T0100491

METHODS IN MOLECULAR BIOLOGY™

Series Editor
John M. Walker
School of Life Sciences
University of Hertfordshire
Hatfield, Hertfordshire, AL10 9AB, UK

For further volumes:
http://www.springer.com/series/7651

Molecular Profiling

Methods and Protocols

Edited by

Virginia Espina and Lance A. Liotta

Center for Applied Proteomics and Molecular Medicine, George Mason University, Manassas, VA, USA

Editors
Virginia Espina
Center for Applied Proteomics
and Molecular Medicine
George Mason University
Manassas, VA, USA
vespina@gmu.edu

Lance A. Liotta
Center for Applied Proteomics
and Molecular Medicine
George Mason University
Manassas, VA, USA
lliotta@gmu.edu

ISSN 1064-3745 e-ISSN 1940-6029
ISBN 978-1-60327-215-5 e-ISBN 978-1-60327-216-2
DOI 10.1007/978-1-60327-216-2
Springer New York Dordrecht Heidelberg London

Library of Congress Control Number: 2011940328

Printed on acid-free paper

Humana Press is part of Springer Science+Business Media (www.springer.com)

Dedication

This book is proudly dedicated to Mary Anne and Len Schiff. Len's encouragement to pursue a life-long interest in research, and Mary Anne's personal sacrifices so I could pursue a graduate degree, were the catalysts that initiated numerous collaborations resulting in the publication of this book.

Virginia Espina

Preface

The next revolution in molecular medicine is the application of molecular profiling to individualized patient therapy. Molecular profiling technology has advanced dramatically, particularly in the field of cancer tissue biomarkers. It is now possible to gather complex genomic and proteomic information from a routine clinical needle biopsy or surgical specimen. This means that translational research scientists can finally begin to address urgent applied research questions that were not possible in the past: (a) How can tissue molecular information be gathered in a reliable and reproducible fashion that is suitable for routine application to the clinic? (b) How does the molecular signature of diseased tissues provide insights into pathogenesis, prognosis, and therapeutic options? (c) What is the best means of combining molecular data with other classes of clinical data (imaging, pathologic staging, clinical chemistry panels) to optimize the treatment plan for the individual patient? (d) How can new classes of clinical research trials be created that are biomarker guided, hypothesis driven, and individualized? The purpose of this volume is to provide an accelerated tutorial to assist students, entrepreneurs, new investigators, and established investigators who want to quickly become versed in, and immersed in, the entire process from discovery to clinical trial validation and commercial public benefit. Our internationally recognized chapter authors have provided the background science, the vision, and the practical experimental protocols, with tips and troubleshooting guides. Since the aim is to span the full process from discovery to commercialization, our practical guides are not limited to experimental methods. We have included tutorials on patents and intellectual property, small business development, FDA review guidelines for molecular profiling, and grant writing tips for investigators seeking funding in translational research.

Molecular Profiling is designed to step the reader through a project/experiment in molecular medicine. The protocol chapters describe detailed techniques for evaluating tissue samples, tissue collection and storage, analytical platforms, and bioinformatics/biostatistics. The narrative chapters are designed to provide the reader with a well-rounded discussion of intellectual property issues in biotechnology, human subjects research requirements, regulatory agency approval processes, and an overview of technology transfer (patent) issues. Although other books have been published covering the topics of genomic profiling, or protein chemistry, we believe this is the first book dedicated to incorporating genomics, proteomics, and bioinformatics with experimental protocols and detailed discussions of future requirements and challenges for commercialization and practical use in the field. An emphasis is placed on tissue-based molecular profiling, a rapidly emerging field that is not covered in pathology text books.

Molecular Profiling covers eleven topics in relation to human disease: Cancer medicine and medical ethics relevant to individualized therapy, genomics, proteomics, microscopic imaging, bioinformatics, tissue preservation/biobanking, individualized therapy regimens, intellectual property, regulatory approval, business development, and grant funding in translational research. A set of core Chapters 1–24 covering genomics, proteomics, imaging, and bioinformatics, illustrate current laboratory protocols for generating data relevant to molecular medicine. Each of these disciplines is complementary and the grouping simply provides a means for differentiating the classes of molecular analytes. The four topics covered in Chapters 25–28 are unique aspects of this volume of the Methods in Molecular Biology series. These latter chapters discuss, in a narrative or tutorial style, future real-world needs in personalized molecular medicine. Important points are highlighted in the Notes section for each chapter.

Although many of the techniques discussed in this volume use commercially available reagents and instrumentation, it is imperative for the user/reader to understand the principles and nuances of these techniques, because they are designed for use with irreplaceable human tissue specimens. In an attempt to provide basic assay information, we have included overview principles in the introduction to each analytical chapter as well as providing troubleshooting tips and tricks for the experienced scientist.

We hope that the readers of this volume will use it as a practical guide at the lab bench as well as the boardroom. The intended readership spans the range of scientists, pathologists, oncologists, residents, biotechnologists, medical students, and nurses involved in clinical trial research. We have a personal hope that this volume will attract new investigators who can apply their creative talents to realize the promise of individualized molecular medicine.

We thank our esteemed chapter authors for their valuable contributions to this volume.

Manassas, VA, USA

Virginia Espina
Lance A. Liotta

Contents

Contributors

ANTHANASIA ANAGOSTOU • *Tatari Design Center and Zorba Marketing Incorporated, Alexandria, VA, USA*
AMY C. ANDERSON • *School of Pharmacy, University of Connecticut, Storrs, CT, USA*
DAGANIA BARZHAGI • *Medestea Research and Production S.p.A, Colleretto Giacosa (TO), Italy*
SVEN BILKE • *Cancer Genetics Branch, National Institutes of Health/ National Cancer Institute, Bethesda, MD, USA*
STEVEN R. BLAKELY • *Applied Biosystems/Life Technologies Corporation, Foster City, CA, USA*
VALERIE CALVERT • *Center for Applied Proteomics and Molecular Medicine, George Mason University, Manassas, VA, USA*
SERENA CAMERINI • *Istituto Superiore di Sanità, Rome (RO), Italy*
WALTER CARBONE • *Merck Serono, Colleretto Giacosa (TO), Italy*
CRISTINA CATTANEO • *Department of Biology, University of Milan, Milan (MI), Italy*
YIDONG CHEN • *Cancer Genetics Branch, National Institutes of Health/ National Cancer Institute, Bethesda, MD, USA*
ROBERT CORNELISON • *Cancer Genetics Branch, National Institutes of Health/ National Cancer Institute, Bethesda, MD, USA*
ROBIN D. COUCH • *Department of Chemistry and Biochemistry, George Mason University, Manassas, VA, USA*
STACY M. COWHERD • *Internal Medicine, University of North Carolina School of Medicine, Chapel Hill, NC, USA*
DANIEL N. COX • *Krasnow Institute for Advanced Study, George Mason University, Fairfax, VA, USA*
THOMAS N. DARLING • *Department of Anatomy, Physiology and Genetics, Uniformed Services University School of Medicine, Bethesda, MD, USA*
SEAN DAVIS • *Cancer Genetics Branch, National Institutes of Health/ National Cancer Institute, Bethesda, MD, USA*
LUCA DEL GIACCO • *Division of Functional and Reproductive Biology, Department of Biology, University of Milan, Milan (MI), Italy*
JIANGHONG DENG • *Center for Applied Proteomics and Molecular Medicine, George Mason University, Manassas, VA, USA*
KIRSTEN H. EDMISTON • *Inova Fairfax Hospital Cancer Center, Fairfax, VA, USA*
GABRIEL S. EICHLER • *InnoCentive Inc., Waltham, MA, USA*
OFER EIDELMAN • *Department of Anatomy, Physiology and Genetics, Uniformed Services University School of Medicine, Bethesda, MD, USA*
VIRGINIA ESPINA • *Center for Applied Proteomics and Molecular Medicine, George Mason University, Manassas, VA, USA*
QINGYUAN FAN • *Department of Neurosciences, Cleveland Clinic, Cleveland, OH, USA*

EOIN F. GAFFNEY • *Department of Histopathology, Biobank Ireland Trust, St James's Hospital, Dublin, Ireland*
FRANCESCA GALDI • *Center for Applied Proteomics and Molecular Medicine, George Mason University, Manassas, VA, USA*
ROSA I. GALLAGHER • *Center for Applied Proteomics and Molecular Medicine, George Mason University, Manassas, VA, USA*
HAROLD R. GARNER • *Virginia Polytechnic Institute and State University, Virginia Bioinformatics Institute, Blacksburg, VA, USA*
FRANCESCO GORRETA • *Genetics and Biomarkers – Exploratory Medicine, Merck Serono Ivrea, Colleretto Giacosa (TO), Italy*
JONATHAN HANSEN • *Center for Applied Proteomics and Molecular Medicine, George Mason University, Manassas, VA, USA*
MICHAEL HEIBY • *University of Virginia School of Medicine, Charlottesville, VA, USA*
PAUL C. HERRMANN • *Department of Pathology and Human Anatomy, Loma Linda University, Loma Linda, CA, USA*
E. CLIFFORD HERRMANN • *Department of Biochemistry and Microbiology, Loma Linda University, Loma Linda, CA, USA*
STEPHEN M. HEWITT • *Tissue Array Research Program/Laboratory of Pathology, National Institutes of Health/National Cancer Institute, Bethesda, MD, USA*
MICHAEL L. HUEBSCHMAN • *McDermott Center for Human Genetics, UT Southwestern Medical Center, Dallas, TX, USA*
CATHERINE E. JOZWIK • *Department of Anatomy, Physiology and Genetics, Uniformed Services University School of Medicine, Bethesda, MD, USA*
YASMIN KAMAL • *Children's National Medical Center, Washington, DC, USA*
J. KEITH KILLIAN • *Cancer Genetics Branch, National Institutes of Health/ National Cancer Institute, Bethesda, MD, USA*
MICHAEL LASKOFSKI • *Office of Sponsored Programs, George Mason University, Fairfax, VA, USA*
JAMIN M. LETCHER • *Krasnow Institute for Advanced Study, George Mason University, Fairfax, VA, USA*
LANCE A. LIOTTA • *Center for Applied Proteomics and Molecular Medicine, George Mason University, Manassas, VA, USA*
CATERINA LONGO • *Department of Dermatology, University of Modena and Reggio Emilia, Modena (MO), Italy*
DEIRDRE MADDEN • *Faculty of Law, University College, Cork, Ireland*
PAUL S. MELTZER • *Cancer Genetics Branch, National Institutes of Health/National Cancer Institute, Bethesda, MD, USA*
BRYAN A. MILLIS • *National Center for Biodefense and Infectious Disease, George Mason University, Manassas, VA, USA*
NOEMI MORONI • *Center for Applied Proteomics and Molecular Medicine, George Mason University, Manassas, VA, USA*
BRYAN T. MOTT • *Department of Chemistry, Johns Hopkins University, Baltimore, MD, USA*
CLAUDIUS MUELLER • *Center for Applied Proteomics and Molecular Medicine, George Mason University, Manassas, VA, USA*

David W. Murray • *Department of Physiology and Medical Physics, Royal College of Surgeons in Ireland, Dublin, Ireland*
Emanuel F. Petricoin III • *Center for Applied Proteomics and Molecular Medicine, George Mason University, Manassas, VA, USA*
Mariaelena Pierobon • *Center for Applied Proteomics and Molecular Medicine, George Mason University, Manassas, VA, USA*
Maria Letizia Polci • *Istituto Superiore di Sanità-Ministero della Salute, Rome (RO), Italy*
Harvey B. Pollard • *Department of Anatomy, Physiology and Genetics, Uniformed Services University School of Medicine, Bethesda, MD, USA*
K. Alex Reeder • *Center for Applied Proteomics and Molecular Medicine, George Mason University, Manassas, VA, USA*
Kevin P. Rosenblatt • *The Brown Foundation Institute of Molecular Medicine/UT Health Science Center CCTS Proteomics Core, Houston, TX, USA*
G. Terry Sharrer • *Inova Health System, Inova Fairfax Hospital, Fairfax, VA, USA*
Michele Signore • *Department of Hematology, Oncology, and Molecular Medicine, Istituto Superiore di Sanità, Rome (RO), Italy*
William I. Smith • *Department of Pathology, Suburban Hospital, Bethesda, MD, USA*
Sinéad M. Smith • *Institute of Molecular Medicine, Trinity College Dublin, Dublin, Ireland*
Alexander Spira • *Virginia Cancer Specialists, Fairfax, VA, USA*
Meera Srivastava • *Department of Anatomy, Physiology and Genetics, Uniformed Services University School of Medicine, Bethesda, MD, USA*
Alessandra Tessitore • *Department of Experimental Medicine, University of L'Aquila, L'Aquila (AQ), Italy*
Geraldine A. Thomas • *Department of Surgery and Cancer, Hammersmith Hospital/ Wales Cancer Bank, London, UK*
J. Lille Tidwell • *Office of Technology Transfer, George Mason University, Fairfax, VA, USA*
Nishant Trivedi • *University of Virginia, Charlottesville, VA, USA*
Amy J. VanMeter • *Center for Applied Proteomics and Molecular Medicine, George Mason University, Manassas, VA, USA*
Robert L. Walker • *Cancer Genetics Branch, National Institutes of Health/National Cancer Institute, Bethesda, MD, USA*
Lindsay Wescott • *Office of Sponsored Programs, George Mason University, Fairfax, VA, USA*
Pamela L. Zeitlin • *Department of Pediatrics, Johns Hopkins School of Medicine, Baltimore, MD, USA*
Weidong Zhou • *Center for Applied Proteomics and Molecular Medicine, George Mason University, Manassas, VA, USA*

Chapter 1

Tumor Staging and Grading: A Primer

Stacy M. Cowherd

Abstract

Cancer staging and grading are used to predict the clinical behavior of malignancies, establish appropriate therapies, and facilitate exchange of precise information between clinicians. The internationally accepted criteria for cancer staging, the tumor-node-metastasis (TNM) system, includes: (1) tumor size and local growth (T); (2) extent of lymph node metastases (N); and (3) occurrence of distant metastases (M). Clinical stage is established before initiation of therapy and depends on the physical examination, laboratory findings, and imaging studies. Pathologic stage is determined following surgical exploration of disease spread and histological examination of tissue. The TNM classification system has evolved over 50 years to accommodate increasing knowledge about cancer biology. Efforts are ongoing to keep the system both synchronized with the most sophisticated cancer technology and simple for ease of clinician/patient use. Upcoming molecular technologies, such as genomic and proteomic profiling of tumors, microRNA profiling, and even ex vivo living tumor tissue treatment, could improve the current TNM staging system. This chapter describes the current TNM system using breast, lung, ovarian, and prostate cancer examples.

Key words: Breast cancer, Grade, Lung cancer, Lymph node, Metastasis, Prostate cancer, Ovarian cancer, Stage, Tumor

1. Introduction

Tumor staging and grading is integral to the practice of clinical oncology because these classifications are the starting point for patient care. Cancer staging and grading are used to predict the clinical behavior of malignancies, establish appropriate therapies, and facilitate exchange of precise information between clinicians. During the staging/grading process, patients are placed in standardized categories according to the anatomical location of dissemination and the pathologic characteristics of their tumors. Therefore, clinicians, pathologists, and radiologists must work together to achieve the most precise classification of neoplasms.

Virginia Espina and Lance A. Liotta (eds.), *Molecular Profiling: Methods and Protocols*, Methods in Molecular Biology, vol. 823, DOI 10.1007/978-1-60327-216-2_1, © Springer Science+Business Media, LLC 2012

Cancer staging refers to the anatomic extent of the disease spread. The internationally accepted criteria for cancer staging, the tumor-node-metastasis (TNM) system, include: (a) tumor size and local growth (T); (b) extent of lymph node metastases (N); and (c) occurrence of distant metastases (M). Cancers are categorized as primary tumor size between T1 and T4, nodes between N0 and N3, and metastases between M0 and M1. Generally, as the size of the primary untreated cancer (T) increases, regional lymph node involvement (N) and distant metastasis (M) become more frequent. The most common sites of metastases are lung, bone, bone marrow, liver, and brain (1).

Different cancer types have unique anatomical patterns of spread, and therefore require distinct TNM classification systems. Following is a general TNM schema for all cancers (1, 2):

Primary Tumor (T)

TX: Tumor cannot be assessed

T0: No evidence of primary tumor

Tis: Carcinoma in situ

T1, T2, T3, T4: Increasing size and/or local extent of tumor

Regional Lymph Nodes (N)

NX: Regional lymph nodes cannot be assessed

N0: No evidence of disease in lymph nodes

N1, N2, N3: Increasing disease involvement of regional lymph nodes

Distant Metastasis (M)

MX: Distant metastasis cannot be assessed

M0: No distant metastasis

M1: Distant metastasis

TNM values are used to classify a patient's cancer into stages between I and IV. Cancers can be assigned both a clinical stage and a pathologic stage. Clinical stage is established before initiation of therapy and depends on the physical examination, laboratory findings, and imaging studies. Pathologic stage is determined following surgical exploration of disease spread and histological examination of tissue (2). Pathologic stage is particularly significant for cancers which are not easily accessible in a clinical setting, such as pancreatic or ovarian carcinoma (1). Both clinical and pathologic stages should be recorded in a patient's permanent medical record, with the clinical stage guiding initial therapy and the pathologic stage guiding adjuvant therapy.

2. Typical Descriptions of the Different Stages and Grades (*1, 2*)

2.1. Tumor Stages

Stage I: Tumor limited to organ of origin, without nodular or vascular spread.

Stage II: Local spread of tumor into surrounding tissue and regional lymph nodes. The lesion is resectable, but there is uncertainty about completeness of removal due to tumor microinvasion of surrounding tissue.

Stage III: Extensive primary tumor with invasion into deeper structures, bone, and lymph nodes. The lesion is operable but not resectable, and gross disease is left behind.

Stage IV: Evidence of distant metastasis beyond tumor organ of origin; primary tumor is inoperable.

2.2. Tumor Grades

Cancer grade is a subjective scoring by the pathologist based on tumor histology and cytomorphology of the tumor lesion. Histopathologic grading is of equal importance to anatomic staging for predicting patient prognosis and guiding treatment (1, 2). Therefore, surgical biopsy or excision of suspicious lesions is essential to confirm cancer diagnosis and classify tumor cellular architecture (1). By definition, malignant tumors invade the basement membrane and extracellular matrix to invade surrounding tissue with indistinct borders (3). Additional microscopic evidence of abnormal, or malignant, aggressiveness includes giant tumor cells, high numbers of mitoses, nucleoli and chromatin morphology, unusual mitoses, aneuploidy, and nuclear pleomorphism (3, 4).

In general, low-grade cancers are well differentiated, resembling healthy cellular counterparts, and high-grade cancers are anaplastic, and disorderly, not resembling normal tissue at that site. The most poorly differentiated part of the tumor determines overall tumor grade (1). In general, high-grade cancers are more clinically aggressive than low-grade cancers. Most grading systems divide tumors into three or four grades according to cellular differentiation (2):

GX: Grade cannot be evaluated

G1: Well differentiated

G2: Moderately differentiated

G3–G4: Poorly differentiated

Using cancer grading and staging in addition to other clinical data, clinicians can construct nomograms to predict treatment outcomes, cure rates, and disease-free survival times. Following is a discussion of specific cancer staging and grading for lung, prostate, breast, and ovarian cancers. Clinical staging information is from the sixth (2002) edition of the American Joint Committee on Cancer's (AJCC) Staging Manual (2).

3. Cancer Classification Examples

3.1. Lung Cancer Clinical Staging Work-Up

Lung cancer is one of the most common malignancies in the Western hemisphere, and the leading cause of cancer death in men and women (2, 5). The stage of lung carcinomas at diagnosis is the single most important prognostic factor for patients (2, 6). Patients with clinically suspected lung carcinoma should receive detailed history and physical exam, chest radiograph, chest computed tomography (CT) scan with intravenous contrast, and laboratory tests including complete blood count, electrolytes, liver function tests, and serum calcium. CT scan is the most important radiologic procedure to visualize lung cancer because the image shows central masses and peripheral nodules, and may suggest tumor invasion into the pleura, mediastinum, and regional lymph nodes (5). Lung cancer spreads locally to intrathoracic, scalene, and supraclavicular lymph nodes (2).

Primary tumor tissue must be procured for confirmation of pathology and definition of histology. Tissue may be collected either through bronchoscopy for central lesions or CT-guided needle biopsy for peripheral lesions. Thoracentesis should also be performed in patients with pleural effusions to determine whether the effusion cells are malignant or paramalignant, exudative with negative cytology. Scalene and intrathoracic lymph nodes, which appear irregular or enlarged on CT scan, should be sampled using mediastinoscopy. This regional lymph node sampling is important for lung cancer patients because therapy is different for N2 versus N3 disease (5, 7).

More extensive CT scanning is useful to detect lung cancer metastases to the liver, adrenal glands, contralateral lung, and brain. Patients with potentially resectable disease may also undergo fluorodeoxyglucose positron emission tomography (FDG-PET) scan to identify occult lymph node infiltrations or distant metastases (5, 7). Metastatic cancer cells appear on PET scan, in addition to infection and severe inflammation, because these cells take up proportionately higher volumes of glucose analog. Any suspicious lesions at distant sites should be sampled (5). See Table 1 for specific TMN staging criteria.

Small cell lung carcinoma (SCLC), a common subtype of lung cancer, is most commonly categorized using a two-stage system rather than TNM staging (5, 6, 8). SCLC tends to be disseminated at the time of diagnosis, with only 25% of patients presenting with "limited" disease (5). SCLC is considered "limited" when it is confined to one radiotherapy port and "extensive" when it extends beyond the ipsilateral hemithorax, or is not confined to a single radiation port. "Limited" SCLC corresponds to stages I through III in the TNM system, and "extensive" SCLC corresponds to stage IV disease (2). All patients diagnosed with SCLC must have a bone scan and brain CT because disease most commonly metastasizes to these sites (5).

Table 1
TNM classification and stage grouping for NSCLC

Stage	Grouping	Descriptions
Occult carcinoma	TX, N0, M0	TX: Primary tumor cannot be assessed, or tumor proven by the presence of malignant cells in sputum or bronchial washings but not visualized with imaging N0: No regional lymph node metastasis M0: No distant metastasis
Stage 0	Tis, N0, M0	Tis: Carcinoma in situ
Stage IA	T1, N0, M0	T1: Tumor 3 cm or less in greatest dimension, surrounded by lung or visceral pleura, without evidence of invasion more proximal than the lobar bronchus
Stage IB	T2, N0, M0	T2: Tumor with any of the following features: • More than 3 cm in greatest dimension • Involves main bronchus, 2 cm or more distal to the carina • Invades the visceral pleura • Associated with atelectasis or obstructive pneumonia that extends to the hilar region but does not involve the entire lung
Stage IIA	T1, N1, M0	N1: Metastasis to ipsilateral peribronchial and/or ipsilateral hilar lymph nodes, and intrapulmonary nodes including involvement by direct extension of the primary tumor
Stage IIB	T2, N1, M0 T3, N0, M0	T3: Tumor of any size that invades any of the following: chest wall, diaphragm, mediastinal pleura, parietal pericardium; or tumor in the main bronchus less than 2 cm distal to the carina, but without involvement of the carina; or associated atelectasis or obstructive pneumonia of the entire lung
Stage IIIA	T1, N2, M0 T2, N2, M0 T3, N1, M0 T3, N2, M0	N2: Metastasis to ipsilateral mediastinal and/or subcarinal lymph node(s)
Stage IIIB	Any T, N3, M0 T4, Any N, M0	N3: Metastasis to contralateral mediastinal, contralateral hilar, ipsilateral or contralateral scalene, or supraclavicular lymph node(s) T4: Tumor of any size that invades any of the following: mediastinum, heart, great vessels, trachea, esophagus, vertebral body, carina; or separate tumor nodules in the same lobe; or tumor with malignant pleural effusion
Stage IV	Any T, Any N, M1	M1: Distant metastasis present

3.1.1. Histology and Grading

Lung cancers are classified into histological types using light microscopy with routinely stained preparations. The two most common subtypes of lung carcinoma are bronchogenic NSCLC and neuroendocrine carcinoma. Bronchogenic NSCLC encompasses several variants, including squamous cell carcinoma, adenocarcinoma, and anaplastic large cell carcinoma. Neuroendocrine carcinomas are further subdivided according to cellular differentiation, with SCLC a poorly differentiated variant of neuroendocrine tumor (6, 8).

Lung cancer grading usually applies to squamous cell carcinoma and adenocarcinoma, the two most common types of NSCLC (6). Large cell carcinomas are inherently high-grade, containing sheets of poorly differentiated cells that do not show differentiation toward either squamous cell carcinoma or adenocarcinoma (6, 8). SCLC is also high-grade by definition (Fig. 1A). Microscopically, SCLC cells are primitive appearing with scant cytoplasm, granular chromatin, and high mitotic activity. Encrustation, with basophilic deposition of DNA within blood vessel walls, is a distinctive histological feature of SCLC (6, 8).

Squamous cell cancers commonly arise from epithelial cells in the proximal tracheobronchial tree, and may therefore present with signs of airway obstruction (5, 6). These carcinomas are graded based on proportion of intercellular bridges and other characteristics of keratinization (Fig. 1B) (6, 8). Most pathologists use a three-tiered grading system to distinguish well, moderately, and poorly differentiated tumors. Grade one or well-differentiated tumors have sheets of cells with ample eosinophilic cytoplasm, round nuclei, prominent nucleoli, and well-defined cellular borders with intercellular bridges. These well-differentiated tumors may contain concentric laminated deposits of amorphous, keratinous material called "squamous pearls" (6, 8). Areas of comedo-like necrosis characterize grade two tumors. Grade three tumors are poorly differentiated and cells tend to grow in

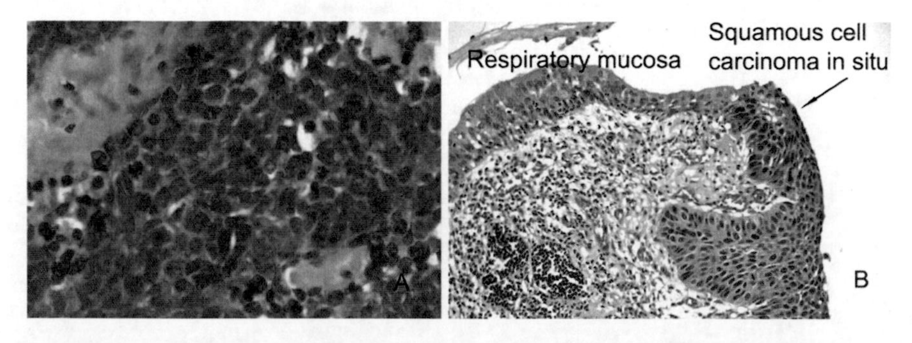

Fig. 1. Lung carcinoma histology. (**A**) Small cell lung carcinoma (SCLC) histology demonstrating poor differentiation. SCLC cells are primitive appearing with scant cytoplasm, granular chromatin, and high mitotic activity. (**B**) Area of normal lung epithelium adjacent to squamous cell carcinoma in situ. Squamous cell carcinomas are graded based on proportion of intercellular bridges and other characteristics of keratinization. Courtesy of William Funkhouser, MD.

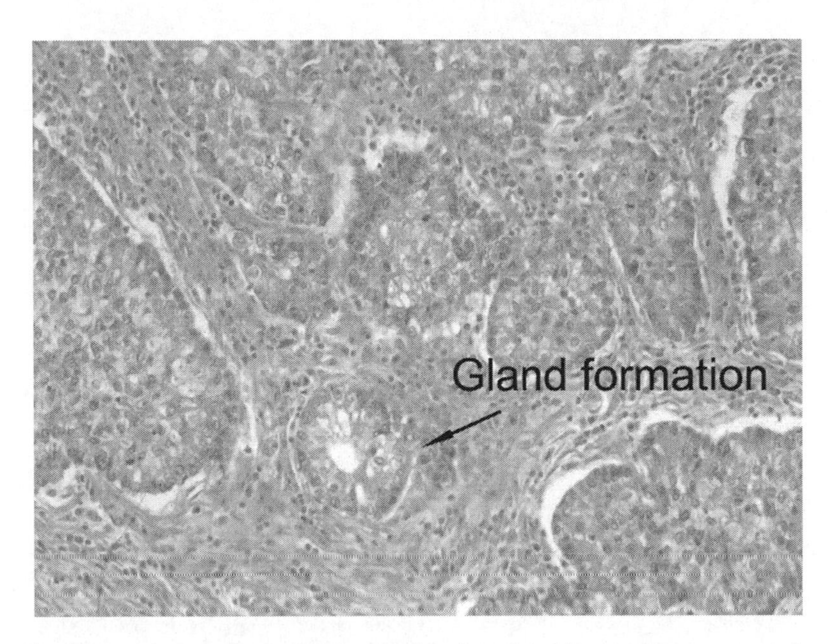

Fig. 2. Lung adenocarcinoma showing glandular formation. Adenocarcinomas are usually located at the periphery of the lung and graded according to number and appearance of glandular structures. Courtesy of William Funkhouser, MD.

confluent sheets. Cells are characterized by bizarre nuclei, cytological atypia, increased mitotic figures, and areas of necrosis and/or hemorrhage (6).

Adenocarcinomas are glandular tumors, usually located at the periphery of the lung and graded according to number and appearance of glandular structures (Fig. 2). As compared to squamous cell carcinomas, adenocarcinomas are more likely to be widely metastatic at the time of diagnosis (5, 8). Grade one or well-differentiated tumors consist of distinctive gland structures throughout 90% of the tumor mass. The glands resemble healthy lung tissue, with tall columnar or mucinous epithelium, eosinophilic cytoplasm, basal nuclei, and prominent nucleoli. A key variant of well-differentiated adenocarcinoma is the bronchioalveolar type, characterized by bland-appearing tumor cells growing continuously along alveolar walls (8). Grade two adenocarcinomas, which are moderately differentiated, consist of glandular or acinar structures throughout at least 50% of the tumor mass. These glands display either poorly formed lumens or atypical, anaplastic cells lining the lumens. Grade three, or poorly differentiated adenocarcinomas, show 5–50% glandular proliferation. The majority of grade three tumors contain solid stromal areas with atypical mucinous cells (6, 8).

3.2. Prostate Cancer Clinical Staging Work-Up

Widespread screening for prostate cancer, with digital rectal exam (DRE) and/or serum prostate-specific antigen (PSA) serum levels, has allowed the majority of cancers to be diagnosed in an asymptomatic and localized stage (9). Prostate biopsy, which is usually performed transrectally with ultrasound guidance, is necessary for a definitive diagnosis of prostate adenocarcinoma following abnormal PSA or DRE. Approximately, ten core needle biopsies are necessary to sample all potentially affected lobes of the prostate (9, 10).

Staging prostate cancer is unique because of the nonspecific nature of PSA and prevalence of equivocal biopsies, many which show histology suggesting "increased risk" for adenocarcinoma. The risk of newly diagnosed prostate cancer progressing over the short-term is low (9, 10); therefore, many patients with "increased risk" or premalignant biopsies choose watchful waiting rather than treatment. Physicians help patients make decisions regarding extensive staging and treatment with mathematical nomograms like the Partin model, the Kattan nomogram, or the D'Amico model. These models are rough predictors of treatment outcome and disease-free survival given an individual patient's PSA level, biopsy Gleason grade, and clinical T stage (9). The Gleason grade is a qualitative assessment of the loss of normal glandular prostate tissue architecture as observed from a stained prostate tissue section (11, 12). The score is based on a grading system of 1–5, with 5 being the worst, or showing the largest loss of normal glandular morphology.

Definitively high-grade biopsy specimens require additional staging work-up and either radical prostatectomy, external beam radiation therapy, or brachytherapy (radioactive seed implants). The primary staging goal of surgery is to determine whether adenocarcinoma extends beyond the prostate capsule. In general, tumors confined to the prostate gland are curable but tumors with extraprostatic extension to the seminal vesicles and regional lymph nodes are not curable (9, 10).

Men with elevated PSA, high Gleason grade, and/or clinical tumor stage greater than T2 may have additional preoperative studies to determine the extent of cancer spread. These studies include radionucleotide bone scan, axial skeleton magnetic resonance imaging (MRI), and CT scan of the abdomen and pelvis. Regional prostate lymph nodes, which are located in the true pelvis below the bifurcation of the common iliac arteries, are difficult to visualize with CT scan (2). Therefore, MRI of the prostate gland with endorectal probe is used to specifically examine seminal vesicle and/or regional lymph node spread (10). Prostate adenocarcinoma most commonly metastasizes to distant lymph nodes and bone, but lung and liver metastases are common in late-stage disease (2). See Table 2 for specific TMN staging criteria.

3.2.1. Histology and Grading

The vast majority of prostate cancers are epithelial adenocarcinomas, although variants include neuroendocrine tumors, stromal tumors, and mesenchymal tumors such as leiomyosarcoma or sarcomatoid carcinoma (13, 14). Putative premalignant lesions

Table 2
TNM classification and stage grouping for prostate adenocarcinoma

Stage	Grouping	Descriptions
I	T1a, N0, M0, G1	T1a: Tumor clinically inapparent. Tumor incidental histological finding in 5% or less of prostate tissue resected N0: No regional lymph node metastasis M0: No distant metastasis G1: Gleason grade 1
II	T1a, N0, M0, G2-4 T1b, N0, M0, Any G T1c, N0, M0, Any G T2, N0, M0, Any G	T1b: Tumor clinically inapparent. Tumor incidental histological finding in more than 5% of tissue resected T1c: Tumor clinically inapparent. Tumor identified by needle biopsy T2: Tumor confined within prostate G2-4: Gleason grades 2, 3, and 4
III	T3, N0, M0, Any G	T3: Tumor extends through the prostate capsule
IV	T4, N0, M0, Any G Any T, N1, M0, Any G Any T, Any N, M1, Any G	T4: Tumor is fixed or invades adjacent structures other than seminal vesicles: bladder neck, external sphincter, rectum, levator muscles, and/or pelvic wall N1: Metastasis in regional lymph node(s) M1: Distant metastasis

include prostatic intraepithelial neoplasia (PIN) and atypical adenomatous hyperplasia (AAH); however, AAH has not proved a true premalignant lesion (14). Low-grade PIN is characterized by a slight increase in cellularity with irregular spacing of epithelial cells. High-grade PIN displays a marked increase in cellularity with nuclear enlargement and hyperchromasia. Both low- and high-grade PIN have preservation of the basal cell layer (9, 14).

Prostate adenocarcinomas are graded using the Gleason system, which classifies specimens between one and five based on glandular architecture and cellular cytomorphology (9, 10, 13). Several clinical trials have validated the prognostic value of the Gleason system, with higher scores predicting widespread disease and worse prognosis (9, 14). Pathologists report both primary and secondary scores, with the primary score representing the most common histological grade in the specimen and the secondary score reflecting the second most common grade. The primary and secondary scores are added to yield overall Gleason score. Thus, Gleason scores between one and three represent well-differentiated adenocarcinomas and scores between eight and ten represent poorly differentiated cancers (9, 13).

Gleason score pattern one cannot be reliably distinguished from adenomatous hyperplasia lesions using needle biopsies (13).

Generally, however, simple round glands with uniform size, shape, and spacing characterize pattern one. The nuclei and nucleoli are markedly enlarged. Gleason pattern score two tumors show more variation in glandular size and shape, and appear incompletely circumscribed. Haphazardly separated glands among bands of fibrous stroma characterize pattern three, the most common microscopic pattern of prostate adenocarcinoma. Pattern four tumor cells are organized into closely packed or fused glands, which invade the stroma with ragged infiltrative edges. Gleason pattern score five tumors contain solid sheets of anaplastic cells with comedo-like necrosis in cribriform nests (13, 14).

3.3. Breast Cancer Clinical Staging Work-Up

Breast cancer, the most common solid-organ malignancy diagnosed among North American women, is usually discovered either through screening mammography or detection of a breast lump (2, 15). Abnormal mammogram findings include breast masses, microcalcifications, asymmetries between the breasts, and architectural distortions. Malignant breast lumps typically present in women over 30 years old as asymptomatic, painless masses which are fixed to surrounding tissue (16). Patients with an abnormal mammogram and/or suspicious breast mass must undergo large core needle biopsies for pathologic diagnosis. Approximately ten core biopsies are usually necessary, each with diameter between 14 and 18 gauge and length between 1 and 3 cm. For women without palpable masses, mammogram or ultrasound guidance is used to precisely localize the lesion (15, 16).

Extensive use of screening mammography has lead to increased diagnoses of noninvasive breast carcinoma or ductal carcinoma in situ (DCIS) (15, 17). DCIS encompasses a wide spectrum of disease with multiple staging and treatment options. In general, DCIS has low metastatic potential but must be completely excised with either radical mastectomy or lumpectomy to prevent local recurrence (17, 18). The most important prognostic factors influencing local recurrence of DCIS include lesion size, adequacy of resection, patient age, and histological grade (15). It is therefore important for surgeons to obtain a wide surgical margin, preferably 10 mm in each dimension. In addition, pathologists must examine biopsy tissue for areas of microscopic stromal invasion or microinvasion. The *AJCC Cancer Staging Manual* classifies microinvasion as T1mic, a subset of T1 breast cancer (2). Sentinel lymph node biopsy and axillary lymph node dissections (ALNDs) are not necessary with DCIS unless the patient has high-grade disease, documentation of microinvasion, or high risk for invasive disease (18).

Invasive breast cancers require complete operative excision plus sentinel lymph node biopsy and/or ALND. Important operative findings for staging include size of the primary tumor and presence of chest wall invasion. If the primary tumor is invasive and sentinel lymph node biopsy is positive, ALND must be performed to evaluate for metastases. Patients with palpable axillary lymph nodes must have ALND regardless of sentinel node status (15, 16). Breast lymphatics drain by the way of three major routes: axillary, transpectoral, and internal mammary. Any other lymph node metastasis, with the exception of supraclavicular spread, is considered metastatic disease (M1) (2).

Additional work-up for suspected breast carcinoma might include supplementary breast imaging, chest imaging, and laboratory work with complete blood count and liver function tests. Breast ultrasound is useful to assess primary lesions in women with dense breasts, precisely locate breast masses, and evaluate ipsilateral axillary lymph nodes. Breast MRI should be used to evaluate for occult disease, either in the ipsilateral or contralateral breast, and to screen for synchronous breast lesions (15, 16). Women with potentially advanced stage cancer should receive abdominopelvic CT scan, radionuclide bone scan, and/or PET scan to evaluate for metastases (15). The most common sites of breast cancer metastasis are bone, brain, liver, and lung (2).

As compared to the fifth edition, the sixth edition of the *AJCC Cancer Staging Manual* includes many significant revisions in breast cancer staging. These changes reflect more specific characterization of tumors using new technologies like sentinel lymph node biopsy, immunohistochemical staining, and reverse transcriptase-polymerase chain reaction. Significantly, the sixth edition distinguishes between tumor micrometastases and isolated tumor cells on the basis of lesion size (19). See Table 3 for specific TMN staging criteria.

3.3.1. Histology and Grading

Adenocarcinoma, which may be either noninvasive or invasive, is the most common histological type of breast cancer. The noninvasive adenocarcinomas include DCIS and lobular carcinoma in situ (LCIS) (15). The most common histological type of invasive breast adenocarcinoma is ductal carcinoma, NOS (not otherwise specified); however, other subtypes of invasive disease include infiltrating lobular, mucinous, medullary, and papillary carcinoma. Familiar subtypes of breast cancer that do not represent special pathologic categories include Paget's disease and inflammatory carcinoma. Paget's disease of the nipple is a variant of high-grade DCIS in subareolar breast ducts (15).

Inflammatory carcinoma is considered an aggressive variant of infiltrating ductal carcinoma, NOS (17).

Breast cancer grading is applicable for DCIS and all invasive carcinomas (2, 20). DCIS is graded on a three-tiered system based on nuclear characteristics. Grade one or low-grade DCIS cells contain small, round, and uniform nuclei with evenly dispersed chromatin (Fig. 3A). Cribriform and micropapillary architectures are common, and neoplastic cells form geometric bulbous projections around which the cells are polarized (17). Grade three or high-grade

Table 3
TNM classification and stage grouping for breast adenocarcinoma

Stage	Grouping	Descriptions
0	Tis, N0, M0	Tis: Ductal or lobular carcinoma in situ N0: No regional lymph node metastasis M0: No disant metastasis
I	T1, N0, M0	T1: Tumor 2 cm or less in greatest dimension
IIA	T0, N1, M0 T1, N1, M0 T2, N0, M0	T2: Tumor more than 2 cm but not more than 5 cm in greatest dimension N1: Metastasis to movable ipsilateral axillary lymph node(s)
IIB	T2, N1, M0 T3, N0, M0	T3: Tumor more than 5 cm in greatest dimension
IIIA	T0, N2, M0 T1, N2, M0 T2, N2, M0 T3, N1, M0 T3, N2, M0	N2: Metastases to ipsilateral axillary lymph nodes fixed or matted; or in clinically apparent ipsilateral internal mammary nodes in the absence of clinically evident axillary lymph node metastasis
IIIB	T4, N0, M0 T4, N1, M0 T4, N2, M0	T4: Tumor of any size with direct extension to chest wall or skin
IIIC	Any T, N3, M0	N3: Metastasis in ipsilateral infraclavicular lymph node(s) with or without axillary lymph node involvement; or in clinically apparent ipsilateral internal mammary lymph node(s) and in the presence of clinically evident axillary lymph node metastasis; or metastasis in ipsilateral supraclavicular lymph node(s) with or without axillary or internal mammary lymph node involvement
IV	Any T, Any N, M1	M1: Distant metastasis

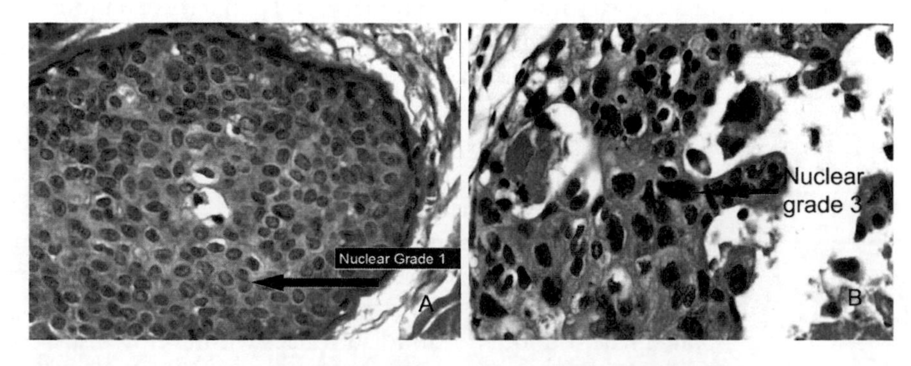

Fig. 3. Breast ductal carcinoma in situ (DCIS). (**A**) Grade one solid breast DCIS cells contain small, round nuclei with evenly dispersed chromatin. (**B**) Grade three breast DCIS cells show nuclear pleomorphism and coarse chromatin. Courtesy of Chad Livasy, MD.

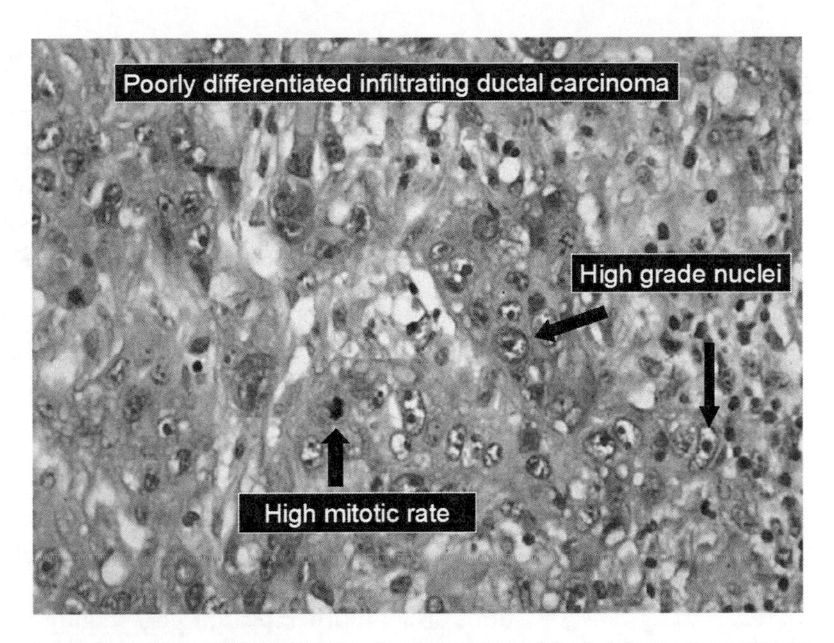

Fig. 4. Poorly differentiated, infiltrating ductal carcinoma of the breast showing cells with high mitotic rates and high-grade nuclei. Courtesy of Chad Livasy, MD.

DCIS tumor cells are pleomorphic with high nuclear cytoplasmic ratio, coarse chromatin, and large nucleoli (Fig. 3B). Mitoses are frequent and necrosis often occurs in the center of ducts surrounded by a solid pattern of neoplastic cells (17). The presence of necrosis within DCIS automatically qualifies the specimen as grade two or three (20).

Invasive breast carcinoma (Fig. 4) is most commonly graded using Elston and Ellis's method, which was endorsed by WHO experts in 2003. This grading technique is based on three histological components: (1) extent of gland and tubule formation, (2) degree of nuclear pleomorphism, and (3) number of mitotic figures. Pathologists assign between one and three points in each of these dimensions, with one point for the most differentiated histology and three points for the least differentiated histology. In the tubule/gland formation category, one point is given for tubule formation in more than 75% of the tumor mass and three points are given for tubule formation in less than 10% of the tumor mass (17, 20). Likewise, nuclear pleomorphism is assessed with one point for small, regular, and uniform nuclei and three points for marked variation among nuclei. It should be noted that, in addition to contributing to calculation of histological grade, nuclear grade is also an independent prognostic marker for invasive breast carcinoma (15). Low numbers of mitotic figures receive one point and high numbers receive three points. The final grade for invasive breast carcinoma is determined by totaling points, with low-grade carcinomas between three and five total points and high-grade carcinomas between eight and nine total points (20).

3.4. Ovarian Cancer Clinical Staging Work-Up

Ovarian cancer is notoriously asymptomatic until it metastasizes; yet there is no reliable method to screen for this disease. Therefore, the majority of patients present with symptoms of stage III or IV cancer such as bloating, abdominal, or pelvic pain, and early satiety (21, 22). Routine pelvic exams may reveal earlier stage carcinomas, which most commonly present as solid, irregular, and fixed adnexal masses in postmenopausal women. Patients with pelvic lymphadenopathy, cul-de-sac nodularity, ascites, or pleural effusion are at high risk for metastatic disease. Suspected ovarian cancer cases should have a transvaginal ultrasound to more accurately determine tumor size and consistency, solid versus cystic (23).

Although sonographic and other clinical data may strongly indicate malignancy, surgery is necessary for ovarian cancer diagnosis, staging, and therapy development (21, 22). Prior to surgery, women with suspected ovarian cancer should have serum glycoprotein CA-125 measurement. CA-125 levels are useful in evaluating the success of future treatments including surgery, radiation, or chemotherapy. Surgical staging must include the following: (1) cytological evaluation of ascites or washings; (2) examination and biopsy of the diaphragm; (3) examination and biopsy of abdominal and pelvic peritoneum; (4) para-aortic and pelvic lymph node sampling; (5) total abdominal hysterectomy, bilateral salpingo-oopherectomy, and partial omentectomy (21, 23). An experienced gynecologic oncologist must perform this extensive surgery, with a pathologist available for specimen interpretation.

Additional work-up for metastatic ovarian cancer includes chest radiograph, bilateral mammogram, barium enema or colonoscopy, and laboratory work including blood chemistry and liver function tests (23). Abdominopelvic CT or MRI may also demonstrate sites of metastatic disease. The most common sites for ovarian spread include the peritoneum, diaphragmatic, and liver surfaces; however, peritoneal ovarian lesions are not classified as distant metastases. M1 disease is characterized by metastases to the liver, lung, or skeleton parenchyma, supraclavicular nodes, and axillary nodes (2).

Epithelial ovarian cancer may spread using any of three primary pathways. First, the tumor can penetrate the ovarian capsule and directly invade adjacent structures such as the uterus, bladder, rectum, or pelvic peritoneum. Second, tumors can spread via lymphatics to the pelvic or para-aortic lymph nodes. Finally, ovarian tumor cells can escape into the peritoneal cavity and spread through the abdomen using peristalsis and the diaphragm's respiratory motions (21). See Table 4 for specific TMN staging criteria.

3.4.1. Histology and Grading

Ovarian tumor histology and grading are especially valuable for predicting prognosis and planning treatment of early cancers (23). Pathologic subtypes of ovarian cancer include epithelial cell tumors, germ cell tumors, and sex cord-stromal tumors (22, 24). Epithelial

Table 4
TNM classification and stage grouping for epithelial ovarian carcinoma

Stage	Grouping	Descriptions
I	T1, N0, M0	T1: Tumor limited to ovaries N0: No regional lymph node metastasis
IA	T1a, N0, M0	T1a: Tumor limited to one ovary; capsule intact, no tumor on ovarian surface. No malignant cells in ascites or peritoneal washings
IB	T1b, N0, M0	T1b: Tumor limited to both ovaries; capsules intact, no tumor on ovarian surface. No malignant cells in ascites or peritoneal washings
IC	T1c, N0, M0	T1c: Tumor limited to one or both ovaries with any of the following: capsule ruptured, tumor on ovarian surface, malignant cells in ascites or peritoneal washings
II	T2, N0, M0	T2: Tumor involves one or both ovaries with pelvic extension and/or implants
IIA	T2a, N0, M0	T2a: Extension and/or implants on uterus and/or tube(s). No malignant cells in ascites or peritoneal washings
IIB	T2b, N0, M0	T2b: Extension to and/or implants on other pelvic tissues. No malignant cells in ascites or peritoneal washings
IIC	T2c, N0, M0	T2c: Pelvic extension and/or implants (T2a or T2b) with malignant cells in ascites or peritoneal washings
III	T3, N0, M0	T3: Tumor involves one or both ovaries with microscopically confirmed peritoneal metastasis outside the pelvis
IIIA	T3a, N0, M0	T3a: Microscopic peritoneal metastasis beyond pelvis (no macroscopic tumor)
IIIB	T3b, N0, M0	T3b: Macroscopic peritoneal metastasis beyond pelvis 2 cm or less in greatest dimension
IIIC	T3c, N0, M0 Any T, N1, M0	T3c: Peritoneal metastasis beyond pelvis more than 2 cm in greatest dimension and/or regional lymph node mestastasis N1: Regional lymph node metastasis
IV	Any T, Any N, M1	M1: Distant metastasis (excludes peritoneal metastases)

Tables 1–4 used with the permission of the American Joint Committee on Cancer (AJCC), Chicago, Illinois. The original source for this material is the AJCC Cancer Staging Manual, Sixth Edition (2002) published by Springer Science and Business Media LLC, http://www.springerlink.com

tumors are the most common ovarian malignancies but benign cystic teratomas, from germ cell origin, are the most common ovarian tumors. Sex cord-stromal tumors, including granulosa cell tumors, are extremely rare and arise from the ovarian stroma and/or sex cord derivatives (24). While epithelial tumors usually occur in postmenopausal women, many types of germ cell and sex

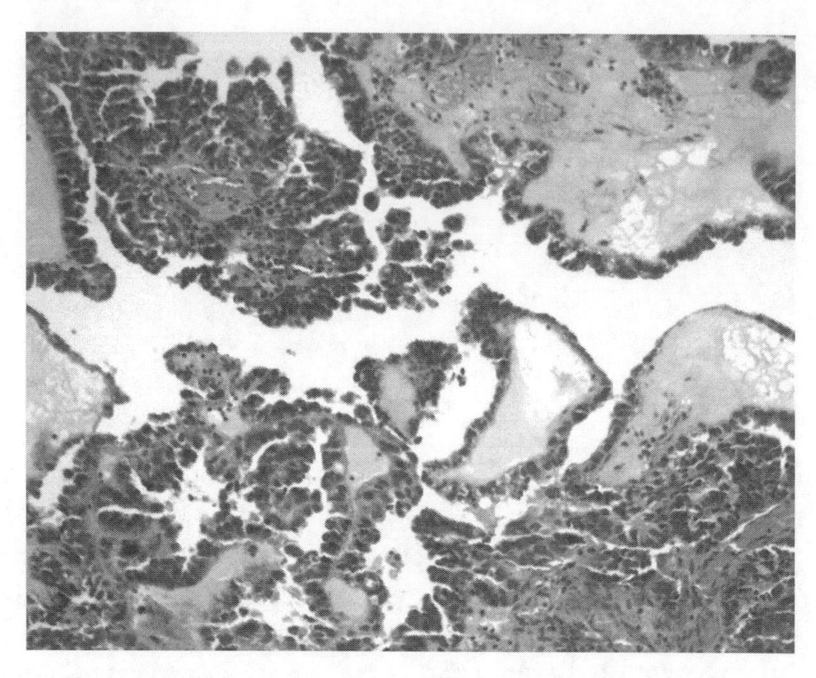

Fig. 5. Epithelial cell ovarian carcinoma of serous type. Epithelial cells display high-grade nuclei and complex papillary growth with tree-like pattern. Courtesy of Chad Livasy, MD.

cord-stromal tumors occur in children and young women. For example, juvenile granulosa cell tumors present in children with symptoms of high estrogen exposure, or precocious pseudopuberty (23, 24). Different types of ovarian cancer present with unique symptoms and histologies; however, tumor grading criteria are usually applied to epithelial tumors (24, 25).

Epithelial tumors arise as adenocarcinomas from the transformation of ovarian coelomic epithelium and surrounding stroma (21, 22). Histological subtypes of epithelial tumors include papillary serous, mucinous, endometrioid, clear cell, and transitional cell or Brenner tumors (22). Serous carcinoma, the most common subtype of epithelial ovarian tumor, has a mixture of cystic, papillary, or solid growth patterns which infiltrate surrounding fibrotic stroma (Fig. 5). Psammoma bodies or small areas of calcification around products of cellular breakdown are common in serous carcinomas. Mucinous carcinomas are large, multilocular cystic tumors composed of columnar cells with stratified nuclei and coarse chromatin. Endometrioid carcinomas have glandular or papillary architecture and resemble endometrial adenocarcinomas. Cells that are cuboidal or polygonal with abundant cytoplasmic glycogen and central vesicular nuclei characterize clear cell carcinomas. Finally, Brenner tumors are solid, fibrous, and composed of epithelial cells resembling transitional cell carcinoma of the bladder (23, 24). Although all malignant epithelial tumors have metastatic potential, serous and clear cell tumors carry a worse prognosis than the other subtypes (24, 25).

All epithelial ovarian tumors are assigned both nuclear and architectural grades on a scale from one to three (25). Nuclear grade one cells have mildly enlarged, uniform nuclei with evenly dispersed chromatin. Nuclear grade three cells have enlarged and pleomorphic nuclei with irregular coarse chromatin and prominent nuclei. Architectural grade one ovarian carcinomas are well differentiated and composed predominantly of glands. Architectural grade three cells are poorly differentiated and composed mostly of solid areas rather than glandular structures (25).

3.5. Future Directions for Cancer Classification

The TNM classification system has evolved over 50 years to accommodate increasing knowledge about cancer biology. Efforts are ongoing to keep the system both synchronized with the most sophisticated cancer technology and simple for ease of clinician/patient use. Debate continues to surround the issue of incorporating cancer grade into the TNM system. Prostate cancer grade is already included in the staging schema, with G1 versus G2-4 distinguishing Stage I (T1a) from Stage II (T1a) disease (2). For many other cancers including lung, breast, and ovarian, however, tumor grading has shown prognostic significance but is not included in the TNM system. This is partially due to variation in tumor grading practices, which leads to lack of quantitative data demonstrating how exactly tumor grade should be incorporated.

Upcoming molecular technologies, such as genomic and proteomic profiling of tumors, microRNA profiling, and even ex vivo living tumor tissue treatment, could improve the current TNM staging system. Some investigators believe that identification and isolation of cancer stem cells has the potential to dramatically shift the paradigm for molecular profiling, since the profile of the bulk tumor cells may not reflect the molecular characteristics of the rare cancer stem cell. Whatever the profiling method, clinicians may eventually tailor cancer therapy to an individual patient's molecular tumor type, rather than anatomic distribution of disease, to maximize treatment success and minimize toxicity. At the same time, rigorous efforts must be made to track treatment effectiveness among groups of patients with similar tumor types. This data is necessary to show statistically significant advantages in prognosis and survival among people receiving individualized cancer therapy. Molecular tumor profiling could direct the future practice of clinical oncology.

Acknowledgments

Special thanks to William Funkhouser, MD, and Chad Livasy, MD, pathologists at the University of North Carolina's School of Medicine, for including their tissue images in this chapter.

References

1. Rubin, P., Williams, J. P., Okunieff, P., Rosenblatt, J. D., Sitzmann, J. V. (2001) Statement of the Clinical Oncologic Problem, in *Clinical Oncology: A Multidisciplinary Approach for Physicians and Students* (Rubin, P., ed.), W.B. Saunders Company, Philadelphia, PA, pp. 1–31.
2. Greene F.L., Page D.L., Fleming, I.D., et al. *AJCC Cancer Staging Manual,* Sixth Edition. New York: Springer, 2002.
3. Spitalnik, P. F., Santagnese, P. A. (2001) The Pathology of Cancer, in *Clinical Oncology: A Multidisciplinary Approach for Physicians and Students* (Rubin, P., ed.), W.B. Saunders Company, Philadelphia, PA, pp. 47–61.
4. Rubin, E., Rubin, R., Aaronson, S. (2005) Neoplasia, in *Rubin's Pathology: Clinicopathologic Foundations of Medicine* (Rubin, E., ed.), Lippincott Williams & Wilkins, Baltimore, MD, pp. 165–213.
5. Houtte, P., McDonald, S., Chang, A. (2001) Lung Cancer, in *Clinical Oncology: A Multidisciplinary Approach for Physicians and Students* (Rubin, P., ed.), W.B. Saunders Company, Philadelphia, PA, pp. 823–844.
6. Moran, C., Suster, S. (2007) Tumors of the lung and pleura, in *Diagnostic Histopathology of Tumors* (Fletcher, C., ed.), Elsevier Limited, Philadelphia, PA, pp. 181–214.
7. Mandel, J., Thomas, K., Weinberger, S. (2007) Overview of non-small cell lung cancer staging, in *UpToDate* (Rose, B., ed.), UpToDate, Waltham, MA, pp.
8. Suster, S., Cesar, M. (2007) Tumors of the Lungs and Pleura, in *Cancer Grading Manual* (Damjanov, I., ed.), Springer, New York, NY, pp. 23–30.
9. Routh, J. C., Leibovich, B. C. (2005) Adenocarcinoma of the prostate: epidemiological trends, screening, diagnosis, and surgical management of localized disease. *Mayo Clin Proc* 80, 899–907.
10. Roach, M., Small, E., Reese, D., Carroll, P. (2001) Urologic and Male Genital Cancers, in *Clinical Oncology: A Multidisciplinary Approach for Physicians and Students* (Rubin, P., ed.), W.B. Saunders Company, Philadelphia, PA, pp. 523–564.
11. Gleason, D. F. (1966) Classification of prostatic carcinomas. *Cancer Chemother Rep* 50, 125–8.
12. Gleason, D. F., Mellinger, G. T. (1974) Prediction of prognosis for prostatic adenocarcinoma by combined histological grading and clinical staging. *J Urol* 111, 58–64.
13. Damjanov, I., Mikuz, G. (2007) Tumors of the Kidney and the Male Urogenital System, in *Tumor Grading Manual* (Damjanov, I., ed.), Springer, New York, NY, pp. 55–63.
14. Ro, J., Amin, M., Kim, K., Ayala, A. (2007) Tumors of the male genital tract: Prostate and seminal vesicles, in *Diagnostic Histopathology of Tumors* (Fletcher, C., ed.), Elsevier Limited, Philadelphia, PA, pp. 749–811.
15. Prosnitz, L., Iglehart, J., Winer, E. (2001) Breast Cancer, in *Clinical Oncology: A Multidisciplinary Approach for Physicians and Students* (Rubin, P., ed.), W.B. Saunders Company, Philadelphia, PA, pp. 252–266.
16. Esserman, L. J., Esserman, L. E. (2007) Diagnostic evaluation and initial staging work-up of women with suspected breast cancer, in *UpToDate* (Rose, B., ed.), UpToDate, Waltham, MA, pp.
17. Ellis, I., Pinder, S., Lee, A. (2007) Tumors of the breast, in *Diagnostic Histopathology of Tumors* (Fletcher, C., ed.), Elsevier Limited, Philadelphia, PA, pp. 903–970.
18. Lagios, M., Silverstein, M. (2007) Breast ductal carcinoma in situ and microinvasive carcinoma, in *UpToDate* (Rose, B., ed.), UpToDate, Waltham, MA, pp.
19. Singletary, S. E., Connolly, J. L. (2006) Breast cancer staging: working with the sixth edition of the AJCC Cancer Staging Manual. *CA Cancer J Clin* 56, 37–47; quiz 50–31.
20. Fan, F., Thomas, P. (2007) Tumors of the Breast, in *Tumor Grading Manual* (Damjanov, I., ed.), Springer, New York, NY, pp. 75–81.
21. Bhoola, S., Hoskins, W. J. (2006) Diagnosis and management of epithelial ovarian cancer. *Obstet Gynecol* 107, 1399–410.
22. Chen, L., Berek, J. (2007) Epithelial ovarian cancer: Clinical manifestations, diagnostic evaluation, staging, and histopathology, in *UpToDate* (Rose, B., ed.), UpToDate, Waltham, MA, pp.
23. Perez, C., Grigsby, P., Mutch, D., Chao, K., Basil, J. (2001) Gynecologic Tumors, in *Clinical Oncology: A Multidisciplinary Approach for Physicians and Students* (Rubin, P., ed.), W.B. Saunders Company, Philadelphia, PA, pp. 462–522.
24. Zaloudek, C. (2007) Tumors of the female genital tract: Ovary, fallopian tube and broad and round ligaments, in *Diagnostic Histopathology of Tumors* (Fletcher, C., ed.), Elsevier Limited, Philadelphia, PA, pp. 567–651.
25. Fan, F., Damjanov, I. (2007) Tumors of the Female Genital Organs, in *Tumor Grading Manual* (Damjanov, I., ed.), Springer, New York, NY, pp. 64–74.

Chapter 2

Clinical Trial Design in the Age of Molecular Profiling

Alexander Spira and Kirsten H. Edmiston

Abstract

The accelerating science of molecular profiling has necessitated a rapid evolution in clinical trial design. Traditional clinical research begins with Phase I studies to characterize dose-limiting toxicities and defines maximally tolerated doses of drugs in small numbers of patients. Traditional Phase II studies test these drugs at the doses discovered during Phase I drug development in small numbers of patients evaluating efficacy and safety. Phase III studies test new therapies to demonstrate improved activity or improved tolerability compared with a standard of care regimen or a placebo. The rapid advances in the understanding of signal transduction, and the identification of new potential diagnostic and therapeutic targets, now require the design and implementation of molecular clinical trials that are very different than traditional Phase I, II, or III trials. The main differentiating factor is the use of a molecular end point to stratify a subset of patients to receive a specific treatment regimen. This chapter focuses on the issues surrounding (a) the definition of clinical end points and the assessment of tumor response; (b) clinical trial design models to define the targeted pathway; and (c) the need for appropriate biomarkers to monitor the response.

Key words: Biomarkers, Clinical trial, Design, End points, Imaging, Pharmacodynamics, Tumor

1. Introduction

1.1. The Definition of Clinical End Points and Assessment of Tumor Response

Establishing the efficacy of a cancer drug is obviously the most important goal of oncology drug development. The gold standard end point is overall survival (OS), i.e., the time from diagnosis to death. Progression-free survival (PFS) is a shorter end point related more directly to the quality of life. Traditionally, response rates (RRs) as determined by tumor volume changes from imaging studies, and physical exams, have been used by investigators and patients to measure drug efficacy (e.g., "What are the chances of Drug X will cause a meaningful shrinkage in my cancer?"). Due to a multitude of reasons, tumor RR is not the best end point, and is not generally accepted in obtaining regulatory approval. Instead, the

Virginia Espina and Lance A. Liotta (eds.), *Molecular Profiling: Methods and Protocols*, Methods in Molecular Biology, vol. 823,
DOI 10.1007/978-1-60327-216-2_2, © Springer Science+Business Media, LLC 2012

more stringent measurement of overall survival as the "defining" end point reflects what is viewed as the most significant clinical benefit. While overall survival is frequently utilized by the US Food and Drug Administration (FDA) as a major end point in the drug approval process, it presents significant problems in molecular target-based therapy (1). Phase II and III studies can often take one or more years to accrue adequately large numbers of patients. An additional number of years must pass before the appropriate numbers of patients live out the natural course of their disease. The prolonged time frame required to assess OS is the reason that this end point may not be useful for phase (Phase II) studies. Overall survival is also affected by the use of subsequent lines of therapies as well as improvements in supportive care that may obscure the effects of a particular therapeutic agent. Moreover, any prolonged time courses add to the cost of clinical trial development as one gets further along in drug development. With the increase in costs to bring novel agents to the market, pharmaceutical manufacturers have reduced incentive to evaluate novel risky therapies.

Determining the efficacy of a therapy by OS alone ignores reduction of symptoms and improvements in the quality of life. Based on all the drawbacks of using the traditional OS end point, clinicians are now utilizing alternative measurement end points that are reached earlier in the treatment course and may be more meaningful to the patient. The approval of gemcitabine in advanced pancreatic cancer was precisely based on symptomatic improvement as the survival benefit was negligible (as were the response rates). Newer clinical trials are incorporating quality-of-life (QOL) measurements, such as the "Lung Cancer Symptom Scale" which is a subjective QOL questionnaire filled out by patients and nurses that reports subjective symptoms, in addition to the standard measurements of response rate and overall survival.

Clinical trial end points, such as time to progression (TTP), which is the time from randomization to the time of progressive disease, may be used as a surrogate for OS. TTP and PFS have been correlated with OS in patients with rectal cancer (2). These end points offer several advantages over the traditional OS in clinical trials design. Both TTP and PFS permit smaller sample sizes and shorter study durations (e.g., months as compared to years). In addition, TTP and PFS do not require demonstration of tumor mass shrinkage. Thus, these end points are useful in trials designed to evaluate cytostatic agents that arrest growth but do not shrink the tumor. TTP and PFS may be measured in real time after a single line of therapy. Consequently, when these measures are used, the designation of a response is not confounded by subsequent events. The disadvantage of TTP and PFS, as compared to OS, in clinical trials is the requirement for costly, frequent, and careful imaging assessment for progression (3, 4).

In order to rapidly assess response, particularly in Phase I and II studies, surrogate measurement techniques have been adopted and standardized to assess results and to compare results across trials. Since 2000, the Response Evaluation Criteria in Solid Tumors (RECIST) criteria have been used to assess response (5). These criteria use the premise that tumor shrinkage reflects a positive outcome of antineoplastic therapy. Measurements for RECIST are based on two-dimensional measurements derived from computed tomography (CT) scans or magnetic resonance imaging (MRI). The RECIST criteria require the radiologist to measure the aggregate of all target tumors in their longest dimensions. Complete resolution of all tumors for at least 4 weeks is a complete response (CR); a 30% shrinkage is a partial response (PR); a 20% increase is progressive disease (PD); and everything else is stable disease (SD) (Table 1). As an example, the drug sorafenib was recently approved for use in patients with advanced hepatocellular carcinoma (HCC) (6, 7). The use of sorafenib was found to prolong survival in patients with HCC by 3 months and was associated with a 31% increase in OS at 1 year. Nevertheless, the RECIST response rate was only 2% (6). The same drug has been studied in patients with advanced renal cell cancer (8). The response rate was only 10%; but this constituted a significant prolongation in survival that led to FDA approval of this agent in advanced renal cell cancer.

There are a number of reasons why RECIST-defined response rates may be problematic end points. First, tumor cytotoxicity may not result in rapid shrinkage, especially in tumors that induce large amounts of stroma (rather than cellular elements that may "die" with chemotherapy). Stroma-rich tumors include sarcomas, lung cancers, and pancreatic cancers. According to RECIST criteria, a tumor that shrinks 29% and a tumor that grows 19% are both considered stable disease while these two responses are clearly different (9). Furthermore, even with modern imaging, there is some scan variability, especially with lesions close to 10 mm, the minimal slice size used in modern CT imaging. Not all cancers are amenable to accurate imaging on scans. For example, a lung cancer can cause adjacent lung atelectasis (collapse) that is difficult to discern from adjacent tumor and hence make accurate calculations difficult. Lastly,

Table 1
Summary of RECIST response criteria (*9*)

Stage	Definition
Complete response (CR)	Complete resolution of the tumor for a least 4 weeks
Partial response (PR)	Greater than 30% decrease in tumor sustained for a least 4 weeks
Progressive disease (PD)	At least 20% increase in tumor size with no CR, PR, or SD documented before the increase of disease
Stable disease (SD)	Neither PR nor PD criteria are met

RECIST criteria cannot be used in tumors with primarily bone lesions since these lesions do not shrink, making it difficult in the assessment of diseases, such as prostate cancer, as well as in many hematologic malignancies that cannot be measured with tumor size.

Recent advances in positron emission tomography (PET), MRI, and dynamic contrast-enhanced MRI (DCE-MRI) have yielded functional methods for tumor staging and assessment of tumor response. PET scans use (18)F-fluorodeoxyglucose (FDG) which accumulates in metabolically active tumor cells and can be described as a "functional" assessment of malignancies. After treatment, decrease in FDG uptake and metabolism correlates with a reduction in tumor viability. To achieve pharmacodynamic end points of tumor response, measurements are made by combining size and tracer uptake activity (standardized uptake value, "SUV"). FDG-PET scans have been very effective in measuring therapeutic responses in gastrointestinal stromal tumors, head and neck, breast, lung, esophageal lymphomas, and high-grade sarcomas (3). Despite many studies demonstrating its usefulness, FDG-PET still has not been utilized in most clinical research due to cost and difficulties in reimbursement as well as lack of an "official" reference standard to date, akin to RECIST (10). Rather than adopting FDG-PET, the US National Cancer Institute uses traditional CT scans as a method of evaluation for assessing response.

MRI and DCE-MRI are increasingly being utilized within clinical trials particularly to measure pharmacodynamic end points for novel antiangiogenic and antivascular targeting agents (11). MRI can provide a functional assessment of tumor physiology through the pattern of progressive enhancement and the change in washout kinetics. In this manner, MRI can provide a functional impression of tumor response superior to tumor shrinkage (Fig. 1).

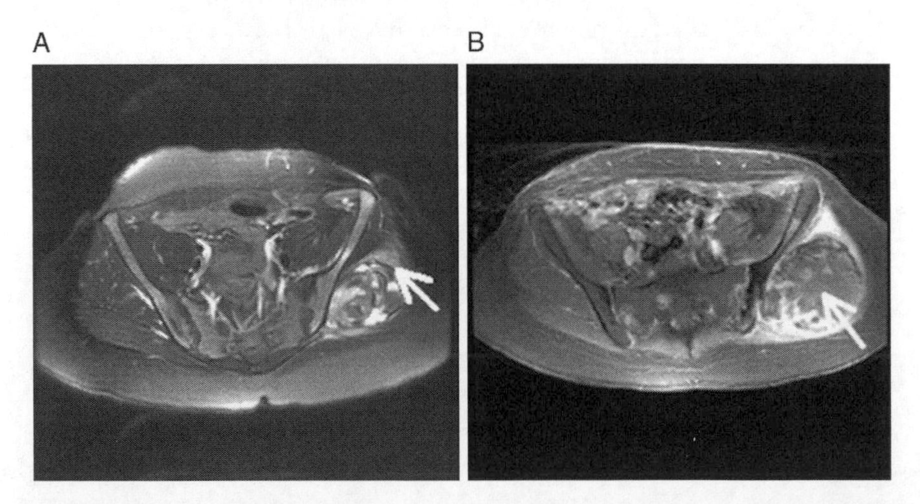

Fig. 1. Example of RECIST criteria for determining tumor size following treatment. Cystosarcoma phylloides of the breast metastatic to the left pelvis before (**a**) and after (**b**) chemotherapy. The size of the tumor increased by approximately 20% by RECIST criteria, yet was felt to be largely necrotic on imaging. At the time of surgery, the tumor was largely necrotic consistent with a good response to chemotherapy. *Arrowheads* indicate the mass. Photo courtesy of A. Spira, Virginia Oncology Services.

This has been well-demonstrated in high-grade soft-tissue sarcomas with the correct prediction of tumor response in 80% of evaluable patients after isolated limb perfusion (12). In addition, MRI is particularly useful for assessing the central nervous system and the bone marrow infiltration that are not well-seen by FDG-PET. MRI is also useful for assessing metastasis and disease progression.

Overall, the accurate assessments of clinical and imaging response rates are critical to establish the pharmacodynamic response of the potential therapy within the target organ. Many clinical trials were done in the time period (1980s and 1990s) prior to the adoption of RECIST criteria. Moreover, the resolution, sensitivity, and specificity of imaging technology have advanced tremendously in the last 10 years. In this light, the use of historical comparisons as reference points to older studies is fraught with potential for error. Additional work needs to be done to integrate these newer noninvasive imaging modalities within clinical trials to provide the "pharmacological audit trail" necessary for drug validation and approval (13).

2. Models for Clinical Trial Design

In many respects, the concept of patient stratification in clinical trials has been around for many years. Many clinical trials have been designed to compare Drug A against Drug B, or placebo, for all patients with a particular stage of a general type of cancer. An example is carboplatin and paclitaxel that were judged to be active in lung cancer (14). The overall observed response was the net combination of the number of beneficial outcomes offset by the number of patients receiving no benefit/toxic outcomes. Traditional chemotherapeutic agents are typically classified as cytotoxic drugs, targeting many pathways of cellular replication and division. Over the past 10 years, there have been rapid advances in molecular profiling technologies (as described in the subsequent chapters) and greater understanding of the specific pathways and upstream regulating molecules responsible for the malignant process. In parallel, there has been an expansion of national and international research collaborations, an increase in data sharing among clinical groups, and an enlargement in the size of patient cohorts in major trials (15). Consequently, researchers are now able to accurately profile patient tumors and design tailored therapy clinical trial models with a high degree of sophistication.

2.1. Clinical Trial Design for Subgroups of Patients with Specific Clinical Attributes

Trials that subdivide patients into defined groups are designed to answer specific treatment questions based on clinical or histologic characteristics (16, 17). These trials often require large numbers of patients and strong research collaborations. The following example demonstrates the value of patient stratification by histology.

2.1.1. Pemetrexed in Lung Cancer

Lung cancer is frequently called "non-small-cell lung cancer (NSCLC)" due to the fact that years of drug development never demonstrated that the identification of a subtype (usually, squamous versus adenocarcinoma) had any bearing on the use of a particular chemotherapeutic choice, hence the broad definition of NSCLC that is usually treated with platinum-based therapy (in the USA, usually carboplatin and paclitaxel) (14).

Pemetrexed is an intravenous methotrexate analog that was FDA approved in 2004 for pleural mesothelioma and as second-line NSCLC therapy due to its tolerability and safety. In an attempt to bring pemetrexed into the first-line setting, gemcitabine/pemetrexed was compared with gemcitabine/cisplatin. Based upon previous leads, histology was identified prestudy to be part of a subgroup analysis (18). Of note, patients with squamous histology did worse and lived for a shorter time when treated with pemetrexed (9.4 months vs. 10.8 months), but patients with adenocarcinoma lived longer when treated with pemetrexed (12.6 months vs. 10.9 months). This data led to the subsequent FDA labeling change specifying that pemetrexed could only be used in nonsquamous (and mainly adenocarcinoma) histologies. To date, this is the only study using traditional chemotherapy (vs. targeted chemotherapy) that demonstrated an impact of histology on outcome.

2.2. Clinical Trial Design Using Prognostic Biomarkers

Within the treatment of breast cancer, a number of different multigene marker sets have been developed and validated to predict clinical outcomes more accurately than traditional clinicopathologic features. These include the 21 gene set (19, 20) and a 70 gene set (21, 22). The 21 gene biomarker (oncotype DX®) has been established and well-validated in hormone receptor-positive, lymph node-negative patients. It is used to predict the low, intermediate, and high recurrence risks. The oncotype DX® score identifies patients who would benefit from hormonal therapy only or would benefit from chemotherapy and hormonal therapy in combination. To further evaluate this prognostic multigene biomarker in clinical decision making, the North American Breast Cancer Intergroup developed the TAILORx trial in 2006 (23). Using a two-way stratified design model, the study first stratifies the patients based on their oncotype DX® recurrence risk score (RRS). Patients with a low RRS (<18) get hormonal therapy alone. Patients with a high RRS (>31) receive chemotherapy and hormonal therapy. Those patients with an intermediate RRS (18–30) are randomized to receive either hormonal therapy only or a combination of chemotherapy and hormonal therapy. This design enabled the Intergroup to take advantage of what was already known about the biomarker in clinical practice to address an important question in a practical manner. This study also utilizes a noninferiority design for the intermediate group and has the statistical power to detect a 3% or greater difference between the randomized arms (24). This design also provides further, although indirect, validation of the biomarker.

To clinically validate the 70 gene set biomarker (originally identified by Van't Veer and coworkers (22)), the EORTC and TRANSBIG used a classifier randomization design. The multigene biomarker score was compared to a common clinical pathological prognostic tool (Adjuvant! Online (25, 26)) to identify patients for adjuvant chemotherapy in node-negative breast cancer (Microarray In Node Negative Disease may Avoid Chemotherapy, MINDACT, trial) (27, 28). This study sought to confirm that patients with a "low-risk" molecular prognosis and "high-risk" clinical prognosis could be safely spared chemotherapy without affecting disease-free survival. In addition, this study compared anthracycline-based chemotherapy to a docetaxel–capecitabine regimen and evaluates the efficacy and safety of 7 years of single-agent Letrozole to sequential 2 years of tamoxifen followed by 5 years of Letrozole (see Note 1).

2.3. Clinical Trial Design to Validate Biomarkers to Predict Clinical Response

In the development of "–omic" (genomic, proteomic, and metabolomic)-targeted therapies, it is critical to test both the utility of predictive/target biomarker as well as the utility of particular therapeutic agent, whether they are inhibitor- or biomarker-directed antibodies (29). A predictive biomarker is one that predicts the differential efficacy of a particular therapy based on the biomarker status (15). Identifying those patients who benefit and those patients who experience toxicity without efficacy is essential given the potential morbidity and cost of these targeted therapies. Many molecular targeted drugs can reach $10,000 per month of treatment, hence making the selection of targeted therapy even more appropriate in the setting of rising health costs.

The earliest example of patient stratification was the demonstration that estrogen-receptor (ER)-positive breast cancers could be treated with hormonal manipulation by tamoxifen and subsequently aromatase inhibitors. The case study of HER-2/neu and the subsequent development of trastuzumab are instructive in development of clinical trials to validate biomarkers.

2.3.1. Breast Cancer and Trastuzumab

HER-2/neu belongs to a family of transmembrane receptor tyrosine kinases that influence cell growth, differentiation, and survival. It is expressed in many normal cells as well as breast cancer cells in particular. Amplification of the HER-2/neu gene, overexpression of the HER-2/neu protein, or both occurs in approximately 25–30% of patients with breast cancer. HER-2/neu-positive patients have a high rate of recurrence and a short disease-free interval after adjuvant conventional anthracycline-based chemotherapy (30–32).

The first antibody discovered against HER-2/neu was trastuzumab, which is a very active intravenous antibody that targets the HER2 extracellular domain (33). The only clinically available biomarker predictor of responsiveness to trastuzumab is HER-2/neu status (30–32). During the initial development and testing of trastuzumab, there were two methods to demonstrate HER-2/neu status – immunohistochemistry (IHC) and fluorescence in situ hybridization (FISH). Traditional pathology looks for activity by the amount of immunohistochemical staining a tumor specimen has for a particular agent. Typically, the pathologist applies the anti-HER2/NEU antibodies, developing agents, and then scores what they see under the microscope (typically, 0, 1+, 2+, or 3+ staining). The disadvantage of IHC scoring is the variability in the quality of the staining, subjectivity by the scoring pathologist, and tumor sampling error. With the advent of FISH testing for HER2 amplification, a much more accurate measurement of HER2 expression can be done without the aforementioned subjectivity (34). FISH measures the fluorescence of the tagged HER2 gene. The amount of fluorescence is proportional to the amplification (number of copies) of the gene. The reproducibility of ICH and FISH testing at the local labs was compared to centralized labs during the Breast Intergroup Trial N9831 on the role of trastuzumab in breast cancer (35). This trial reported strong concordance between central IHC and FISH testing (92%) but poor concordance (74%) between local and central testing for HER-2/neu, thus underscoring the need for standardized testing for biomarkers. To establish consistency in tumor marker prognostic studies across preclinical and clinical trials, the US National Cancer Institute and the European Organization for Research and Treatment came together in July 2000 to publish the REporting recommendations for tumour MARKer prognostic studies (REMARK) (Table 2) (36).

To demonstrate clinical utility, Phase I clinical trials showed that the antibody is safe and confined to the tumor (unpublished

Table 2
REporting recommendations for tumour MARKer prognostic studies (REMARK) (*36*)

Introduction
1. State the marker examined, study objectives, and any prespecified hypotheses
Materials and Methods
Patients
2. Describe the characteristics (e.g., disease stage or comorbidities) of the study patients, including their source and inclusion and exclusion criteria
3. Describe treatments received and how chosen (e.g., randomized or rule based)
Specimen characteristics
4. Describe the type of biological material used (including control samples), and methods of preservation and storage
Assay methods
5. Specify the assay method used and provide (or reference) a detailed protocol, including specific reagents or kits used, quality control procedures, reproducibility assessments, quantitation methods, and scoring and reporting protocols. Specify whether and how assays were performed blinded to the study end point

(continued)

Table 2
(continued)

Study design
6. State the method of case selection, including whether prospective or retrospective and whether stratification or matching (e.g., by stage of disease or age) was employed. Specify the time period from which cases were taken, the end of the follow-up period, and the median follow-up time
7. Precisely define all clinical end points examined
8. List all candidate variables initially examined or considered for inclusion in models
9. Give rationale for sample size; if the study was designed to detect a specified effect size, give the target power and effect size
Statistical analysis methods
10. Specify all statistical methods, including details of any variable selection procedures and other model-building issues, how model assumptions were verified, and how missing data were handled
11. Clarify how marker values were handled in the analyses; if relevant, describe methods used for cut-point determination
Results
Data
12. Describe the flow of patients through the study, including the number of patients included in each stage of the analysis (a diagram may be helpful) and reasons for dropout. Specifically, both overall and for each subgroup extensively examined report the numbers of patients and the number of events
13. Report distributions of basic demographic characteristics (at least age and sex), standard (disease specific) prognostic variables, and tumor marker, including numbers of missing values
Analysis and presentation
14. Show the relation of the marker to standard prognostic variables
15. Present univariate analyses showing the relation between the marker and outcome, with the estimated effect (e.g., hazard ratio and survival probability). Preferably provide similar analyses for all other variables being analyzed. For the effect of a tumor marker on a time-to-event outcome, a Kaplan–Meier plot is recommended
16. For key multivariable analyses, report estimated effects (e.g., hazard ratio) with confidence intervals for the marker and, at least for the final model, all other variables in the model
17. Among reported results, provide estimated effects with confidence intervals from an analysis in which the marker and standard prognostic variables are included, regardless of their significance
18. If done, report results of further investigations, such as checking assumptions, sensitivity analyses, internal validation
Discussion
19. Interpret the results in the context of the prespecified hypotheses and other relevant studies; include a discussion of limitations of the study
20. Discuss implications for future research and clinical value

data). Subsequent Phase II trials demonstrated that many women with HER2-positive metastatic disease who had relapsed after chemotherapy had a response to trastuzumab; as suggested by the preclinical data, the efficacy of trastuzumab when given with chemotherapy was superior to its effectiveness when used alone (37). The central issues in the Phase I and II studies were to identify the optimal biologic dose and the best dosing schedule to optimize target binding and inhibition (38). Slamon et al. reported the results of the Phase III trial in which women with cancers that overexpressed HER2 who had not previously received chemotherapy for metastatic disease were randomly assigned to receive chemotherapy alone or chemotherapy plus trastuzumab (39). The primary end points of the study were the time to disease progression and the incidence of adverse effects. Secondary end points were the rates and the duration of responses, the time to treatment failure, and overall survival. The addition of trastuzumab to chemotherapy was associated with a longer time to disease progression (median, 7.4 vs. 4.6 months; $P<0.001$), a higher rate of objective response (50% vs. 32%, $P<0.001$), a longer duration of response (median, 9.1 vs. 6.1 months; $P<0.001$), a lower rate of death at 1 year (22% vs. 33%, $P=0.008$), longer survival (median survival, 25.1 vs. 20.3 months; $P=0.046$), and a 20% reduction in the risk of death (39). Further work is being conducted to refine the clinical utility for trastuzumab and additional biomarkers to identify the further subset of HER-2/neu-positive patients that respond to trastuzumab.

2.4. Clinical Trial Design to Test Specific Pathway/Targeted Molecules

The next generation of therapies is designed to suppress specific cellular protein signal pathways (rather than a general cytotoxic approach), driving the cancer cell. Clinically, the most successful example of this approach involves the highly selective tyrosine kinase inhibitor, imatinib, in the treatment of chronic myelogenous leukemia and gastrointestinal stromal tumors. The new generation of agents is small molecules with selective tyrosine kinase inhibitory activity (e.g., erlotinib, imatinib, sutininib, sorafenib) or antibodies to epidermal growth factor receptors (EGFRs), CD20, or vascular endothelial growth factor (cetuximab, bevacizumab, panitumumab, rituximab). The evolving role of *K-ras* mutations as a predictor of the lack of the effectiveness of EGFR inhibitors underscores the importance of understanding the comprehensive pathway in targeted therapies.

2.4.1. EGFR Inhibitors and K-ras Mutations

The EGFR is transmembrane receptor that is a family of molecules both required for normal cell development as well as for the proliferation and growth of malignant cells, particularly lung, pancreas, head and neck, and colorectal cancers (40). Both small-molecule inhibitors of the receptor are currently available (erlotinib, gefitinib) as well as monoclonal antibodies (panitiumimab and cetuximab) that inhibit this family of receptors.

At best, the efficacy of these drugs are modest, with an RR of 8% with cetuximab in colorectal cancer, a prolonged stable disease rate of 31% compared with 11% given the best supportive care alone, and an improvement in survival of 1.5 months; more than 50% of patients did not go beyond a single disease assessment due to early progressive disease (41). Even though the EGFR signaling pathway is important in cancer cell proliferation, it is not the only step in this pathway and can easily be bypassed by alternative or downstream pathways. Subsequent to one of the pivotal studies that led to the approval of cetuximab, the role of *K-ras* was studied (42). *K-ras* is a small G-protein downstream of EGFR, and mutations in exon 2 can become "activating," hence isolating *K-ras* activity from upstream EGFR signaling. Karpetis et al. found that patients with wild-type (wt) *K-ras* treated with cetuximab had nearly double the survival (4.8 vs. 9.5 mos) compared with BSC; yet those patients with mutant *K-ras* treated with survival had the identical median survival (4.5 vs. 4.6 months) (Fig. 2) (41, 42). Hence, by understanding pathway activation, one could identify the appropriate treatment for a patient. Similar results were also seen for the monoclonal antibody panitumimab in colorectal cancer (43). It is expected that there will be other downstream molecules that are likely to be just as important as *K-ras*, as well as other parallel pathways as well, independent of the EGFR pathway.

2.4.2. EGFR and Lung Cancer

Small-molecule EGFR inhibitors have been available for several years. They are associated with modest response rates, on the order of 10%. While these drugs are modestly effective for the overall population, a small percentage (8–15%) of patients have a dramatic response that lasted far longer than the average (i.e., years vs. months). Soon after the responder subgroup was first identified, clinical and epidemiologic criteria were found that predicted a higher-than-normal likelihood of benefit. These criteria included the following set: Asian race, adenocarcinoma histology, female, nonsmokers. Researchers went on to identify the molecular basis for the dramatic responses by studying changes in the EGFR molecule that would correlate with these responses. Two groups of researchers in Boston, MA, identified mutations in the ATP-binding pocket of EGFR that strongly correlated with tumor response to gefitinib (44, 45). A subsequent study looked at the use of gefitinib as front-line therapy in patients with EGFR-activating mutations, and found a striking response rate of 55% (46). In this study, interestingly, only two patients with EGFR-activating mutations (out of 34) demonstrated resistance (i.e., early progression) on gefitnib, and subsequent analysis demonstrated that one had a subsequent mutation in EGFR associated with gefitinib resistance (T790M) while the other had MET amplification, which is also associated with gefitnib resistance (46). This work further underscores the need to understand the targeted pathway as well as the impact of various receptor mutations on determining responsiveness to specific targeted therapies.

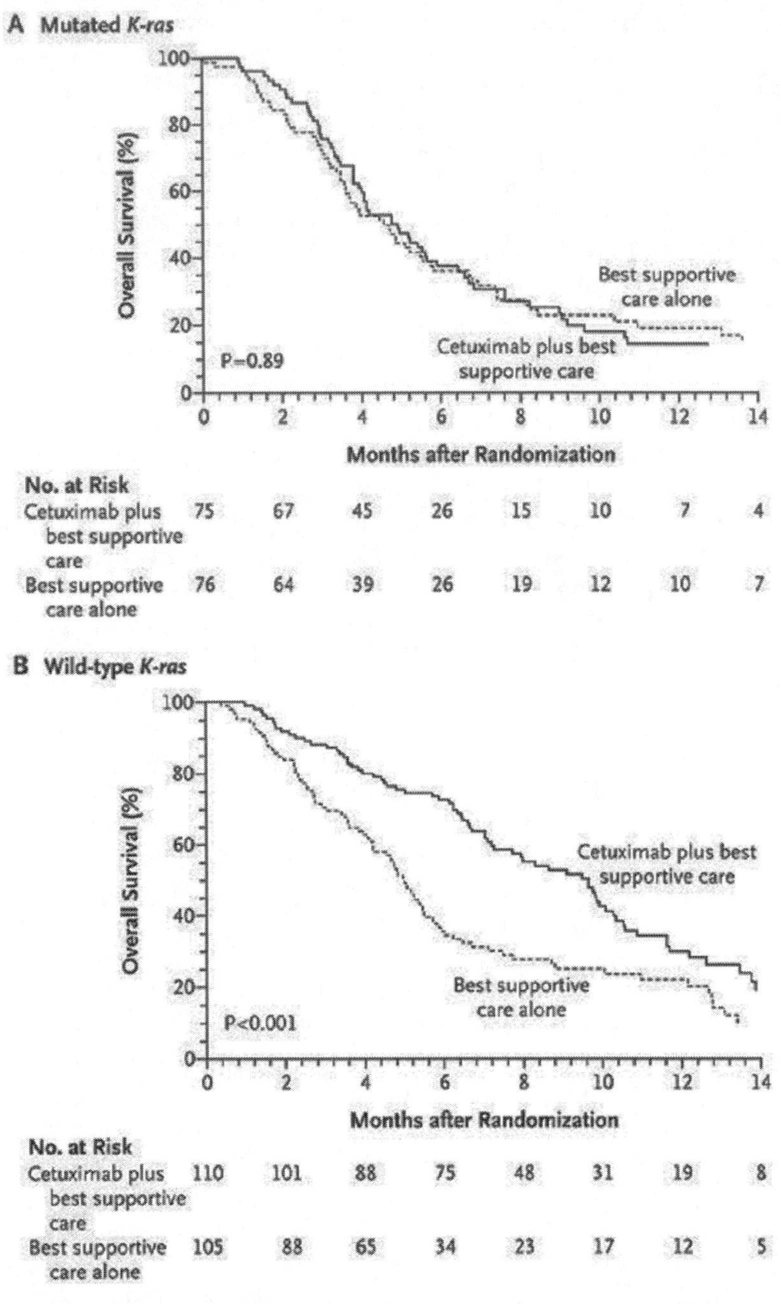

Fig. 2. Comparison of overall survival for patients post cetuximab treatment based on *K-ras* mutation status. Non-small-cell lung cancer patients harboring wild-type or mutated *K-ras* were treated with cetuximab. Panel (**A**) shows results for patients with mutated *K-ras* tumors, and panel (**B**) for patients with wild-type *K-ras* tumors. Cetuximab as compared with best supportive care alone was associated with improved overall survival among patients with wild-type *K-ras* tumors but not among those with mutated *K-ras* tumors. The difference in treatment effect according to mutation status was significant (test for interaction, $P = 0.01$). Reprinted with permission from NEJM (2008).

3. Conclusions

We hope to leave the reader with several thoughts going forward. First, trial design must ideally be hypothesis driven, biomarker guided, and individualized based on the biology of tumors. As good as modern imaging is, the use of response rate and overall survival has many limitations and should be applied only in the context of the overall clinical picture of the patient's disease course and QOL. The traditional Phase II response rate end point based on RECIST is being supplanted by OS end points and "TTP", i.e., the waiting time before the tumor begins to grow again after initial growth arrest. The thought of "prolonged stable disease" emerged, based on the above understanding, as a viable end point and surrogate for survival.

In the future, it is imperative to design studies that match a therapy with a specific molecular correlate of response: e.g., companion diagnostic. Techniques, such as genomic microarrays, reverse-phase protein microarray analysis, DNA mutation studies, and even IHC, are very important in identifying drug target pathways. Targeted therapy, by definition, treats a pathologic signaling pathway that drives the cancer. The responsibility falls to the designers of the clinical trial to identify the nodes in the pathway affected by a particular drug and to use this information to predict what surrogates or molecular end points can be used to stratify patients. In the past, patients were stratified by histology alone. In the future, DNA mutations in a particular gene, RNA transcript profiles, or proteomic profiles, including the activation state or phosphorylation of a protein in the target signal pathway, will constitute molecular "theranostics" (47). Stratification of patients by molecular profiling increases the likelihood of response while sparing toxicity with no treatment benefit. Molecular stratification may allow drugs with significant activity to be detected in small populations, obviating the time delay to accrue and study large populations of patients. Ideally, the oncologist of the future will not treat patients based on their specific organ category of disease, i.e., adenocarcinoma of the colon. Instead, they will treat the molecular pathway defect itself, which may be independent of histology.

4. Notes

1. A summary of the MINDACT trial is available at http://www.eortc.be/services/unit/mindact/documents/MINDACT_trial_outline.pdf.

References

1. Pazdur, R. (2008) Endpoints for assessing drug activity in clinical trials. *Oncologist* **13 Suppl 2**, 19–21.
2. Glynne-Jones, R., Mawdsley, S., Pearce, T., Buyse, M. (2006) Alternative clinical end points in rectal cancer--are we getting closer? *Ann Oncol* **17**, 1239–48.
3. Schuetze, S. M., Baker, L. H., Benjamin, R. S., Canetta, R. (2008) Selection of response criteria for clinical trials of sarcoma treatment. *Oncologist* **13 Suppl 2**, 32–40.
4. Scott, J., McGettigan, G. (2005) Regulatory approvals for oncology products based on accelerated clinical development and limited data packages-2. *Regulatory Rapporteur* **2**, 6–15.
5. Therasse, P., Arbuck, S. G., Eisenhauer, E. A., Wanders, J., Kaplan, R. S., Rubinstein, L. et al. (2000) New guidelines to evaluate the response to treatment in solid tumors. European Organization for Research and Treatment of Cancer, National Cancer Institute of the United States, National Cancer Institute of Canada. *J Natl Cancer Inst* **92**, 205–16.
6. Llovet, J. M., Ricci, S., Mazzaferro, V., Hilgard, P., Gane, E., Blanc, J. F. et al. (2008) Sorafenib in advanced hepatocellular carcinoma. *N Engl J Med* **359**, 378–90.
7. Alves, R. C., Alves, D., Guz, B., Matos, C., Viana, M., Harriz, M. et al. (2011) Advanced hepatocellular carcinoma. Review of targeted molecular drugs. *Ann Hepatol* **10**, 21–7.
8. Escudier, B., Eisen, T., Stadler, W. M., Szczylik, C., Oudard, S., Siebels, M. et al. (2007) Sorafenib in advanced clear-cell renal-cell carcinoma. *N Engl J Med* **356**, 125–34.
9. Eisenhauer, E. A., Therasse, P., Bogaerts, J., Schwartz, L. H., Sargent, D., Ford, R. et al. (2009) New response evaluation criteria in solid tumours: revised RECIST guideline (version 1.1). *Eur J Cancer* **45**, 228–47.
10. Hillner, B. E., Siegel, B. A., Liu, D., Shields, A. F., Gareen, I. F., Hanna, L. et al. (2008) Impact of positron emission tomography/computed tomography and positron emission tomography (PET) alone on expected management of patients with cancer: initial results from the National Oncologic PET Registry. *J Clin Oncol* **26**, 2155–61.
11. Evelhoch, J., Garwood, M., Vigneron, D., Knopp, M., Sullivan, D., Menkens, A. et al. (2005) Expanding the use of magnetic resonance in the assessment of tumor response to therapy: workshop report. *Cancer Res* **65**, 7041–4.
12. van Rijswijk, C. S., Geirnaerdt, M. J., Hogendoorn, P. C., Peterse, J. L., van Coevorden, F., Taminiau, A. H. et al. (2003) Dynamic contrast-enhanced MR imaging in monitoring response to isolated limb perfusion in high-grade soft tissue sarcoma: initial results. *Eur Radiol* **13**, 1849–58.
13. Reid, A. H., Baird, R., Workman, P. (2008) Emerging molecular therapies: Drugs interfering with signal transduction pathways, in Principles of Molecular Oncology (Bronchud, M. H., Foote, M., Giaccone, G., Olopade, O., and Workman, P., ed.), Humana Press, Totowa, NJ, pp. 325–329.
14. Weissman, C. H., Reynolds, C. H., Neubauer, M. A., Pritchard, S., Kobina, S., Asmar, L. (2011) A phase III randomized trial of gemcitabine-oxaliplatin versus carboplatin-paclitaxel as first-line therapy in patients with advanced non-small cell lung cancer. *J Thorac Oncol* **6**, 358–64.
15. Therasse, P., Carbonnelle, S., Bogaerts, J. (2006) Clinical trials design and treatment tailoring: general principles applied to breast cancer research. *Crit Rev Oncol Hematol* **59**, 98–105.
16. Mandrekar, S. J., Sargent, D. J. (2009) Clinical trial designs for predictive biomarker validation: theoretical considerations and practical challenges. *J Clin Oncol* **27**, 4027–34.
17. Mandrekar, S. J., Sargent, D. J. (2010) Predictive biomarker validation in practice: lessons from real trials. *Clin Trials* **7**, 567–73.
18. Scagliotti, G. V., Parikh, P., von Pawel, J., Biesma, B., Vansteenkiste, J., Manegold, C. et al. (2008) Phase III study comparing cisplatin plus gemcitabine with cisplatin plus pemetrexed in chemotherapy-naive patients with advanced-stage non-small-cell lung cancer. *J Clin Oncol* **26**, 3543–51.
19. Paik, S., Shak, S., Tang, G., Kim, C., Baker, J., Cronin, M. et al. (2004) A multigene assay to predict recurrence of tamoxifen-treated, node-negative breast cancer. *N Engl J Med* **351**, 2817–26.
20. Paik, S., Tang, G., Shak, S., Kim, C., Baker, J., Kim, W. et al. (2006) Gene expression and benefit of chemotherapy in women with node-negative, estrogen receptor-positive breast cancer. *J Clin Oncol* **24**, 3726–34.
21. van de Vijver, M. J., He, Y. D., van't Veer, L. J., Dai, H., Hart, A. A., Voskuil, D. W. et al. (2002) A gene-expression signature as a predictor of survival in breast cancer. *N Engl J Med* **347**, 1999–2009.

22. van 't Veer, L. J., Dai, H., van de Vijver, M. J., He, Y. D., Hart, A. A., Mao, M. et al. (2002) Gene expression profiling predicts clinical outcome of breast cancer. *Nature* **415**, 530–6.
23. Sparano, J. A. (2006) TAILORx: trial assigning individualized options for treatment (Rx). *Clin Breast Cancer* 7, 347–50.
24. Sparano, J. A., Paik, S. (2008) Development of the 21-gene assay and its application in clinical practice and clinical trials. *J Clin Oncol* **26**, 721–8.
25. Oakman, C., Santarpia, L., Di Leo, A. (2010) Breast cancer assessment tools and optimizing adjuvant therapy. *Nat Rev Clin Oncol* 7, 725–32.
26. Ross, J. S., Hatzis, C., Symmans, W. F., Pusztai, L., Hortobagyi, G. N. (2008) Commercialized multigene predictors of clinical outcome for breast cancer. *Oncologist* **13**, 477–93.
27. Bogaerts, J., Cardoso, F., Buyse, M., Braga, S., Loi, S., Harrison, J. A. et al. (2006) Gene signature evaluation as a prognostic tool: challenges in the design of the MINDACT trial. *Nat Clin Pract Oncol* **3**, 540–51.
28. Cardoso, F., Van't Veer, L., Rutgers, E., Loi, S., Mook, S., Piccart-Gebhart, M. J. (2008) Clinical application of the 70-gene profile: the MINDACT trial. J Clin Oncol 26, 729–35.
29. Schaeybroeck, S. V., Allen, W. L., Turkington, R. C., Johnston, P. G. (2011) Implementing prognostic and predictive biomarkers in CRC clinical trials. *Nat Rev Clin Oncol* doi:10.1038/nrclinonc.2011.15
30. Jacobs, T. W., Gown, A. M., Yaziji, H., Barnes, M. J., Schnitt, S. J. (1999) Specificity of HercepTest in determining HER-2/neu status of breast cancers using the United States Food and Drug Administration-approved scoring system. *J Clin Oncol* **17**, 1983–7.
31. Elkin, E. B., Weinstein, M. C., Winer, E. P., Kuntz, K. M., Schnitt, S. J., Weeks, J. C. (2004) HER-2 testing and trastuzumab therapy for metastatic breast cancer: a cost-effectiveness analysis. *J Clin Oncol* **22**, 854–63.
32. Bartlett, J. M., Going, J. J., Mallon, E. A., Watters, A. D., Reeves, J. R., Stanton, P. et al. (2001) Evaluating HER2 amplification and overexpression in breast cancer. *J Pathol* **195**, 422–8.
33. Goldenberg, M. M. (1999) Trastuzumab, a recombinant DNA-derived humanized monoclonal antibody, a novel agent for the treatment of metastatic breast cancer. *Clin Ther* **21**, 309–18.
34. Wolff, A. C., Hammond, M. E., Schwartz, J. N., Hagerty, K. L., Allred, D. C., Cote, R. J. et al. (2007) American Society of Clinical Oncology/College of American Pathologists guideline recommendations for human epidermal growth factor receptor 2 testing in breast cancer. *Arch Pathol Lab Med* **131**, 18–43.
35. Roche, P. C., Suman, V. J., Jenkins, R. B., Davidson, N. E., Martino, S., Kaufman, P. A. et al. (2002) Concordance between local and central laboratory HER2 testing in the breast intergroup trial N9831. *J Natl Cancer Inst* **94**, 855–7.
36. McShane, L. M., Altman, D. G., Sauerbrei, W., Taube, S. E., Gion, M., Clark, G. M. (2005) REporting recommendations for tumour MARKer prognostic studies (REMARK). *Br J Cancer* **93**, 387–91.
37. Baselga, J., Tripathy, D., Mendelsohn, J., Baughman, S., Benz, C. C., Dantis, L. et al. (1996) Phase II study of weekly intravenous recombinant humanized anti-p185HER2 monoclonal antibody in patients with HER2/neu-overexpressing metastatic breast cancer. *J Clin Oncol* **14**, 737–44.
38. Arteaga, C. L., Baselga, J. (2003) Clinical trial design and end points for epidermal growth factor receptor-targeted therapies: implications for drug development and practice. *Clin Cancer Res* **9**, 1579 89.
39. Slamon, D. J., Leyland-Jones, B., Shak, S., Fuchs, H., Paton, V., Bajamonde, A. et al. (2001) Use of chemotherapy plus a monoclonal antibody against HER2 for metastatic breast cancer that overexpresses HER2. *N Engl J Med* **344**, 783–92.
40. Ciardiello, F., Tortora, G. (2008) EGFR antagonists in cancer treatment. *N Engl J Med* **358**, 1160–74.
41. Jonker, D. J., O'Callaghan, C. J., Karapetis, C. S., Zalcberg, J. R., Tu, D., Au, H. J. et al. (2007) Cetuximab for the treatment of colorectal cancer. *N Engl J Med* **357**, 2040–8.
42. Karapetis, C. S., Khambata-Ford, S., Jonker, D. J., O'Callaghan, C. J., Tu, D., Tebbutt, N. C. et al. (2008) K-ras mutations and benefit from cetuximab in advanced colorectal cancer. *N Engl J Med* **359**, 1757–65.
43. Amado, R. G., Wolf, M., Peeters, M., Van Cutsem, E., Siena, S., Freeman, D. J. et al. (2008) Wild-type KRAS is required for panitumumab efficacy in patients with metastatic colorectal cancer. *J Clin Oncol* **26**, 1626–34.
44. Lynch, T. J., Bell, D. W., Sordella, R., Gurubhagavatula, S., Okimoto, R. A., Brannigan, B. W. et al. (2004) Activating mutations in the epidermal growth factor receptor underlying responsiveness of non-small-cell lung cancer to gefitinib. *N Engl J Med* **350**, 2129–39.

45. Paez, J. G., Janne, P. A., Lee, J. C., Tracy, S., Greulich, H., Gabriel, S. et al. (2004) EGFR mutations in lung cancer: correlation with clinical response to gefitinib therapy. *Science* **304**, 1497–500.
46. Sequist, L. V., Martins, R. G., Spigel, D., Grunberg, S. M., Spira, A., Janne, P. A. et al. (2008) First-line gefitinib in patients with advanced non-small-cell lung cancer harboring somatic EGFR mutations. *J Clin Oncol* **26**, 2442–9.
47. Sheehan, K. M., Calvert, V. S., Kay, E. W., Lu, Y., Fishman, D., Espina, V. et al. (2005) Use of reverse phase protein microarrays and reference standard development for molecular network analysis of metastatic ovarian carcinoma. *Mol Cell Proteomics* **4**, 346–55.

Chapter 3

Personalized Medicine: Ethics for Clinical Trials

G. Terry Sharrer

Abstract

Modern ethical codes in medicine were developed following World War II to provide respect for persons, beneficence, and justice in clinical research. Clinical trial medicine involves greater scrutiny than most research activities. In every instance, clinical trials have institutional review boards to ensure the medical procedure under study complies with regulatory requirements, privacy, informed consent, good practices, safety monitoring, adverse events reporting, and is free of conflicting interests. Mandatory training in medical ethics for all clinical staff is becoming more common, and at some institutions, knowledgeable patient advocates play a watchdog role. In personalized medicine, each patient becomes a clinical trial of one, based on the uniqueness of the person's illness and the relatively tailored treatment. These features imply a shared responsibility between the patient and the researchers because uncertainty exists over the outcome for each individual patient. This chapter introduces ethical considerations using case studies, with historical context, and describes general ethical guidelines for initiating a clinical trial.

Key words: Belmont Report, Biomedical testing, Clinical trials, Declaration of Helsinki, Ethics, Gene therapy, Human subjects protection, Informed consent, Institutional review board

1. Introduction

Goldstein proclaimed at the opening of the 2010 decade that genomic sequencing will lead to the identification of an increasing number of mutations that will form the basis for accurate predictions of disease risk as part of the accelerating revolution in personalized medicine (1). Despite this initial enthusiasm, however, current thought leaders are worried that personalized genomic medicine should not be implemented until key ethical and technological issues are solved. "If genomic innovations have great power to help, they also have power to harm, and patients and the public deserve a rigorous evaluation of what scientists bring to the table"

Virginia Espina and Lance A. Liotta (eds.), *Molecular Profiling: Methods and Protocols*, Methods in Molecular Biology, vol. 823, DOI 10.1007/978-1-60327-216-2_3, © Springer Science+Business Media, LLC 2012

cautions Khoury et al. (2–4). Ng et al. (5) found that only 50% of the disease predictions agreed across the tests offered by two genomics companies evaluated over five individuals. Reflecting on the danger of false predictions, and the fear expressed by a public worried that confidential genetic information can lead to discrimination by insurance carriers and employers, many believe that "it will take 10–20 years for personalized medicine, based on genetic information, to become commonplace" (6). While we can expect that the accuracy of genomic and genetic testing will continue to improve in the future based on emerging technology and further research, this does not address the ethical issues of conducting and implementing personalized medicine. This chapter will explore the ethical side of personalized medicine, because this is the most fundamental challenge for society and the biotechnology industry.

2. Ethics in Clinical Research

Because personalized medicine extols the individual, let us begin our discussion of ethics in clinical research with one individual in particular: 18-year old Jesse Gelsinger, who in September 1999 was the first person to die from gene therapy – one form of personalized medicine.

Physicians first diagnosed Jesse Gelsinger at age 2 with the X-linked urea cycle disorder, ornithine transcarbamylase (OTC, EC 2.1.3.3) deficiency (7). It is a rare disease – perhaps occurring in 1:80,000 live births – and is an odd one at that (8, 9). In clear instances, hyperammonemia presents from any of 341 known alterations in the 73 kb OTC gene on the X-chromosome (Xp.21.1) (10). In healthy individuals, OTC in hepatocytes carries out the second of five reactions that convert ammonia, from protein metabolism, into excretable urea (11). Without or with too little of this enzyme, ammonia levels in the blood become lethal – for newborns, in as little as 72 h. While this seems straight-forward as X-linked traits go, some hemizygous males have no disease onset until their fourth or fifth decade (12). Heterozygous females can be completely asymptomatic or severely affected, with random inactivation of the gene presenting early or late onset. About 20% of proven OTC deficiency cases have no known genetic mutation. In Jesse Gelsinger's instance, the disease apparently arose from a spontaneous mutation during gestation or sometime before his second birthday, when it first presented (7, 13). Some of his liver cells expressed OTC; most did not. By carefully regulating his diet – minimizing protein and compensating with carbohydrates and fats for energy, he had remained relatively healthy until his 18th birthday when, at the age of consent, he planned to enroll in a clinical trial for partial OTC deficiency (7).

Neither Jesse nor his family needed much explanation of the disease; they had lived it, beyond what even the most informed nonafflicted expert knew. They did need to understand, though, that this phase one trial was aimed at determining safety rather than efficacy. This, they clearly grasped. Perhaps murkier was the protocol itself, in which the investigators proposed to insert the OTC gene into a weakened adenovirus, somewhat like National Institutes of Health (NIH) researchers had done with the human gene for adenosine deaminase and a mouse retrovirus to conduct the first approved gene therapy clinical trial 9 years earlier (14–18). Adenovirus vectors had advantages of being highly efficient for delivering genes to most bodily tissues, whether in dividing or quiescent cells; they did not integrate their double-stranded DNA into the nucleus of the target cells; and a decade of live adenovirus vaccinations for military recruits had shown no propensity for causing cancer (19). On the other hand, because the adenovirus vector did not integrate, any therapeutic benefit would be transitory. Repeated applications could induce a specific immune response to any transformed cells, and while investigators knew how to weaken the virus, it could not be made completely harmless because some of its essential genes might produce immune reactivity. By the mid-1990s, the National Institutes of Health's (NIH) Recombinant DNA Advisory Committee (RAC) had approved two dozen gene therapy clinical trials that relied on adenoviral vectors: 9 for cystic fibrosis, 14 for cancers, and 1 for OTC deficiency, at the University of Pennsylvania, where Jesse Gelsinger signed up on June 22nd, 1999 (20).

Gastrointestinal surgeon Steve Raper ordered the blood and liver function tests that showed Jesse to be fit enough for the trial. A year earlier, he had experienced a crisis – liver failure and coma – since then a strict diet and medication had maintained his health. Indeed, it was his good overall condition and the fact that he had partial OTC deficiency that made him a desirable candidate for the trial. University of Pennsylvania ethicist Arthur Kaplan persuaded the gene therapists that desperately sick patients would have trouble reaching a cool assessment of the risks involved prior to signing the informed consent (see Note 1). Jesse was hopeful but not desperate. In the procedure Dr. Raper laid out, a catheter would be inserted into Jesse's groin and advanced to the hepatic artery. Then, slowly, 30 ml of the weakened and recombined adenovirus vector would be injected. He might have flu-like symptoms afterwards, but nothing more serious seemed in store. Jesse was scheduled to be the eighteenth patient in this trial, and in an earlier case, a woman had well tolerated the same dose. Because the NIH and Food and Drug Administration (FDA) had not only reviewed this protocol but had funded the trial, the Gelsingers reasonably could assume that close oversight prevailed even if the informed consent agreement noted that no patient benefit was anticipated and a chance of death existed.

With everything in place, Jesse Gelsinger's gene therapy began on September 13. Over night, he developed a high fever and by the following morning began to show signs of liver failure, which worsened. His blood ammonia level rose to more than ten times the normal level. Neither dialysis nor an induced coma, to allow extracorporeal membrane oxygenation, stemmed the organ failure that proceeded. On September 14, Jesse died, from either a severe immune reaction to the adenoviral vector, or from losing a complicated, calculated gamble in which a large cadre of physicians, ethicists, and institutional regulators had, in their best judgment, bet with him (7). In either case, he lost his life because of the treatment he received.

Drs. James Wilson, who directed Penn's Human Gene Therapy Institute, surgeon Steven Raper, and pediatric geneticist Mark Batshaw, who was the Principal Investigator at the collaborating Children's National Medical Center for the OTC deficiency trial, immediately reported the death to the Food and Drug Administration as required. In reviewing events of the case, FDA investigators identified several deficiencies in the way the study had been handled in general and regarding Jesse Gelsinger in particular. First, they asserted that Gelsinger should have been excluded from the trial because his ammonia levels were high on the day of his gene therapy. Wilson responded that the researchers acted within the protocol the FDA had approved, that Gelsinger's ammonia levels were under medical control, and the FDA had been kept informed. Second, the investigators in the clinical trial should have stopped if patients suffered serious adverse events. Wilson said all serious side effects had been reported and the FDA had not ordered a cessation of the trial. And, third, the investigators alleged that the deaths of 2 out of 11 monkeys in preceding animal studies were not mentioned in the informed consent agreement. Wilson countered that only one monkey had died and only after receiving a dose of the adenovirus that was 17 times higher than Gelsinger received. Wilson, Raper, and Bratshaw felt that they had acted responsibly in every instance, and in testimony before the Senate Committee on Health, Education, Labor and Pensions (Subcommittee on Public Health) in February 2000, the Director of the NIH's Office of Biotechnology Activities, Dr. Amy Patterson, seemingly vouched for the Penn investigators, after pointing out the inherent risk in clinical research (21). Jesse Gelsinger's father, at that point, did not feel that the gene therapists had done anything less than the best they could (7, 21).

The FDA, however, did not let the matter go, and, perhaps, responded to a lawsuit the Gelsinger family filed in September 2000. The FDA's Office of Criminal Investigations and the Health and Human Services-Office of Inspector General sent Dr. Batshaw a warning letter on November 30, 2000 about false statements made between July 1998 and September 1999, and subsequently turned the investigation over to the Justice Department for prosecution. Assistant US Attorney for the Eastern District of Pennsylvania, David Hoffman, handled the case, alleging that toxicities which should have terminated the clinical experiment did not; reports to the FDA, NIH, and Penn's own institutional review board (IRB) misrepresented actual clinical findings – violating the civil False Claims Act (an 1863 law which allows the government to join private plaintiffs who file fraud charges against federal contractors); and that the informed consent agreement did not sufficiently disclose possible toxicity. On February 9, 2005, the Justice Department announced that the case had been settled: both the University of Pennsylvania and the Children's National Medical Center agreed to pay more than $500,000 to the government to resolve the allegations. Drs. Wilson, Raper, and Bratshaw also agreed to being restricted for 3–5 years in FDA-regulated clinical trials, to undertaking various training/education courses dealing with human research subjects, and to accepting a "special monitor" who would oversee their research – with semiannual reports on compliance to the FDA and NIH. Yet, they were excused from admitting to the government's charges and continued to assert that they had acted properly at all times (22).

3. Significance in Medical Ethics

What, then, is the significance of the Gelsinger case in the genre of medical ethics? In the broadest sense, there is an ethical continuum that goes back to Hippocrates – not to the Hippocratic Oath, which pledges doctors to abide by community morals, but to the corpus on epidemics: "As to diseases, make a habit of two things – to help, or at least to do no harm." Jesse was harmed, even if unintentionally so. Interestingly, Hippocrates pointed out that diseases do not exist in nature. They are intellectual constructions to guide treatments. What exist naturally, he felt, were sick people and no two people are sick in exactly the same way. Jesse Gelsinger had a rare form of a rare disease. Considering what's now known about human genetic variation – single nucleotide polymorphisms, copy-number differences, overlapping genes, and significant disparities in base pair numbers between individuals (12, 23, 24) – Gelsinger's disease was unlike any other patient's OTC deficiency. Further, following that reasoning, no two people – not even conjoined identical twins – ever had exactly the same "disease." Taking this into account, personalized medicine – its practice and ethics – probably faces greater complexity and uncertainty than has ever existed in medicine before – with attending propensity for causing harm.

3.1. Modern Ethical Codes of Conduct

Modern ethical codes in medicine are not very old; they largely follow World War II. Earlier, standards aimed to ameliorate relations between physicians. Though the Hippocratic Oath mentioned abstaining from taking sexual advantage of patients, its first rule was that a medical student should show perpetual devotion to the teacher. In both of those frames of reference, the sick person was virtually a passive object (see Note 2). Medicine always has had a sacred feature – justly deserved in matters of life and death – but more so before 1945 than after. Change resulted from Nazi medical atrocities (25). At the so-called "Doctors' Trial" of the Nuremberg Tribunals in 1946–1947, 23 defendants stood accused of war crimes and crimes against humanity; 20 were physicians. Chief Counsel Telford Taylor said: "These defendants did not kill in hot blood, nor for personal enrichment. Some of them may be sadists who killed and tortured for sport, but they are not all perverts. They are not ignorant men. Most of them are trained physicians and some of them are distinguished scientists. Yet these defendants, all of whom were fully able to comprehend the nature of their acts, and most of whom were exceptionally qualified to form a moral and professional judgment in this respect, are responsible for wholesale murder and unspeakably cruel tortures." (26). Six were acquitted, ten sentenced to life in prison, and seven – all "practicing" physicians – were hanged. The resulting "Nuremberg Code" (1947) established ten principles for medical experiments, starting with "voluntary consent of the human subject is absolutely essential." While these principles (and their subsequent revision into the "Declaration of Helsinki" in 1964 (http://www.wma.net/e/policy/b3.htm) (27)) were intended to be "universal," they were not applied to investigations which US government scientists carried out for nuclear, biological, and chemical warfare between 1945 and the early 1970s. Nor were they in place for US Public Health Service's (USPHS) Syphilis Study at Tuskegee, AL between 1932 and 1972 (28).

3.1.1. Belmont Report

It was reasonable for medical investigators to wonder if syphilis followed different clinical courses along racial lines – the Tuskegee study's original justification. African-Americans in the southern states infected with syphilis often presented with hypertension, diabetes, sickle cell anemia, and malnutrition. Over four decades, researchers enrolled 623 men of whom 200 were healthy "controls;" the others had syphilis (see Note 3). The study team felt that the patients were unable to understand the medical aspects of their disease, but benefited from free exams, free meals, burial costs, and the opportunity to participate in a process that benefited "mankind." (see Note 1). Viewing the participants as clinical subjects rather than sick men was dreadful, but the moral crime of this experiment arose from not treating the disease when a good remedy came along – penicillin. This drug, developed in 1943 and generally available after World War II, cured syphilis in most cases, yet the USPHS researchers in Tuskegee withheld it from the study participants so they could continue following the natural course of the disease. Finally, in July 1972, the Washington Star ran a story that brought a huge public outcry (29). Senator Edward Kennedy held hearings about medical experiments involving human subjects that resulted in the National Research Act in 1974. This law created a National Commission to protect people in medical research, and subsequently, the Commission produced "Ethical Principles and Guidelines for the Protection of Human Subjects of Research," also known as "the Belmont Report" (1979, named for a Maryland conference center) (30).

The Belmont Report specified three principles: respect for persons, beneficence, and justice (31). Respect and beneficence largely recapitulated the Nuremberg Code; the justice provision was new, and a response to the syphilis study, observing that injustice arose from social, racial, sexual, and cultural biases that had become institutionalized in American society. It also noted "One special instance of injustice results from the involvement of vulnerable subjects. . . . Given their dependent status and their frequently compromised capacity for free consent, they should be protected against the danger of being involved in research solely for administration convenience, or because they are easy to manipulate as a result of their illness or socioeconomic condition."(31) (http://www.hhs.gov/ohrp/humansubjects/guidance/belmont.htm).

3.2. Patient Privacy

Neither Nuremberg nor Belmont addressed patient privacy, largely because there was not a need at this point in time. Hippocrates mentioned the need to keep secret forever any knowledge arising from medical practice (not limited to patients) and that standard prevailed for centuries until acquired immunodeficiency syndrome (AIDS) and development of the Internet forced new concerns regarding confidentiality. AIDS hit right at Belmont's justice principle. AIDS sufferers became politically active, pushing for more government research funding, while demanding greater protection for their privacy. In the early 1980s, several jurisdictions in the US still required passing a syphilis test for marriage; so with reason, AIDS sufferers worried about privacy invasion and discrimination. While AIDS grew to pandemic proportions, the Internet created both an opportunity and a privacy concern in medicine. Electronic medical files, mainly for billing and reimbursement, offered a solution to healthcare's paper-handling blizzard, yet they also allowed insurers to exert greater scrutiny over their subscribers. Computer viruses and hackers' intrusions added security matters to a growing list of privacy issues. Patients did have some privacy rights, but those that existed by the early 1990s arose mainly from case law, not statutes.

3.2.1. Health Insurance Portability and Accountability Act

Eight years after the US Supreme Court upheld the privacy of garbage (California v. Greenwood, 486 US 35 (1988) (32), Congress passed the first medical privacy law of consequence – the Health Insurance Portability and Accountability Act (HIPAA, http://www.hhs.gov/ocr/hipaa/) of 1996. It mandated certain electronic records, using numerical identifiers for patients, doctors, hospitals, insurers, etc. To protect an individual's healthcare information, HIPAA required patient privacy standards. For hospitals, the Joint Commission on Accreditation of Healthcare Organizations (JCAHO) oversees privacy compliance. Now, when a sick person enters a hospital or doctor's office, the registrar is obliged to ask for the patient's consent in sharing personal health information ("PHI") with the thousand or so workers who may see a record as it goes through the billing chain.

3.3. Guidelines for Human Subjects Protection in Clinical Research

3.3.1. Shared Responsibility

The Nuremberg Code, the Belmont Report, and HIPPA, despite their shortcomings, are pillars of modern medical ethics, for now and the future. But even adhering more stridently to those principles would not have saved Jesse Gelsinger's life. He was not a victim of torture, or injustice, or invasion of privacy. He died from disputable errors of judgment. Still, his story may have more to teach about ethics in the age of personalized medicine than any other example.

The context of Jesse's experience was a clinical trial. Clinical trial medicine involves greater scrutiny than most research activities. In every instance, trials have IRBs to ensure the medical procedure under study complies with regulatory requirements, privacy, informed consent, good practices, safety monitoring, adverse events reporting, and is free of conflicting interests. Mandatory training in medical ethics for all clinical staff is becoming more common, and at some institutions, knowledgeable patient advocates play a watchdog role. These features imply shared responsibility for the patient because exceptional uncertainty exists over the outcome. In personalized medicine, each patient becomes a clinical trial of one – based on the uniqueness of the person's illness and the relatively tailored treatment. The ethical issue for the providers, then, is how seriously each person takes their oversight responsibility, forming a group assurance of best judgment (see Note 4).

While healthcare has become less doctor-centric – giving rise to the interests of hospitals, corporations, and governments – personalized medicine puts the doctor–patient relationship foremost in determining all ethical considerations, including issues of the greater social good in most instances. In the doctor–patient exchange, clinicians will have knowledge of how a protocol worked in tissue culture or animal studies, and how various treatment options effected individual cases. Patients, besides knowing intimate health details that may not be in their medical records, also have expectations that effect their treatment choices and thereby influence their assessment of risks and benefits. Once a physician and patient have reached agreement or "belief" about how to proceed with an uncertain situation, new information derived from the treatment feeds back to confirm, modify, or reject the original protocol. Thus, the patient has both greater autonomy than has ever existed in clinical practice, and a heavier burden in becoming informed about the nature of their health and health care.

3.4. Personalized Medicine in a Primary Care Facility

Greater oversight and greater patient autonomy could imply that personalized medicine will remain limited to major medical facilities and to patients with enough background in science to understand some level of biochemistry. But that is not the case for at least one primary care facility already doing personalized medicine: the Clinic for Special Children, founded in 1989 as a pediatric service for the Amish and Old Order Mennonites in Lancaster County, PA. In many ways, this clinic is an exception to how healthcare is now practiced in the USA. Its patient base is unusual in having many illnesses that are rare in the general population; it is also unusual that this locality has been a major hearth for the study of Mendelian inheritance in man. Few other communities sense that the survival of their religious faith depends, in part, on dealing with inherited disorders. The clinic is run on a modest fee for service basis; patient families do not subscribe to health insurance. Episodes are not rated for patient satisfaction; they are seen as God's will. "Special children" is a euphemism for birth defects, which the Amish and Old Order Mennonites interpret as divine gifts to teach love.

Yet, in other ways, the clinic reflects conventional medical research and practice (33–35). It conforms to research protocols of its collaborating partners: Lancaster General Hospital, Children's Hospital of Philadelphia, and Johns Hopkins University Hospital. It's daily caseload centers around children with profound inherited disorders, such as Crigler–Najjar syndrome (OMIM #218800) (36), maple syrup urine disease (OMIM #248600) (37), and adenosine deaminase deficiency (OMIM #102700) (38), which are gene-specific disorders that frequently are described nonspecifically as cerebral palsy, mental retardation, and lethal infection. Annually, two physicians and a staff geneticist serve over 800 patients presenting more than 100 genetic disorders. Not every patient is treatable, but most are, and many are highly treatable. Incurable does not equate to untreatable. Early diagnosis, using genetic, genomic, and proteomic technologies, is key to effective management. And, the clinic's outcomes show impressive results: for example, in managing Crigler–Nijjar syndrome five international patient cohorts (1952–1998), totaling 180, reported 41% experienced brain damage and 9% untimely death. The clinic handled 20 Crigler–Nijjar patients, between 1989 and 2006, with none dying or suffering brain damage. In some instances, treatment after diagnosis seems simple. With maple syrup urine disease – where sufferers are enzyme deficient for branched-chain alpha-keto acid dehydrogenase – a dietary formula lacking the branched-chain amino acids leucine, isoleucine, and valine sustains normal growth and development. Management becomes more difficult when the usual childhood illnesses impact on these metabolically fragile patients, but close and careful monitoring of serum proteins allows the storm to pass (39).

In conducting its work, the Clinic for Special Children aspires: "to make medical care for special children accessible, affordable, and culturally acceptable; to identify genetic causes of childhood disability and death with Plain communities; to use state-of-the-art technologies to improve the accuracy and economy of diagnosis; and to use presymptomatic diagnosis and preventative therapy to improved child health and reduce medical costs."(http://www.clinicforspecialchildren.org). These goals, along with its sensitivity in seeing patients as individuals who need help rather than interesting medical problems, are, with little modification, compelling ethical standards for personalized medicine wherever it is practiced.

How personalized medicine proceeds depends on four fundamental elements: accumulated knowledge, apparent need, available capital, and social acceptability. Ethics largely comes under the social acceptability heading, but in fact, all depend on moral principles in one way or another. It is through high ethical standards that researchers are drawn to create knowledge, that needs are defined, that money is invested in the transition from clinical research to clinical practice, and that society comes to see medicine more as benefit to mankind than a chronic social problem.

As with all healing arts, the second ethical law of personalized medicine will remain: "at least, do no harm." Its first, as ever, is "to help."

3.5. General Guidelines for Initiating Clinical Research

Prior to conducting clinical research, the responsible physician and researchers should seek specific guidance from each physician's institution, federal and professional governing bodies, or international research entities regarding their particular intended clinical research project. The guidelines listed herein are intended to provide an overview of key issues/steps in clinical research (Fig. 1).

3.5.1. Preliminary Research Activities

1. Prior to performing clinical research on living subjects, the scientific basis of the research must have been derived from laboratory and/or animal experiments or other scientifically established facts (27).
2. Clinical research should be supervised only by a qualified physician and conducted only by scientifically qualified persons.
3. Comparison of the foreseeable benefits to the inherent risks of the research must be assessed prior to initiation of clinical research.

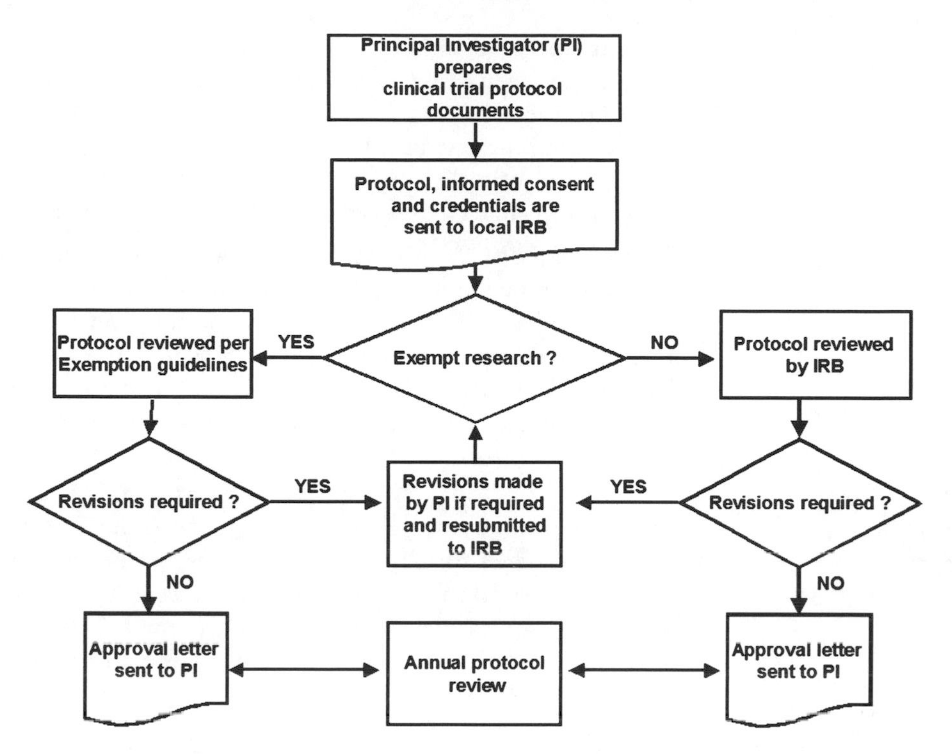

Fig. 1. Outline of processes required for obtaining human subjects protection approval and institutional review board approval. Research and clinical trials involving human subjects must undergo appropriate review board approval prior to initiating the study.

3.5.2. Research Institution Approval Process and Documentation

Documents generally required for IRB approval include: complete clinical trial protocol, copy of informed consent, and biosketches/credentials for persons performing the research.

1. The clinical trial protocol should include, but is not limited to, the following topics: (a) scientific basis for the study, (b) intended research populations, (c) inclusion and exclusion criteria, (d) treatment regimens, (e) anticipated outcome, (f) adverse event reporting instructions, (g) research locations and personnel, and (h) pertinent scientific references.
2. The informed consent document should contain: (a) a description in lay terms of the research, (b) voluntary participation/withdrawal statement, (c) description of the anticipated benefits, (d) description of the inherent/foreseeable risks, (e) funding sources for the research, (f) description of the information/samples/data to be collected, (g) contact information regarding the study, and (h) confidentiality.

3.5.3. Exempt and Nonexempt Clinical Research

Research activities that may be exempt from IRB approval include, but are not limited to the following activities (see Note 5):

1. Research involving the collection or study of existing data, documents, records, pathological specimens, or diagnostic specimens, if these sources are publicly available or if the

information is recorded by the investigator in such a manner that subjects cannot be identified, directly or through identifiers linked to the subjects.

Research involving the use of educational tests (cognitive, diagnostic, aptitude, achievement), survey procedures, interview procedures, or observation of public behavior, unless:

2. Information obtained is recorded in such a manner that human subjects can be identified, directly or through identifiers linked to the subjects; and any disclosure of the human subjects' responses outside the research could reasonably place the subjects at risk of criminal or civil liability or be damaging to the subjects' financial standing, employability, or reputation.

3.6. Tissue Collection Guidelines for Clinical Research

Tissue, blood, or body fluids (collectively referred to as tissue) collected for clinical research studies must be done according to the above guidelines for clinical research. Tissue collection should incur minimal risk to the patient, should not be required for diagnosis/prognosis/treatment, should be anonymized/de-identified to the researcher, with no commercial development from the specimen (see Note 6).

4. Notes

1. All research studies must include informed consent. IRB approval must be obtained for all studies involving human subjects even if the IRB determines that the study would qualify for an "exempt" status. The human subjects review board (IRB) can waive the requirement for a subject's signature on their name on a document or where the research will be conducted electronically.
2. A human subject is defined as a living individual about whom an investigator conducting research obtains (1) data through intervention or interaction with the individual or (2) identifiable private information.
3. Research means a systematic investigation, including research development, testing, and evaluation, designed to develop or contribute to generalizable knowledge. Investigations designed to develop or contribute to generalizable knowledge are those designed to draw general conclusions (i.e., knowledge gained from a study may be applied to populations outside of the specific study population), inform policy, or generalize findings.
4. It is the duty of the doctor to remain the protector of the life and health of the person on whom the research is conducted (27) per the Declaration of Helsinki. Clinical research cannot

be conducted on a living human subject without his/her free consent after he/she has been informed of the nature of the project, anticipated risks, and/or potential benefits, if any.

5. Research activities associated with any group of persons considered "at risk" such as prisoners, children, or mentally impaired individuals do not qualify for "exempt" status. These groups of people may not be able to make informed, nonbiased judgments regarding their participation in research activities.
6. A discarded specimen is that portion of a collected specimen that is not needed for assessment of diagnostic, prognostic, and other parameters in the diagnosis and treatment of the patient. Autopsy tissue is not from a living human being; therefore IRB regulations do not apply. Unless the tissue is anonymous to the investigator, HIPAA regulations may apply and the investigator may need to request a HIPAA waiver from their IRB.

References

1. Goldstein, D. B. (2010) Personalized medicine. *Nature* 463, 10.
2. Khoury, M. J., Evans, J. P., Burke, W. (2010) A reality check for personalized medicine. *Nature* 464, 680.
3. Burke, W., Burton, H., Hall, A. E., Karmali, M., Khoury, M. J., Knoppers, B. et al. (2010) Extending the reach of public health genomics: what should be the agenda for public health in an era of genome-based and "personalized" medicine? *Genet Med* 12, 785–91.
4. Evans, J. P., Burke, W., Khoury, M. (2010) The rules remain the same for genomic medicine: the case against "reverse genetic exceptionalism". *Genet Med* 12, 342–3.
5. Ng, P. C., Murray, S. S., Levy, S., Venter, J. C. (2009) An agenda for personalized medicine. *Nature* 461, 724–6.
6. Butler, D. (2010) Science after the sequence. *Nature* 465, 1000–1001.
7. Stolberg, S. G. (28 November 1999) The Biotech Death of Jesse Gelsinger, in *New York Times*, Sunday Magazine, New York.
8. McCullough, B. A., Yudkoff, M., Batshaw, M. L., Wilson, J. M., Raper, S. E., Tuchman, M. (2000) Genotype spectrum of ornithine transcarbamylase deficiency: correlation with the clinical and biochemical phenotype. *Am J Med Genet* 93, 313–9.
9. Tuchman, M., McCullough, B. A., Yudkoff, M. (2000) The molecular basis of ornithine transcarbamylase deficiency. *Eur J Pediatr* 159 Suppl 3, S196-8.
10. Lindgren, V., de Martinville, B., Horwich, A. L., Rosenberg, L. E., Francke, U. (1984) Human ornithine transcarbamylase locus mapped to band Xp21.1 near the Duchenne muscular dystrophy locus. *Science* 226, 698–700.
11. Horwich, A. L., Kalousek, F., Fenton, W. A., Pollock, R. A., Rosenberg, L. E. (1986) Targeting of pre-ornithine transcarbamylase to mitochondria: definition of critical regions and residues in the leader peptide. *Cell* 44, 451–9.
12. Tuchman, M., Jaleel, N., Morizono, H., Sheehy, L., Lynch, M. G. (2002) Mutations and polymorphisms in the human ornithine transcarbamylase gene. *Hum Mutat* 19, 93–107.
13. Drogari, E., Leonard, J. V. (1988) Late onset ornithine carbamoyl transferase deficiency in males. *Arch Dis Child* 63, 1363–7.
14. Anderson, W. F., Blaese, R. M., Culver, K. (1990) The ADA human gene therapy clinical protocol: Points to Consider response with clinical protocol, July 6, 1990. *Hum Gene Ther* 1, 331–62.
15. Anderson, W. F. (1984) Prospects for human gene therapy. *Science* 226, 401–9.
16. Belmont, J. W., Henkel-Tigges, J., Chang, S. M., Wager-Smith, K., Kellems, R. E., Dick, J. E. et al. (1986) Expression of human adenosine deaminase in murine haematopoietic progenitor cells following retroviral transfer. *Nature* 322, 385–7.
17. Belmont, J. W., Henkel-Tigges, J., Wager-Smith, K., Chang, S. M., Caskey, C. T. (1986)

Towards gene therapy for adenosine deaminase deficiency. *Ann Clin Res* 18, 322–6.

18. Eglitis, M. A., Kantoff, P., Gilboa, E., Anderson, W. F. (1985) Gene expression in mice after high efficiency retroviral-mediated gene transfer. *Science* 230, 1395–8.
19. Gaydos, C. A., Gaydos, J. C. (1995) Adenovirus vaccines in the U.S. military. *Mil Med* 160, 300–4.
20. Culver, K. W. Gene Therapy: A Primer for Physicians, 2nd. Mary Ann Liebert, Inc, Larchmont, NY, 1996.
21. Stillman, B. A., Tonkinson, J. L. (2000) FAST slides: a novel surface for microarrays. *Biotechniques* 29, 630–5.
22. U.S. Settles Case of Gene Therapy Study that ended with Teen's Death. (2005) University of Pennsylvania Almanac 51:21: Philadelphia.
23. Wong, L. J., Dimmock, D., Geraghty, M. T., Quan, R., Lichter-Konecki, U., Wang, J. et al. (2008) Utility of oligonucleotide array-based comparative genomic hybridization for detection of target gene deletions. *Clin Chem* 54, 1141–8.
24. Ogino, W., Takeshima, Y., Nishiyama, A., Okizuka, Y., Yagi, M., Tsuneishi, S. et al. (2007) Mutation analysis of the ornithine transcarbamylase (OTC) gene in five Japanese OTC deficiency patients revealed two known and three novel mutations including a deep intronic mutation. *Kobe J Med Sci* 53, 229–40.
25. Barondess, J. A. (1996) Medicine against society. Lessons from the Third Reich. *Jama* 276, 1657–61.
26. Trials of War Criminals before the Nuremberg Military Tribunals under Control Council Law No. 10. Nuremberg, October 1946-April 1949, U.S. Government Printing Office, Washington, DC, 1949–1953.
27. Williams, J. R. (2008) The Declaration of Helsinki and public health. *Bull World Health Organ* 86, 650–2.
28. Vollmann, J., Winau, R. (1996) Informed consent in human experimentation before the Nuremberg code. *Bmj* 313, 1445–9.
29. Heller, J. (July 25, 1972) Syphilis Patients Died Untreated, in *Washington Star*, Washington, DC.
30. Jones, J. H. Bad Blood: The Tuskegee Syphilis Experiment, Free Press, New York, 1981.
31. Office of the Secretary (1979) Ethical Principles and Guidelines for the Protection of Human Subjects of Research. The National Commission for the Protection of Human Subjects of Biomedical and Behavioral Research: Washington, DC.
32. California V. Greenwood, et al. No. 86–684, Supreme Court of the United States, 486 U.S. 35, 1988.
33. Strauss, K. A., Puffenberger, E. G., Robinson, D. L., Morton, D. H. (2003) Type I glutaric aciduria, part 1: natural history of 77 patients. *Am J Med Genet C Semin Med Genet* 121 C, 38–52.
34. Morton, D. H., Morton, C. S., Strauss, K. A., Robinson, D. L., Puffenberger, E. G., Hendrickson, C. et al. (2003) Pediatric medicine and the genetic disorders of the Amish and Mennonite people of Pennsylvania. *Am J Med Genet C Semin Med Genet* 121 C, 5–17.
35. Morton, D. H., Bennett, M. J., Seargeant, L. E., Nichter, C. A., Kelley, R. I. (1991) Glutaric aciduria type I: a common cause of episodic encephalopathy and spastic paralysis in the Amish of Lancaster County, Pennsylvania. *Am J Med Genet* 41, 89–95.
36. Crigler, J. F., Jr., Najjar, V. A. (1952) Congenital familial nonhemolytic jaundice with kernicterus. *Pediatrics* 10, 169–80.
37. Westall, R. G., Dancis, J., Miller, S. (1957) Maple syrup urine disease. *Am. J. Dis. Child* 94, 571–572.
38. Giblett, E. R., Anderson, J. E., Cohen, F., Pollara, B., Meuwissen, H. J. (1972) Adenosine-deaminase deficiency in two patients with severely impaired cellular immunity. *Lancet I* 1067–1069.
39. Morton, D. H., Strauss, K. A., Robinson, D. L., Puffenberger, E. G., Kelley, R. I. (2002) Diagnosis and treatment of maple syrup disease: a study of 36 patients. *Pediatrics* 109, 999–1008.

Chapter 4

Reduction of Preanalytical Variability in Specimen Procurement for Molecular Profiling

Virginia Espina and Claudius Mueller

Abstract

Despite the tremendous perceived value, and the predicted high abundance, of disease-associated tissue biomarkers, the number of biomarkers that have been validated for routine clinical use is very low. The major roadblock has been the sample-to-sample variability and perishability of biomolecules in tissue. A chief source of variability is biomolecule perturbation caused by sample handling, the time delays following procurement, and the method of preservation. Living tissue that has been separated from its blood supply during surgical procurement goes through defined stages of reactive changes preceding death, beginning with oxidative, hypoxic, and metabolic stress. These reactive fluctuations in the tissue biomolecules can occur within 20 min postexcision, and can significantly distort the levels of critical diagnostic and prognostic biomolecules. Depending on the delay time ex vivo, and manner of handling, protein biomarkers such as signal pathway phosphoproteins will be elevated or suppressed in a manner that does not represent the biomarker levels at the time of excision. Based on analysis of phosphoproteins, one of the most labile tissue protein biomarkers, we set forth tissue procurement guidelines for clinical research. We further propose the future use of a multipurpose fixative solution designed to stabilize, preserve and maintain proteins, nucleic acids, and tissue architecture.

Key words: Fixation, Histomorphology, Phosphoprotein, Preanalytical variability, Preservative, Stability, Tissue

1. Introduction

1.1. Preanalytical Variability: A Major Roadblock to Translational Research

George Poste in a 2011 Comment in Nature (1) warned that research in biomarkers to diagnosis disease and guide therapy has "not yet delivered on its promise," even though "the ability of biomarkers to improve health and reduce health care costs is greater than any other area of biomedical research." While 150,000 papers have been published claiming new biomarkers, only 100 molecular biomarkers are routinely used in the clinic, and these 100 were largely identified more than 10 years ago. Poste states,

Virginia Espina and Lance A. Liotta (eds.), *Molecular Profiling: Methods and Protocols*, Methods in Molecular Biology, vol. 823, DOI 10.1007/978-1-60327-216-2_4, © Springer Science+Business Media, LLC 2012

"A major impediment to progress in the hunt for biomarkers is the lack of standardization in how specimens are collected" (1). He proposes that funding agencies should only support research programs that have stringent characterization of specimens, impose rigorous quality control in specimen acquisition handling and storage, and possess the full spectrum of cross-disciplinary capabilities needed to translate laboratory findings to the clinic (1). The purpose of this chapter is to (a) provide a physiologic rationale supporting the paramount importance of tissue handling, and (b) establish guidelines for tissue procurement for high-quality translational research.

As technologies advance from initial conception to clinical applications, one must consider the entire spectrum of clinical assay variability, including preanalytical as well as postanalytical events, which could potentially impact the final result (Fig. 1). Cells within a tissue biopsy react and adapt to the trauma of excision, ischemia, hypoxia, acidosis, accumulation of cellular waste, absence of electrolytes, and temperature changes (2, 3). It would be expected that a large surge of stress, hypoxia, and wound repair-related protein signal pathway proteins and transcription factors would be induced in the tissue immediately following procurement (4, 5). Further complicating the tissue fidelity issue is the type of analysis to be performed on the tissue. Nucleic acid stabilization has been accomplished with a variety of commercial products as long as the product is used promptly post sample collection (6), while protein

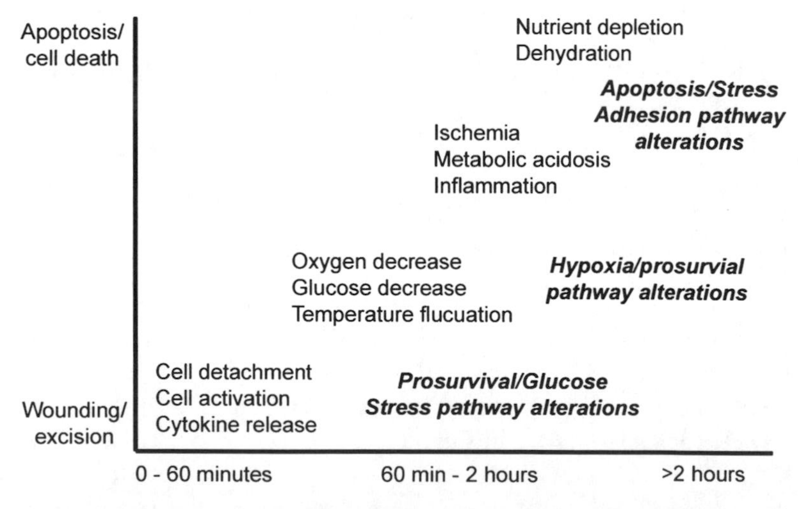

Fig. 1. Cellular responses to wounding/excision may contribute to preanalytical variability. Tissue samples undergo multiple reactive changes postexcision, each of which could potentially contribute to preanalytical variability. The timeline is illustrative and each sample may exhibit more or less protein pathway variation based on specific environmental stresses/conditions as well as on preexcision in vivo conditions, such as warm ischemia time, subject hydration status, anesthesia, etc. Adapted from Espina et al., *Mol Cell Proteomics* 7, 2008 (*3*).

post-translational modifications require different stabilization strategies. Although investigators have worried about the effects of vascular clamping and anesthesia (warm ischemia times) prior to excision on the fidelity of molecular data in tissues (7), a much more significant and underappreciated issue is the fact that excised tissue is alive and reacting to ex vivo stress (3, 8). During the ex vivo time period, because the tissue cells are alive and reactive, phosphorylation/dephosphorylation of certain kinase substrates may transiently increase due to the persistence of functional signaling, activation by hypoxia, or other stress-response signals (3, 9–11). Without molecular stabilization, imbalances of kinases/phosphatases will significantly distort the tissue's molecular signature compared to the in vivo state.

2. Phosphoprotein Stability: The Balance Between Kinases and Phosphatases

Measurement of phosphoprotein signal pathway epitopes in tumor biopsy samples is crucial for individualizing the new generation of molecular therapy aimed at targeting kinase pathways. Phosphoprotein epitopes, generated by the action of a kinase on a substrate protein, provide a minute by minute record of ongoing signal pathways relevant to therapeutic target selection and the prediction of toxicity. Kinases phosphorylate the substrate amino acid and phosphatases remove the phosphorylation of the amino acid. At any point in time within the tissue cellular microenvironment, the phosphorylated state of a protein is a function of the local stoichiometry of associated kinases and phosphatases specific for the phosphorylated residue. Thus, in the absence of kinase activity a phosphorylated residue will be removed by a phosphatase and this could reduce the level of a phosphoprotein analyte and cause a false negative result.

A variety of chemical- and protein-based inhibitors of phosphatases exist (12, 13). However if the kinase remains active, then the addition of a phosphatase alone will result in augmentation of the phosphoepitope, generating a false positive (see Note 1) (3). Optimally, the stabilizing chemistry should arrest both sides of the kinase/phosphatase balance, in order to prevent positive or negative fluctuations in phosphorylation events as the living excised tissue reacts to the ex vivo conditions (3) (see Note 2). An important long-term need for the clinical implementation of phosphoprotein biomarkers will be the design of stabilizers for the preservation of phosphoproteins while maintaining histomorphology.

2.1. Time Delay in Clinical Tissue Procurement

Tissue is removed for diagnostic pathology under three main scenarios: (a) surgery in a hospital-based operating room, (b) biopsy conducted in an outpatient clinic, and (c) image-directed needle biopsies or needle aspirates conducted in a radiologic suite.

The time delay from patient excision to pathologic examination is often not recorded and may vary from 30 min to many hours depending on the time of day, the length of the procedure, and the number of concurrent cases.

The postexcision delay time, or EDT, is the time from excision to the time that the specimen is placed in a stabilized state, e.g., immersed in fixative or snap-frozen in liquid nitrogen. Given the complexity of patient-care settings, during the EDT the tissue may reside at room temperature in the operating room or on the pathologist's cutting board, or it may be refrigerated in a specimen container. In addition to uncertainty about delays following procurement and preservation, a host of known and unknown variables can influence the stability of tissue molecules during the postprocurement period (14). These include (1) temperature fluctuations prior to fixation or freezing, (2) preservative chemistry and rate of tissue penetration by the preservative, (3) size of the tissue specimen which lengthens the time it takes for the fixative to penetrate the internal regions of the tissue, and (4) extent of cutting, washing, and crushing of the tissue during excision or diagnostic evaluation.

Given current practices, in the face of these uncertainties, it is imperative that a standardized procedure is followed for collecting specimens utilized for translational research and clinical profiling. Espina et al. have developed guidelines based on a detailed timecourse study of phosphoprotein fluctuations in tissue samples procured from a variety of organs (3, 8). Phosphoprotiens were utilized in this study because (a) they are one of the most sensitive indicators of the state of the tissue, and (b) they comprise one of the most important new categories of tissue analytes for molecular profiling.

2.2. Importance of Preserving Phosphoproteins

An urgent clinical goal is to identify functionally important molecular networks associated with subpopulations of cancer patients that may respond individually to molecular targeted inhibitors that target the pathogenic signaling pathways. The majority of the current molecular targeted therapeutics is directed at protein targets, and these targets are often protein kinases and/or their substrates. The human "kinome," or full complement of kinases encoded by the human genome, comprises the molecular networks and signaling pathways of the cell. The activation state of these proteins and these networks fluctuate constantly depending on the cellular microenvironment. Consequently, the source material for molecular profiling studies needs to shift from in vitro models to the use of actual diseased human tissue and the application of molecular profiling to provide individually tailored therapy should include direct proteomic pathway analysis of patient material. Moreover, because the kinome represents a rich source for new molecular targeted therapeutics, technologies that can broadly profile and assess the activity of the human kinome will be critical for the realization of patient-tailored therapy. The intracellular information network delicately regulates, by internal feedback processes, cellular homeostasis. This intracellular balance is carefully maintained by constant rearrangements of the post-translational modification of proteins through the activity of a series of kinases and phosphatases. As a consequence, the study of the these kinase and phosphatase events is a fundamental aspect for understanding and characterizing cellular activities in a variety of normal and/or disease processes (15). The phosphorylation status of proteins can be detected and measured using specific antiphosphoprotein antibodies (16, 17). Antibodies have been developed to specifically recognize the phosphorylated isoform of kinase substrates. Through the use of these phospho-specific antibodies, it is now possible to evaluate the state of entire portions of a signaling pathway or cascade, even though the cell is lysed.

3. Tissue Procurement Guidelines for Molecular Analysis (3)

The paramount requirement of any human tissue procurement is an accurate histopathologic diagnosis. The pathologic diagnosis is the determining factor for irrevocable clinical decisions about mode and extent of surgery, the successful attainment of clean surgical margins, and the administration of toxic therapies. Surgeons, pathologists, and institutional review boards are frequently concerned that tissue procured for exploratory research will compromise the accuracy of the histopathologic diagnosis. Their concern is based on the fact that tissue used for molecular profiling may not contain the same pathologic changes present in the tissue sample used for primary diagnosis. Adherence to the following guidelines for tissue procurement will aid in providing quality specimens suitable for diagnosis and/or research.

1. Tissue is alive and reactive to ex vivo stress. Kinase pathways are active and reactive until the tissue cells are stabilized. RNA transcription is reactive to ex vivo stress and RNA stability rapidly declines over time in a highly temperature-dependent manner. Tissue procurement protocols must recognize these facts and provide methods for limiting exposure to extreme environmental conditions.
2. Reactive changes occurring in tissue postexcision can generate false elevation as well as false declination in protein and nucleic acid analytes. This may be a significant source of bias in the analysis of protein or nucleic acid as potential biomarkers.
3. Kinase pathway stabilization methods should block both sides of the kinase/phosphatase kinetic reaction. Blocking only phosphatases can cause false elevation of an analyte's phosphorylation level.

4. Tissue should be stabilized as soon as possible after excision. Taking into consideration the average time for procurement in a community hospital, the recommended maximum elapsed time is 20 min from excision to stabilization (e.g., flash freezing, thermal denaturation, or chemical stabilization).
5. Tissue stabilization and preservation methods should be compatible with the intended downstream analysis. Preservation of tissue histology and morphology is essential for verification of tissue type and cellular content.
6. Documentation of the sample excision/collection time, elapsed time to preservation/stabilization, and length of fixation time and type of preservation are critical data elements for sample quality assessments.

3.1. Formalin Fixation Pitfalls

Formalin fixation, the standard for 100 years, is considered by many molecular scientists to be outmoded due to its slow penetration and the cross links it generates (14). The slow penetration can deprive inner regions of the tissue from receiving adequate fixation at the time they are undergoing hypoxic stress. The cross links produced by formalin hinder biomolecule extraction and negatively effect the quality of proteins and nucleic acids, therefore existing formalin-fixed paraffin-embedded (FFPE) tissue archives are not suitable for the next generation of molecular analyses required for individualized therapy and the prediction of aggressiveness.

Even if a strict protocol is followed, there is no ultimate assurance that processing variables are free from compromise up to the time that the molecular profile data is collected. Thus, a major requirement for stabilization technology of the future is to provide immediate stabilization of molecular and histologic tissue signatures at the time of procurement.

Formalin penetrates tissue at a variable rate, reported within the range of mm/h (18–20). During this time delay period the portion of the living tissue deeper than several millimeters would be expected to undergo significant fluctuations in regards to phosphoprotein analytes. Cellular molecules in the depth of the tissue will have significantly degraded by the time formalin permeates the tissue (18, 21). Penetration is not synonymous with fixation. In aqueous solution formaldehyde becomes hydrated, forming methylene glycol (18, 20). The small percentage of formaldehyde in solution forms the actual covalent cross-links with proteins and nucleic acids. Methylene glycol penetrates the tissue, yet it is the carbonyl formaldehyde component that causes tissue fixation (18, 20). Formalin cross-linking, the formation of methylene bridges between amide groups of protein, blocks analyte epitopes as well as decreases the yield of proteins extracted from the tissue. Often the dimensions of the tissue and the depth of the block from which samples are prepared are unknown variables for the translational researcher. Consequently, formalin fixation would be expected to cause significant variability in protein and phosphoprotein stability for molecular diagnostics (18, 22, 23).

3.2. Formalin Fixation Alternatives

New chemistries are needed for preserving proteins and post-translationally modified proteins (24–26) and nucleic acids. Although new fixatives have been developed for preservation and/or extraction of RNA from formalin-fixed tissue, these same chemistries have not been thoroughly evaluated as a timecourse analysis of phosphoproteins or other post-translationally modified proteins (21, 24, 25, 27–30), and they may not be suitable for diagnostic histopathology and immunohistology. Thermal/pressure inactivation of RNAses protein kinases and phosphatases has been proposed for rapid protein stabilization/inactivation (31). Rapid thermal inactivation of enzymes ensures stabilization of kinetic reactions but fails to maintain the tissue morphology. Preservation of tissue morphology is an absolute requirement for acceptance by pathologists. Ultrasound rapid fixation (21, 27, 28, 30, 32) and nonformalin based fixatives (24, 29) processed with or without microwave assistance, are emerging technologies that can maintain histomorphology and maximize antigen retrieval for immunohistochemistry. Nevertheless, the contribution of preanalytical variables (prior to immersion in the nonformalin fixative or delay in processing) and the preservation of phosphoproteins, must still be addressed before these technologies gain widespread clinical utility.

3.3. Preserving Tissue Histomorphology for Shared Diagnostic/Research Samples

It is widely recognized that phosphoproteins are the emerging molecular markers that provide key information about the state of cellular signaling pathways that drive cancer pathogenesis. Measurement of phosphoprotein signal pathway epitopes in tumor biopsy samples is crucial for individualizing the new generation of molecular therapy aimed at targeting kinase pathways.

Future tissue preservation technology must take into consideration the requirement for an accurate histopathologic diagnosis, and recognize limitations in the availability of specialized processing instruments, or liquid nitrogen for flash freezing of tissue. The pathologic diagnosis is the determining factor for irrevocable clinical decisions about mode and extent of surgery, the successful attainment of clean surgical margins, and the administration of toxic therapies. Surgeons, pathologists, and institutional review boards, are concerned that tissue procured for molecular diagnostics, or research, will compromise the accuracy of the histopathologic diagnosis. The ideal future technology is a new class of one-step, multipurpose fixation chemistry. At the time of tissue procurement, the tissue would be immediately immersed in the stabilizing chemistry to arrest all reactive fluctuations in protein and nucleic acid macromolecules.

This proposed chemistry would retain histomorphology in multiple tissues without compromising nuclear size, preserve phosphorylated antigens for immunohistochemistry, and inhibit phosphatases that may be active in tissue postexcision and during tissue processing. In addition, multiple sections from the same paraffin block could be cut and distributed for (a) histopathologic diagnosis, (b) immunohistochemistry, (c) microdissection, (d) proteomic analysis, and (e) nucleic acid analysis. Finally, the tissue could be stored indefinitely at room temperature as a paraffin block in the standard fashion of an anatomic pathology archive. Such a technology will transform clinical molecular medicine because (a) all molecular profiling can become standardized, (b) preanalytical variables caused by tissue preservation delays will be eliminated, (c) no additional equipment or costly freezers will be required, and most importantly (d) there will be no chance that a molecular profile will be conducted on a piece of tissue that does not have histopathologic verification.

4. Notes

1. The addition of a phosphatase inhibitor to a native (nondenatured) protein sample may result in false elevation of phosphorylated proteins due to the uninhibited action of kinases. It is important to consider inhibiting both phosphatases and kinases in a sample, or using protein denaturing conditions which are likely to denature both phosphatases and kinases.
2. Phosphoprotein antigen epitopes are not adequately preserved by formalin fixation and paraffin embedding.

References

1. Poste, G. (2011) Bring on the biomarkers. *Nature* **469**, 156–7.
2. Spruessel, A., Steimann, G., Jung, M., Lee, S. A., Carr, T., Fentz, A. K. et al. (2004) Tissue ischemia time affects gene and protein expression patterns within minutes following surgical tumor excision. *Biotechniques* **36**, 1030–7.
3. Espina, V., Edmiston, K. H., Heiby, M., Pierobon, M., Sciro, M., Merritt, B. et al. (2008) A portrait of tissue phosphoprotein stability in the clinical tissue procurement process. *Mol Cell Proteomics* 7, 1998–2018.
4. Li, J., Gould, T. D., Yuan, P., Manji, H. K., Chen, G. (2003) Post-mortem interval effects on the phosphorylation of signaling proteins. *Neuropsychopharmacology* **28**, 1017–25.
5. Li, X., Friedman, A. B., Roh, M. S., Jope, R. S. (2005) Anesthesia and post-mortem interval profoundly influence the regulatory serine phosphorylation of glycogen synthase kinase-3 in mouse brain. *J Neurochem* **92**, 701–4.
6. Imbeaud, S., Auffray, C. (2005) 'The 39 steps' in gene expression profiling: critical issues and proposed best practices for microarray experiments. *Drug Discov Today* **10**, 1175–82.
7. Dash, A., Maine, I. P., Varambally, S., Shen, R., Chinnaiyan, A. M., Rubin, M. A. (2002) Changes in differential gene expression because of warm ischemia time of radical prostatectomy specimens. *Am J Pathol* **161**, 1743–8.
8. Espina, V., Mueller, C., Edmiston, K., Sciro, M., Petricoin, E. F., Liotta, L. A. (2009) Tissue is alive: New technologies are needed to address the problems of protein biomarker pre-analytical variability. *Proteomics Clin Appl* **3**, 874–882.

9. Grellner, W., Vieler, S., Madea, B. (2005) Transforming growth factors (TGF-alpha and TGF-beta1) in the determination of vitality and wound age: immunohistochemical study on human skin wounds. *Forensic Sci Int* **153**, 174–80.
10. Grellner, W. (2002) Time-dependent immunohistochemical detection of proinflammatory cytokines (IL-1beta, IL-6, TNF-alpha) in human skin wounds. *Forensic Sci Int* **130**, 90–6.
11. Grellner, W., Madea, B. (2007) Demands on scientific studies: vitality of wounds and wound age estimation. *Forensic Sci Int* **165**, 150–4.
12. Goldstein, B. J. (2002) Protein-tyrosine phosphatases: emerging targets for therapeutic intervention in type 2 diabetes and related states of insulin resistance. *J Clin Endocrinol Metab* **87**, 2474–80.
13. Neel, B. G., Tonks, N. K. (1997) Protein tyrosine phosphatases in signal transduction. *Curr Opin Cell Biol* **9**, 193–204.
14. Lim, M. D., Dickherber, A., Compton, C. C. (2011) Before you analyze a human specimen, think quality, variability, and bias. *Anal Chem* **83**, 8–13.
15. Baker, A. F., Dragovich, T., Ihle, N. T., Williams, R., Fenoglio-Preiser, C., Powis, G. (2005) Stability of phosphoprotein as a biological marker of tumor signaling. *Clin Cancer Res* **11**, 4338–40.
16. Paweletz, C. P., Charboneau, L., Bichsel, V. E., Simone, N. L., Chen, T., Gillespie, J. W. et al. (2001) Reverse phase protein microarrays which capture disease progression show activation of pro-survival pathways at the cancer invasion front. *Oncogene* **20**, 1981–9.
17. Nishizuka, S., Charboneau, L., Young, L., Major, S., Reinhold, W. C., Waltham, M. et al. (2003) Proteomic profiling of the NCI-60 cancer cell lines using new high-density reverse-phase lysate microarrays. *Proc Natl Acad Sci U S A* **100**, 14229–34.
18. Fox, C. H., Johnson, F. B., Whiting, J., Roller, P. P. (1985) Formaldehyde fixation. *J Histochem Cytochem* **33**, 845–53.
19. Helander, K. G. (1994) Kinetic studies of formaldehyde binding in tissue. *Biotech Histochem* **69**, 177–9.
20. Srinivasan, M., Sedmak, D., Jewell, S. (2002) Effect of fixatives and tissue processing on the content and integrity of nucleic acids. *Am J Pathol* **161**, 1961–71.
21. Nassiri, M., Ramos, S., Zohourian, H., Vincek, V., Morales, A. R., Nadji, M. (2008) Preservation of biomolecules in breast cancer tissue by a formalin-free histology system. *BMC Clin Pathol* **8**, 1.
22. Devireddy, R. V. (2005) Predicted permeability parameters of human ovarian tissue cells to various cryoprotectants and water. *Mol Reprod Dev* **70**, 333–43.
23. He, Y., Devireddy, R. V. (2005) An inverse approach to determine solute and solvent permeability parameters in artificial tissues. *Ann Biomed Eng* **33**, 709–18.
24. Bellet, V., Boissiere, F., Bibeau, F., Desmetz, C., Berthe, M., Rochaix, P. et al. (2007) Proteomic analysis of RCL2 paraffin-embedded tissues. *J Cell Mol Med*
25. Delfour, C., Roger, P., Bret, C., Berthe, M. L., Rochaix, P., Kalfa, N. et al. (2006) RCL2, a new fixative, preserves morphology and nucleic acid integrity in paraffin-embedded breast carcinoma and microdissected breast tumor cells. *J Mol Diagn* **8**, 157–69.
26. Perlmutter, M. A., Best, C. J., Gillespie, J. W., Gathright, Y., Gonzalez, S., Velasco, A. et al. (2004) Comparison of snap freezing versus ethanol fixation for gene expression profiling of tissue specimens. *J Mol Diagn* **6**, 371–7.
27. Nadji, M., Nassiri, M., Vincek, V., Kanhoush, R., Morales, A. R. (2005) Immunohistochemistry of tissue prepared by a molecular-friendly fixation and processing system. *Appl Immunohistochem Mol Morphol* **13**, 277–82.
28. Chu, W. S., Furusato, B., Wong, K., Sesterhenn, I. A., Mostofi, F. K., Wei, M. Q. et al. (2005) Ultrasound-accelerated formalin fixation of tissue improves morphology, antigen and mRNA preservation. *Mod Pathol* **18**, 850–63.
29. Stanta, G., Mucelli, S. P., Petrera, F., Bonin, S., Bussolati, G. (2006) A novel fixative improves opportunities of nucleic acids and proteomic analysis in human archive's tissues. *Diagn Mol Pathol* **15**, 115–23.
30. Vincek, V., Nassiri, M., Nadji, M., Morales, A. R. (2003) A tissue fixative that protects macromolecules (DNA, RNA, and protein) and histomorphology in clinical samples. *Lab Invest* **83**, 1427–35.
31. Svensson, M., Skold, K., Nilsson, A., Falth, M., Nydahl, K., Svenningsson, P. et al. (2007) Neuropeptidomics: MS applied to the discovery of novel peptides from the brain. *Anal Chem* **79**, 15–6, 18–21.
32. Morales, A. R., Nassiri, M., Kanhoush, R., Vincek, V., Nadji, M. (2004) Experience with an automated microwave-assisted rapid tissue processing method: validation of histologic quality and impact on the timeliness of diagnostic surgical pathology. *Am J Clin Pathol* **121**, 528–36.

Chapter 5

The Human Side of Cancer Biobanking

Eoin F. Gaffney, Deirdre Madden, and Geraldine A. Thomas

Abstract

The future success of translational research is critically dependent on the procurement and availability of high-quality tissue specimens linked to accurate histopathologic and clinical information about the individual banked specimen. The international community has awakened to this critical need only recently. Three major roadblocks have hindered the success of previous biobank consortiums: (1) Ethical issues surrounding patient consent and ownership of intellectual property, (2) Failure to properly preserve the molecular content of the tissue, and failure to reliably document clinical data linked to the specimen, and (3) Management issues: inadequate funding, competition for use of the tissue, inadequate personnel and facilities, and absence of dedicated database software. This chapter reviews these critical roadblocks and discusses international efforts to provide strategies to implement high-quality biobanks.

Key words: Biobanking, Database, DNA, Ethic, Pathology, Protein, Quality control, RNA, Specimen, Tissue

1. Introduction

The development of more effective, targeted treatment for cancer depends on increased understanding of the molecular mechanisms involved in the initiation of the tumour and its progression to metastatic disease. Research studies depend on the availability of high-quality, well-preserved biological material with full pathologic diagnosis and linked clinical information. Collections of human biospecimens ("biobank") obtained according to stringent, but practical, standard operation procedures (SOPs) linked with detailed clinical information, updated over time, will be required to deliver the promise of personalised medicine (1, 2) (see Note 1).

High-quality biobanks are also crucial for validating disease biomarkers and translating them to clinical benefit. George Poste in a 2011 Comment in Nature (3) proclaimed that research in

Virginia Espina and Lance A. Liotta (eds.), *Molecular Profiling: Methods and Protocols*, Methods in Molecular Biology, vol. 823, DOI 10.1007/978-1-60327-216-2_5,

biomarkers to diagnosis disease and guide therapy has not yet delivered on its promise, even though "the ability of biomarkers to improve health and reduce health care costs is greater than any other area of biomedical research. While thousands of papers have been published claiming new biomarkers, only 100 biomarkers are routinely used in the clinic today. In this chapter, Poste asserts that the major roadblock preventing the successful discovery and validation of new biomarkers is the lack of high-quality biobanks (3). He proposes that stringent international guidelines would improve the practice of tissue storage and banking. While the need for high-quality biobanking is not a new idea, it is an idea that has finally begun to resonate at the national and international level (4). Time magazine recognised Biobanks as one of the "ten ideas changing the world now" (5). The NIH and funding agencies, such as the European Framework Programme, the Innovative Medicines Initiative (Europe), and the UK Welcome Trust, are now investing in new biobanking initiatives that are based on international cooperation and follow strict quality control protocols.

There is no shortage of new molecular techniques for investigating cancer, but unfortunately there is a real shortage of high quality frozen and fresh samples to investigate in large numbers. Rapid access to large sample numbers is an important requirement for translational research, and a lack of samples is regarded as the greatest roadblock. It is ironic that potentially useable tissue is discarded daily in pathology departments throughout the world. If pathology departments were to receive (nominally increased) resources, and biobank personnel, and were supported in a cancer biobank network, samples from selected tumour types (or from all resected tumours) could be collected and stored as "standard of care." This scenario is beginning to happen in many countries of the world, following the lead of the Spanish Tumour Bank Network (6, 7).

The authors believe that cancer research should be patient-focused and therapy should be individualised. Consequently, if cancer biobanking is to be of real benefit to those with cancer, and indeed to researchers, it must be carried out in interinstitutional collaborations in regional or national networks, with easy linkage to international networks. International harmonisation is being promoted by the US National Cancer Institute (NCI), which is appropriate because cancer is a global health issue, and the difficult questions are better tackled by international consortiums. Barriers to achieving this goal include severely reduced funding for biomedical research and difficulties in exchanging research material between countries, due to different legislation (6, 8). Study sections may be inclined to fund original focused experimental research at a priority greater than biobank funding. Assuming high quality fully annotated specimens, a final, and underappreciated, barrier to a successful biobanking is how to decide who is permitted to receive samples from the precious

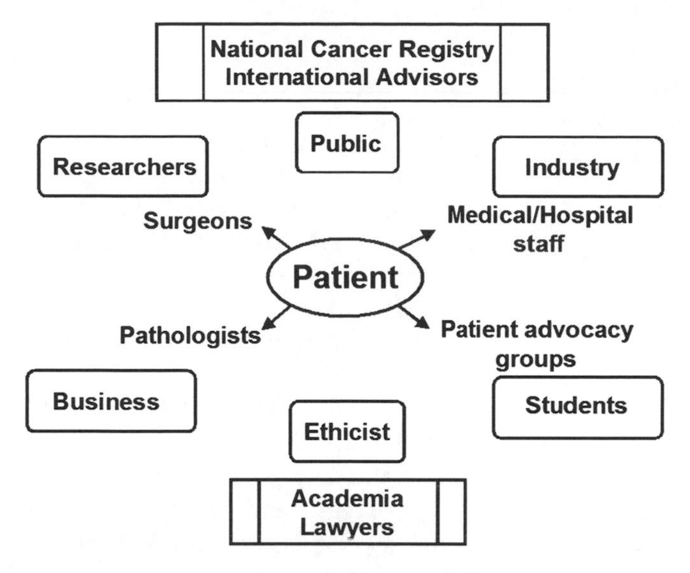

Fig. 1. Biobank network stakeholders. Patients are essential to the success of a biobank because tissue donation begins with an approved protocol and an informed, consenting patient. The medical community must support the patient and provide opportunities for tissue donation. Businesses and pharmaceutical industry leaders may provide financial support for maintaining the biobank. Scientists benefit from biobanks due to the availability of samples in either large numbers or from patients with rare conditions.

resource of specimens. Many past biobank specimens have remained fallow while the biobank owners have waited for the ideal request.

Biobanking shares more stakeholder groups than any other area of biomedical research (Fig. 1). Apart from patient donors, there are stakeholders within and outside the hospital and the research laboratory. These include government ministers and health agency officials, hospital management, hospital staff (nurses obtaining consent and clinical follow-up, surgeons, pathologists, biobank technicians, data managers, network co-ordinators, information technology staff, and scientific and ethics review committees [Institutional review boards (IRBs)]. Hospital staff and review boards constitute the essential links in the "chain of trust" between the patient and the researcher (see Note 2) (9). Patient advocacy groups and an informed public are strong advocates of biobanking. So too are professional medical societies, scientific organisations, charities, and the biotechnology and pharmaceutical industry, albeit for different reasons. The Genetic Alliance Biobank (10) and PXE International (11) are two examples of US-based advocacy organisations that partner with researchers and industry to drive translational research. Although the components of biobanking infrastructure are straightforward, putting it together and maintaining it require excellent communication and an attitude of genuine commitment and collaboration between many individuals.

2. Purpose and Goals of a Biobank Network

The purpose and goals of the cancer biobank network must be defined at the outset. Although no two biobanks are, or will be, the same (12), the goals will vary depending on the bank's mission. For example, the biobank network may focus on specific tumour types, and may or may not collect samples for industry. Other goals might be to collect rare tumours or tumours in unusual sites or in atypical clinical circumstances (familial cancer syndromes) and to obtain samples to enhance clinical trials. Goals may be modified or additional goals added over time, although the purpose should be unchanged.

The purpose of cancer biobank networks is much broader than that of the "stand-alone" biobank. Biobanks were synonymous with freezers for storing "researchers' samples" for use by the researcher in hypothesis-driven research projects. Biobanks were – and many still are – managed by researchers themselves, samples being withdrawn when researchers are ready to use a particular technique to investigate individual aspects of the project. The emphasis is now on neutral biobank personnel or "honest brokers" who not only collect samples prospectively for specific projects (13), but who in many cases also bank all tumours for future use in undefined projects and for industry. Nevertheless, biobank personnel must not lose sight of the fact that samples have been donated by patients in the expectation that they will be used. Access to samples must be fair and based on a project's objective scientific merit and ethical review, and should not be restricted to an institution's researchers – although they often have first claim to the material.

2.1. Benefits

The benefits of a biobank network are potentially far-reaching. A successful biobank network benefits all, including the economy (potential for employment and increased industry research and development), but above all, the biobank network helps future patients as it is the bridge between research and care (Fig. 2).

A biobank network has the greatest chance of success if researchers' needs are addressed in the planning and development process. The network can present immediate advantages: (a) management of the logistics of specimen collection and preservation, (b) availability of online data protection-compliant virtual database cataloguing high-quality samples, and (c) storage of restricted patient data in many different institutions. Availability of enough sample aliquots, or preferably RNA, DNA, protein, or paraffin sections, will facilitate collaboration in the project. Researchers must provide feedback to the biobank on sample quality and acknowledge the bank in disseminations. This helps promote awareness of the bank, which should be a service to the scientific community. The biobank should communicate directly with researchers, clearly outlining

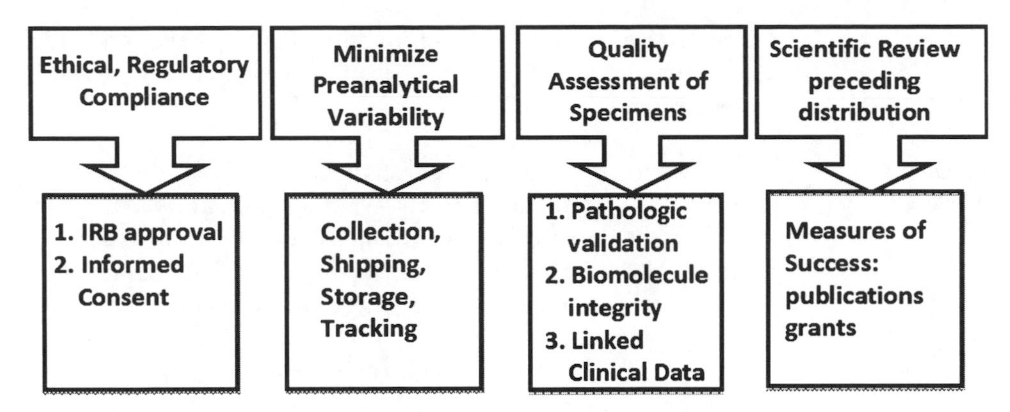

Fig. 2. Biobank specimen processes. Compliance with regulatory agencies ensures ethical tissue acquisition. Preanalytical variability between samples is minimised through the development and implementation of standard operating procedures. Quality control and quality assurance procedures provide information regarding molecular integrity and suitability of the sample. Biobank success is measured by the number of literature citations and grants that were awarded based on samples from the biobank.

what can be provided and the requirements for access (13). Grants have begun to provide resources for biobanking administrative expenses. This will shorten the time needed to acquire preliminary results and the time needed to complete a project.

Quality assured samples, all treated according to uniform operating procedures, and linked to pathology and clinical data, unquestionably offer the greatest chance of high-quality research (see Note 2). The principal advantage of pathologist input is quality assurance – selecting the most suitable tissue areas to store in biobanks and not required for standard of care diagnosis. Pathologists can also assist researchers in study design and feasibility and in interpretation of microscopic findings (14). They can indicate common data elements for particular tumour sites. Pathologists may be in a position to offer molecular pathology training and support, such as laser capture microdissection and tissue microarrays, to complement techniques used for frozen material (15). This type of collaboration and attention to detail greatly benefits research projects and characterises the sharing attitude of modern biobanking. But the biobank must be resourced adequately, particularly for pathology, in order to obtain pathologist buy-in.

Hospitals and patients benefit too, in the medium and long-term. The need for biobanking ensures prompt examination and fixation of the fresh cancer resection and hence more rapid pathology reporting. Location of biobanks in hospitals focuses the emphasis on translational research for better patient care. In contrast, universities do not have a primary role in cancer resection or sample collection, or in patient diagnosis or treatment, and therefore their role in biobanking is secondary. The capacity to obtain samples from clinical trials patients offers the opportunity to assess efficacy and toxicity of new drugs and ensures a uniform standard of cancer care for those patients. This represents added value for pharmaceutical companies sponsoring clinical trials, and may accelerate the development of real benefits for patients – new and better patient interventions.

2.2. Ethical Issues in Biobanking

This section focuses on two main ethical issues that arise in the context of biobanking, namely, whether human tissue is the property of the person from whom it is removed and what form of control should exist in relation to the uses of the tissue; and what kind of consent is appropriate where stored tissue is being used for research purposes.

Other issues for consideration in this area include the question of what measures should be taken to ensure that the privacy of the individual is not infringed by the research; whether there should be feedback of research results to those whose tissue was used; what form of ethical oversight is appropriate; and whether there ought to be a system of benefit-sharing where commercial research is involved. These latter issues pertain more broadly to many different aspects of medical research and will not be the focus of this chapter.

2.3. Ownership of Human Bodies and Tissue

2.3.1. Corpses

There has been much academic discussion about whether the law recognises the human body as property that is capable of being owned. Historically, the human body after death was seen as *res derelicta*, a thing with no value and no ownership rights attaching to it. Although there is some dispute regarding the exact provenance of this rule in the common law system, it now seems to be so enshrined in legal doctrine that it is now unlikely to be overturned (16). Most jurisdictions have traditionally prohibited human bodies having a property value on the basis that it would violate the fundamental principle of respect for human dignity.

2.3.2. Body Parts

In relation to separated body parts, the law is less clear. In the UK, hair, blood, and urine have all been held to be property for the purposes of the criminal offence of theft. In recent years, the English courts have also held that parts of a corpse are capable of being property if they have acquired different attributes by virtue of the application of skill (17, 18). The cases do not clearly set out what must be done to a part of a body in order for it to acquire the status and nature of property. Some indicate that dissection or preservation may be sufficient, in which case the person applying the skill would have a property right in respect of it. There have been no Irish cases on this point but given the common law tradition shared by both jurisdictions, it is unlikely that an Irish court would reach a different conclusion.

2.3.3. Tissue from Living Donors

The law relating to the removal and control of tissue, which has been excised from living persons for therapeutic purposes, is similarly unsatisfactory in not providing clear guidance to patients, clinicians, and researchers. A number of cases in the USA have held that the person from whom tissue is excised for therapeutic purposes has no property right in that tissue. In the well-known case of *Moore v Regents of the University of California* (19, 20), John Moore had his spleen removed as part of treatment for hairy cell leukaemia. The treating physician discovered that Moore's spleen cells contained potentially beneficial properties and he developed a cell line that he eventually sold for $15 million. His research was carried out without Moore's knowledge or consent. A legal action was initiated by Moore on a number of grounds, namely, conversion (the use of another's property without their consent), breach of fiduciary duty, and failure to obtain informed consent. The Californian Supreme Court rejected the property rights claim on the basis that there was no precedent on which it could be said that people had property rights in their bodies and it would be inappropriate for the law to recognise such a right as it would hinder medical research and lead to the sale of bodily parts.

In a more recent case, *Washington University v Catalona* (17, 21), the defendant worked for many years as Chief of the Division of Urologic Surgery in a private research university with a medical school. His research focused on prostate cancer and over many years of surgery, he collected research samples from the excised cancerous tissue of his patients. The samples were stored in a biobank operated by the University and were used strictly for research purposes. Research participants were asked to sign informed consent forms that stated that they could not assert any ownership rights in respect of the products resulting from the research. In 2003, Catalona left the university to work elsewhere and wrote to the research participants whose samples were stored at the biobank asking for their consent to have their samples released to him. A legal wrangle subsequently ensued in relation to whether the sample was owned by the University or the research participants, such that they could decide to withdraw their samples from the biobank in favour of Catalona. Under relevant legislation in Missouri, the court held that the University had satisfied the two pronged test for ownership, i.e., exclusive possession and control. The next issue was whether the research participants had made a gift of their samples, and if so was the gift made to the University or to Catalona. Based on the wording of the informed consent forms, the court held that they had intended to make a gift to the University.

In both cases, the courts were concerned with the potential adverse effect on research if a property right were held to exist over excised tissue. They were also opposed to the notion that people would be able to sell their tissue to the highest bidder. There have been no Irish cases on this point but it is likely that an Irish court would have similar objections to the property argument.

In addition to the question of whether the management of human tissue should be governed by property laws or another regulatory framework (22), biobanking raises issues about how to strike the appropriate balance between respect for persons and society's interest in promoting research.

2.4. Consent

Respect for individual autonomy is usually expressed through the concept of informed consent. In medical ethics and law, the doctrine requires that patients be given full information regarding their treatment so that they can make an informed choice as to whether or not to have the treatment (23, 24). Informed consent is commonly set as the gold standard to be achieved in relation to participation in research also. However, there is no international consensus on how the doctrine of informed consent applies in relation to archival human tissue or future research uses of stored tissue (25).

It will usually be the case that, in the case of donation of biological samples for research where the donor–patient remains anonymous to the researchers, no potential risk arises other than in relation to the taking of extra blood samples preoperatively. Although there is no anticipated therapeutic benefit to the donor–patient, the donation may generate a positive benefit for future patients. Until recently, international statements and guidelines did not address the specific issue of archived samples. The main question here is whether it is in keeping with respect for autonomy that persons could be asked to give a general or broad consent to all future uses of their tissue, or is it necessary that a separate consent be given in relation to each research project as it arises?

2.5. International Guidelines

In 2003, the World Health Organisation stated that archived material removed in the course of medical care should be capable of being used for research once anonymised (21). The Human Genome Organisation (HUGO) (26), the Council for International Organisations of Medical Sciences (CIOMS) (27), and United Nations Educational, Scientific and Cultural Organisation (UNESCO) (28) (see Note 3) have been a little less restrictive by recommending that samples may be used for research not only in anonymised form but potentially also in coded form without re-consent, provided certain conditions are met such as notification of such a policy to patients, no objection by patients, approval by an ethics committee, minimal risk, potential significant scientific knowledge benefits, and the impracticability of obtaining consent in the circumstances.

Many European guidelines permit general or broad consent for unspecified future research use, for example, Germany (29), UK (30), Sweden, Iceland, and Estonia (31). The Council of Europe's Steering Committee on Bioethics states that "when biological materials of human origin and personal data are collected, it is best practice to ask the sources for their consent to future use, even in cases where the specifics of the future research projects are unknown" (32–34). This form of general consent is seen as acceptable if two conditions are met, namely, research ethics review, and the right of the participant to withdraw at any time. This is a relaxation of the classical research ethics position in relation to biobanks, and a less strict standard of consent (35). While the former condition does not appear to pose significant problems, other than perhaps slowing the progress of the research protocol, the second condition is perhaps more potentially troubling as the withdrawal of samples in any significant number from the biobank may lead to problems in checking the validity of the research at a later date.

In the USA, the prevailing opinion until recent years was in favour of the classical standard of informed consent. In relation to the difficulties of obtaining consent in respect of future uses, multi-layered consent was seen as the most acceptable means of ensuring that participants were informed of the relevant information and choices. This form of consent allows for different choices to be presented on a detailed form, enabling participants to choose to limit their consent to use for research into a specific disease or a specific research project. This approach was seen as imposing a burden on research (35).

In 1999, the guidelines of the US National Bioethics Advisory Committee proposed a strategy of waivers where informed consent would not be necessary where the research involves no more than minimal risk to the subjects; the waiver will not adversely affect the rights or welfare of the subjects; the research could not be practicably be carried out without the waiver; and whenever appropriate the subjects will be provided with additional information following their participation. Despite the granting of a waiver it was also recommended that consent should be sought unless it was impracticable to locate the subjects in question (36–38).

In 2004, the US Office for Human Research Protection (OHRP) proposed a different solution (39). It was recommended that research on unidentifiable specimens should not be classed as research involving human subjects. As a result, it was not necessary to obtain informed consent or ethics review. This opinion was due to the interpretation of "research involving human subjects" as involving an interaction with a living person. The OHRP now considers private information or specimens not to be individually identifiable when they cannot be linked to specific individuals by the investigators either directly or indirectly through coding systems. This is the case when the investigators and the holder of the code enter into an agreement prohibiting the release of the code to the investigators under any circumstances until the individuals are deceased. The advantage of this approach is clearly that it facilitates research by avoiding the requirement for informed consent and ethics review simply by entering into an agreement prohibiting the researchers accessing the code. This means there will be no delays, no costs, and no ethics review.

2.6. Views of Patients and Participants

Studies indicate that a majority of patients favour the requirement of consent to the use of their tissue even where the samples are anonymous, particularly where the research involves genetic testing of the samples (40). It is also noteworthy that a substantial majority of patients and research participants find general consent adequate, largely because they do not wish to be re-contacted repeatedly (41). Because of their unique insight, it is highly appropriate that patient advocate groups participate in the design of consent forms and sit on ethics committees (IRBs). Patients are aware of the importance of research, and may wish to actively participate in directing the biobank's operations (42). Less than 1% of patients approached for consent by the Wales Cancer Bank (http://www.walescancerbank.com) refuse to donate material, and only one out of more than 2000 so far consented refused on the grounds that he did not agree with generic (enduring) consent. Although there is some antipathy to pharmaceutical companies, a New Zealand study reported that over 97% of donors consented to commercial research (43).

A survey carried out in 2005 in Ireland indicates that 86% of the population would be willing to allow the use of their excised tissue for research purposes (44). The primary motivation for their willingness to donate was that it could be potentially beneficial for members of the patient's family (96%), or their own future health (92%), or for potential benefit to society generally (80%). There was an 89% preference for linked donation once the benefit of linkage was explained. In relation to the form of consent to be adopted, 36% preferred specific consent, 44% chose general consent, and 16% opted for the personal choice model which leaves it to the individual to choose specific or general consent.

2.7. Ethical Issues

Consent is regarded as key to all medical treatment and research. This applies both to the use of samples stored as part of medical care and to secondary uses of research samples. However, there appears to be a gradual move in other jurisdictions and in international and professional guidelines toward allowing biobanks to obtain a broad consent for future secondary research using either anonymous or coded samples and data once there is sufficient public notification and ethics review processes in place.

The current point of view is that persons ought not to be regarded as private owners of tissue, and an encouragement of the altruistic and communitarian approach that would recognise the societal and public health benefit to be gained by such a model.

2.8. The Patient, Hospital Staff, and the Biobank

Successful biobanks communicate with and receive excellent collaboration from hospital staff. A successful biobank reflects a good corporate image and the hospital can help disseminate information for patients and on the biobank's progress. The central location, use of one system, and avoidance of duplication with uniform safe storage, constitute an infrastructure to facilitate translational research in a first class institution. The biobank may lead to recruitment of more clinical trial patients. All of the above potentially serve to reinforce management's positive view of the value of the biobank to their hospital and to the wider community. The most important long-term stakeholder is the individual patient. The patient voluntarily donates surplus tissue from his/her cancer resection specimen or biopsy as a gift to the biobank, which then becomes custodian for the tissue.

From a day-to-day perspective, it is the smooth working relationship between cancer research nurses, operating theatre nurses, surgeons, porters, pathologists, pathology technicians, and biobank personnel that enables biobank staff to perform their job to the best of their ability. It is important therefore to meet with the above groups to explain why their role is important, to discuss any work practice changes and why the biobank needs the specimen fresh and quickly to minimise RNA degradation. Training of medical staff in aspects of biobanking may be helpful in their appreciation that it is an important but standard hospital activity. The surgeon must be informed and agree that the cancer resection he is going to perform will need to be sent to pathology quickly. Even surgeons not directly involved in research have demonstrated the capacity to adapt their practices to facilitate the rapid transfer of the resection specimen. Acting on notification by the operating theatre nurse, resection specimens are taken on ice by a porter or biobank personnel to pathology. This requires the biobank to be made aware in advance that the operation is almost finished. Time is precious – time from resection of the cancer to freezing of samples should not be more than 30 min (45). Certainly, if RNA and protein–protein interactions are being investigated, this time should ideally be less. Of paramount importance, sampling for research must not jeopardise full pathology examination, and if the tumour is too small or not visible grossly, either a frozen section can be carried out to confirm tumour or to identify tumour type, or the specimen should not be sampled at all. Sometimes the frozen section shows that the "tumour" is an inflammatory mass or other benign process. Ideally, the pathologist who sampled the specimen should be the individual to examine and report it. If enough aliquots are taken or if the biobank only releases RNA, DNA, or sections rather than blocks (see below), there will be material for several academic research projects and for academic–industry partnerships. Patients must consent to the above, and be reassured that their samples will not be sold. Unless investigated with numerous other samples and altered by research, their sample has no commercial value, and no financial gain will accrue to them. However, after research, intellectual property can be held by researchers, their institution, a company, and the nation.

For uniform collection, processing and storage of samples, satisfactory performance of SOPs should be assessed regularly, both internally and externally if possible, in each biobank of the network. Although failure to observe SOPs is likely to lead to poor sample quality, a high level of RNA degradation might also be due to as yet poorly studied patient factors. These include intaoperative procedural factors, warm ischaemia time, sample handling postprocurement, and differences in tissue type (4). Tissue to be sampled is subject to heterogeneous immunological and metabolic insults in the region of the tumour and in the cancer resection specimen. Until recently, efforts to control these factors have been based on anecdotal evidence. This new field of research – biospecimen science – is a current research programme at NCI, and developments will clearly be of great interest to the international biobanking community.

3. Centralised or Decentralised Biobank Network

A centralised biobank network – centralised sample storage and informatics – may seem attractive and appropriate. However, tissue biobank networks worldwide with very few exceptions use a decentralised model: samples are stored in the centres in which the patients donate. Each hospital has its own database(s) too, but data is relayed to a central co-ordinating office for the central common database. Decentralised storage is preferable for ethical and legal reasons and when a patient's previous sample is unexpectedly needed for diagnostic purposes. Significant territorialism has always been associated with biobanks, and has the potential to block all sharing, unless researchers perceive clear benefits for them. Therefore, the system should be the most suitable, advantageous, and not necessarily the least expensive. (A centralised biobank structure would be expected to be cheaper – like a central warehouse for stock distribution – but the experience worldwide is that buy-in is much more difficult, potentially undermining the operation.)

3.1. Database, Common Data Elements, and Data Entry

Each hospital biobank retains its own database for sample inventory and for importing pathology and clinical data. Identical common data elements are agreed by the participating biobanks, but often not without considerable discussion. Data entry is performed at different points in time, by designated persons only. Restricted data from each biobank database is exported at regular intervals to the common database in the central office. The common database should be CaBIG-compliant, for greatly added value. Clearly and extensively annotated samples are essential for addressing specific research questions (4, 46). Alternatively, if resources are suboptimal, the biobank database might contain only very basic data, that can be supplemented with longitudinal clinical follow-up later during a project.

3.1.1. Data Security

Patient identifiers must never be available to researchers. This can be achieved by separating databases on which patient-identifiable data is stored from databases on which information on the storage and use of the specimen donated is collected. This is best achieved by providing a code that is unrelated to any other form of patient identifier (e.g., hospital number, national insurance number, etc.). This can then provide the link between the clinical and tissue bank database. Only those who are bound by clinical contracts that state that the individual must maintain patient confidentiality at all times should have access to any data that contains patient identifiers and the tissue bank database code. All databases should be password protected, with individuals only having access to areas of the database that are necessary for their day-to-day work. Ideally the patient-identifiable data should remain within the healthcare system, and be subject to the usual stringent controls that operate to preserve patient confidentiality. The data associated only with a tissue bank number should still be protected by a high level of password-based security, but may not be required to be maintained within the healthcare system. The benefit of using this type of system is that clinical data on the tissue bank database can be updated either by individuals who have access to the clinical database or by electronic transfer using the tissue bank number as the identifier.

3.2. Sample Processing

In order to maximise the use of material available to a biobank from a single consented patient, one should endeavour to collect material in a variety of formats. A single blood sample taken in an EDTA tube can be used for the isolation of germline DNA. One sample can be split into many aliquots that can be issued to different research projects. Similarly, a single blood sample can be centrifuged and the serum divided into a number of different aliquots. Where possible a biobank should endeavour to obtain samples of frozen material and formalin-fixed paraffin-embedded material. In addition, RNA, DNA, and protein should be extracted from the same block of tissue. This enables direct comparison of gene expression at the mRNA level with chromosomal changes at the DNA level, and proteomic studies. Providing a whole tissue block to one study is wasteful of the donated material as most researchers will not use the entire sample. Biobanks should be very reluctant to distribute "the last sample." By providing further processing to DNA and RNA, biobanks can provide material to many different projects from the same patient, thereby maximising their resource. Perhaps more importantly, provisions for DNA and RNA extraction by the biobank permits the bank to provide quality assured material that is fit for purpose (4). For example, if the tissue RNA is degraded, it still may be suitable for DNA analysis or immunohistochemistry.

3.3. Pathology Quality Assurance

It is important to ensure that the material stored in a biobank is "what is says on the pot", and that it is suitable for purpose (see Note 4). Cancer samples are composed of a mixture of different tissue and cell types, and in many cases the genes that are being studied by researchers are present in only one cell type (e.g., epithelial cells). However, the connective tissue stroma surrounding a carcinoma may also be specifically examined. The sensitivity of molecular biological techniques varies and it is therefore important that researchers realise what proportion of the material contains tumour tissue. Also, the presence of necrotic cells can have a deleterious effect on some assays. Frozen sections should be cut from each tissue block prior to extraction of nucleic acids because the percentage of tumour cells can vary considerably between blocks from the same operative specimen and in different types of cancer.

3.4. Molecular Biology Quality Control

Providing nucleic acids extracted from frozen tissue blocks rather than the frozen blocks themselves enables researchers to be given material that is of the correct quality standard for their projects. Different types of molecular biological techniques demand differing standards. For high-throughput technologies that study the expression of a large number of genes simultaneously, higher molecular weight RNA or DNA is required (i.e., RNA or DNA that is present in its full length, not degraded into smaller fractions). Other technologies (e.g., those that look at the expression of a single or a small number of genes simultaneously) can use samples of a lower quality. An assessment of extract purity should be made, using spectrophotometry to compare ratios of DNA:protein (260/280 wavelength ratio), the amount of nucleic acid extracted (the yield), and microfluidics for RNA using an Agilent Bioanalyser to determine RNA fidelity (the RNA Integrity Number – or RIN).

3.5. Central Management

Central management provides physical structure as well as online database and handling, applications for materials, training, and reviews compliance with procedures. It oversees re-evaluation and modifications to procedures. Developing a common ethics policy, communications, PR, media contact, monitoring expenditure, and maintaining a Web site for researchers, patient advocates, and all stakeholders – are other important functions of the central office.

The level of funding determines the extent of the biobank services. Costs recouped by administrative charges are insufficient to balance the books. Funding mechanisms differ between biobanks: many biobank networks outside the USA receive a significant government contribution. Additional funding may be derived from universities or other institutions, industry, or philanthropists. Securing adequate funding is always a concern: despite the importance of biobanking, it does not loosen purse-strings. Fundraising for revenue shortfalls may be combined with events and public awareness campaigns.

3.6. Communication, Trust, and Collaboration

Transparency, trust, collaboration, and excellent communication at all levels are essential for a successful biobank network. Terminology makes a difference: the word "biobank" is much more likely to be remembered by the public than "biorepository", provided that the difference between a monetary bank and a biobank is clarified. Biobanking procedures, and in particular, database privacy, security, and patient confidentiality must be transparent for patient donors' trust. The potential benefits of biobank networks should not be exaggerated. Only successful research will lead to the development of new drugs and biomarkers and their introduction into the clinic, but biobank networks improve the chance of this happening. Patients need to know that their samples are being used and in the way intended – "we believed it was going to be used for something useful" (47). They should be informed where to find answers to questions, e.g., How will researchers use my samples? How will the samples help those with cancer? Researchers' trust is gained by their perception that the biobank network represents a distinct advantage for them compared to how they operated previously, and that access to samples and the approval process are fair and not overly complicated. Criteria for rules of access and decisions to approve sample drawdown must be equitable, consistent, and clearly communicated. If there is prioritisation of applicants, this should be clear. Some banks rely on external reviewers for applications. Research data and updated clinical data add value to the biobank for all, and researchers must acknowledge the biobank in publications.

3.7. Applications for Samples and Data

One of the functions of the central co-ordinating office is triaging applications for samples and data. Researchers with scientifically and ethically approved projects need equitable access. Access to an online database may be restricted, but should not exclude those working in institutions that have no access to human tissue. Decisions on applications for materials from the biobank network must be equitable (7). Applications to the Wales Cancer Bank are examined by outside reviewers. Researchers from the host institution may be given priority over industry or international groups, but in practice, reasonable requests from most applicants should be satisfied, provided there are enough samples in the bank. Decisions should be prompt: one of the purposes of the network is to fast-track the process from research concept to results that may lead to better patient interventions.

3.8. Academic–Industry Partnerships

Academic–industry partnerships represent a vital collaboration for researchers, industry, and the patient. In the past, industry was assigned a low priority for sample access. Large pharmaceutical companies have excellent biobanking facilities, procedures and personnel, but many state they have no guarantee of sample quality. The biobank network, by acquiring multiple aliquots, will have adequate sample quantities for both academic and industry's research. Indeed, it is arguable which should have priority. Both are important, and if a new biomarker or drug is to be manufactured, high-quality samples should be used because the products are for patient care. Some patients may opt out of their samples being used by industry, but market research indicates that most agree, on the altruistic basis that the samples may eventually be used to help others. Cancer patients and the public alike need to understand that pharmaceutical companies need patient donors' biobanked blood and tissue samples for developing improved new cancer drugs. Therefore, pharmaceutical and biotechnology companies, as end-users of samples should be encouraged to contribute to the biobank's running expenses. Transparency by industry has been lacking in the past, but the boundaries of academic research and research in industry have become blurred (48). This may represent an advantage for the exchange of ideas, and ultimately for better patient interventions.

3.9. Patient, Public Awareness, Web site, Meetings, Fundraising, PR, Media Coverage

Consultation with and improved awareness for patient groups and the public is essential. A realistic future goal is that a knowledgeable public would understand that it will become standard procedure to biobank samples from cancer resection specimens, unless they opt out. This level of trust requires that easily understood research results are presented clearly, on a regular basis and on a Web site. Biobank personnel and scientists must learn how to communicate with nonscientists, using simple language and an uncomplicated format. Web sites have qualified usefulness, but can illustrate biobank progress. Carefully planned public meetings are well appreciated and have the advantage of potential media coverage, which greatly expands the audience base. In British Columbia, a process of deliberative public engagement has been reported to shape the governance of a biobank (49, 50).

Funding shortfalls are overcome in different ways, through philanthropy, industry, granting agencies, and advocacy organisations. In Ireland, a charitable trust – Biobank Ireland (http://www.biobankireland.ie) holds fundraisers and has celebrity endorsement for promoting development of a network. Biobank Ireland has received funds from philanthropists and more recently from the corporate sector. Of all the stakeholders consulted in Ireland, members of public and patient groups have been by far the most supportive. It is very important that the public understands the role of biobanking: an informed public has the potential to form a powerful lobby group for obtaining vital government funding.

4. Notes

1. Investigators intending to use biobanked specimens to investigate the molecular mechanisms of disease, or to discover a new blood or tissue biomarker, must recognise that the weakest link in the project can be the uncertain quality of the specimen. Unfortunately, there is a graveyard of previously published biomarkers or disease correlations that never made any clinical impact because they were based on poor quality specimens (4, 51). In order to maximise the reliability of biobank specimens, the new investigator must verify that the specimen comes from a biobank with stringent public standards and that the biobank contains tissue and linked clinical data that are appropriate for the proposed study.
2. Quality metrics for an acceptable tissue biobank specimen are: (a) The tissue was collected under full consent under an IRB approved protocol that is available for review. Informed patient consent must be obtained prior to collection of tissue for research/biobanking purposes. IRB or ethics committee approval is also required prior to initiation of tissue collection for research. (b) Within 20 min or less, following excision, the tissue was frozen or otherwise preserved in a manner that stabilises the molecular analytes of interest. Clinical data at a minimum should include: tissue type, age, sex, diagnosis, preservation method, date/time collected, date/time preserved, and collecting institution. (c) The histopathology of the tissue is known (evaluated by a qualified pathologist), and the tissue is in the proper form for microdissection to isolate the cell population of interest. (d) Proper control tissue for the disease of interest was collected under exactly the same protocol as the diseased tissue. (e) Statistically powered numbers of matched disease and control groups are available in the bank for discovery and for separate independent validation. (f) The clinical data linked to the specimen must be all-inclusive as it relates to the disease question under study. For example, if the tissue is breast cancer, the clinical information should include stage, grade, and menopausal status, as well as treatment history and concurrent medical conditions (e.g., diabetes).
3. UNESCO guidance document regarding human genetic data can be accessed at http://www.portal.unesco.org/en/ev.php-URL_ID=17720&URL_DO=DO_TOPIC&URL_SECTION=201.html.
4. An epidemiologist and a biostatistician should be consulted prior to initiating the study. They should be involved in (a) the evaluation of the biobank that will supply the specimens and (b) the drafting of the specimen request proposal that is submitted to the biobank.

References

1. Robb, J. (2010) 2010 Revised NCI Best Practices for Biospecimen Resources. National Cancer Institute: Bethesda, MD.
2. Pitt, K. E. (2008) 2008 Best Practices for Repositories: Collection, Storage, retrieval and Distribution of Biological Materials for Research. *Cell Preservation Technology* 6, 3–58.
3. Poste, G. (2011) Bring on the biomarkers. *Nature* 469, 156–7.
4. Lim, M. D., Dickherber, A., Compton, C. C. (2011) Before you analyze a human specimen, think quality, variability, and bias. *Anal Chem* 83, 8–13.
5. Park, A. (2009) 10 Ideas Changing the World Right Now - Biobanks. *TIME*, Inc. Thursday March 12.
6. Mager, S. R., Oomen, M. H., Morente, M. M., Ratcliffe, C., Knox, K., Kerr, D. J. et al. (2007) Standard operating procedure for the collection of fresh frozen tissue samples. *Eur J Cancer* 43, 828–34.
7. Morente, M. M. (2011) Tumour Bank: The CNIO Tumour Bank Network. 2011. Spanish National Cancer Research Centre, http://www.cnio.es/ing/index.asp.
8. Riegman, P. H., Bosch, A. L. (2008) OECI TuBaFrost tumor biobanking. *Tumori* 94, 160–3.
9. Sebire, N. J., Dixon-Woods, M. (2007) Towards a new era of tissue-based diagnosis and research. *Chronic Illn* 3, 301–9.
10. Editorial (2006) The advocates. *Nat Genet* 38, 391.
11. Terry, S. F., Terry, P. F., Rauen, K. A., Uitto, J., Bercovitch, L. G. (2007) Advocacy groups as research organizations: the PXE International example. *Nat Rev Genet* 8, 157–64.
12. Eiseman, E., Bloom, G., Brower, J., Clancy, N., Olmsted, S. S. Case Studies of Existing Human Tissue Repositories: "Best Practices" for a Biospecimen Resource for the Genomic and Proteomic Era, RAND Corporation, Santa Monica, CA, 2003.
13. Qualman, S. J., France, M., Grizzle, W. E., LiVolsi, V. A., Moskaluk, C. A., Ramirez, N. C. et al. (2004) Establishing a tumour bank: banking, informatics and ethics. *Br J Cancer* 90, 1115–9.
14. Grizzle, W. E., Aamodt, R., Clausen, K., LiVolsi, V., Pretlow, T. G., Qualman, S. (1998) Providing human tissues for research: how to establish a program. *Arch Pathol Lab Med* 122, 1065–76.
15. Becich, M. J. (2000) The role of the pathologist as tissue refiner and data miner: the impact of functional genomics on the modern pathology laboratory and the critical roles of pathology informatics and bioinformatics. *Mol Diagn* 5, 287–99.
16. (2004) AB v Leeds Teaching Hospital NHS Trust. *Fam Law Rep* 2, 365–439.
17. Quigley, M. (2009) Property: The Future Of Human Tissue? 10.1093/medlaw/fwp021 *Medical Law Review* 17 457–466.
18. Various (1998 Summer) Commentary. 10.1093/medlaw/6.2.247. *Medical Law Review* 6, 247–261.
19. Matthews, P. (1995) The Man Of Property. 10.1093/medlaw/3.3.251. *Medical Law Review* 3, 251–274.
20. Price, D. (2011 Spring) From Cosmos and Damian to Van Velzen: The Human Tissue Saga Continues. 10.1093/medlaw/11.1.1. *Medical Law Review* 11, 1–47.
21. Nwabueze, R. N. (2008) Donated Organs, Property Rights And The Remedial Quagmire. 10.1093/medlaw/fwn004 *Medical Law Review* 16 201–224
22. Glantz, L., Roche, P., Annas, G. J. (2008) Rules for donations to tissue banks--what next? *N Engl J Med* 358, 298–303.
23. Knoppers, B. M. (2005) Consent revisited: points to consider. *Health Law Rev* 13, 33–8.
24. Clayton, E. W. (2005) Informed consent and biobanks. *J Law Med Ethics* 33, 15–21.
25. Knoppers, B. M. (2005) Biobanking: international norms. *J Law Med Ethics* 33, 7–14.
26. Knoppers, B. M., Hirtle, M., Lormeau, S., Laberge, C. M., Laflamme, M. (1998) HUGO Ethics Committee Statement on DNA sampling: control and access. *Genetic Resour* 11, 43–4.
27. (2002) International ethical guidelines for biomedical research involving human subjects. *Bull Med Ethics* 17–23.
28. (2003) [The UNESCO international declaration about human genetic data]. *Law Hum Genome Rev* 239–53.
29. Simitis, S. (2004) Biobaken fur die Forschung. Nationaler Ethikrat: Hamburg.
30. HGC (2002) Inside information:Balancing interests in the use of personal genetic data. Human Genetics Commission: London, 183.
31. Kaye, J., Helgason, H. H., Nomper, A., Sild, T., Wendel, L. (2004) Population genetic databases: a comparative analysis of the law in Iceland, Sweden, Estonia and the UK. *TRAMES* 8, 15–33.
32. Bale, M. (2009) Key principles relating to genetic testing and insurance. *Law Hum Genome Rev* 203–7.

33. (2004) Draft additional protocol to the Convention on Human Rights and Biomedicine, on biomedical research. *J Int Bioethique* 15, 107–22.
34. Doppelfeld, E. (2002) Good medical research--the view of the CDBI/Council of Europe. *Sci Eng Ethics* 8, 283–6.
35. Elger, B. S., Caplan, A. L. (2006) Consent and anonymization in research involving biobanks: differing terms and norms present serious barriers to an international framework. *EMBO Rep* 7, 661–6.
36. Meslin, E. M. (1999) The National Bioethics Advisory Commission (NBAC) report. *Biol Psychiatry* 46, 1011–2.
37. Crigger, B. J. (2001) National Bioethics Advisory Commission Report: Ethical and policy issues in international research. *Irb* 23, 9–12.
38. Childress, J. F. (1998) The National Bioethics Advisory Commission: bridging the gaps in human subjects research protection. *J Health Care Law Policy* 1, 105–22.
39. (2008) OHRP - Guidance on Research Involving Coded Private Information or Biological Specimens. US Department of Health and Human Services, Office of Human Research Protection: Bethesda, MD,
40. Wendler, D. (2006) One-time general consent for research on biological samples: is it compatible with the health insurance portability and accountability act? *Arch Intern Med* 166, 1449–52.
41. Kettis-Lindblad, A., Ring, L., Viberth, E., Hansson, M. G. (2007) Perceptions of potential donors in the Swedish public towards information and consent procedures in relation to use of human tissue samples in biobanks: a population-based study. *Scand J Public Health* 35, 148–56.
42. Borisch, B. (2007) Tissue banking in a regulated environment--does this help the patient? Part 2--Patient views and expectations (including the EUROPA DONNA Forum UK position). *Pathobiology* 74, 223–6.
43. Morrin, H., Gunningham, S., Currie, M., Dachs, G., Fox, S., Robinson, B. (2005) The Christchurch Tissue Bank to support cancer research. *N Z Med J* 118, U1735.
44. Cousins, G., McGee, H., Ring, L., Conroy, R., Kay, E., Croke, D. T. et al. (2005) Public perceptions of biomedical research: a survey of the general population in Ireland., Health Research Board: Dublin,
45. Espina, V., Edmiston, K. H., Heiby, M., Pierobon, M., Sciro, M., Merritt, B. et al. (2008) A portrait of tissue phosphoprotein stability in the clinical tissue procurement process. *Mol Cell Proteomics* 7, 1998–2018.
46. Patel, A. A., Gilbertson, J. R., Parwani, A. V., Dhir, R., Datta, M. W., Gupta, R. et al. (2006) An informatics model for tissue banks--lessons learned from the Cooperative Prostate Cancer Tissue Resource. *BMC Cancer* 6, 120.
47. Dixon-Woods, M., Wilson, D., Jackson, C., Cavers, D., Pritchard-Jones, K. (2008) Human Tissue and the Public: The Case of Childhood Cancer Tumour Banking. *BioSocieties* 3, 57–80.
48. Wilan, K. (2007) From bench to business and back again. *Cell* 131, 211–3.
49. Burgess, M. M., O'Doherty, K. C., Secko, D. M. (2009) Biobanking in British Columbia: discussions of the future of personalized medicine through deliberative public engagement. *Per Med* 5, 285–296.
50. O'Doherty, K. C., Burgess, M. M. (2009) Engaging the public on biobanks: outcomes of the BC biobank deliberation. *Public Health Genomics* 12, 203–15.
51. Imbeaud, S., Auffray, C. (2005) 'The 39 steps' in gene expression profiling: critical issues and proposed best practices for microarray experiments. *Drug Discov Today* 10, 1175–82.

Chapter 6

Introduction to Genomics

Luca Del Giacco and Cristina Cattaneo

Abstract

The Science of Genomes: Only within the past few decades have scientists progressed from the analysis of a single or a small number of genes at once to the investigation of thousands of genes, going from the study of the units of inheritance to the investigation of the whole genome of an organism. The science of the genomes, or "genomics," initially dedicated to the determination of DNA sequences (the nucleotide order on a given fragment of DNA), has promptly expanded toward a more functional level – studying the expression profiles and the roles of both genes and proteins. The aim of the chapter is to review some basic assumptions and definitions that are the fabric of genomics, and to elucidate key concepts and approaches on which genomics rely.

Key words: Chromosome walking, Contig, Genetic map, Genomic annotation, Genomics, Physical map, Polymorphism, Restriction pattern analysis, RFLPs, Sequencing, SSLPs, SSR, VNTR

1. Introduction

The leading power behind genomics has been, without a doubt, the Human Genome Project (HGP). During the second half of the 1980s, the US National Institute of Health (NIH) and the Department of Energy (DOE) joined forces, establishing the HGP. The common effort led to a multiphase plan with a series of goals (1–3), most of which have been accomplished more than 2 years ahead of schedule and at a cost less than originally expected (3).

One of the primary goals of the HGP was the creation of genetic and physical high-resolution maps of each one of the human chromosomes (4). The goal for developing such maps was to greatly enhance the ability to localize and identify genes for inherited disorders.

The idea of a genetic (linkage) map is based on the relative position along the chromosome of known specific markers (a genetic marker can be any inheritable distinguishable trait), and the distance

Virginia Espina and Lance A. Liotta (eds.), *Molecular Profiling: Methods and Protocols*, Methods in Molecular Biology, vol. 823,
DOI 10.1007/978-1-60327-216-2_6, © Springer Science+Business Media, LLC 2012

between two markers is expressed in units of recombination. Polymorphic markers (in biology, the term polymorphism describes the existence of multiple forms, also known as morphs) can be phenotype-based (identifiable characters of the individual, essentially determined by the expressed regions of the genome, the genes) or polymorphism-based (DNA polymorphisms that characterize both coding and noncoding DNA). Genetic maps represented an invaluable tool for the construction of physical maps. Physical maps depict the order of DNA sequences on the chromosome, specifying physical distances between landmarks. These maps are typified by different degrees of resolution, from the lowest resolution chromosomal map (based on the banding pattern of stained chromosomes) to the definitive highest resolution physical map, represented by the entire genomic DNA sequence itself, that allows the measurement of physical distances in term of single nucleotides. Completion of the human genome sequence ("the perfect" physical map) represented the fulfillment of the HGP primary goal. Following the announcement, in 2000, that a first working draft of the human genome had been completed, the scientific world celebrated the event when, in 2001, the sequence of the human genome was published (3, 5). The HGP is only one of several genome projects dedicated to the identification of the DNA sequences of a wide variety of organisms, including animal models (fruit flies (6), nematodes (7), mice (8–10), zebrafish (11)), bacteria (12, 13), and plants (14).

The elucidation of genomic DNA sequences represents only the very first step toward the identification and understanding of mechanisms that link the mere distribution of nucleotides along chromosomes with the specification of the living matter. The accumulation of enormous amount of sequence data, that triggered the rapid growth of bioinformatic sciences, now provide the resources to identify all the genes present in a genome and approaches to analyze their expression and functions. The umbrella term Functional Genomics covers all approaches that use DNA sequence and structure data to elucidate molecular profiles such as gene and protein expression, and protein–protein interactions (15). The introduction of DNA microarray technology, for example, has made feasible the gene expression analysis on a large scale, allowing the study of the transcriptional activity of several thousands of genes at once (potentially all the genes represented in a genome), in any biological and medical context (16). Proteomics (*proteome* is the set of proteins synthesized by a *genome*) implemented the expression studies focusing on the array of proteins expressed by a genome rather than transcripts, and approached the analysis of post-translational modifications and protein–protein interactions (17–19).

Genomes of individuals belonging to the same species are not identical. The slight variations in genomes are called single nucleotide polymorphisms, or SNPs. These single nucleotide differences contribute to diversity among individuals and, in some cases, can predispose one to disease or influence the response to substances or agents. For instance, pharmacogenomics correlates these and other genetic variations with the variable effects of drugs in different patients, with the definitive aim of designing "the perfect drug" (or "the perfect combination of drugs") for the specific genetic background of the individual (20, 21).

Genome projects have generated an exceptional amount of information unparalleled by anything else in biology. The complete genome sequence of several organisms is revolutionizing biology and medicine by providing answers for many important issues, from species evolution to discovery of novel and more efficient therapies for the treatment of individual patients.

2. Genetic and Physical Mapping of Genomes

2.1. Genetic Maps

A genetic map is based on the relative location of specific markers along the chromosome. Genetic markers on the same chromosome are physically connected (genetically linked) and segregate together during meiosis. On the other hand, markers on different chromosomes (not genetically linked) segregate in an independent fashion, as a consequence of the independent chromosomes re-assortment during meiosis.

The distance between two markers, expressed in centiMorgan (cM, units of recombination), named after the American geneticist Thomas Hunt Morgan, is defined by their tendency to recombine, so that it does not correspond to an actual physical distance, but simply reflects the probability that the markers can be separated by a meiotic crossing-over. The rate of recombination along the chromosome is not uniform due to interference, which is the phenomenon by which "*the occurrence of one crossing-over interferes with the coincident occurrence of another crossing-over in the same pair of chromosomes*" (22). Thus, a high degree of discrepancy exists between the genetic and physical distances. Nevertheless, the general concept on which genetic maps rely is that the further apart the markers are on the chromosome, the higher the chance that they will segregate separately (23, 24).

In model organisms, genetic markers are typically phenotypes associated to genes inherited in a Mendelian fashion. The simple crossing of individuals carrying these different traits establishes whether or not two genes are on the same chromosome. Genes located on the same chromosome are called syntenic.

Obviously, the recombination of genes in controlled mating is not a realistic option to map human chromosomes, also in the light of the fact that the availability of human genetic markers, such as red cell enzymes and serum protein polymorphisms, are limited in number and not sufficient to generate a complete linkage map. In 1978, the discovery of a DNA sequence polymorphism near the beta globin gene (25) paved the way for the construction of the first human genetic map (26). DNA-based polymorphisms are available in high number since they are distributed not only in the genes, but also in the noncoding regions of the genome, where DNA is subjected to far less selective pressure and allowed to mutate.

2.1.1. Restriction Fragment Length Polymorphisms

Restriction fragment length polymorphisms (RFLPs), the first identified type of DNA-based polymorphism, correspond to the different electrophoresis restriction patterns obtained from digesting the DNA of two individuals using a particular restriction enzyme. Differences in the length of DNA fragments originate from the presence or absence of a specific restriction site recognized by the enzyme (26). The major disadvantage of using RFLPs is their bi-allelic nature. Indeed, an RFLP can exist in just two forms, cut and uncut, so that it might not always be informative. Nevertheless, the first human genetic map was realized by means of RFLPs (27).

2.1.2. Simple Sequence Length Polymorphisms

Simple sequence length polymorphisms (SSLPs) are an additional class of multiallelic DNA markers that greatly accelerated the construction of more accurate genetic maps. SSLPs are repetitive DNA sequences, or blocks of DNA, that form tandem repeats in the genome of organisms. SSLPs are subdivided into repeat units, and each SSLP allele is characterized by its own length, defined on the basis of how many times the repeat unit recurs in the allele itself. Both minisatellites (or Variable number of tandem repeats, VNTR) and microsatellites (or Simple Sequence Repeats, SSRs) belong to the SSLPs (reviewed in ref. 28). Minisatellites and microsatellites differ essentially by the length of the repeats (each repeat unit is 10–100 bp in length in VNTRs versus 1–9 bp in SSRs) and length of the clusters (VNTRs range from 1 to 20 kb, while in SSRs the whole repetitive region spans less than 150 bp). Both minisatellites and microsatellites are unstable (SSRs more than VNTRs) for several possible reasons such as DNA replication slippage, DNA recombination, and their expansions–contractions make them highly polymorphic and informative markers. A serious drawback with the minisatellites is their uneven allocation in the genome. VNTRs are distributed almost exclusively in close proximity or within the telomeric regions. On the other hand, microsatellites such as SSRs (di-, tri-, and tetra-nucleotide repeats) are dispersed throughout the genome at a minimum of once in 10 kb in eukaryotes. Microsatellites are considerably shorter than minisatellites and their length can be more precisely characterized using polymerase chain reaction (PCR) techniques. Microsatellites significantly accelerated the achievement of one of the main goals of the HGP – the construction of a comprehensive, high-density genetic map – that marked the moment of the switch from genetic to physical mapping (23).

2.2. Physical Maps

Chronologically speaking, the very first human physical map available was the cytogenetic map, allowing researches to have a direct look (through a microscope!) of the particular banding patterns of each stained chromosome (29). Cytogenetic maps permitted elucidation of chromosomal alterations, such as deletions and translocations, thus confirming the diagnosis of several diseases. Currently, the *Fluorescence In Situ Hybridization* (FISH) technique visually maps the position of a gene on the chromosomes by hybridizing the gene of interest to a cloned DNA sequence that has been fluorescently labeled (30). FISH performed with two gene-specific probes, for example, provides the ability to compare the gene of interest to a control gene.

2.2.1. Contig Maps

Several other physical maps have been constructed each having different levels of resolution. The most important is the clone contig (from contiguous) map that consists of DNA clones with overlapping inserts. Contig assembly relies on genomic libraries; these are generated through partial DNA restriction digest in which only some of the restriction sites will be digested by the enzyme in order to obtain a collection of fragments that partially overlap each other. The DNA fragments are then inserted into proper vectors based on the kilobase size of the fragment. Yeast Artificial Chromosomes (YACs) host molecules up to thousands of kilobases in size, whereas Bacterial Artificial Chromosomes (BACs) or P1 Artificial Chromosomes (PACs) host up to hundreds of kb, while plasmids or phages only host a few kb. Unfortunately, by performing this crucial step of DNA cloning, the original information regarding the reciprocal position of the fragments is missed. Numerous approaches are in place to recover the lost data, from the identification of common restriction fragments among clones to the chromosome walking technique (31, 32), to name a few. In the restriction pattern analysis, isolated clones are digested with a restriction enzyme and their electrophoresis profiles are compared to identify possible common patterns (DNA fingerprinting). The chromosome walking technique identifies a genomic region surrounding a known DNA fragment (the starting probe). A genomic library can be screened with the probe, and all the positive clones can be characterized to identify which one extends furthest into the neighboring DNA. The terminal sequence of the clone is then used as a new probe to repeat the procedure and extend the cloning to more distal regions.

High-resolution contig maps represent an indispensable tool for the assembly of a complete genome sequence (the definitive physical map), providing a scaffold in which the markers can be allocated to their actual position.

3. Genome Sequencing and DNA Assembly

As soon as genetic and physical maps of the organism's genome are available, the time to accomplish the definitive goal of the genome project has come: to complete the genomic DNA sequence of that organism. In order to achieve such a task, there are two fundamental aspects that need to be considered: (1) availability of techniques for DNA sequencing, to reveal the actual nucleotide order in a DNA clone (i.e., a BAC), that represents just a little fraction of the chromosome it belongs to, and (2) accessibility to the methodologies for the correct assembly of the hundreds of thousands of discrete clones into the final sequences of the chromosomes.

DNA sequencing has been available since the 1970s with the chain-termination or Sanger method the most popular one as a result of its practicality (33). The advent of PCR and the introduction of improved new chemistries (fluorescent dyes, thermal stable DNA polymerases specifically customized for sequence reactions) increased the efficiency and the speed of sequencing in the 1990s, and the introduction of the dye-terminator cycle sequencing methodology, together with automated high-throughput DNA sequence analyzers, made genome-scale sequencing projects possible.

Assembly of the enormous number of clones into a contiguous DNA sequence representing the whole chromosome can be achieved with two different strategies. The first one, known as clone contig approach, has been partially addressed in the *Physical Maps* paragraph (see above). Briefly, genomic DNA libraries are generated by partial digestion of the DNA, the fragments are cloned into proper high capacity vectors (i.e., BAC or YAC) and the order of the clones is determined with different methods, such as DNA fingerprinting and chromosome walking. A minimum number of overlapping clones can be chosen in order to cover the entire chromosome length (this scaffold of clones is defined as the minimum tiling path, MTP). Once this physical "sub-map" has been generated, each one of the clones of the MTP is fragmented, and each fragment is processed for sequencing. The sequences of all the subcontigs composing the individual clone of the MTP are assigned to their original position (i.e., comparing their in silico restriction patterns with the restriction pattern of the clone from which they originated).

The second approach, known as shotgun, is based on the random breakage of the high-molecular genomic DNA into fragments that will be later cloned and processed for sequencing. Unlike the clone contig approach and its pre-ordered scenario (MTP), that allows the positioning of the incoming sequences directly on the chromosome, the shotgun approach is totally independent from any previous information regarding the genome so that it can be performed in the absence of maps, either genetic or physical. For this reason, the shotgun approach is strongly reliant on computer algorithms that are necessary to organize the numerous overlapping sequences into contigs.

The two approaches described are not mutually exclusive but complement each other. Genome projects (including the HGP) employ both methods. Shotgun sequencing is quicker than the traditional clone contig method, but is less precise in the following assembly process, especially in regards to complex genomes of higher eukaryotes that contain large numbers of identical sequences (for instance, interspersed repeats represent ~50% of the human genome).

3.1. Genomic Annotation

Genomic annotation makes sense of the fascinating, yet "cryptic" endless line of nucleotides by labeling the sequences with biological information. Once a genome is sequenced, the very first step is the identification of the genomic elements and the determination of their potential function. Computerized annotation tools (GeneFinder, Genscan, and several others) play a fundamental role in this aspect of genomics and are essential in recognizing genomic elements, such as genes and their structure (i.e., introns, exons, regulatory elements), but human intervention is still irreplaceable, and inputs from researchers are necessary for fine-tuning annotations. Once a genomic component has been identified, the next step is comprehension of the function of the component itself (functional annotation). For instance, the ultimate demonstration that a genomic element is an actual gene is its expression (a gene is expressed when transcribes mRNA or, more generally, as in the case of the genes whose product is not a protein, when the corresponding RNA in synthesized). For this reason, virtually all genome projects branch into Expressed Sequence Tags (ESTs) "sub"-projects (34). mRNAs from a certain biological sample (i.e., tissues, organs, embryonic developmental stages) are employed to generate, through reverse-transcription, a complementary DNA (cDNA) library (cDNA is a double-stranded DNA copy of an mRNA molecule). ESTs are obtained through the single sequencing of a few hundred base pairs of the clones from the library, and represent the transcriptome (the set of all mRNAs expressed, for example, by a population of cells) of that particular biological sample. An individual EST can be used for the in silico identification of the active gene in the genome by sequence alignment using the BLAST (Basic Local Alignment Search Tool) bioinformatics tool (http://www.ncbi.nlm.nih.gov/blast/Blast.cgi). BLAST is useful for determining regions of similarity between biological sequences. At the time of writing, more than 68 million ESTs are available in the GenBank dbEST public repository, 8 million of the ESTs are human (http://www.ncbi.nlm.nih.gov/projects/dbEST/) (35).

Genomic annotation does not end with the identification of the protein coding genes. Besides traditional genes, several other features of biological relevance characterize a genome, such as noncoding RNAs (ncRNAs), promoters, regulatory elements, or structural aspects intrinsic in the architecture of genomes (i.e., sequence repeats). All these features represent annotations that can be attached to a sequence.

3.2. Next Generation Sequencing

DNA sequencing by the traditional Sanger method (33) has not been referred to as "first generation sequencing," therefore it may not be obvious that next generation sequencing is simply an improved high-throughput method(s). Next-generation sequencing enables high throughput, massively parallel sequencing of either DNA or RNA, in a matter of weeks, with single nucleotide resolution (36, 37).

Gigabase pairs of DNA sequences can be generated, producing terabytes of data. Although data generation can be accomplished in a matter of weeks, data analysis requires a longer timeframe due to the sheer volume and complexity of data. Next generation sequencing, also referred to as ultra-sequencing or deep sequencing, is rapidly surpassing traditional microarray analysis.

Pyrosequencing technology is an example method for massively parallel DNA sequencing. A complementary DNA strand is synthesized enzymatically, one base pair at a time, to an immobilized DNA template (38–40). RNA can also be subjected to high-throughput sequencing in a method called RNA-seq. RNA-seq provides a global view of the entire transcriptome, in static systems as well as under differing conditions (41–43).

Pyrosequencing has been used in numerous applications including sequencing SNPs (44) and the p53 tumor suppressor gene (45). Next generation sequencing has been used to identify translocations in solid tumors from individual patients (46). Deep sequencing is unraveling the intricacies of specific enzyme-mediated oncogenesis (47). These sequencing studies are leading to the development of personalized tumor biomarkers (46).

Fascinatingly, the more we gain insights into the nature of biological systems, the more our skills in reading the messages hidden in the DNA sequences increase, making the process of genomic annotation an intriguing, never-ending, challenge.

Acknowledgment

The authors would like to thank Dr. Anna Pistocchi for critical reading of the manuscript.

References

1. Collins, F., and Galas, D. (1993) A new five-year plan for the U.S. Human Genome Project. *Science* **262**, 43–6.
2. Collins, F. S., Patrinos, A., Jordan, E., Chakravarti, A., Gesteland, R., *et al.* (1998) New goals for the U.S. Human Genome Project: 1998–2003. *Science* **282**, 682–9.
3. Venter, J. C., Adams, M. D., Myers, E. W., Li, P. W., Mural, R. J., *et al.* (2001) The sequence of the human genome. *Science* **291**, 1304–51.
4. Collins, F. S., Morgan, M., and Patrinos, A. (2003) The Human Genome Project: lessons from large-scale biology. *Science* **300**, 286–90.
5. Lander, E. S., Linton, L. M., Birren, B., Nusbaum, C., Zody, M. C., *et al.* (2001) Initial sequencing and analysis of the human genome. *Nature* **409**, 860–921.
6. Adams, M. D., Celniker, S. E., Holt, R. A., Evans, C. A., Gocayne, J. D., *et al.* (2000) The genome sequence of Drosophila melanogaster. *Science* **287**, 2185–95.
7. (1998) Genome sequence of the nematode C. elegans: a platform for investigating biology. *Science* **282**, 2012–8.
8. Kawai, J., Shinagawa, A., Shibata, K., Yoshino, M., Itoh, M., *et al.* (2001) Functional annotation of a full-length mouse cDNA collection. *Nature* **409**, 685–90.
9. Hayashizaki, Y. (2003) The Riken mouse genome encyclopedia project. *C R Biol* **326**, 923–9.
10. Hayashizaki, Y. (2003) Mouse Genome Encyclopedia Project. *Cold Spring Harb Symp Quant Biol* **68**, 195–204.
11. Jekosch, K. (2004) The zebrafish genome project: sequence analysis and annotation. *Methods Cell Biol* **77**, 225–39.
12. Winsor, G. L., Van Rossum, T., Lo, R., Khaira, B., Whiteside, M. D., *et al.* (2009) Pseudomonas Genome Database: facilitating user-friendly, comprehensive comparisons of microbial genomes. *Nucleic Acids Res* **37**, D483-8.
13. Winsor, G. L., Khaira, B., Van Rossum, T., Lo, R., Whiteside, M. D., *et al.* (2008) The Burkholderia Genome Database: facilitating flexible queries and comparative analyses. *Bioinformatics* **24**, 2803–4.
14. Chandler, V. L., and Brendel, V. (2002) The Maize Genome Sequencing Project. *Plant Physiol* **130**, 1594–7.
15. Ganesan, A. K., Ho, H., Bodemann, B., Petersen, S., Aruri, J., *et al.* (2008) Genome-wide siRNA-based functional genomics of pigmentation identifies novel genes and pathways that impact melanogenesis in human cells. *PLoS Genet* **4**, e1000298.
16. Lockhart, D. J., and Winzeler, E. A. (2000) Genomics, gene expression and DNA arrays. *Nature* **405**, 827–36.
17. Paweletz, C. P., Charboneau, L., Bichsel, V. E., Simone, N. L., Chen, T., *et al.* (2001) Reverse phase protein microarrays which capture disease progression show activation of pro-survival pathways at the cancer invasion front. *Oncogene* **20**, 1981–9.
18. Petricoin, E. F., 3 rd, Espina, V., Araujo, R. P., Midura, B., Yeung, C., *et al.* (2007) Phosphoprotein pathway mapping: Akt/mammalian target of rapamycin activation is negatively associated with childhood rhabdomyosarcoma survival. *Cancer Res* **67**, 3431–40.
19. VanMeter, A. J., Rodriguez, A. S., Bowman, E. D., Jen, J., Harris, C. C., *et al.* (2008) Laser capture microdissection and protein microarray analysis of human non-small cell lung cancer: differential epidermal growth factor receptor (EGPR) phosphorylation events associated with mutated EGFR compared with wild type. *Mol Cell Proteomics* **7**, 1902–24.
20. Kalow, W. (2006) Pharmacogenetics and pharmacogenomics: origin, status, and the hope for personalized medicine. *Pharmacogenomics J* **6**, 162–5.
21. Vosslamber, S., van Baarsen, L. G., and Verweij, C. L. (2009) Pharmacogenomics of IFN-beta in multiple sclerosis: towards a personalized medicine approach. *Pharmacogenomics* **10**, 97–108.
22. Mueller, H. (1916) The mechanism of crossing over. *Am Nat* **50**, 421–34.
23. Murray, J. C., Buetow, K. H., Weber, J. L., Ludwigsen, S., Scherpbier-Heddema, T., *et al.* (1994) A comprehensive human linkage map with centimorgan density. Cooperative Human Linkage Center (CHLC). *Science* **265**, 2049–54.
24. Eichler, E. E., Nickerson, D. A., Altshuler, D., Bowcock, A. M., Brooks, L. D., *et al.* (2007) Completing the map of human genetic variation. *Nature* **447**, 161–5.
25. Kan, Y. W., and Dozy, A. M. (1978) Polymorphism of DNA sequence adjacent to human beta-globin structural gene: relationship to sickle mutation. *Proc Natl Acad Sci U S A* **75**, 5631–5.
26. Botstein, D., White, R. L., Skolnick, M., and Davis, R. W. (1980) Construction of a genetic linkage map in man using restriction fragment length polymorphisms. *Am J Hum Genet* **32**, 314–31.
27. Donis-Keller, H., Green, P., Helms, C., Cartinhour, S., Weiffenbach, B., *et al.* (1987) A genetic linkage map of the human genome. *Cell* **51**, 319–37.

28. Thomas, E. E. (2005) Short, local duplications in eukaryotic genomes. *Curr Opin Genet Dev* **15**, 640–4.
29. Caspersson, T., Zech, L., Modest, E. J., Foley, G. E., Wagh, U., *et al.* (1969) DNA-binding fluorochromes for the study of the organization of the metaphase nucleus. *Exp Cell Res* **58**, 141–52.
30. Trask, B. J., Massa, H., Kenwrick, S., and Gitschier, J. (1991) Mapping of human chromosome Xq28 by two-color fluorescence in situ hybridization of DNA sequences to interphase cell nuclei. *Am J Hum Genet* **48**, 1–15.
31. Jones, C. W., and Kafatos, F. C. (1981) Linkage and evolutionary diversification of developmentally regulated multigene families: tandem arrays of the 401/18 chorion gene pair in silkmoths. *Mol Cell Biol* **1**, 814–28.
32. Bender, W., Spierer, P., and Hogness, D. S. (1983) Chromosomal walking and jumping to isolate DNA from the Ace and rosy loci and the bithorax complex in Drosophila melanogaster. *J Mol Biol* **168**, 17–33.
33. Sanger, F., and Coulson, A. R. (1975) A rapid method for determining sequences in DNA by primed synthesis with DNA polymerase. *J Mol Biol* **94**, 441–8.
34. Boguski, M. S. (1995) The turning point in genome research. *Trends Biochem Sci* **20**, 295–6.
35. Boguski, M. S., Lowe, T. M., Tolstoshev, C. M. (1993) dbEST--database for "expressed sequence tags". *Nat Genet* **4**, 332–3.
36. Mardis, E. R. (2008) The impact of next-generation sequencing technology on genetics. *Trends Genet* **24**, 133–41.
37. Ding, L., Wendl, M. C., Koboldt, D. C., Mardis, E. R. (2010) Analysis of next-generation genomic data in cancer: accomplishments and challenges. *Hum Mol Genet* **19**, R188-96.
38. Nordstrom, T., Nourizad, K., Ronaghi, M., Nyren, P. (2000) Method enabling pyrosequencing on double-stranded DNA. *Anal Biochem* **282**, 186–93.
39. Ronaghi, M., Karamohamed, S., Pettersson, B., Uhlen, M., Nyren, P. (1996) Real-time DNA sequencing using detection of pyrophosphate release. *Anal Biochem* **242**, 84–9.
40. Ronaghi, M., Uhlen, M., Nyren, P. (1998) A sequencing method based on real-time pyrophosphate. *Science* 281, **363**, 365.
41. Wilhelm, B. T., Marguerat, S., Watt, S., Schubert, F., Wood, V., Goodhead, I. et al. (2008) Dynamic repertoire of a eukaryotic transcriptome surveyed at single-nucleotide resolution. *Nature* **453**, 1239–43.
42. Wilhelm, B. T., Marguerat, S., Goodhead, I., Bahler, J. (2010) Defining transcribed regions using RNA-seq. *Nat Protoc* **5**, 255–66.
43. Marguerat, S., Wilhelm, B. T., Bahler, J. (2008) Next-generation sequencing: applications beyond genomes. *Biochem Soc Trans* 36, 1091–6.
44. Nordstrom, T., Ronaghi, M., Forsberg, L., de Faire, U., Morgenstern, R., Nyren, P. (2000) Direct analysis of single-nucleotide polymorphism on double-stranded DNA by pyrosequencing. *Biotechnol Appl Biochem* **31** (Pt 2), 107–12.
45. Ahmadian, A., Lundeberg, J., Nyren, P., Uhlen, M., Ronaghi, M. (2000) Analysis of the p53 tumor suppressor gene by pyrosequencing. *Biotechniques* **28**, 140–4, 146–7.
46. Leary, R. J., Kinde, I., Diehl, F., Schmidt, K., Clouser, C., Duncan, C. et al. (2010) Development of personalized tumor biomarkers using massively parallel sequencing. *Sci Transl Med* **2**, 20ra14.
47. Yamane, A., Resch, W., Kuo, N., Kuchen, S., Li, Z., Sun, H. W. et al. (2010) Deep-sequencing identification of the genomic targets of the cytidine deaminase AID and its cofactor RPA in B lymphocytes. *Nat Immunol* **12**, 62–9.

Chapter 7

Genomic Profiling: cDNA Arrays and Oligoarrays

Francesco Gorreta, Walter Carbone, and Dagania Barzaghi

Abstract

The introduction of microarray technology, which is a multiplexed hybridization-based process, allows simultaneous analysis of a large number of nucleic acid transcripts. This massively parallel analysis of a cellular genome will become essential for guiding disease diagnosis and molecular profiling of an individual patient's tumor. Nucleic acid based microarrays can be used for: gene expression profiling, single-nucleotide polymorphisms (SNPs) detection, array-comparative genomic hybridizations, comparisons of DNA methylation status, and microRNA evaluation.

A multitude of commercial platforms are available to construct and analyze the microarrays. Typical workflow for a microarray experiment is: preparation of cDNA or gDNA, array construction, hybridization, fluorescent detection, and analysis. Since many sources of variability can affect the outcome of one experiment and there is a multitide of microarray platforms available, microarray standards have been developed to provide industry-wide quality control and information related to each microarray. In this chapter, we review array construction, methodologies, and applications relevant to molecular profiling.

Key words: cDNA, DNA methylation, Genomics, Hybridization, miRNA, Oligoarray, PCR RNA

1. Introduction

1.1. DNA Microarray History

Until recently, transcription analysis suffered significant limitations and only few genes at a time could be followed during a single experiment. This limitation was due to the elaborate experimental protocols available and the small number of identified genes. The introduction of the array technology, a multiplexed hybridization-based process, and completion of the Human Genome Mapping project allowed studies on a large number of transcripts simultaneously (1–4).

A microarray is defined as an ordered array of nucleotides (or in other cases proteins or small molecules) that enables parallel analysis of complex biochemical samples. The nucleotide, referred

Virginia Espina and Lance A. Liotta (eds.), *Molecular Profiling: Methods and Protocols*, Methods in Molecular Biology, vol. 823, DOI 10.1007/978-1-60327-216-2_7, © Springer Science+Business Media, LLC 2012

to as a "probe", is immobilized as a small discrete spot on a solid support. The probe contains known amounts of individual, specific sequences that can be used to hybridize to the free nucleic acid in the sample (1, 5). The arrays are constructed using different techniques, the most widespread being spotting of the probe with robotic systems and in situ synthesis of oligonuleotides (2, 3, 6). The sample in analysis is hybridized to the probes after being labeled; detection and quantitation are then obtained with a scanning confocal microscope and ad hoc software.

The first arrays were fabricated in the mid-1980s by spotting probes (bacteria carrying cDNA inserts, DNA clones, and PCR products) on supports such as nylon membranes and were called macroarrays; due to the size of the spots, they enabled analyses up to approximately 2,000 genes (6).

The growing availability of information about genomic sequences of different species further increased the "need" for high-throughput analysis of gene expression patterns, better representing, in a snapshot, the status of the analyzed sample (7). The advent of microarray technology, in the 1990s, made this possible (3, 6).

The microarray technology was immediately recognized as truly revolutionary and soon many applications, other than gene expression analyses, were developed for its exploitation: SNPs detection, array-comparative genomic hybridizations, DNA-methylation status, etc.

2. Fabrication and Characteristics of DNA Microarrays

DNA microarrays can be classified according to the fabrication and probe technology employed. Arrays can be arranged either by printing (or spotting) or in situ synthesizing DNA on a solid surface (8–13).

2.1. Printed or Spotted Microarrays

Spotted or printed microarrays are prepared by printing probes of oligonucleotides or cDNA that is prepared from known, selected probes. The probes hybridize to cDNA or cRNA derived from experimental samples. Using the printed microarray, it is possible to transfer onto a glass surface an existing library, composed, for example, of presynthesized oligonucleotides or by double-stranded cDNA. When using oligonucleotides, generally gene-specific long oligos (between 50 and 70-mers) are designed in order to have similar annealing temperature (see Note 1) (14). In the case of cDNA microarrays, the double-stranded DNA to be printed can be generated by PCR amplification of a selected cDNA library of an organism (15). The probes can then be transferred onto the glass surface by several kinds of robotic stations or arrayers (16). One of the most common methods is called "contact printing"; it involves a robotic arm loaded with pins which harvest the sample by immersion and then deposit pico- to nanoliter volumes of it (spot) on the slide by touching the glass surface (17). Printing pins resemble needles with a terminal groove (10–100 μm in diameter) so that the sample is loaded by capillary action. Size and regularity of the spot depend on the characteristics of the probes (i.e., resuspension buffer and DNA concentration) and on the pin material, shape, and diameter.

Alternative approaches are the noncontact printing methods. Typically, they employ piezoelectric inkjet printing devices to dispense pico- to nanoliter volumes of probes on the array surface (17). Electrical pulses are used to deform a transducer material, creating a pressure inside a capillary; this determines the dispensation of small volumes from a nozzle.

Slides used for the production of printed arrays are coated with chemicals with active groups (e.g., Aminopropylsilane, providing hydroxyl groups) that allow the spotted DNA to bind covalently to the surface (18).

2.2. High-Density In Situ Synthesis of Oligonucleotide Arrays

Alternatively, microarrays can be fabricated by in situ synthesis of desired oligonucleotide sequences. Affymetrix (Santa Clara, CA, USA) was the first company to commercialize this type of technology (GeneChip®), basing the production on photolithography and solid-phase chemical synthesis, with a technique that enables the creation of high-density oligonucleotide arrays (19). This light-based synthesis approach has been for many years the gold standard method for microarray industrial manufacturing. In this method, oligonucleotides are synthesized directly on the array surface via a series of chemical coupling reactions. The procedure utilizes a solid support derivatized with a covalent linker molecule terminating with a photolabile protection group. Ultraviolet light causes deprotection and activates these molecules for chemical coupling with new protected nucleotides. UV rays are precisely localized using serial checkerboard photomasks with clear regions at desired positions. This process allows the activation of specific positions in order to stepwise synthesize oligos. It is important to note that, with this method, only relatively short, 25–30 base long probes can be synthesized. To overcome low specificity of short oligos, multiple probes are present for the same target gene. Moreover for each probe complementary to the target gene (perfect match), there is a probe with one mismatching nucleotide to monitor nonspecific binding. The advantage of this technology is the ability to manufacture high-density arrays containing up to 500,000 probes/cm^2.

2.2.1. Oligo DNA Microarrays

Another approach for industrial scale production of DNA oligo microarrays is available from Agilent (Santa Clara, CA, USA) (20, 21). It combines in situ synthesis of oligonucleotides with inkjet printing technology. With this method, 60-mers are synthesized by stepwise inkjet printing of DNA nucleotides followed by coupling based on phosphoramidite chemistry. Needing no mask, this system is extremely flexible.

In general, long oligonucleotide arrays have advantages over cDNA microarrays, specifically sequences are known and designed to avoid cross-hybridization, while spotted cDNA might present more specificity problems (22, 23). Different genes are usually represented by cDNA with different lengths and often the sequence printed is not known entirely in a spotted microarray. In addition, mistakes in cDNA libraries often require performing sequence validation of clones.

On the other hand, the use of cDNA in spotted microarrays enables the production of arrays from organisms without the need to know their entire genome sequence; they are extremely flexible and easy to customize. For these reasons and for their low cost, their manufacture is the most widespread in academic settings (24).

One of the main advantages of industrial short-oligo arrays is their reproducibility and the standardization of the array. The entire Affymetrix platform for example (commercial arrays, standardized protocols, controlled fluidic station, and calibrated scanner) is designed to provide reproducible results, allowing comparison of data from different laboratories. The main disadvantages in this case are represented by the considerable costs and the difficulty to customize arrays based on specific researchers' needs. Translational research applications of expression profiling and transcript validation encompass scenarios in which an investigator acquires frozen tumor tissue samples and desires to develop a genomic profile associated with biology or outcome (25). In such a case, RNA expression profiling is often implemented (Table 1).

2.3. Gene Expression Validation Platforms

It is appropriate to validate expression profile findings using an independent quantitative or semiquantitative method. There are a variety of methods for the quantitation of mRNA. These include: Northern blotting, ribonuclease protection assays (RPAs), in situ hybridization, and PCR (26–28).

PCR is the most sensitive method and can discriminate closely related mRNAs. Northern blotting and RPAs are the gold standards, since no amplification is involved, whereas in situ hybridization is qualitative rather than quantitative. Techniques such as Northern blotting and RPAs require more RNA than is available in a tissue sample. PCR methods are therefore particularly valuable when amounts of RNA are low. In contrast to regular reverse transcriptase-PCR and analysis by agarose gels, real-time PCR provides quantitative results.

Examples of commonly used platforms for RT-PCR are (see Note 2): Applied Biosystems (ABI7300), Roche (LightCycler), Stratagene (Mx4000), Cephid (SmartCycler), Corbett (Rotor-Gene 6000), Eppendorf (Mastercycler), and Bio-Rad (MiniOpticon, MyiQ, Opticon2, Chromo4, iQ5).

Table 1
Commercially available examples of RNA expression profiling platforms for correlating expression profiles with clinical outcome

Platform	Number of oligo-nucleotide probes	Number of genes	Method	Input material (μg Total RNA)	Detection/scanner
Applied Biosystems expression array	31,700 60-mer	27,868	Digoxigenin-UTP labeled cRNA	2	Chemiluminescence Applied Biosystems 1700 Chemiluminescent microarray analyzer
Stanford human cDNA microarray	42,000 Features	24,271 Unique cluster IDs (UniGene Build Number 173)	Van Gelder et al. (26) method using MessageAmp amplification kit (Ambion)	3	Cy5 for tumor RNA Cy3 for reference RNA Agilent DNA microarray scanner Images analyzed by GenePix Pro v 4.1
Agilent whole human genome oligo microarray	44,000 60-mer	41,000	T7 RNA polymerase incorporation of Cy3-CTP Cy5-CTP	500	Cy5 for tumor RNA Cy3 for reference RNA Agilent DNA microarray scanner Agilent Feature Extraction Software A.7.5.1

3. Gene Expression Analyses: Sample Preparation, Workflow, and Experimental Design for a Microarray Experiment

For gene expression analysis, the transcriptome of the sample is analyzed through hybridization with the complementary probes present on the array. It is important to note that target sequences (from the unknown query sample) need to be labeled to allow signal detection by scanning confocal microscopy. A signal is detected when the target hybridizes to its corresponding probe sequence on the array.

3.1. Sample Preparation

Labeling of RNA samples can be achieved through a reverse transcription reaction by direct incorporation of nucleotides linked to fluorochromes, typically Cy3 or Cy5, or by incorporation of aminoallyl-dUTP followed by coupling to the Cy3 or Cy5 dyes (indirect labeling). The resulting cDNA can be then hybridized to the array.

It is important to note that, since the labeled antisense strand of the starting RNA is hybridized in oligo microarrays, the probes on the array are synthesized in the sense orientation. In cDNA microarrays, each probe is generated by PCR amplification, therefore both strands are present on the array. Those differences need to be taken into account for some applications (e.g., double-stranded cDNA spots can be stained with intercalating dyes such as POPO-3).

Another widely used method for sample preparation is the linear amplification of RNA to generate labeled antisense cRNA. This procedure was first described by Eberwine and coworkers (29) and provides a means for analyzing RNA samples extracted from limited sources of biological materials. Briefly, with this method, a first-strand cDNA synthesis is obtained by reverse transcription using oligo(dT) primers linked to the T7 promoter sequence followed by second-strand synthesis. The presence of the promoter allows the linear amplification of the template through an in vitro transcription reaction using the T7 RNA polymerase. During this step, labeled or modified nucleotides are incorporated in the cRNA sequences. In any case, the resulting labeled targets are then hybridized to the DNA microarrays under stringent conditions. After overnight incubation, the array is washed to remove any nonspecific hybridization and to reduce background signals. Laser confocal devices are next used to scan the arrays and acquire images and spot intensities.

3.2. Experimental Design

Though both cDNA spotted microarrays and oligo microarrays are widely employed for differential gene expression analyses, it is important to take into account that different kinds of microarrays may require different experimental designs (Table 2) (30). The simplest experiment to perform involves two samples only: a query and its control. Printed arrays (cDNA and long oligo arrays) allow

Table 2
Comparison of commercially available example DNA array platforms

Platform	Technology summary	Application	References
Affymetrix	906.6k SNPs and 946k copy number probes; array based	Gene copy number and association	(51, 52)
Agilent	Oligonucleotide probes covering coding and noncoding regions	Genetic detection Genetic-based research	(38)
ROMA/ NimbleGen	Oligonucleotides Use *Bgl II* cutting sites	Chromosomal or microsatellite genetic instability	(37) (39)
Illumina	109k SNP markers; bead based	Gene copy number LOH/allelic ratios	(9)

the contemporaneous hybridization of the two different samples, labeled with different fluorochromes, so that the test and control can be directly compared on the same slide. For this reason, printed arrays are also called "two-color arrays." Direct comparison of the samples on the same array reduces artifacts due to uneven hybridization, local background differences, and slide-to-slide variations; on the other hand, dye bias requires performing dye-swap replicates. When considering an experiment involving multiple samples with two-color arrays, particular attention has to be paid to its design. Often the lack of control RNA, the presence of large number of samples, possibly made available during a long stretch of time, does not allow a direct comparison among multiple samples. In this situation, all samples are usually compared indirectly through a reference, i.e., a common arbitrary "control." A reference can be prepared by pooling samples within the study or it can be obtained from external sources: RNA extracted from a combination of different cell lines and pooled, random genomic amplification (DNA-based reference), or by PCR amplification of the library printed on the array in use (31–33).

Short-oligo arrays support the hybridization of only one sample per chip; therefore they are also called one-color arrays. In this case, the intensity of each spot is measured and the comparison of different samples is achieved by comparing data from different chips. For this reason, consistency between sample preparation and hybridization procedures needs to be under tight control, but artifacts like uneven hybridization, etc. may still constitute a significant problem. An advantage of one-color arrays is that the experimental design can be very simple. Indeed it is possible to analyze a large number of samples, to include ex novo samples during the experiment or even to merge easily different data sets together, possibly from different laboratories. In these situations, though, external references are still useful, but mostly they are employed to check reproducibility among analysis runs or bias due to different laboratory sites.

3.3. Applications of DNA Microarrays to Molecular Profiling

Analysis of somatic genetic alterations in primary tumors and their metastasis provides information about cancer etiology, progression, and drug sensitivity. The information derived is different from that obtained by RNA transcript profiling. A common application is the association of copy number changes with cancer phenotypes or critical genes. DNA, for many investigators, is a preferred class of molecular analyte because DNA is highly stable and can be readily amplified by PCR. There are many DNA microarray platforms from which to choose. For array-comparative genomic hybridization (array CGH), several methods and platforms have been developed. Microarray copy number detection systems differ in their probe derivation (BAC, cDNA, or oligonucleotides) and array fabrication (spotting, polymerization, or microbeads), and the density of probes per gene within the genome (20). Hybridization and labeling techniques (single- or two-color systems) differ among DNA microarray platforms. A particular class of platform may be best suited to specific biological questions (34–36). The choice of the platform is further influenced by the experimental designs, sample size restrictions, and resolution (density of information per gene or chromosome) or data processing challenges. Three major example platforms are: (a) Agilent Human Genome CGH Microarray 44k, (b) ROMA/NimbleGen Representational Oligonucleotide Microarray 82k, and (c) Illumina Human-1 Genotyping 109k BeadChip (Table 2) (37).

3.3.1. Agilent Human Genome CGH Microarray 44k

The oligonucleotide probes used for the Agilent array cover both coding and noncoding sequences, and most reporters are located in genes (gene-oriented arrangement). Agilent 44k CGH arrays have a gene-oriented arrangement being enriched particularly for cancer relevant genes with local high variation of number of reporters (38). The combination of these two features may be the reason for the high number of specific small high frequency amplification or deletion peaks particular in the Agilent platform.

3.3.2. ROMA/NimbleGen Representational Oligonucleotide Microarray 82k

In contrast to the Agilent platform, oligonucleotides in the ROMA/NimbleGen technology are based on *Bgl* II cutting sites, so the reporters are more or less randomly distributed across the entire genome. The ROMA technology was invented at Cold Spring Harbor Laboratory and by NimbleGen (39). Since the oligonucleotide probes are designed for cutting site derived fragments of the human genome sequence, which are more or less randomly distributed across the genome, the ROMA/NimbleGen arrays provide a gene-independent arrangement of the structure of the complete genome at a high resolution. This broad distribution of reporters provides a detailed picture of the structure and organization of the complete genome (genome orientated arrangement).

3.3.3. Illumina Human-1 Genotyping 109k BeadChip

A third type of DNA mircroarray technology, the Illumina platform provides a dense, exon-centric view of the genome presented in a molecular cytogenetics genotyping arrangement (9). The Illumina bead-based Human-1 109k arrays query the genome through their 109k SNP markers, of which 70% are located in gene exons or within 10 kb of transcripts. This platform employs an allele-specific primer extension assay using two probes (bead types) in one-color channel to score SNPs.

Which platform is best for which application? Baumbusch et al. (37) conclude that the Agilent or the Illumina system is best for gene detection and gene-oriented research. The ROMA/NimbleGen approach, revealing a compact picture of the entire genome structure, is the method of choice for exploration of genomic instability mechanisms such as chromosomal instability (CIN) or microsatellite instability (MIN). Finally, the Illumina SNP-CGH arrays can be readily applied to detecting loss of heterozygosity (LOH) and allelic ratio aberrations associated with copy number changes (40).

3.4. MammaPrint® and Oncotype DX® for Personalized Therapy of Breast Cancer

Gene expression profiling for personalized therapy is now accessible to physicians treating patients with breast cancer. Onco*type* Dx® and MammaPrint® (Agendia) are currently the leading commercial examples for which a body of clinical experience has been acquired. In the current marketplace, each test has competitive advantages and disadvantages, summarized below.

3.4.1. Test Description and Indicated Use

Onco*type* DX®, created by Genomic Health, is a diagnostic test that scores the likelihood of disease recurrence in women with early stage breast cancer. The score is provided to clinicians who wish to determine whether the patient is likely to benefit from certain types of chemotherapy (41–44). Onco*type* DX® analyzes a panel of 21 genes within a tumor to determine a Recurrence Score. The Recurrence Score is a number between 0 and 100 that corresponds to a specific likelihood of breast cancer recurrence within 10 years of the initial diagnosis.

Patients eligible for the test harbor invasive breast cancer who are ER+ and whose lymph nodes are negative. Typically, in these cases, treatment with hormonal therapy, such as tamoxifen, is planned. Onco*type* DX® is not for use in patients with carcinoma in situ (precancerous) or metastatic breast cancer (cancer that has spread beyond the breast), or stages III and IV disease.

MammaPrint® is marketed by Agendia and is based on the Amsterdam 70-gene breast cancer gene signature as published by van't Veer et al. (25). MammaPrint® is designated for use in the USA for lymph node negative breast cancer patients under 61 years of age with tumors less than 5 cm. The test requires a fresh breast specimen. MammaPrint® is an expression profile microarray-based test. The score classifies analyzed tumors as low or high risk for recurrence of the disease (45–47).

3.4.2. Comparison of Technical Differences Between Oncotype Dx® and Mammaprint®

Validation Comparison

Based on currently published studies, the onco*type* DX® test has been validated as a stand-alone prognostic test and has been interpreted as a predictive test for response to tamoxifen and to the cytoxan, methotrexate, and fluorouracil adjuvant chemotherapy regimen (although concurrent with tamoxifen) (41–44). In contrast, the MammaPrint® assay is validated as a prognostic test only and has not been formally validated as a predictive test for specific endocrine or cytotoxic therapy regimens.

Comparison of Gene Coverage

The two assays only have one overlapping gene among the list of genes in each test. Nevertheless the important subsets of genes used for the scoring for both tests cover transcripts relevant to proliferation, ER and HER2. Thus, it can be argued that there is a functional overlap in the biologic basis of the two tests.

Comparison of Sample Requirements: Frozen Versus Formalin Fixed

The onco*type* DX® test can use formalin fixed tissue, while the MammaPrint® assay requires frozen specimens. The majority of newly diagnosed breast cancers in the USA and Europe are formalin processed for microscopic examination. The onco*type* DX® test therefore provides an ease of use for FFPE-based assays. Frozen tissue is associated with a higher yield of transcript information. This difference in input sample requirements translates to a larger gene number of the MammaPrint® assay that could potentially provide additional pharmacogenomic information missed by the 21-gene onco*type* DX® assay.

Patient Coverage

The MammaPrint® test accepts both ER-positive and -negative patients, which also allows for inclusion of a greater number of younger patients, whereas the original onco*type* DX® test was restricted to ER-positive patients. Neither test accepts stages III and IV patients, or patients with Ductal Carcinoma In Situ.

Regulatory Approval

The MammaPrint® assay has received 510(k) clearance by the US Food and Drug Administration. The onco*type* DX® test has been exempt. The Genomic Health Inc.'s central testing laboratory has been approved under the US Clinical Laboratory Improvement Amendment (CLIA) regulations to offer the onco*type* DX test as a homebrew assay. More recently, onco*type* DX has been designated as "recommended for use" by the American Society of Clinical Oncology Breast Cancer Tumor Markers Update Committee, whereas the MammaPrint® assay was classified by the group as "under investigation."

3.4.3. Clinical Validation Studies

In collaboration with several independent investigators, onco*type* DX® was evaluated in numerous studies involving over 3,300 patients (41–44). Onco*type* DX® was clinically validated in a large, independent multicenter trial of patient samples from the National Surgical Adjuvant Breast and Bowel (NSABP) Study B-14.2. The study determined that onco*type* DX® was a prognostic assay to determine the likelihood of recurrence in tamoxifen-treated patients with node-negative, estrogen receptor-positive breast cancer (43). A second study, NSABP Study B-20, determined that onco*type* DX® can predict chemotherapy benefit (44). Archived tumor samples from a large cohort of patients who did not receive chemotherapy and had been diagnosed with node negative breast cancer were assayed using the 21-gene onco*type* DX® assay (42). This US community health care-based patient study confirmed that the recurrence score was significantly associated with risk of death among ER-positive, tamoxifen-treated and -untreated patients (42).

There have been three papers validating and testing the use of the MammaPrint® 70-gene profile in clinical conditions (45–47). van't Veer et al. developed the 70-gene multivariate profile for breast cancer prognosis using a population of 78 patients (6.4% of which had adjuvant treatment) (25). van de Vijver et al. validated the 70-gene profile in consecutive series of breast cancer patients ($n = 151$; 5.2% of which had adjuvant treatment) (47). In a study of 302 patients without adjuvant treatment, the 70-gene signature was validated by the European TRANSBIG consortium (45). To date, there are no clinical trials assessing the utility of adjuvant therapy in high-risk versus low-risk individuals. This means that the MammaPrint® is prognostic not predictive. In these studies, researchers could not conclude that the high-risk patients would be more likely than the low-risk patients to benefit from chemotherapy.

3.5. Standardization and Comparison Among Different Microarray Platforms: MIAME Standards and MAQC Project

Microarray experiments soon generated large amounts of data. The initial lack of standardization represented a major problem for data exchange, interpretation, and for repeatability of results. To fill this gap, in 2001, a consortium created by the "Microarray and Gene Expression Data Society" (MGED) proposed the Minimum Information About a Microarray Experiment (MIAME) that should be made available to facilitate the sharing and interpretation of data (48). In particular, since many variables could affect the outcome of one experiment, MIAME requested the availability of at least the following elements: spot intensity raw data, normalized data, sample annotations, experimental design, annotation of the array probes, laboratory protocols, and data processing procedures. MIAME standards have been adopted by most scientific journals and this information must be submitted for review with any manuscript.

Another consortium was coordinated by the US Food and Drug Administration (FDA) with the aim to assess the performance of several microarray platforms, the proficiency of individual laboratories, and data analysis procedures (49). This project, named MicroArray Quality Control (MAQC), involved different FDA centers (e.g., NCTR, CBER, and CDER), the National Institutes of Health (NIH), providers of array platforms and RNA samples (e.g., Affymetrix, Agilent, Applied Biosystems, Illumina, and Ambion), the National Institute of Standards and Technology (NIST), academic laboratories (e.g., Stanford University and Duke University), and other institutions (e.g., Novartis and Cold Spring Harbor Laboratory). The scope was to develop standards and quality control to support the use of microarray data in drug discovery, clinical practice, and regulatory decision-making. The MAQC project evaluated and compared data deriving from several sources of RNA, different microarrays platforms using a common control RNA sample (five commercial whole-genome microarray platforms and custom arrays), and various data analysis approaches. Moreover, reproducibility within test sites and across different sites has been investigated. Finally, the study showed that microarray results are repeatable within test site and reproducible between sites. Different platforms provide comparable results when considering differences in probe annotations (50–52). Ultimately, the project confirmed that microarray technology has the capabilities required to be considered for regulatory applications.

Today regulatory agencies encourage pharmacogenomic data submissions and several guidelines have been issued for industry to provide information regarding how to perform a microarray experiment, the use that can be made of the data in drug development, and when the data can be used in regulatory decision-making (53).

3.6. Other Microarray Applications

Microarray technology offers several applications beyond gene expression analysis. Indeed a genome-wide platform capable of analyzing sequence specificity and abundance raised interest in different areas of biotechnology and biomedical research.

Many biological questions can be simply investigated by varying DNA sequences present on the array and by hybridizing different sources of query target. It is worth noticing that oligonucleotide arrays find more application than cDNA arrays, as they allow better control sequence specificity. Below is an overview of different applications.

3.6.1. Alternative Splicing Analysis

Alternative splicing can produce several gene products from a single locus; each product may represent different or even opposite functions. Therefore, the importance of analyzing a complete transcriptome and discriminating among different gene variants is extremely important (54).

Usually probes on oligo arrays employed for traditional gene expression analysis are designed to represent the 3′-untranslated region (3′ UTR) of the genes. The main reason for this choice is that reverse transcription of RNA using poly(T) primers generates cDNA enriched in that region. Moreover, sequence divergence is typically greater in 3′-UTR. On the other hand, it is not possible to discriminate among different splicing variants in this region. To retrieve this information, oligo arrays were made with combinations of exons and junction probes. For this application, it is important to label the entire length of transcripts. Therefore, the reverse transcription step is usually performed using random primers or a combination of random and poly(T) primers. Examples of these arrays are SpliceArrays™ from Agilent and GeneChip® Exon Arrays provided by Affymetrix.

3.6.2. miRNA Analysis

microRNAs (miRNAs) are short noncoding, single-stranded RNAs (about 21–23 nucleotides), involved in the regulation of many biological processes. Being partially complementary to some mRNAs, they regulate the expression of many genes by inhibiting protein translation and by mediating the cleavage of their mRNA. To better investigate gene expression profiling and global regulation, oligo microarrays have been designed with specific miRNA probes. Oligo arrays composed of miRNA have been used to evaluate the microRNA expression of single embryonic stem cells, and identify novel microRNAs in colorectal tumor cells (55–57).

An example is provided by the mirVana Array System from Ambion (Austin, TX) which includes arrays and miRNA Labeling Kit (56). Target labeling is achieved by adding a tail of 20–50 nuclueotides to the 3′ end of each miRNA. Amine-modified nucleotides present in the tail are then used to couple CyDye fluorophores.

3.6.3. Genome Variation Analysis

Other varieties of oligo arrays have been designed to investigate genome variations; in particular, SNP and copy number variation (CNV) such as deletion or duplication (9). These analyses are particularly useful in cancer research, in linkage analysis, or to genotype drug-metabolism enzymes in order to predict drug exposure.

As target, the whole-genome DNA sequence is used for hybridization. The labeling can be achieved by genomic DNA digestion with restriction enzymes followed by ligation of an adaptor; a single primer, complementary to the adaptor, is then used to amplify DNA fragments by PCR. Amplicons are then fragmented and end-labeled with biotin using terminal deoxynucleotidyl transferase enzyme.

Among arrays to discriminate SNPs, Affymetrix manufactured chips with 24–40 different 25-mers per locus; some of these oligos represent the perfect match of one allele while another set is complementary to the alternative allele (58, 59). Both kinds have the corresponding control mismatch. In addition, probes targeting known CNV regions and probes evenly spaced along the genome allow monitoring gene deletions or amplification. Another platform for the analysis of copy variants is the Agilent Human Genome CGH Microarray (20).

3.6.4. Methylation and DNA-Binding Analysis

Other oligo microarrays allow monitoring epigenetic changes of the genome. In particular, there is a lot of interest in investigating methylation events in CpG islands in gene promoter regions. Indeed these events regulate gene expression and they can be linked to cancer onset or have an important role during embryonic development.

To study methylation on a large scale (methylome), two-color microarrays can be employed in combination with methyl-DNA immunoprecipitation approach. With this method, methylated DNA is enriched from a genome using antibody against 5-methylcytosine (21). The fraction is then Cy5-labeled and hybridized versus the nonenriched DNA, labeled with Cy3. For this application, probes on the array can be designed on regulatory and promoter regions of specific genes of interest or on known CpG Islands.

Another application allows studying global regulation through the investigation of binding events of DNA-binding proteins to the genome. For example, it is possible to monitor transcription factor's binding activities and to identify DNA target sequences. This method employs chromatin immunoprecipitation (ChIP) of DNA fragments using specific antibodies against the protein of interest followed by PCR amplification of the enriched DNA (ChIP on chip) (60). The PCR products are then fragmented, labeled, and hybridized. For this application, tiling arrays are commonly used, which are designed by dividing the nonrepetitive sequence of genomic regions into contiguous probes.

4. Notes

1. Extensive literature exists for troubleshooting microarray processes. An excellent reference describing the required steps for quality microarray design and analysis is: Imbeaud and Auffray (61).
2. For a full comparison of these example RT-PCR platforms, see ref. 28.

References

1. Southern, E., Mir, K., and Shchepinov, M. (1999) Molecular interactions on microarrays. *Nat Genet* **21**, 5–9.
2. Ramsay, G. (1998) DNA chips: state-of-the art. *Nat Biotechnol* **16**, 40–4.
3. Marshall, A., and Hodgson, J. (1998) DNA chips: an array of possibilities. *Nat Biotechnol* **16**, 27–31.
4. Venter, J. C., Adams, M. D., Myers, E. W., Li, P. W., Mural, R. J., *et al.* (2001) The sequence of the human genome. *Science* **291**, 1304–51.
5. Phimister, B. (1999) Going global. *Nature Genetics* **21**, 1.
6. Lockhart, D. J., and Winzeler, E. A. (2000) Genomics, gene expression and DNA arrays. *Nature* **405**, 827–36.
7. Baldwin, D., Crane, V., and Rice, D. (1999) A comparison of gel-based, nylon filter and microarray techniques to detect differential RNA expression in plants. *Curr Opin Plant Biol* **2**, 96–103.
8. Smith, D. R., Quinlan, A. R., Peckham, H. E., Makowsky, K., Tao, W., *et al.* (2008) Rapid whole-genome mutational profiling using next-generation sequencing technologies. *Genome Res* **18**, 1638–42.
9. Quail, M. A., Kozarewa, I., Smith, F., Scally, A., Stephens, P. J., *et al.* (2008) A large genome center's improvements to the Illumina sequencing system. *Nat Methods* **5**, 1005–10.
10. Mardis, E. R. (2008) Next-generation DNA sequencing methods. *Annu Rev Genomics Hum Genet* **9**, 387–402.
11. Mardis, E. R. (2008) The impact of next-generation sequencing technology on genetics. *Trends Genet* **24**, 133–41.
12. Kozarewa, I., Ning, Z., Quail, M. A., Sanders, M. J., Berriman, M., *et al.* (2009) Amplification-free Illumina sequencing-library preparation facilitates improved mapping and assembly of (G+C)-biased genomes. *Nat Methods* **6**, 291–5.
13. Bentley, D. R. (2006) Whole-genome re-sequencing. *Curr Opin Genet Dev* **16**, 545–52.
14. Rouillard, J. M., Zuker, M., and Gulari, E. (2003) OligoArray 2.0: design of oligonucleotide probes for DNA microarrays using a thermodynamic approach. *Nucleic Acids Res* **31**, 3057–62.
15. Hegde, P., Qi, R., Abernathy, K., Gay, C., Dharap, S., *et al.* (2000) A concise guide to cDNA microarray analysis. *Biotechniques* **29**, 548–50, 52–4, 56 passim.
16. Holloway, A. J., van Laar, R. K., Tothill, R. W., and Bowtell, D. D. (2002) Options available – from start to finish – for obtaining data from DNA microarrays II. *Nat Genet* **32 Suppl**, 481–9.
17. Barbulovic-Nad, I., Lucente, M., Sun, Y., Zhang, M., Wheeler, A. R., *et al.* (2006) Bio-microarray fabrication techniques – a review. *Crit Rev Biotechnol* **26**, 237–59.
18. Grainger, D. W., Greef, C. H., Gong, P., and Lochhead, M. J. (2007) Current microarray surface chemistries. *Methods Mol Biol* **381**, 37–57.
19. Fodor, S. P., Read, J. L., Pirrung, M. C., Stryer, L., Lu, A. T., *et al.* (1991) Light-directed, spatially addressable parallel chemical synthesis. *Science* **251**, 767–73.
20. Barrett, M. T., Scheffer, A., Ben-Dor, A., Sampas, N., Lipson, D., *et al.* (2004) Comparative genomic hybridization using oligonucleotide microarrays and total genomic DNA. *Proc Natl Acad Sci USA* **101**, 17765–70.
21. Pokholok, D. K., Harbison, C. T., Levine, S., Cole, M., Hannett, N. M., *et al.* (2005) Genome-wide map of nucleosome acetylation and methylation in yeast. *Cell* **122**, 517–27.
22. Nielsen, H. B., and Knudsen, S. (2002) Avoiding cross hybridization by choosing nonredundant targets on cDNA arrays. *Bioinformatics* **18**, 321–2.
23. Taylor, E., Cogdell, D., Coombes, K., Hu, L., Ramdas, L., *et al.* (2001) Sequence verification as quality-control step for production of cDNA microarrays. *Biotechniques* **31**, 62–5.
24. Hager, J. (2006) Making and using spotted DNA microarrays in an academic core laboratory. *Methods Enzymol* **410**, 135–68.
25. Veer, L. J., and De Jong, D. (2002) The microarray way to tailored cancer treatment. *Nat Med* **8**, 13–4.
26. VanGuilder, H. D., Vrana, K. E., and Freeman, W. M. (2008) Twenty-five years of quantitative PCR for gene expression analysis. *Biotechniques* **44**, 619–26.
27. Udvardi, M. K., Czechowski, T., and Scheible, W. R. (2008) Eleven golden rules of quantitative RT-PCR. *Plant Cell* **20**, 1736–7.
28. Logan, J., Edwards, K., and Saunders, N. (2009) Real-Time PCR: Currnet Technology and Applications, Caister Academic Press, Norwich, UK.
29. Van Gelder, R. N., von Zastrow, M. E., Yool, A., Dement, W. C., Barchas, J. D., *et al.* (1990)

Amplified RNA synthesized from limited quantities of heterogeneous cDNA. *Proc Natl Acad Sci USA* **87**, 1663–7.

30. Lee, N. H., and Saeed, A. I. (2007) Microarrays: an overview. *Methods Mol Biol* **353**, 265–300.
31. Gorreta, F., Barzaghi, D., VanMeter, A. J., Chandhoke, V., and Del Giacco, L. (2004) Development of a new reference standard for microarray experiments. *Biotechniques* **36**, 1002–9.
32. Kim, H., Zhao, B., Snesrud, E. C., Haas, B. J., Town, C. D., *et al.* (2002) Use of RNA and genomic DNA references for inferred comparisons in DNA microarray analyses. *Biotechniques* **33**, 924–30.
33. Sterrenburg, E., Turk, R., Boer, J. M., van Ommen, G. B., and den Dunnen, J. T. (2002) A common reference for cDNA microarray hybridizations. *Nucleic Acids Res* **30**, e116.
34. Li, C., Li, M., Long, J. R., Cai, Q., and Zheng, W. (2008) Evaluating cost efficiency of SNP chips in genome-wide association studies. *Genet Epidemiol* **32**, 387–95.
35. Li, M., Li, C., and Guan, W. (2008) Evaluation of coverage variation of SNP chips for genome-wide association studies. *Eur J Hum Genet* **16**, 635–43.
36. Barrett, J. C., and Cardon, L. R. (2006) Evaluating coverage of genome-wide association studies. *Nat Genet* **38**, 659–62.
37. Baumbusch, L. O., Aaroe, J., Johansen, F. E., Hicks, J., Sun, H., *et al.* (2008) Comparison of the Agilent, ROMA/NimbleGen and Illumina platforms for classification of copy number alterations in human breast tumors. *BMC Genomics* **9**, 379.
38. Brennan, C., Zhang, Y., Leo, C., Feng, B., Cauwels, C., *et al.* (2004) High-resolution global profiling of genomic alterations with long oligonucleotide microarray. *Cancer Res* **64**, 4744–8.
39. Ylstra, B., van den Ijssel, P., Carvalho, B., Brakenhoff, R. H., and Meijer, G. A. (2006) BAC to the future! or oligonucleotides: a perspective for micro array comparative genomic hybridization (array CGH). *Nucleic Acids Res* **34**, 445–50.
40. Dong, L. M., Brennan, P., Karami, S., Hung, R. J., Menashe, I., *et al.* (2009) An analysis of growth, differentiation and apoptosis genes with risk of renal cancer. *PLoS One* **4**, e4895.
41. Hornberger, J., Cosler, L. E., and Lyman, G. H. (2005) Economic analysis of targeting chemotherapy using a 21-gene RT-PCR assay in lymph-node-negative, estrogen-receptor-positive, early-stage breast cancer. *Am J Manag Care* **11**, 313–24.
42. Habel, L. A., Shak, S., Jacobs, M. K., Capra, A., Alexander, C., *et al.* (2006) A population-based study of tumor gene expression and risk of breast cancer death among lymph node-negative patients. *Breast Cancer Res* **8**, R25.
43. Paik, S., Shak, S., Tang, G., Kim, C., Baker, J., *et al.* (2004) A multigene assay to predict recurrence of tamoxifen-treated, node-negative breast cancer. *N Engl J Med* **351**, 2817–26.
44. Paik, S., Tang, G., Shak, S., Kim, C., Baker, J., *et al.* (2006) Gene expression and benefit of chemotherapy in women with node-negative, estrogen receptor-positive breast cancer. *J Clin Oncol* **24**, 3726–34.
45. Buyse, M., Loi, S., Veer, L., Viale, G., Delorenzi, M., *et al.* (2006) Validation and clinical utility of a 70-gene prognostic signature for women with node-negative breast cancer. *J Natl Cancer Inst* **98**, 1183–92.
46. Glas, A. M., Floore, A., Delahaye, L. J., Witteveen, A. T., Pover, R. C., *et al.* (2006) Converting a breast cancer microarray signature into a high-throughput diagnostic test. *BMC Genomics* **7**, 278.
47. van de Vijver, M. J., He, Y. D., Veer, L. J., Dai, H., Hart, A. A., *et al.* (2002) A gene-expression signature as a predictor of survival in breast cancer. *N Engl J Med* **347**, 1999–2009.
48. Brazma, A., Hingamp, P., Quackenbush, J., Sherlock, G., Spellman, P., *et al.* (2001) Minimum information about a microarray experiment (MIAME)-toward standards for microarray data. *Nat Genet* **29**, 365–71.
49. Shi, L., Tong, W., Su, Z., Han, T., Han, J., *et al.* (2005) Microarray scanner calibration curves: characteristics and implications. *BMC Bioinformatics* **6 Suppl 2**, S11.
50. Shi, L., Reid, L. H., Jones, W. D., Shippy, R., Warrington, J. A., *et al.* (2006) The MicroArray Quality Control (MAQC) project shows inter- and intraplatform reproducibility of gene expression measurements. *Nat Biotechnol* **24**, 1151–61.
51. Shi, L., Tong, W., Fang, H., Scherf, U., Han, J., *et al.* (2005) Cross-platform comparability of microarray technology: intra-platform consistency and appropriate data analysis procedures are essential. *BMC Bioinformatics* **6 Suppl 2**, S12.
52. Patterson, T. A., Lobenhofer, E. K., Fulmer-Smentek, S. B., Collins, P. J., Chu, T. M., *et al.* (2006) Performance comparison of one-color and two-color platforms within the MicroArray Quality Control (MAQC) project. *Nat Biotechnol* **24**, 1140–50.
53. Goodsaid, F., and Frueh, F. W. (2007) Implementing the U.S. FDA guidance on

pharmacogenomic data submissions. *Environ Mol Mutagen* **48**, 354–8.

54. Lee, C., and Roy, M. (2004) Analysis of alternative splicing with microarrays: successes and challenges. *Genome Biol* **5**, 231.
55. Tang, F., Hajkova, P., Barton, S. C., Lao, K., and Surani, M. A. (2006) MicroRNA expression profiling of single whole embryonic stem cells. *Nucleic Acids Res* **34**, e9.
56. Tang, F., Hajkova, P., Barton, S. C., O'Carroll, D., Lee, C., *et al.* (2006) 220-plex microRNA expression profile of a single cell. *Nat Protoc* **1**, 1154–9.
57. Cummins, J. M., He, Y., Leary, R. J., Pagliarini, R., Diaz, L. A., Jr., *et al.* (2006) The colorectal microRNAome. *Proc Natl Acad Sci USA* **103**, 3687–92.
58. Kathiresan, S., Voight, B. F., Purcell, S., Musunuru, K., Ardissino, D., *et al.* (2009) Genome-wide association of early-onset myocardial infarction with single nucleotide polymorphisms and copy number variants. *Nat Genet* **41**, 334–41.
59. Zheng, W., Long, J., Gao, Y. T., Li, C., Zheng, Y., *et al.* (2009) Genome-wide association study identifies a new breast cancer susceptibility locus at 6q25.1. *Nat Genet* **41**, 324–8.
60. Collas, P., and Dahl, J. A. (2008) Chop it, ChIP it, check it: the current status of chromatin immunoprecipitation. *Front Biosci* **13**, 929–43.
61. Imbeaud, S., and Auffray, C. (2005) 'The 39 steps' in gene expression profiling: critical issues and proposed best practices for microarray experiments. *Drug Discov Today* **10**, 1175–82.

Chapter 8

Genome-Wide Methylation Profiling in Archival Formalin-Fixed Paraffin-Embedded Tissue Samples

J. Keith Killian, Robert L. Walker, Sven Bilke, Yidong Chen, Sean Davis, Robert Cornelison, William I. Smith, and Paul S. Meltzer

Abstract

New technologies allow for genome-scale measurement of DNA methylation. In an effort to increase the clinical utility of DNA methylation as a biomarker, we have adapted a commercial bisulfite epigenotyping assay for genome-wide methylation profiling in archival formalin-fixed paraffin-embedded pathology specimens. This chapter takes the reader step by step through a biomarker discovery experiment to identify phenotype-correlated DNA methylation signatures in routine pathology specimens.

Key words: Biomarker, DNA methylation, Epigenetics, FFPE, Pathology

1. Introduction

Epigenetic modifications include covalent and noncovalent modifications that map to the primary DNA sequence, a subset of which are established and maintained with cell-lineage specificity (1). Cytosine 5-methylation is a vital cellular process that requires DNA methyltransferase enzymes; it is the only known physiologic covalent modification of DNA, and some regard 5-methyl-cytosine (5mC) as a fifth nucleotide, along with adenine (A), cytosine (C), guanine (G), and thymine (T). Cytosine methylation adds significant value to the genome for biomarker discovery because uniform patterns of cytosine methylation can be identified which correlate with cell phenotype (neuron vs. lymphocyte) and biologic potential (benign vs. malignant). Thus, DNA methylation patterns are akin to histologic and immunophenotypic cytologic and histologic classifiers.

Virginia Espina and Lance A. Liotta (eds.), *Molecular Profiling: Methods and Protocols*, Methods in Molecular Biology, vol. 823, DOI 10.1007/978-1-60327-216-2_8, © Springer Science+Business Media, LLC 2012

In addition, in contrast with RNA and protein, DNA methylation markers appear to be significantly more stable within, and recoverable from, diagnostic specimens collected in current medical practice.

DNA methylation profiling is capable of identifying phenotype-specific signatures for both normal and cancer tissues (Fig. 1) (2–5). Such biomarkers are readily recoverable on a large scale from archival formalin-fixed paraffin-embedded (FFPE) samples. Large-scale retrospective clinical studies can be performed on archival patient samples to identify DNA methylation correlates of clinical endpoints, and further the goal of patient-tailored cancer diagnosis and treatment.

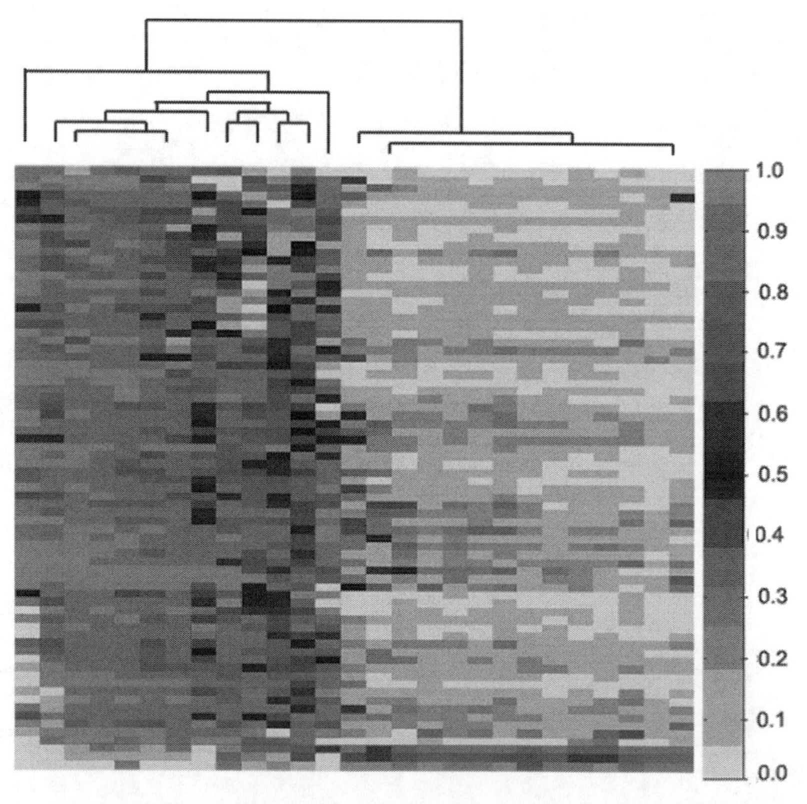

Fig. 1. Representative heatmap of differential methylation between cancer and normal archival lymph node tissues. BeadStudio (Illumina) targets were pre-selected based on group average difference >50% methylation between benign and malignant groups, and then unsupervised hierarchical clustering (Euclidean) was performed on all individual samples. All cancers segregated to the left main branch, and all benign samples to the right.

2. Materials

2.1. Human Subjects Research Approval

Archival FFPE pathology samples: tissue blocks and matched H&E stained slides.

2.2. Tissue Blocks and Slides

Microscope (Optional: digital slide scanner).

Sharpie-type pen to mark slides and block.

2.3. Sample Acquisition

20-gauge needle or 0.6 mm Tissue MicroArray (TMA) coring needle.

0.2 mL tubes or 96-well microtiter plate.

2.4. Dewaxing Tissue

Xylene or Xylene substitute (Histochoice, Sigma).

50% (v/v) Ethanol:Water (DNAase-free water).

Centrifuge with 96-well plate adaptor.

2.5. Sample Lysis and Genomic DNA Release

High pH antigen retrieval solution (ARS) (Dako target retrieval solution).

Tissue lysis buffer (Qiagen ATL buffer with proteinase K): 120.0 μL Qiagen ATL buffer plus 30.0 μL Qiagen proteinase K (ratio of 80:20 ATL buffer:proteinase K for a final lysis volume 150 μL).

Proteinase K.

Heat blocks at 65°C, 105°C, and 37°C.

NanoDrop UV spectrophotometer or spectrophotometer (260/280 nm) and cuvettes.

PicoGreen dsDNA fluorometric assay (Invitrogen).

Fluorescence plate reader.

2.6. Sodium Bisulfite Mutagenesis of DNA

Bisulfite DNA conversion kit (Zymo Research or Qiagen).

Oligonucleotide A, G, T-primer sequence.

Ethanol (molecular biology grade 96–100%).

2.7. bisDNA Quantitation

OliGreen® reagent (Invitrogen).

TE Buffer: 10 mM Tris–HCl, 1 mM EDTA, pH 7.5.

Sterile, distilled, DNase-free water.

Fluorescence plate reader (excitation 480 nm, emission 520 nm).

Plastic tubes.

96-Well microtiter plate.

SpeedVac Concentrator (Thermo).

2.8. Methylation Analysis

GoldenGate methylation kit (Illumina, Inc.) (see Note 1).

BeadArray reader, iScan, or HiScanSQ (Illumina, Inc.).

3. Methods

3.1. Human Subjects Research Approval

1. Obtain approval for human subjects research from the local Institutional Review Board (IRB). One potential advantage of archival pathology material for biomarker research is that anonymized materials, from which patient identifiers are removed, may be exempted from IRB review (6).

3.2. Acquire Formalin-Fixed Paraffin-Embedded Tissue and H&E-Stained Slides

1. Obtain FFPE tissue blocks and slides corresponding to the samples to be analyzed.
2. If possible, obtain matched tissue blocks with extant pathology slides, to avoid unnecessarily recutting the tissue block (see Notes 2 and 3).
3. Review the slides to verify the archival sample pathologic classification. Optimal study design usually requires histologic control/verification of the samples (see Note 4).
4. It may not be possible to have histologic control for specimens. In such nonideal cases, a surrogate molecular marker for percent tumor cellularity may be used, such as a genotyping assay capable of semiquantitative measurement of DNA nucleotide mutation or DNA copy number abnormalities. A genome-scale analysis of DNA copy number alterations often can support a pathologic diagnosis as well (7–9).

3.3. Pathology Review and Sample Procurement

1. Review the set of pathology slides for regions of relevant pathology, and mark the areas of interest. If slides have been scanned, digital pathology review is possible over the internet, which permits collaboration with expert pathologists at remote locations. The region of the FFPE block corresponding to the designated region on the slide is then located, and may be dotted with a Sharpie pen.
2. Use a 20-gauge needle or a TMA (tissue microarray) coring needle, to take a targeted micropunch from the FFPE block and eject it into a recipient microplate well or microcentrifuge tube (see Note 5). The precise number of cores taken from the tissue block to obtain sufficient DNA for array analysis depends on the tissue cellularity, age of the block, and needle diameter, and needs to be empirically determined (see Note 6).
3. Proceed with block micropunches/tissue procurement until the complete set of tissue samples has been obtained, as all samples will be processed together during subsequent steps.

3.4. Sample Dewaxing

1. Dewax all the cores after all the blocks have been cored and prior to tissue lysis. Fill the sample microcentrifuge tubes/ microtiter plate with liquid dewaxing reagent (xylene or xylene-substitute) (i.e., Histochoice clearing agent), and incubate for 20 min at 60–70°C (see Note 7).
2. After adding dewaxing reagent, spin the tubes/plate in a centrifuge to pellet the tissue.
3. Discard the supernatant containing dissolved wax.
4. Wash the tissue with 50% ethanol/water solution, spin the tubes/plate in a centrifuge to pellet the tissue, and discard the supernatant.

3.5. Sample Lysis and Genomic DNA Release

1. Add 50 μL of 1× antigen retrieval solution (Dako, ARS) to each sample (using a multichannel pipette if samples are in a plate format), and heat the samples at 105°C for 20 min. We do not consider the ARS step optional when extracting gDNA from FFPE samples (see Note 8).
2. After the ARS incubation, cool the plate to 37°C and add 100 μL of freshly mixed tissue lysis buffer (30 μL proteinase K plus 120 μL Qiagen ATL buffer) directly to the samples containing ARS. At this point, the tissue cores are in a 50:80:20 solution of ARS:ATL buffer: proteinase K.
3. Incubate samples overnight at 65°C in a heat block or thermomixer.
4. Next day, quick-spin the samples in the centrifuge to collect any evaporate from the lid, and add an additional 10.0 μL proteinase K (see Note 9).
5. Add additional proteinase K as needed every 6 h until tissue is completely lysed (see Note 10). Usually within 1 h of the first proteinase K booster the tissues will be completely lysed, without visually detectable tissue fragments.
6. Analysis of gDNA yield and quality: Use both NanoDrop UV absorbance (or another spectrophotometric method) and PicoGreen dsDNA fluorometry (Invitrogen) to measure DNA levels in the lysates (see Notes 11 and 12).

3.6. Sodium Bisulfite Mutagenesis of DNA

The sodium bisulfite conversion is a chemical modification of DNA that results in the conversion of unmethylated cytosine to uracil, while methyl-cytosine remains unmodified (10). This chemical reaction is not impeded by impurities in typical tissue lysates, and so the lysates do not need to be further purified prior to bisulfite conversion in our experience. We have used both Zymo and Qiagen bisulfite modification kits with comparable results. Thus, the tissue lysates can be directly inoculated into the bisulfite conversion reaction, using at least 500 ng gDNA. The bisulfite modification is carried out in a 96-well plate using a single batch of freshly prepared bisulfite reagent, prepared according to the kit directions. This method minimizes intersample variability in bisulfite conversion.

1. Use the following thermal incubation steps for the sodium bisulfite conversion: 98°C × 10 min (to denature the DNA into single strands, as the bisulfite conversion requires ssDNA), 64°C × 2.5 h, and ending with room temperature soak. If the samples have not been dewaxed, it is preferable to end at room temperature to prevent solidification of the paraffin wax.
2. Apply the samples to the DNA binding columns that come with the bisulfite kit, in the presence of supplied binding buffer.
3. Either centrifuge the columns. As a greater filtration force can be achieved by centrifuge, this is a more effective method for samples that contain carryover wax.
4. Wash the bound DNA samples once with an ethanol-based supplied reagent, filter by centrifuge, and then apply the provided desulfonation reagent as directed.
5. Wash the samples again and elute the bisulfite DNA from the columns in elution buffer. Elution efficiency may be improved by heating the elution buffer, and incubating for 5 min, prior to centrifugation to collect the eluant.

3.7. Measure Yield of Bisulfite Converted DNA with OliGreen® Reagent

At this point, one has a 96-well plate of purified bisDNA samples, where greater than 99% of unmethylated cytosines have been converted to uracil. The next step is to determine the bisDNA yield of the procedure. The target goal is 250 ng for input into the Illumina GoldenGate methylation assay (see Note 1).

After the bisulfite conversion, DNA is no longer self-complementary, and is essentially single stranded. Thus, the PicoGreen reagent cannot be used to assess DNA quantity (see Note 12). We have used the OliGreen reagent, which is similar to PicoGreen except it binds and fluoresces upon contact with ssDNA (11). A critical detail for measurement with these fluorescent reagents is the use of an appropriate control DNA for generation of a standard curve to which the unknown samples can be mapped for concentration determination. The OliGreen reagent binds differentially to the different nucleotides in DNA; the presence of uracil in the bisulfite converted DNA relative to DNA standards could also have a confounding effect on the sample measurement. Because the relative amounts of uracil, cytosine, and thymine in bisDNA are cell-lineage and phenotype-dependent, and OliGreen differentially binds these bases, this creates a real theoretical concern for absolute quantification of bisDNA using OliGreen (see Note 13).

We have compared concentration measurements based upon two different oligonucleotide standards, one composed of A, G,

and C, the other composed of A, G, and T. The latter DNA standard gives a lower estimate of sample DNA concentration by as much as one order of magnitude, consistent with the greater fluorescence response of OliGreen to T than C. We err on the side of overestimation of DNA concentration for the Illumina methylation assay, and therefore base our calculation of bisDNA yield on a standard curve generated using an A, G, T-primer (see Note 14). Regarding bisDNA stability, samples stored at –20°C have remained amplifiable by PCR for at least 1.5 years in our experience.
Preparation of working OliGreen Reagent (Invitrogen)

1. Dilute OliGreen reagent 1:200 in TE buffer (10 mM Tris–HCl, 1 mM EDTA, pH 7.5). 10 µL OliGreen plus 1.99 mL TE buffer (see Note 15). Use plastic tubes because OliGreen may bind to glass. Reagent is sufficient for analyzing 20 samples. Protect from light. Use within 3 h of preparation.

Prepare oligonucleotide standard curve

1. Prepare a standard curve of an oligonucleotide primer, such as the 18-mer M13 sequencing primer provided with the OliGreen® reagent (Invitrogen).
2. Dilute the oligonucleotide primer to 2 µg/mL in TE buffer.
3. Verify the concentration of the oligonucleotide stock. Measure the absorbance at 260 nm (A_{260}) in a cuvetter with a 1.0-cm pathlength. A_{260} 1.0 is equivalent to 30–35 µg/mL oligonucleotide.
4. Prepare oligonucleotide standards in the range of 100 pg/mL to 50 ng/mL. Dilute the stock 2 µg/mL oligonucleotide 1:20 to make a 100-ng/mL oligonucleotide solution. Prepare additional standards and blank (Table 1).

Table 1
Oligonucleotide standard preparation for quantifying bisDNA (ssDNA)

Volume (µL) of 100 ng/mL oligonucleotide solution	Volume (µL) of TE buffer	Final oligonucleotide concentration in OliGreen assay
1,000	0	50 ng/mL
100	900	5 ng/mL
10	990	500 pg/mL
2	998	1,000 pg/mL
0	1,000	Blank

5. Add 100 μL of each standard or blank to a 96-well microtiter plate.
6. Add 100 μL OliGreen® reagent to each well.
7. Incubate 2–5 min at room temperature. Protect from light.
8. Measure fluorescence with a microplate reader (excitation 480 nm, emission 520 nm) (see Note 16).
9. Subtract the fluorescence of the blank from the fluorescence of each standard.
10. Create a standard curve of fluorescence (*y*-axis) versus oligonucleotide concentration (*x*-axis).

Measure absorbance of samples

1. Dilute samples in 100 μL TE buffer in a 96-well microtiter plate. At least two different dilutions should be tested.
2. Add 100 μL OliGreen® reagent to each well.
3. Incubate 2–5 min at room temperature. Protect from light.
4. Measure fluorescence with a microplate reader (excitation 480 nm, emission 520 nm).
5. Subtract the fluorescence of the blank from the fluorescence of each sample.
6. Read the concentration of the samples from the standard curve prepared in Subheading 3.7, step 2.

3.8. Illumina GoldenGate Methylation Assay

Technical details of Illumina's GoldenGate methylation assay have been described previously (12). For the Illumina methylation assay, we follow the manufacturer's instructions without modifications (see Note 1). We recently reported that matched pairs of frozen and FFPE lymph node tissues yielded essentially identical results for differential methylation profiling (13).

1. Add 250 ng of bisDNA in a volume of 5 μL (50 ng/μL) for the bisulfite genotyping assay (see Note 17).
2. bisDNA may be diluted with water or concentrated in a SpeedVac Concentrator to obtain the necessary concentration for downstream applications.

3.8.1. Data Analysis

Bisulfite sequencing and genotyping assays provide a measure of C/T ratio with nucleotide-level resolution in the genome. Each DNA allele present in the sample will have an equal representation in the methylation measurement at each target in the assay. Pathologic hypermethylation will tend toward an increased C/T ratio relative to a reference normal sample of otherwise similar cellular composition.

Several points should be considered regarding the analysis of methylation data in tissue samples, which should also be considered at the experimental design stage of the project. For one, in the case of cancer, normal tissue architecture is typically effaced by an increased population of malignant cells. Thus, cancer tissues contain compositional enrichment and depletion of various cell-lineage constituents, in addition to de novo presence of malignant cells. Thus, differential methylation measured in a cancer may be due to alteration in the mixture of normal cells, as well as pathologic departures from normal lineage states. Still, we have not found it necessary to employ tissue fractionation, such as laser capture microdissection (LCM) (14), for example, in order to detect a robust number of differential methylations in the majority of our experiments.

The precise quantitation of methylation levels in cancer cells may benefit from such fractionation, but the detection of differential methylation relative to a reference normal tissue sample from the same tissue site is still achievable. One exception is in the case of heavily inflamed tumor specimens, where a normal lymphoid signature may dominate the tissue epigenetic profile. The rule of thumb is one cell, one vote, so if a requirement of 20% differential methylation difference is the cutoff, one should take care that the tumor sample contains at least 20% tumor cell nuclei. One advantage of DNA methylation as a biomarker, as opposed to gene expression and proteomic biomarkers, is that the one cell, one vote rule diminishes the influence of outlier high-amplitude biomarkers in individual samples.

4. Notes

1. The Illumina GoldenGate Methylation Assay is being replaced by the 450K Infinium Methylation Assay. The concepts presented herein are applicable for both assays. The reader should refer to the vendor's website for the most current protocols (www.illumina.com).
2. Recutting these older blocks often requires wasteful resurfacing, and strips the block of its outer coat of more oxidized paraffin that is potentially protective of deeper tissue biomolecules.
3. We make a digital image archive of research pathology sections using a slide scanner, therefore we are able to return all original slides to pathology departments when requested.
4. Histologic control is important because of potential regional tissue heterogeneity, including lymphoid infiltrates and percent tumor cellularity; moreover, anatomic pathology classification guidelines may have changed since the original diagnosis of the specimen, and histologic review allows confirmation of the original diagnosis. Finally, it is not uncommon for specimens

to be inadvertently incorrectly archived and/or retrieved, and pathology review often resolves such issues when they arise.

5. We prefer to eject the cores directly into a microplate, to facilitate subsequent pipetting steps by permitting a multichannel pipetter to be used and eliminating the need to open and close multiple tubes; however, greater care needs to be taken using this approach than with individual tubes, as the minute cores may more easily be inadvertently dropped or placed in the wrong well of the plate.
6. Low gDNA yield from tissues: The amount of tissue required to obtain sufficient DNA needs to be determined empirically. Highly cellular tissues where the cells have a high nuclear:cytoplasmic ratio, such as lymph node, yield larger amounts of DNA than low-cellularity tissues such as normal breast. FFPE samples typically yield an order of magnitude less DNA than an equivalent amount of fresh tissue. Lymph nodes, for example, yield microgram amounts of DNA from a single 0.6 mm core, while normal breast tissue may require several larger cores to obtain a microgram of DNA. In the case of extremely minute biopsies (gastrointestinal biopsies, for example) we occasionally melt the FFPE block and retrieve the entire tissue specimen for molecular analysis.
7. While dewaxing the cores will facilitate subsequent DNA purification after the bisulfite treatment and may modestly improve total DNA yield, tissue lysis can be achieved without dewaxing the tissue core. The DNA-binding filter plates used to purify the bisulfite-modified DNA tend to become clogged by carryover wax, and the time spent dewaxing the cores is partially recuperated during subsequent steps.
8. Antigen retrieval solutions combined with heat have been shown to improve the reversal of formalin cross-links, and appear to improve the recovery of gDNA from archival pathology specimens (15). This step only adds about 30 min to the tissue lysis process, and samples do not need to be re-purified from the ARS before proceeding to tissue lysis.
9. One advantage of the Qiagen proteinase K is that it can be stored at room temperature, and beyond that somewhat trivial advantage, we have usually obtained excellent tissue lysis with the Qiagen reagents.
10. Hairs or other nondigestable materials may persist, and are not a problem. A total of 24–48 h of digestion is almost always sufficient for complete tissue dissolution.
11. Ideally a determination of DNA yield is made at this point, after tissue lysis and before bisulfite modification. Most commercial kits for bisulfite conversion specify an optimal amount of DNA to convert with their reagents. However, because of

all the impurities in the lysates – which may include residual paraffin wax, proteinaceous material, RNA, antigen retrieval solution and lysis buffers, and other debris – it may be difficult to get an accurate measurement of DNA quantity.

12. Unexpected UV absorbance profile of bisDNA: the 260/280 ratio of bisDNA is more like that of an RNA sample. PicoGreen fluoresces when bound to dsDNA. The PicoGreen assay tends to estimate significantly lower DNA quantity relative to NanoDrop; reasons for the discrepancy include detection of RNA by NanoDrop (agarose gels of tissue lysates often show an RNAse-labile fraction), and the selection of a suitable reference DNA standard for PicoGreen (use of lambda DNA as a reference as opposed to human placental DNA will decrease the PicoGreen-measured DNA test sample concentration by 50%, and theoretically, cancer associated microsatellite instability, copy number changes, and other mutations will subvert the accuracy of using even a human-derived DNA reference). Thus, the exact determination of DNA amount in small tissue lysates is problematic. Fortunately, precise determination of DNA yield from the tissue lysate is not necessary prior to bisulfite conversion; we typically proceed directly with bisulfite modification of tissue lysates prior to assessing DNA yield. Some investigators may wonder about putting too much DNA into the bisulfite reaction, as the manufacturer Zymo, for example, suggests using up to 2 μg DNA; we have converted ten times that amount with excellent conversion and sample performance on downstream applications.
13. Other bisDNA standards could be employed, such as a commercially purchased DNA sample of known quantity that is bisulfite modified in parallel with experimental samples; however, such commercial-quality DNA may provide different bisDNA yields from FFPE samples. As an alternative to fluorescence measurements of bisDNA, we have also used quantitative real-time bisulfite PCR to determine the amount of amplifiable bisDNA in the sample and predict success in downstream applications. The real-time bisulfite PCR approach is of course more expensive and time consuming than fluorescence measurement.
14. A spectrophotometer such as NanoDrop can be used to measure the bisDNA yield; it must be remembered that bisDNA contains a significant proportion of uracil and reduction in cytosine content, so the 260/280 ratio will be closer to an RNA sample than a gDNA sample, that is, 2.4 rather than 1.8. The 260/280 ratio should be approximately 2.4 for bisDNA. Convert the absorbance value to concentration. Overall, the quantitation of bisDNA samples requires careful contemplation, and various approaches can be used.
15. Use sterile, nuclease-free distilled water to prepare the TE buffer.

16. Set the gain on the fluorescent plate reader so the most concentrated standard has the maximum fluorescence. This allows greater sensitivity with the most dilute standards/samples. Keep the time of measurement consistent between all samples.
17. Insufficient labeling/poor performance of bisulfite-treated DNA in Illumina assay: whenever we have input 250 ng of bisulfite-modified DNA into the Illumina assay based on OliGreen measurement, we obtain passing fluorescent intensities on the bead array hybridizations. There are multiple controls included in the GoldenGate methylation assay that monitor parameters including bisulfite conversion efficiency, input DNA amount, sample contamination, and hybridization processes; these controls help identify and troubleshoot possible sample failures.

References

1. Gama-Sosa, M. A., Midgett, R. M., Slagel, V. A., Githens, S., Kuo, K. C., Gehrke, C. W. et al. (1983) Tissue-specific differences in DNA methylation in various mammals. *Biochim Biophys Acta* **740**, 212–9.
2. Yoo, C. B., Jones, P. A. (2006) Epigenetic therapy of cancer: past, present and future. *Nat Rev Drug Discov* **5**, 37–50.
3. Steiner, I., Jung, K., Schatz, P., Horns, T., Wittschieber, D., Lein, M. et al. (2010) Gene promoter methylation and its potential relevance in early prostate cancer diagnosis. *Pathobiology* **77**, 260–6.
4. Kron, K., Pethe, V., Briollais, L., Sadikovic, B., Ozcelik, H., Sunderji, A. et al. (2009) Discovery of novel hypermethylated genes in prostate cancer using genomic CpG island microarrays. *PLoS One* **4**, e4830.
5. Kron, K. J., Liu, L., Pethe, V. V., Demetrashvili, N., Nesbitt, M. E., Trachtenberg, J. et al. (2010) DNA methylation of HOXD3 as a marker of prostate cancer progression. *Lab Invest* **90**, 1060–7.
6. Williams, J. R. (2008) The Declaration of Helsinki and public health. *Bull World Health Organ* **86**, 650–2.
7. Quail, M. A., Kozarewa, I., Smith, F., Scally, A., Stephens, P. J., Durbin, R. et al. (2008) A large genome center's improvements to the Illumina sequencing system. *Nat Methods* **5**, 1005–10.
8. Zheng, W., Long, J., Gao, Y. T., Li, C., Zheng, Y., Xiang, Y. B. et al. (2009) Genome-wide association study identifies a new breast cancer susceptibility locus at 6q25.1. *Nat Genet* **41**, 324–8.
9. Kathiresan, S., Voight, B. F., Purcell, S., Musunuru, K., Ardissino, D., Mannucci, P. M. et al. (2009) Genome-wide association of early-onset myocardial infarction with single nucleotide polymorphisms and copy number variants. *Nat Genet* **41**, 334–41.
10. Frommer, M., McDonald, L. E., Millar, D. S., Collis, C. M., Watt, F., Grigg, G. W. et al. (1992) A genomic sequencing protocol that yields a positive display of 5-methylcytosine residues in individual DNA strands. *Proc Natl Acad Sci U S A* **89**, 1827–31.
11. Borin, M., Siffert, W. (1990) Stimulation by thrombin increases the cytosolic free Na+concentration in human platelets. Studies with the novel fluorescent cytosolic Na+indicator sodium-binding benzofuran isophthalate. *J Biol Chem* **265**, 19543–50.
12. Bibikova, M., Lin, Z., Zhou, L., Chudin, E., Garcia, E. W., Wu, B. et al. (2006) High-throughput DNA methylation profiling using universal bead arrays. *Genome Res* **16**, 383–93.
13. Killian, J. K., Bilke, S., Davis, S., Walker, R. L., Killian, M. S., Jaeger, E. B. et al. (2009) Large-scale profiling of archival lymph nodes reveals pervasive remodeling of the follicular lymphoma methylome. *Cancer Res* **69**, 758–64.
14. Espina, V., Wulfkuhle, J. D., Calvert, V. S., VanMeter, A., Zhou, W., Coukos, G. et al. (2006) Laser-capture microdissection. *Nat Protoc* **1**, 586–603.
15. Shi, S. R., Cote, R. J., Wu, L., Liu, C., Datar, R., Shi, Y. et al. (2002) DNA extraction from archival formalin-fixed, paraffin-embedded tissue sections based on the antigen retrieval principle: heating under the influence of pH. *J Histochem Cytochem* **50**, 1005–11.

Chapter 9

An Overview of MicroRNA Methods: Expression Profiling and Target Identification

Sinéad M. Smith and David W. Murray

Abstract

MicroRNAs (miRNAs) are small, single-stranded RNA molecules encoded by genes that are transcribed from DNA but not translated into protein (noncoding RNA). The ability of miRNA to regulate the expression of, as yet, an unknown quantity of targets has recently become an area of huge interest to researchers studying many different areas in many species. Identifying miRNA targets provides functional insights and strategies for therapy. Furthermore, the recent advent of high-throughput methods for profiling miRNA expression and for the identification of miRNA targets has ushered in a new era in the research of gene regulation. miRNA profiling further adds a new dimension of information for the molecular profiling of disease. Summarized herein are the methods used to query the expression of miRNAs at both an individual and global level. We have also described modern computational approaches to identifying miRNA target transcripts.

Key words: MicroRNA, Expression profiling, Microarray analysis, Quantitative, PCR *In silico*

1. Introduction

MicroRNAs (miRNAs) are small, highly conserved, single-stranded RNA molecules of approximately 21–23 nt in length. miRNAs are encoded by genes that are transcribed from DNA but not translated into protein (noncoding RNA). miRNAs regulate gene expression by binding to the 3′-untranslated regions (3′ UTR) of specific messenger RNAs (mRNAs). Although miRNAs were first described in 1993 (1), it is only in recent years that the diversity and significance of this class of regulatory molecule has been fully appreciated. Several hundred miRNAs have been cloned and sequenced to date. Over 300 different miRNA genes have been identified in the human genome alone, comprising 1–4% of all

Virginia Espina and Lance A. Liotta (eds.), *Molecular Profiling: Methods and Protocols*, Methods in Molecular Biology, vol. 823, DOI 10.1007/978-1-60327-216-2_9, © Springer Science+Business Media, LLC 2012

expressed genes (2), and more that 1,000 predicted miRNA genes are awaiting experimental confirmation (3). miRNA sequences share a great deal of homology among organisms, implying that they represent an important regulatory pathway (4).

The processing of active single-stranded miRNAs from precursor transcripts is a multistep process. miRNAs are first transcribed by RNA polymerase II as large RNA primary transcripts (pri-miRNA) (5) with a cap and poly-A tail (6, 7). The pri-miRNAs are processed into ~70-nt stem-loop structures known as pre-miRNA by the nuclease Drosha and the double-stranded RNA-binding protein Pasha (8). Exportin-5 transports the pre-miRNAs to the cytoplasm where the endonuclease Dicer excises a double-stranded RNA of ~22 nt in length from the pre-miRNA. This short RNA duplex is subsequently incorporated into the RNA-induced silencing complex (RISC), which is responsible for the gene silencing. The mature miRNA strand is preferentially retained in the RISC complex, where it base pairs with its complementary mRNA molecule, resulting in the induction of mRNA degradation, thereby negatively regulating gene expression (2). miRNA molecules therefore represent a novel class of gene regulator.

Bioinformatic data indicate that a single miRNA may bind up to 200 gene targets of varying function. Thus, miRNAs potentially control the expression of one third of human mRNAs, highlighting the potential influence of miRNAs on almost every genetic pathway (2). Several studies have provided evidence that miRNAs may act as key regulators of development (9, 10), cell proliferation, apoptosis (11, 12), and cell differentiation (13, 14). Consequently, dysregulated expression or deletion of miRNAs, or indeed components of the miRNA processing pathway, have been implicated in many diseases. Since numerous miRNA sequences have already been identified, a focus has developed on miRNA expression levels between different tissues, developmental stages, or disease states (15). Expression profiling has implicated miRNAs in various cancers including, B-cell chronic lymphocytic leukemia (16), Burkitt's lymphoma (17), breast cancer (18), Barrett's esophagus (19), colorectal cancer (20), liver cancer (10, 21, 22), and lung cancer (23–25). miRNA genes are thought to function as both tumor suppressors and oncogenes (26, 27). From a clinical standpoint, the differential expression of certain miRNAs in various diseases has the potential to reveal them as diagnostic markers or therapeutic targets (28–30). miRNA expression profiling may therefore become a valuable tool to aid diagnosis and treatment (31, 32).

There has been a rapid advancement in the analysis platforms available for miRNA profiling in recent years. Various companies, including Affymetrix, Applied Biosystems, Invitrogen, and

Illumina, have developed products to robustly detect and quantify expression of miRNAs. These products allow for detection of specific miRNAs, or global miRNA expression profiling in a variety of species using microarray technology or high-throughput quantitative PCR (qPCR). The platform of choice for an investigator depends on the particular aims of the experiment, as well as the budget and access to specific equipment within a given laboratory. Given the availability of excellent commercially available kits and reagents, this chapter describes an example method for miRNA isolation and expression profiling techniques incorporating reverse transcription (RT) and qPCR technology, and is designed to be used in conjunction with the user manuals provided with the commercially available kits (33–35). In addition, computational tools for miRNA sequence and target identification are discussed. These databases provide an excellent resource for today's researcher involved in the field of miRNA analysis.

2. Materials

2.1. Disruption of Tissue or Cultured Cells and Extraction of Total RNA

1. Fresh, frozen, or RNA*later*-stored mammalian tissue or cultured cells.
2. RNA*later* (Ambion P/N 7020). Store at room temperature. Optional reagent – only required if tissue/cell samples are to be stored for processing at a future time.
3. RNase*Zap* solution (Ambion P/N 9780). Store at room temperature.
4. Trypsin – only required for adherent cell line samples.
5. *mir*Vana miRNA Isolation Kit Protocol (34) (Ambion P/N 1560M, Revision C 01/2011).
6. *mir*Vana miRNA Isolation Kit containing acid-phenol:chloroform (Ambion P/N 1560): miRNA wash solution 1, wash solution 2/3, collection tubes, filter cartridges, lysis/binding buffer, miRNA homogenate additive, acid-phenol:chloroform, gel loading buffer, and elution solution. The lysis/binding buffer, miRNA homogenate additive and acid-phenol:chloroform are stored at 4°C. The gel loading buffer is stored at –20°C and all other kit components are stored at room temperature.
7. Nuclease-free 1.5-mL and 0.5-mL microfuge tubes and tips.
8. Mortar and pestle prechilled on dry ice or in liquid nitrogen.
9. Phosphate-buffered saline (PBS).

2.2. Isolation of Total RNA (Containing the miRNA Fraction)

1. *mir*Vana miRNA Isolation Kit Protocol (34) (Ambion P/N 1560M, Revision C 01/2011).
2. *mir*Vana miRNA Isolation Kit containing acid-phenol:chloroform (Ambion P/N 1560): miRNA wash solution 1, wash solution 2/3, collection tubes, filter cartridges, lysis/binding buffer, miRNA homogenate additive, acid-phenol:chloroform, gel loading buffer and elution solution. The lysis/binding buffer, miRNA homogenate additive and acid-phenol:chloroform are stored at 4°C. The gel loading buffer is stored at –20°C and all other kit components are stored at room temperature.
3. Nuclease-free 1.5-mL and 0.5-mL microfuge tubes and tips.
4. 100% ethanol, ACS grade. Room temperature.
5. Nuclease-free water.

2.3. Quantifying Total RNA

1. UV spectrophotometer with cuvettes.
2. Nuclease-free water.

2.4. Detection and Quantifying Individual miRNAs

1. TaqMan® Small RNA Assays Protocol (Applied Biosystems P/N 4364031E, Revision E) (35).
2. Inventoried TaqMan® microRNA Assay for miRNA of interest (Applied Biosystems P/N 4427975). Store at –20°C and protect from light.
3. TaqMan® microRNA Reverse Transcription Kit (Applied Biosystems P/N 4366596). Store at –20°C.
4. Nuclease-free 1.5-mL and 0.2-mL microfuge tubes.
5. TaqMan® Universal PCR Master Mix (Applied Biosystems P/N 4324018). Store at 4°C.
6. Endogenous control assay, e.g., *Homo sapiens* mature miRNA control (P/N 4427975, assay ID 001093 Applied Biosystems). Store at –20°C.
7. Nuclease-free water.
8. MicroAmp Optical 384-well plate (Applied Biosystems P/N 4309849).
9. MicroAmp Optical Adhesive Film (Applied Biosystems P/N 4311971).
10. Applied Biosystems 7900HT Thermal Cycler.

2.5. High-Throughput Expression Profiling of miRNAs

1. Megaplex Pools for microRNA Expression Analysis Protocol (33) (Applied Biosystems P/N 4399721C, Revision C 07/2010, by Life Technologies, Carlsbad, CA, USA).
2. TaqMan microRNA Reverse Transcription Kit (Applied Biosystems P/N 4366596). Store at –20°C.

3. Megaplex RT Primers, Human Pool Set (Pool A and Pool B) v3.0 (Applied Biosystems P/N 4444745). Store at −20°C.
4. Nuclease-free 1.5-mL microfuge tubes and tips.
5. MicroAmp Optical 96-well Reaction Plates (Applied Biosystems P/N N8010560). Store at room temperature.
6. MicroAmp Clear Adhesive Film (Applied Biosystems P/N 4306311).
7. Applied Biosystems 7900HT Thermal Cycler.
8. TaqMan® Universal PCR Master Mix (Applied Biosystems P/N 4324018). Store at 4°C.
9. Nuclease-free water.
10. TaqMan® Arrays Human microRNA A and B Cards Set v3.0 (Applied Biosystems P/N 4444913). Store at 4°C.

3. Methods

Global miRNA expression by microarray analysis is a widely used high-throughput technique for comparing differential miRNA expression between control and disease tissues, and in cell line models. Specific detection and quantitation of an miRNA of interest may be carried out using qRT-PCR, which is also an important tool used to validate changes in miRNA expression identified during microarray analysis. In order to identify the targets of miRNAs many computational approaches have been developed. Furthermore, databases have been assembled which include extensive information on known miRNAs and therefore provide an excellent resource for today's researcher (36–40).

The *mir*Vana miRNA Isolation kit allows for the purification of total RNA suitable for many miRNA analysis techniques (34). This kit employs organic extraction and RNA mobilization on glass fiber filters to purify either total RNA or RNA enriched for small species (≤200 nt), depending on the particular requirements of the investigator. The purified total RNA described here may be used for qRT-PCR analysis of specific miRNA targets (see Subheading 3.4) or for use in high-throughput expression profiling of miRNA (see Subheading 3.5).

3.1. Disruption of Tissue and Cultured Cells and Extraction of Total RNA

3.1.1. Tissue

1. Harvest tissue and remove any extraneous material. Rinse with cold PBS to remove red blood cells.
2. Cut the tissue into smaller pieces for either storage or disruption.
3. For storage, inactivate RNases by placing the tissues in liquid Nitrogen until frozen through. Subsequently, remove tissue

from the liquid nitrogen and store in an airtight container at −70°C. Alternatively, the sample may be placed directly in RNA*later* (see Note 1) and frozen at −70°C.

4. Measure the weight of either fresh or frozen tissue sample.
5. For frozen tissue, grind the tissue to a powder with a prechilled mortar and pestle sitting in dry ice. Place 10 volumes of lysis/binding buffer per tissue mass (i.e., 1 mL per 0.1 g tissue) into a plastic weigh boat on ice. Scrape the powdered frozen tissue into the lysis/binding buffer using a chilled metal spatula, and mix rapidly until all clumps are dispersed (see Note 2).
6. Alternatively, place fresh tissue directly into 10 volumes of lysis/binding buffer per tissue mass in a homogenization tube on ice, and thoroughly disrupt sample.

3.1.2. Cultured Cells

1. For suspension cells, centrifuge 10^2–10^7 cells at low speed and discard the supernatant (see Note 3). Resuspend the cells in 1 mL PBS and centrifuge. Discard supernatant and place the washed cells on ice.
2. For adherent cells, trypsinize the cells and count. Centrifuge cell suspension at low speed and wash pellet in 1 mL PBS. Centrifuge again, discard supernatant, and place the cells on ice.
3. For both cell types, add 300–600 μL lysis/binding solution per 10^2–10^7 cells.
4. Vortex the sample to lyse the cells, or pipette vigorously up and down.

3.1.3. Organic Extraction

1. Add 1/10 volume of miRNA homogenate additive to the tissue or cell lysate (e.g., Add 30 μL homogenate additive to 300 μL lysate). Vortex thoroughly to mix.
2. Leave the mixture on ice for 10 min.
3. Add a volume of acid-phenol:chloroform that is equal to the lysate volume prior to addition of the miRNA homogenate additive (e.g., if original lysate volume before adding the homogenate was 200 μL, add 200 μL acid-phenol:chloroform) (see Note 4). Vortex the sample for 1 min.
4. In order to separate the aqueous and organic phases, centrifuge at 10,000 × *g* for 10 min at room temperature. Following centrifugation, if the interphase layer is not compact, repeat the centrifugation step.
5. Carefully transfer the upper aqueous phase to a fresh tube without disturbing the lower phase and note the volume removed. The aqueous phase sample will be used in Subheading 3.2 to isolate total RNA.

3.2. Isolation of Total RNA (Containing the miRNA Fraction)

1. Prepare the wash buffers from the *mir*Vana miRNA Isolation Kit by adding 21 mL of 100% ethanol to miRNA wash solution 1 and 40 mL of 100% ethanol to wash solution 2/3.
2. Preheat 1-mL nuclease-free water to 95°C for recovering the final eluate (Subheading 3.2, step 12).
3. Add 1.25 volumes of 100% ethanol (room temperature) to the aqueous phase.
4. Isolate total RNA (see Note 5). Place a filter cartridge into a collecting tube for each sample. Transfer the lysate/ethanol mixture into the filter cartridge. For samples >700 μL, apply the mixture in successive steps to the same filter cartridge.
5. Centrifuge at 10,000 × *g* for 15 s at room temperature and discard to the flow-through. The collection tube may be reused during the washing steps.
6. Place 700 μL wash solution 1 in the filter cartridge and centrifuge for 10 s at 10,000 × *g* at room temperature.
7. Discard the flow-through and place the filter cartridge back into the same collection tube.
8. Add 500 μL wash solution 2/3 to the filter cartridge.
9. Centrifuge for 10 s at 10,000 × *g* at room temperature.
10. Discard the flow-through. Add 500 μL wash solution 2/3 and centrifuge for 10 s at 10,000 × *g* at room temperature.
11. Discard the flow-through. In order to remove any remaining fluid from the filter, place the filter cartridge back in the same collecting tube and centrifuge for 1 min at 10,000 × *g*. Discard the flow-through.
12. Transfer the filter cartridge to a new collection tube. Add 100 μL preheated (95°C) nuclease-free water to the center of the filter and close the cap (see Note 6).
13. Centrifuge for 30 s at 10,000 × *g* to recover the total RNA.
14. Remove 1–2 μL of the total RNA sample for quantifying RNA (Subheading 3.3) and place it into a clean nuclease-free tube.
15. Store the remaining eluate at –70°C until required for analysis.

3.3. Quantifying and Quality Assessment of Total RNA

1. In order to determine the concentration of RNA isolated dilute an aliquot of the total RNA sample 1:100 in nuclease-free water.
2. Measure the absorbance at 260 nm and 280 nm on a UV spectrophotometer, using a water sample to blank the spectrophotometer.

3. Based on the fact that a 40 μg/mL sample of pure RNA has an absorbance reading of 1 at 260 nm, calculate the concentration of the sample using the following equation:

$$\text{Concentration of RNA } (\mu g/\mu L) = \frac{\text{Absorbance (260 nm)} \times 40 \times 100 \text{ (dilution factor)}}{1000}$$

4. Calculate the ratio of the absorbance at 260 and 280 nm ($A_{260/280}$) to determine the quality of the preparation. A ratio between 1.8 and 2.1 is indicative of high-quality RNA, suitable for use in microarray experiments.

3.4. Detection and Quantifying Individual miRNAs

This protocol describes detection and quantitation of specific miRNAs using individual TaqMan® miRNA assays (35). This involves a reverse transcription (RT) reaction step in which cDNA is reverse transcribed from total RNA samples using a predesigned stem-looped primer and a sequence-specific TaqMan® qPCR assay to detect mature miRNAs. TaqMan® miRNA assays are available for the majority of miRNAs listed in the Sanger miRBase sequence repository (http://www.mirbase.org) (41–43). Each assay includes a tube containing an miRNA-specific RT primer (5×) and one tube containing a mix of miRNA-specific forward PCR primer, specific reverse PCR primer, and miRNA-specific TaqMan® probe (20×) (35).

3.4.1. Reverse Transcription

1. Thaw total RNA sample (see Note 7) prepared in Subheading 3.2, the miRNA-specific 5× RT primer (from TaqMan® miRNA Assay) and the components of the TaqMan® microRNA Reverse Transcription Kit on ice. Mix the tube contents by flicking the tubes and briefly centrifuge the tubes. 1–10 ng total RNA should be used per 15 μL RT reaction.
2. Prepare the RT Master Mix (see Note 8) in a 1.5-mL polypropylene tube with the components of the TaqMan® microRNA Reverse Transcription Kit (35). The total volume of Master Mix to be prepared is calculated with the formula: volume of Master Mix required = 7.0 μL × (number of reactions).
3. Master Mix for one RT reaction (see Note 8) (35): 0.18 μL 100 mM dNTPs, 1.2 μL MultiScribe™ Reverse Transcriptase (50 U/μL), 1.8 μL 10× reverse transcriptase buffer, 0.23 μL RNase inhibitor (20 U/μL), and 4.99 μL nuclease-free water.
4. Invert the master mix tube to mix and centrifuge briefly.
5. For each 15 μL RT reaction, mix 7 μL of RT Master Mix, 5 μL total RNA (containing 1–10 ng) and 3 μL miRNA-specific 5× RT primer in a 0.2-mL tube. Mix by inverting and centrifuge briefly. Seal the tube and place on ice for 5 min.

6. Place RT reaction tubes in a bench-top thermal cycler and run RT reaction at 16°C for 30 min, 42°C for 30 min, 85°C for 5 min, and 4°C indefinitely.
7. The resulting RT product may be stored at –20°C until further use.

3.4.2. Quantitative PCR

1. Prepare reagents for qPCR by thawing the RT product (from Subheading 3.4.1, step 7) and TaqMan® miRNA assay on ice, and inverting the TaqMan® Universal PCR Master Mix to mix.
2. Calculate the number of reactions per assay, including replicates. Also include an endogenous control (see Note 9) and nontemplate controls (see Note 10).
3. According to the calculated number of reactions for each assay, prepare the qPCR reaction mix in a 1.5-mL tube.
4. qPCR reaction mix (for 1 sample in triplicate): 3.60 μL TaqMan® Small RNA Assay (20×), 4.80 μL product from RT reaction, 36.0 μL TaqMan® Universal PCR Master Mix (2×), 27.61 μL nuclease-free water.
5. Mix the tube by inverting and centrifuge briefly.
6. Transfer 20 μL of the qPCR reaction mix into each well of a 384 well plate, seal with adhesive film, centrifuge briefly and load into the thermal cycler.
7. Run the qPCR program on standard mode, 20 μL volume, enzyme activation 95°C 10 min, 40 cycles of PCR 95°C 15 s, and anneal/extend 60°C 60 s.
8. Analyze differential miRNA expression between samples using the comparative Ct method (http://www.appliedbiosystems.com/dataassist).

3.5. High-Throughput Expression Profiling of miRNAs

Here, we describe the Applied Biosystems platform for high-throughput detection and quantitation of mature miRNAs using a protocol involving a reverse transcription step using Megaplex™ primer pools to create single strand cDNA from RNA (33), followed by qPCR using corresponding TaqMan® miRNA array cards, allowing for rapid analysis of up to 754 unique miRNAs in human total RNA samples. This platform offers comprehensive coverage of human miRNAs listed in the Sanger miRBase database.

3.5.1. Reverse Transcription of Total RNA to Single Strand cDNA

1. Thaw total RNA sample prepared during Subheading 3.2, the components of the TaqMan® microRNA Reverse Transcription Kit and the Megaplex™ primers on ice. Mix the tubes by flicking and briefly centrifuge.
2. Based on the concentration of total RNA calculated in Subheading 3.3, use 500 ng total RNA per RT reaction. Each RT reaction has a final volume of 7.5 μL, comprising 3 μL

total RNA (corresponding to 500 ng) and 4.5 μL RT reaction mix. For a full miRNA profile, run two Megaplex™ RT reactions (pool A and B) and two TaqMan® miRNA arrays (array A and B) per sample.

3. Calculate the number of RT reactions for each Megaplex™ RT primer pool according to the number of total RNA preparations.
4. Prepare a mix of RT reaction components in a 1.5-mL tube (33), for every 10 reactions (including 12.5% excess for volume loss during pipetting, add 9.00 μL Megaplex™ RT Primers, 2.25 μL dNTPs with dTTP (100 mM), 16.88 μL MultiScribe™ Reverse Transcriptase (50 U/μL), 9 μL 10× RT buffer, 10.12 μL $MgCl_2$, 1.12 μL RNase inhibitor (20 U/μL), 2.25 μL nuclease-free water.
5. Mix by inverting the tube and centrifuge briefly.
6. Add 4.5 μL of RT mix to each well of a 96-well MicroAmp® reaction plate, followed by 3 μL (containing 500 ng) of total RNA.
7. Seal the plate using a MicroAmp® Adhesive Film. Invert the plate to mix the components and centrifuge briefly.
8. Place the plate on ice for 5 min.
9. Load the plate into the thermal cycler and run the RT with the following conditions: 16°C 2 min, 42°C 1 min, 50°C 1 s, 85°C 5 min, 4°C infinity.
10. Following RT, the cDNA may be stored at –20°C until further use (see Note 11).

3.5.2. Real-Time PCR and TaqMan® miRNA Array

DNA polymerase amplifies the target cDNA using sequence specific primers in a real-time PCR and probe reaction on the TaqMan® miRNA Array.

1. Prepare reagents for qPCR by thawing the RT product from Subheading 3.5.1, and the PreAmp reagent on ice.
2. Invert the TaqMan® Universal PCR Master Mix.
3. Bring the TaqMan® miRNA Array Cards to room temperature.
4. Prepare the PCR reaction mix in a 1.5-mL tube: 6 μL RT product, 450 μL of TaqMan® Universal PCR Master Mix and 440 μL of nuclease-free water.
5. Invert the tube to mix and centrifuge briefly.
6. Add 100 μL of PCR reaction mix into each port of the TaqMan® miRNA Array. Centrifuge the array briefly and seal (see Note 12).
7. Load the array into the thermal cycler and run the TaqMan low density array default program on the machine.

8. Analyze the data for changes in miRNA expression using the comparative Ct method, as outlined in the Megaplex™ Pools for miRNA Expression Analysis protocol and http://www.affymetrix.com/dataassist. For further downstream analysis, Integromics StatMiner Software may be used (http://www.integromics.com).

3.6. Identification of miRNAs and Their Targets

The identification of miRNA targets is a major challenge since conventional NCBI BLAST searching will not yield satisfactory results as miRNAs usually bind to their targets with incomplete complementarity. miRNAs bind their mRNA targets in the 3′-untranslated region (UTR). The 5′ end of miRNA molecules has been shown to be crucial for the stability and proper loading of the miRNA into the miRISC complex as well as being important for biological function (19, 20). The majority of bioinformatic tools use what is termed the "seed," a region that encompasses a 7–8 bp sequence starting from either the first or second base of the 5′ end of the mature miRNA. This is used to search for complementarity to sequences in the 3′ UTR of mRNA transcripts. Some current *in silico* methodologies for miRNA-mRNA duplex prediction as well as databases of known miRNAs and their targets are summarized in Table 1. Of note is the miRNA registry or miRBase, a database of all known miRNAs, which was established in 2004, and is frequently updated (42, 43). In parallel with this, efforts have also been made in standardizing the nomenclature and classification of miRNA molecules (44). Below we discuss useful Web resources for both miRNA sequence information

Table 1
miRNA databases and identification tools

Tool	Notes	URL
miRBase	Database of microRNA and target sequences	http://www.mirbase.org
Tarbase	Database of experimentally verified miRNA targets	http://www.diana.pcbi.upenn.edu/tarbase.html
miRNAMap	Database of miRNA and target sequences	http://mirnamap.mbc.nctu.edu.tw/
TargetScan	Prediction of vertebrate miRNA targets	http://www.targetscan.org/
miTarget	Nonspecies-specific prediction of miRNA targets	http://cbit.snu.ac.kr/~miTarget/
PicTar	miRNA target prediction for various species	http://pictar.bio.nyu.edu/
MicroInspector	Nonspecies-specific prediction of miRNA targets	http://mirna.imbb.forth.gr/microinspector/

(see Subheading 3.6.1) and miRNA target identification (see Subheading 3.7). An example of an miRNA target identification protocol using the TargetScan program is outlined (see Subheading 3.7.1). In addition, considerations for validating newly identified miRNA targets are discussed (see Subheading 3.8).

3.6.1. miRBase

miRBase is a Web-based tool which can be used to search for known miRNAs in various species. miRBase is a very useful starting point for the investigation of miRNAs. miRBase contains all published miRNA sequences, genomic locations, and associated annotation. It also contains a newly developed database of predicted miRNA target genes and furthermore provides assistance in assigning official names for novel miRNA genes prior to publication of their discovery (42). The latest release of the database, miRBase release 16 (September 2010, http://www.mirbase.org), contains 15,172 entries representing hairpin precursor miRNAs, expressing 17,341 mature miRNA products, in primates, rodents, birds, fish, worms, flies, plants, and viruses (41). miRBase also contains substantial information on potential targets of miRNAs for various species. Utilizing the miRanda software involves scanning all available miRNA sequences for a given genome against 3′ UTR sequences of that genome (45–47). Users can therefore determine whether their gene of interest has miRNA target sites, and also identify all the targets of an miRNA of interest.

3.6.2. miRNAMap

miRNAMap is an online database that integrates experimentally verified miRNAs and experimentally verified miRNA target genes in human, mouse, rat, and dog. miRNAMap collects experimental evidence relating to miRNAs and miRNA targets from other databases as well as published literature. Furthermore, putative miRNA targets are predicted using the built-in miRanda algorithm (48).

3.6.3. Tarbase

Precise and reliable experimental identification and validation of miRNA targets remains a major hurdle for both the analysis of miRNA function and for the construction of computational methods. One example of a database consisting of experimentally validated targets is Tarbase (49). Tarbase collects targets reported in the literature from a variety of experimental techniques. These miRNAs have therefore been experimentally validated to confirm that the miRNA is capable of regulating the target mRNA through its 3′ UTR.

3.7. miRNA and Target Prediction Methodologies

Computational approaches can be employed to predict miRNAs and their targets. miRNA detection relies on (1) conservation of miRNAs in the genomes of related species (2) formation of stable stem-loop in pre-miRNA structures and (3) the presence of mature miRNAs in the stem and not in the loop of pre-miRNAs (50). Following input of a target mRNA sequence and, if relevant, a known miRNA sequence, the majority of miRNA target prediction algorithms incorporate the steps below (51):

1. Determination of complementarity within "seed" region (52, 53).
2. Determination of the strength of interaction based on the thermodynamic properties of the duplex formation (54).
3. Determination of the level of conservation of miRNA sequences between species. This is necessary to increase sensitivity, thereby minimizing the false identification of non-miRNA sequences (55, 56).
4. Determination of multiple miRNA molecules binding one mRNA transcript (47, 57, 58).

3.7.1. TargetScan

TargetScan is an *in silico* tool for identifying targets of vertebrate miRNA (59). TargetScan performs thermodynamic modeling of miRNA:mRNA duplex interactions as well as comparative sequence analysis. For a given an miRNA, TargetScan firstly searches 3′ UTRs for segments of complementarity within a 7 nucleotide "seed" 2–8 nt from the 5′ end of miRNA. It then extends the seed matches to predict the remaining extent of miRNA:mRNA binding, allowing G:U pairing after which it calculates the thermodynamic properties of base pairing between the miRNA target and the extended seed sequence is calculated using the RNAfold program (54, 60). An updated version of TargetScan, TargetScanS introduced several new features. Most target prediction algorithms consider the free energy of complementarity when making predictions. The assumption being that target site occupancy correlates with the base pairing strength (61). The developers of TargetScan, Lewis et al. (62), have shown that free energy calculations did not have much impact on the success of TargetScanS predictions. Furthermore, with TargetScanS, the seed nucleotide match has been reduced to 6 nt (2–7 from the 5′ end of the miRNA molecule), and it does not consider the thermodynamic properties of pairing. Furthermore, in order to reduce noise, pairing in the vicinity immediately up- and downstream of the seed and the presence of multiple-binding sites per UTR are assessed (62, 63).

3.7.2. Protocol for Target Identification Using TargetScan

The following is an example of miRNA target identification using the TargetScan program. For the sake of this methodology chapter, all algorithms mentioned above cannot be described in detail; however, their designated Web locations can be found in Table 1.

1. After going to the TargetScan Web site (http://www.targetscan.org/), select the species under investigation and select the miRNA family of interest, or enter the name of the miRNA if known.

2. Click "submit". The next page displays a table, including all the target genes for that miRNA family. Other information given includes the amount of conserved sites on the target.
3. The user can then link out to NCBI Gene for further information on the targets.

For the identification of miRNA sites within a given mRNA the following steps should be followed:

1. From the TargetScan homepage enter the entrez gene symbol for the transcript of interest and click "submit".
2. An image will display the predicted miRNA-binding sites within the 3′ UTR as well as the level of conservation of those sites across species.

3.7.3. miTarget

miTarget assesses information on structure, thermodynamics, and position by SVM-analysis. miTarget uses what is termed a support vector machines (SVM)-based analysis, a branch of supervised machine learning (64–66). miTarget uses an SVM to learn target rules from positive and negative examples of validated target–miRNA interactions. It then quantitates the target–miRNA interaction using structural, thermodynamic, and positional features. It is therefore very dependent on the quality of the training data.

3.7.4. PicTar

PicTar utilizes powerful cross-species comparative data as criteria for identifying miRNA target genes (58, 67, 68). PicTar also reports the maximum likelihood that a given mRNA is bound by one or more miRNAs. Firstly, target mRNAs are predicted using such criteria as optimal binding thermodynamics, and are then statistically tested using cross-species genome-wide alignment, thereby removing false positives. Krek et al. employed this method to predict that approximately 200 transcripts are regulated by a single miRNA (58). PicTar locates perfect 7 nucleotide matches starting at position 1 or 2 of the 5′ end of the miRNA. The hybridization energy is then calculated and unstable matches are discarded. PicTar calculates the maximum probability that an mRNA is regulated by two or more miRNAs.

PicTar and TargetScan have been described as good performers, both with about 20–30% false-positive rates and despite different search criteria; there is an 80–90% overlap between the two for human targets (56, 69).

3.8. Considerations for Experimental Validation of miRNAs and Their Targets

Computational predictions represent a huge contribution to miRNA gene discovery; however, real experimental validation is required to confirm candidate miRNAs thereby evaluating the performance of the algorithm. Furthermore, experimental confirmation of computationally predicted miRNA targets is crucial for understanding their biological significance. The various computational prediction algorithms discussed above all make use of established miRNA–mRNA interaction rules to identify targets. To maximize their dependability, the majority of these in silico methods require a fully complementary "seed" sequence in the 3′ UTR of the mRNA as well as conservation of this site across several species, thus potentially missing targets that do not conform to these rules. mirTools is a recently described software suite for characterizing the small size RNA transcriptome using multiple levels of computational analyses(70). mirTools can be accessed at http://centre.bioinformatics.zj.cn/mirtools/ and http://59.79.168.90/mirtools (70).

Although experimental validation of miRNA target genes is challenging compared to computational validation, these *in silico* methods are not perfect and independent laboratory-based confirmation of miRNA target interactions is essential therefore in detecting and confirming novel miRNA targets. Highly sensitive techniques must be employed to confirm miRNA targets because of their small size, sequence similarity among members and low expression levels. Reporter-gene assays are commonly used for experimental validation of computationally predicated targets (71, 72). Furthermore, microarray analysis, as discussed above, provides a powerful and high-throughput method for observing cleaved target mRNAs (73, 74). There are no high-throughput techniques to verify the interaction with, and regulation of a target (75, 76). Techniques which are widely used in other areas of biological research and which have also been successfully used to validate miRNA identification data include Northern blot analysis (16, 77, 78), RT-PCR, and qRT-PCR (79, 80), microarray (79–81), and in situ hybridization (82).

4. Notes

1. To prevent RNA degradation by RNases and ensure high-quality RNA isolation, it is important to work in an RNase-free environment. Clean work surfaces and treat with an RNase decontaminating spray, such as RNase*Zap*. Wear latex gloves when carrying out experiments and change gloves frequently. It is recommended to use RNase-free tubes and tips.
2. RNA may be isolated from 0.5 to 250 mg tissue per preparation using the *mir*Vana miRNA Isolation Kit. In order obtain a high yield of good quality RNA, it is important to limit the time between obtaining the samples and inactivating RNases. In order to inactivate RNases, the samples may be placed directly into either RNA*later* (5–10 volumes), the lysis buffer contained in the *mir*Vana miRNA Isolation Kit or into liquid

Nitrogen. It is necessary to cut the tissue sample into pieces small enough to facilitate complete penetration of the storage solution. When using RNA*later*, do not freeze samples immediately. Samples should be stored at 4°C overnight to allow complete penetration of the solution prior to storage at –70°C long term. The RNA*later* supernatant may be removed prior to freezing to expedite thawing. It is very important to process tissue samples immediately upon removal from –70°C freezer, as the ice crystals in partially thawed samples rupture cellular compartments releasing RNases. Tissue samples should be homogenized in 10 volumes of lysis buffer per mass using a motorized rotor-stator homogenizer (e.g., TissueRuptor and disposable probes, Qiagen, Valencia, CA, USA). As foaming may occur during the homogenization step, choose a suitably sized RNase-free tube. Homogenization should be carried out on ice until all clumps of tissue are dispersed.

3. It is recommended to use 10^2–10^7 cells per RNA preparation using the *mir*Vana miRNA Isolation Kit. If possible, cells in culture should be processed fresh as opposed to frozen samples. However, cells stored prior to miRNA isolation should be maintained in RNA*later* at –70°C.

4. Do not mix the acid-phenol:chloroform reagent. The top layer is aqueous buffer. It is important to only use the bottom layer which contains the phenol:chloroform. Insert your pipette tip into the bottom layer, and aspirate from the bottom layer only.

5. Total RNA preparations can be enriched for small RNA molecules (≤200 nt) if necessary for other downstream techniques. If the RNA preparation is to be used for miRNA expression profiling by microarrays, it is recommended to first prepare total RNA containing the miRNA fraction. This allows the preparation to be accurately quantified and the quality analyzed to check the sample is suitable for microarray analysis experiments. It is essential to use only extremely high-quality RNA when performing microarrays. The 2100 Bioanalyzer (Agilent, Santa Clara, CA, USA) is a useful microfluidics-based platform that provides detailed information on the quality and quantity of RNA samples prior to beginning array experiments. Following quality control, the miRNA population can be further purified according to the *mir*VANA miRNA Isolation Kit Protocol prior to labeling samples for miRNA microarray analysis if necessary.

6. At this step, RNA may be eluted in either nuclease-free H_2O or in the elution buffer supplied with the kit, which contains 0.1 mM EDTA. Nuclease-free water is recommended to prevent interference with any downstream applications.

7. In order to preserve endogenous control sequences, it is recommended to use a total RNA preparation in the RT reaction and not a sample enriched for the miRNA fraction.
8. The volumes listed for the Master Mix in this protocol incorporate an extra 20% volume to account for losses of reagent during pipetting. 7.0 μL of the Master Mix is required for each RT reaction. Prepare enough Master Mix for your desired number of reactions.
9. When carrying out qPCR, errors may occur due to variations in the amount of starting material, sample collection, RNA preparations, and the efficiency of the RT reaction. By normalizing gene expression to endogenous control genes, problems due to these variations may be overcome. A robust endogenous control gene is expressed constantly and abundantly across tissues and cell types.
10. Include H_2O as template in some wells as a negative control. If there is amplification in these wells, it is indicative of contamination in the reagents used. When preparing samples for qPCR, ensure the workspace is amplicon-free by preparing the reactions in an area free of artificial templates and separate to where siRNA transfections are performed. If possible, use separate areas and supplies for sample preparation, PCR setup, and PCR amplification. Use aerosol resistant pipette tips and centrifuge tubes before opening.
11. If the yield of total RNA starting material is low (<300 ng), Megaplex PreAmp Primer Sets, available from Applied Biosystems, may be used in order to amplify miRNA cDNA targets prior to the qPCR step.
12. TaqMan miRNA arrays may be prepared in advance of qPCR run and stored at 4°C for up to 48 h until loaded into the thermal cycler.

References

1. Lee, R. C., Feinbaum, R. L., Ambros, V. (1993) The C. elegans heterochronic gene lin-4 encodes small RNAs with antisense complementarity to lin-14. *Cell* 75, 843–54.
2. Esquela-Kerscher, A., Slack, F. J. (2006) Oncomirs - microRNAs with a role in cancer. *Nat Rev Cancer* 6, 259–69.
3. Calin, G. A., Croce, C. M. (2006) MicroRNA signatures in human cancers. *Nat Rev Cancer* 6, 857–66.
4. Grosshans, H., Slack, F. J. (2002) MicroRNAs: small is plentiful. *J Cell Biol* 156, 17–21.
5. Lee, Y., Jeon, K., Lee, J. T., Kim, S., Kim, V. N. (2002) MicroRNA maturation: stepwise processing and subcellular localization. *Embo J* 21, 4663–70.
6. Smalheiser, N. R. (2003) EST analyses predict the existence of a population of chimeric microRNA precursor-mRNA transcripts expressed in normal human and mouse tissues. *Genome Biol* 4, 403.
7. Cai, X., Hagedorn, C. H., Cullen, B. R. (2004) Human microRNAs are processed from capped, polyadenylated transcripts that can also function as mRNAs. *Rna* 10, 1957–66.
8. Denli, A. M., Tops, B. B., Plasterk, R. H., Ketting, R. F., Hannon, G. J. (2004) Processing of primary microRNAs by the Microprocessor complex. *Nature* 432, 231–5.

9. Reinhart, B. J., Slack, F. J., Basson, M., Pasquinelli, A. E., Bettinger, J. C., Rougvie, A. E. et al. (2000) The 21-nucleotide let-7 RNA regulates developmental timing in Caenorhabditis elegans. *Nature* 403, 901–6.
10. Cairo, S., Wang, Y., de Reynies, A., Duroure, K., Dahan, J., Redon, M. J. et al. (2010) Stem cell-like micro-RNA signature driven by Myc in aggressive liver cancer. *Proc Natl Acad Sci U S A* 107, 20471–6.
11. Brennecke, J., Hipfner, D. R., Stark, A., Russell, R. B., Cohen, S. M. (2003) bantam encodes a developmentally regulated microRNA that controls cell proliferation and regulates the proapoptotic gene hid in Drosophila. *Cell* 113, 25–36.
12. Xu, P., Vernooy, S. Y., Guo, M., Hay, B. A. (2003) The Drosophila microRNA Mir-14 suppresses cell death and is required for normal fat metabolism. *Curr Biol* 13, 790–5.
13. Chen, X. (2004) A microRNA as a translational repressor of APETALA2 in Arabidopsis flower development. *Science* 303, 2022–5.
14. Houbaviy, H. B., Murray, M. F., Sharp, P. A. (2003) Embryonic stem cell-specific MicroRNAs. *Dev Cell* 5, 351–8.
15. Rosenfeld, N., Aharonov, R., Meiri, E., Rosenwald, S., Spector, Y., Zepeniuk, M. et al. (2008) MicroRNAs accurately identify cancer tissue origin. *Nat Biotechnol* 26, 462–9.
16. Calin, G. A., Dumitru, C. D., Shimizu, M., Bichi, R., Zupo, S., Noch, E. et al. (2002) Frequent deletions and down-regulation of micro- RNA genes miR15 and miR16 at 13q14 in chronic lymphocytic leukemia. *Proc Natl Acad Sci U S A* 99, 15524–9.
17. Metzler, M., Wilda, M., Busch, K., Viehmann, S., Borkhardt, A. (2004) High expression of precursor microRNA-155/BIC RNA in children with Burkitt lymphoma. *Genes Chromosomes Cancer* 39, 167–9.
18. Iorio, M. V., Ferracin, M., Liu, C. G., Veronese, A., Spizzo, R., Sabbioni, S. et al. (2005) MicroRNA gene expression deregulation in human breast cancer. *Cancer Res* 65, 7065–70.
19. Fassan, M., Volinia, S., Palatini, J., Pizzi, M., Baffa, R., De Bernard, M. et al. (2010) MicroRNA expression profiling in human Barrett's carcinogenesis. *Int J Cancer*
20. Michael, M. Z., SM, O. C., van Holst Pellekaan, N. G., Young, G. P., James, R. J. (2003) Reduced accumulation of specific microRNAs in colorectal neoplasia. *Mol Cancer Res* 1, 882–91.
21. Tomimaru, Y., Eguchi, H., Nagano, H., Wada, H., Tomokuni, A., Kobayashi, S. et al. (2010) MicroRNA-21 induces resistance to the anti-tumour effect of interferon-alpha/5-fluorouracil in hepatocellular carcinoma cells. *Br J Cancer* 103, 1617–26.
22. Ji, J., Yamashita, T., Budhu, A., Forgues, M., Jia, H. L., Li, C. et al. (2009) Identification of microRNA-181 by genome-wide screening as a critical player in EpCAM-positive hepatic cancer stem cells. *Hepatology* 50, 472–80.
23. Takamizawa, J., Konishi, H., Yanagisawa, K., Tomida, S., Osada, H., Endoh, H. et al. (2004) Reduced expression of the let-7 microRNAs in human lung cancers in association with shortened postoperative survival. *Cancer Res* 64, 3753–6.
24. Liu, X., Sempere, L. F., Galimberti, F., Freemantle, S. J., Black, C., Dragnev, K. H. et al. (2009) Uncovering growth-suppressive MicroRNAs in lung cancer. *Clin Cancer Res* 15, 1177–83.
25. Liu, X., Sempere, L. F., Ouyang, H., Memoli, V. A., Andrew, A. S., Luo, Y. et al. (2010) MicroRNA-31 functions as an oncogenic microRNA in mouse and human lung cancer cells by repressing specific tumor suppressors. *J Clin Invest* 120, 1298–309.
26. Nicoloso, M. S., Spizzo, R., Shimizu, M., Rossi, S., Calin, G. A. (2009) MicroRNAs--the micro steering wheel of tumour metastases. *Nat Rev Cancer* 9, 293–302.
27. Ryan, B. M., Robles, A. I., Harris, C. C. (2010) Genetic variation in microRNA networks: the implications for cancer research. *Nat Rev Cancer* 10, 389–402.
28. Junker, A., Hohlfeld, R., Meinl, E. (2010) The emerging role of microRNAs in multiple sclerosis. *Nat Rev Neurol* 7, 56–59.
29. Hassan, M. Q., Gordon, J. A., Beloti, M. M., Croce, C. M., van Wijnen, A. J., Stein, J. L. et al. (2010) A network connecting Runx2, SATB2, and the miR-23a~27a~24-2 cluster regulates the osteoblast differentiation program. *Proc Natl Acad Sci U S A* 107, 19879–84.
30. Nana-Sinkam, S. P., Fabbri, M., Croce, C. M. (2010) MicroRNAs in cancer: personalizing diagnosis and therapy. *Ann N Y Acad Sci* 1210, 25–33.
31. Krol, J., Loedige, I., Filipowicz, W. (2010) The widespread regulation of microRNA biogenesis, function and decay. *Nat Rev Genet* 11, 597–610.
32. Garzon, R., Marcucci, G., Croce, C. M. (2010) Targeting microRNAs in cancer: rationale, strategies and challenges. *Nat Rev Drug Discov* 9, 775–89.
33. (2010) Megaplex Pools for microRNA Expression Analysis. 4399721 Rev C. Applied

Biosystems/Life Technologies: Foster City, CA, 1–30.

34. (2011) *mir*Vana miRNA Isolation Kit Protocol. Revision C Ambion P/N 1560M. Ambion/Life Technologies: Carlsbad, CA, 1–33.
35. (2011) TaqMan Small RNA Assays Protocol. 4364031 Rev E. Applied Biosystems/Life Technologies: Carlsbad, CA, 1–41.
36. Gamazon, E. R., Im, H. K., Duan, S., Lussier, Y. A., Cox, N. J., Dolan, M. E. et al. (2010) Exprtarget: an integrative approach to predicting human microRNA targets. *PLoS One* 5, e13534.
37. Friard, O., Re, A., Taverna, D., De Bortoli, M., Cora, D. (2010) CircuitsDB: a database of mixed microRNA/transcription factor feed-forward regulatory circuits in human and mouse. *BMC Bioinformatics* 11, 435.
38. Bartonicek, N., Enright, A. J. (2010) SylArray: a web server for automated detection of miRNA effects from expression data. *Bioinformatics* 26, 2900–1.
39. Yang, Z., Ren, F., Liu, C., He, S., Sun, G., Gao, Q. et al. (2010) dbDEMC: a database of differentially expressed miRNAs in human cancers. *BMC Genomics* 11 Suppl 4, S5.
40. Zhang, Y., Guan, D. G., Yang, J. H., Shao, P., Zhou, H., Qu, L. H. (2010) ncRNAimprint: a comprehensive database of mammalian imprinted noncoding RNAs. *Rna* 16, 1889–901.
41. Griffiths-Jones, S. (2010) miRBase: microRNA sequences and annotation. *Curr Protoc Bioinformatics* Chapter 12, Unit 12 9 1–10.
42. Griffiths-Jones, S. (2006) miRBase: the microRNA sequence database. *Methods Mol Biol* 342, 129–38.
43. Griffiths-Jones, S., Grocock, R. J., van Dongen, S., Bateman, A., Enright, A. J. (2006) miRBase: microRNA sequences, targets and gene nomenclature. *Nucleic Acids Res* 34, D140-4.
44. Ambros, V., Bartel, B., Bartel, D. P., Burge, C. B., Carrington, J. C., Chen, X. et al. (2003) A uniform system for microRNA annotation. *Rna* 9, 277–9.
45. Enright, A. J., John, B., Gaul, U., Tuschl, T., Sander, C., Marks, D. S. (2003) MicroRNA targets in Drosophila. *Genome Biol* 5, R1.
46. John, B., Enright, A. J., Aravin, A., Tuschl, T., Sander, C., Marks, D. S. (2004) Human MicroRNA targets. *PLoS Biol* 2, e363.
47. Stark, A., Brennecke, J., Russell, R. B., Cohen, S. M. (2003) Identification of Drosophila MicroRNA targets. *PLoS Biol* 1, E60.
48. Hsu, P. W., Huang, H. D., Hsu, S. D., Lin, L. Z., Tsou, A. P., Tseng, C. P. et al. (2006) miRNAMap: genomic maps of microRNA genes and their target genes in mammalian genomes. *Nucleic Acids Res* 34, D135-9.
49. Sethupathy, P., Corda, B., Hatzigeorgiou, A. G. (2006) TarBase: A comprehensive database of experimentally supported animal microRNA targets. *Rna* 12, 192–7.
50. Chaudhuri, K., Chatterjee, R. (2007) MicroRNA detection and target prediction: integration of computational and experimental approaches. *DNA Cell Biol* 26, 321–37.
51. Watanabe, Y., Tomita, M., Kanai, A. (2007) Computational methods for microRNA target prediction. *Methods Enzymol* 427, 65–86.
52. Lin, S. Y., Johnson, S. M., Abraham, M., Vella, M. C., Pasquinelli, A., Gamberi, C. et al. (2003) The C elegans hunchback homolog, hbl-1, controls temporal patterning and is a probable microRNA target. *Dev Cell* 4, 639–50.
53. Doench, J. G., Sharp, P. A. (2004) Specificity of microRNA target selection in translational repression. *Genes Dev* 18, 504–11.
54. Wuchty, S., Fontana, W., Hofacker, I. L., Schuster, P. (1999) Complete suboptimal folding of RNA and the stability of secondary structures. *Biopolymers* 49, 145–65.
55. Rajewsky, N., Socci, N. D. (2004) Computational identification of microRNA targets. *Dev Biol* 267, 529–35.
56. Rajewsky, N. (2006) microRNA target predictions in animals. *Nat Genet* 38 Suppl, S8-13.
57. Watanabe, Y., Yachie, N., Numata, K., Saito, R., Kanai, A., Tomita, M. (2006) Computational analysis of microRNA targets in Caenorhabditis elegans. *Gene* 365, 2–10.
58. Krek, A., Grun, D., Poy, M. N., Wolf, R., Rosenberg, L., Epstein, E. J. et al. (2005) Combinatorial microRNA target predictions. *Nat Genet* 37, 495–500.
59. Lewis, B. P., Shih, I. H., Jones-Rhoades, M. W., Bartel, D. P., Burge, C. B. (2003) Prediction of mammalian microRNA targets. *Cell* 115, 787–98.
60. Hofacker, I. L. (2003) Vienna RNA secondary structure server. *Nucleic Acids Res* 31, 3429–31.
61. Burgler, C., Macdonald, P. M. (2005) Prediction and verification of microRNA targets by MovingTargets, a highly adaptable prediction method. *BMC Genomics* 6, 88.
62. Lewis, B. P., Burge, C. B., Bartel, D. P. (2005) Conserved seed pairing, often flanked by adenosines, indicates that thousands of human genes are microRNA targets. *Cell* 120, 15–20.
63. Friedman, R. C., Farh, K. K., Burge, C. B., Bartel, D. P. (2009) Most mammalian mRNAs are conserved targets of microRNAs. *Genome Res* 19, 92–105.

64. Helvik, S. A., Snove, O., Jr., Saetrom, P. (2007) Reliable prediction of Drosha processing sites improves microRNA gene prediction. *Bioinformatics* 23, 142–9.
65. Hertel, J., Stadler, P. F. (2006) Hairpins in a Haystack: recognizing microRNA precursors in comparative genomics data. *Bioinformatics* 22, e197-202.
66. Kim, S. K., Nam, J. W., Rhee, J. K., Lee, W. J., Zhang, B. T. (2006) miTarget: microRNA target gene prediction using a support vector machine. *BMC Bioinformatics* 7, 411.
67. Grun, D., Wang, Y. L., Langenberger, D., Gunsalus, K. C., Rajewsky, N. (2005) microRNA target predictions across seven Drosophila species and comparison to mammalian targets. *PLoS Comput Biol* 1, e13.
68. Lall, S., Grun, D., Krek, A., Chen, K., Wang, Y. L., Dewey, C. N. et al. (2006) A genome-wide map of conserved microRNA targets in C. elegans. *Curr Biol* 16, 460–71.
69. Doran, J., Strauss, W. M. (2007) Bio-informatic trends for the determination of miRNA-target interactions in mammals. *DNA Cell Biol* 26, 353–60.
70. Zhu, E., Zhao, F., Xu, G., Hou, H., Zhou, L., Li, X. et al. (2010) mirTools: microRNA profiling and discovery based on high-throughput sequencing. *Nucleic Acids Res* 38, W392-7.
71. Miranda, K. C., Huynh, T., Tay, Y., Ang, Y. S., Tam, W. L., Thomson, A. M. et al. (2006) A pattern-based method for the identification of MicroRNA binding sites and their corresponding heteroduplexes. *Cell* 126, 1203–17.
72. Stark, A., Brennecke, J., Bushati, N., Russell, R. B., Cohen, S. M. (2005) Animal MicroRNAs confer robustness to gene expression and have a significant impact on 3 UTR evolution. *Cell* 123, 1133–46.
73. Wang, X. (2006) Systematic identification of microRNA functions by combining target prediction and expression profiling. *Nucleic Acids Res* 34, 1646–52.
74. Lim, L. P., Lau, N. C., Garrett-Engele, P., Grimson, A., Schelter, J. M., Castle, J. et al. (2005) Microarray analysis shows that some microRNAs downregulate large numbers of target mRNAs. *Nature* 433, 769–73.
75. Bentwich, I., Avniel, A., Karov, Y., Aharonov, R., Gilad, S., Barad, O. et al. (2005) Identification of hundreds of conserved and nonconserved human microRNAs. *Nat Genet* 37, 766–70.
76. Vatolin, S., Navaratne, K., Weil, R. J. (2006) A novel method to detect functional microRNA targets. *J Mol Biol* 358, 983–96.
77. Ambros, V., Lee, R. C., Lavanway, A., Williams, P. T., Jewell, D. (2003) MicroRNAs and other tiny endogenous RNAs in C. elegans. *Curr Biol* 13, 807–18.
78. Sempere, L. F., Freemantle, S., Pitha-Rowe, I., Moss, E., Dmitrovsky, E., Ambros, V. (2004) Expression profiling of mammalian microRNAs uncovers a subset of brain-expressed microRNAs with possible roles in murine and human neuronal differentiation. *Genome Biol* 5, R13.
79. Tang, F., Hajkova, P., Barton, S. C., Lao, K., Surani, M. A. (2006) MicroRNA expression profiling of single whole embryonic stem cells. *Nucleic Acids Res* 34, e9.
80. Fu, H. J., Zhu, J., Yang, M., Zhang, Z. Y., Tie, Y., Jiang, H. et al. (2006) A novel method to monitor the expression of microRNAs. *Mol Biotechnol* 32, 197–204.
81. Liu, C. G., Calin, G. A., Meloon, B., Gamliel, N., Sevignani, C., Ferracin, M. et al. (2004) An oligonucleotide microchip for genome-wide microRNA profiling in human and mouse tissues. *Proc Natl Acad Sci U S A* 101, 9740–4.
82. Nelson, P. T., Baldwin, D. A., Kloosterman, W. P., Kauppinen, S., Plasterk, R. H., Mourelatos, Z. (2006) RAKE and LNA-ISH reveal microRNA expression and localization in archival human brain. *Rna* 12, 187–91.

Chapter 10

Antibody Validation by Western Blotting

Michele Signore and K. Alex Reeder

Abstract

Validation of antibodies is an integral part of translational research, particularly for biomarker discovery. Validation is essential to show the specificity of the reagent (antibody) and to confirm the identity of the protein biomarker, prior to implementing the biomarker in clinical studies.

Antibody validation is the procedure in which a single antibody is thoroughly assayed for sensitivity and specificity. Although a plethora of commercial antibodies exist, antibody specificity must be thoroughly demonstrated using a complex biological sample, rather than a recombinant protein, prior to use in clinical translational research. In the simplest iteration, antibody specificity is determined by the presence of a single band in a complex biological sample, at the expected molecular weight, on a western blot.

Numerous western blotting procedures are available, spanning the spectrum of single blots to multiplex blots, with images and quantitation generated by manual or automated systems. The basic principles of western blotting are (a) separation of protein mixtures by gel electrophoresis, (b) transfer of the proteins to a blot, (c) probing the blot for a protein or proteins of interest, and (d) subsequent detection of the protein by chemiluminescent, fluorescent, or colorimetric methods. This chapter focuses on the chemiluminescent detection of proteins using a manual western blotting system and a vacuum-enhanced detection system (SNAP i.d.™, Millipore)

Key words: Alkaline phosphatase, Antibody, Antigen, Chemiluminescence, Horseradish peroxidase, Reverse-phase protein microarray, SDS-PAGE, SNAP i.d., Western blot

1. Introduction

Antibodies are basic reagents used in numerous research grade as well as clinical grade assays. Antibodies are employed in diverse applications such as flow cytometry, immunohistochemistry, ELISA, immunoprecipitation, and protein mircroarrays. Consequently, antibody validation is an integral part of translational research, particularly for biomarker discovery, to show the specificity of the reagent (antibody) and to confirm the identity of the protein biomarker (1, 2). Prior to approval by regulatory agencies, such as

Virginia Espina and Lance A. Liotta (eds.), *Molecular Profiling: Methods and Protocols*, Methods in Molecular Biology, vol. 823, DOI 10.1007/978-1-60327-216-2_10, © Springer Science+Business Media, LLC 2012

the US Food and Drug Administration, most diagnostic assays that measure the concentration, or simply the presence of various types of analytes in clinical samples, undergo sensitivity/specificity analysis (3). Sensitivity is the rate of true positives and specificity is the rate of true negatives. In addition, sensitivity of an assay is dependent on the prevalence of a disease or condition inside a population (4). Usually, a diagnostic test is designed to have a sufficient discriminatory power to support the intended use. Consequently, cutoff values are chosen so that the sensitivity of an assay is sacrificed to a certain extent. Sensitivity is sacrificed because obtaining accurate, specific results may be more important than the ability to detect the analyte of interest at very low concentrations. However, the merits of both sensitivity and specificity must be taken into consideration for an individual analyte/assay prior to use.

Sensitivity and specificity concepts are applicable to the field of translational research especially when large-scale detection is taken into account. It is acknowledged and broadly accepted within the scientific community that, when performing a western blot, proteins other than the protein of interest can be detected (1–3, 5). Typically, these additional bands are seen on a blot as multiple bands of various molecular weights (3, 5). The additional bands may be different proteins altogether, posttranslationally modified forms of the target protein, or splice variants. Without further analysis, a determination of antibody specificity cannot be made between specific (expected molecular weight) and nonspecific (unexpected molecular weight) bands (Fig. 1) (6). Isoforms may appear as protein dimers with molecular weights twice the expected molecular weight (7). Techniques such as peptide competition can show the specificity of the antibody in question as well as identification of degraded, cleaved, or other modified forms of the protein of interest (3, 6, 8) (see Note 1).

Antibodies used in clinical or clinical research applications should be validated, at a minimum, using a standardized western blot procedure in which control lysates, known to contain and/or to be devoid of the protein of interest, are probed with an antibody, together with lysates that are comparable to those which will be analyzed with the intended assay (Fig. 2) (9). In addition to traditional western blotting, a dot-blot style validation can be performed using reverse phase protein microarray techniques (RPMA). RPMA validation entails immobilizing peptides of known sequence, in an array format, on a nitrocellulose-coated slide (8). Phospho-peptides and nonphosphorylated peptides with matching sequences may be printed on one array as a means of determining the specificity of phosphospecific antibodies. The phosphospecific antibody to be validated is used to immunostain the array containing the peptides. By comparing the presence/absence of signal in the corresponding peptides one can verify antibody specificity and/or cross-reactivity (8).

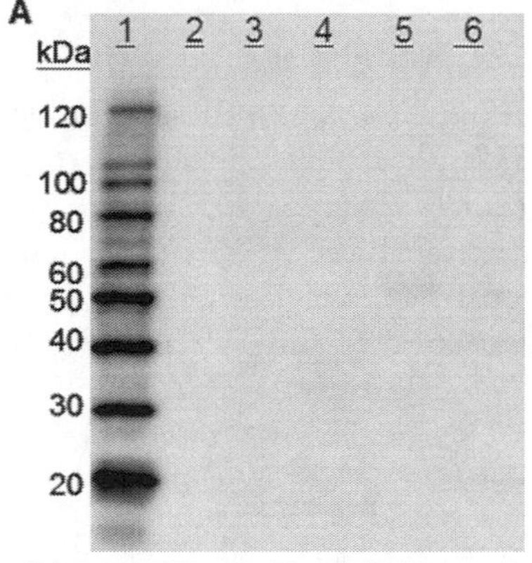

Primary antibody: Numb P-20 (goat)

Expected Molecular Weight: 71 kDa

Secondary antibody: Donkey anti-goat AP conjugate

Lane 1: Chemiluminescent marker
Lane 2: Pre-stained marker
Lane 3: HeLa cell lysate
Lane 4: A431 cell lysate
Lane 5: T98G cell lysate
Lane 6: Ovarian tissue lysate

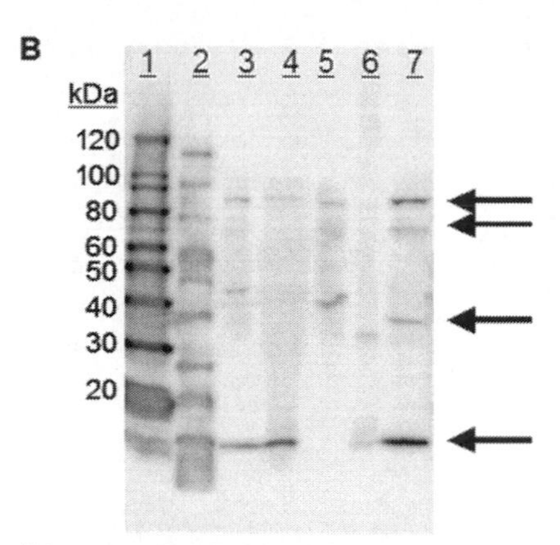

Primary antibody: Numb (rabbit)

Expected Molecular Weight: 75 kDa

Secondary antibody: Goat anti-rabbit HRP conjugate

Lane 1: Chemiluminescent marker
Lane 2: Pre-stained marker
Lane 3: Kelly lysate
Lane 4: SKN cell lysate
Lane 5: Rat brain tissue lysate
Lane 6: A431 cell lysate
Lane 7: T98G cell lysate

Fig. 1. Example western blots of nonvalidated antibodies. (**a**) A blot probed with anti-Numb does not show any bands at the expected molecular weight. (**b**) A blot probed with anti-Numb from a different source has multiple bands at various molecular weights, indicated by *arrows*. The additional bands were not present when the blot was reprobed with secondary antibody only indicating that the bands were due to cross-reactivity with the primary anti-Numb antibody.

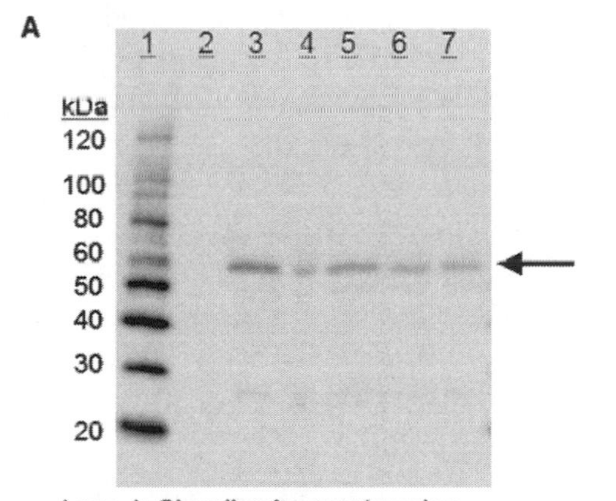

Lane 1: Chemiluminescent marker
Lane 2: Pre-stained marker
Lane 3: Raji cell lysate
Lane 4: HeLa cell lysate
Lane 5: HCT-8cell lysate
Lane 6: U266 cell lysate
Lane 7: U266+Pervanadate

Primary antibody: Beclin

Expected Molecular Weight: 60kDa

Secondary antibody: Goat anti-rabbit HRP conjugate

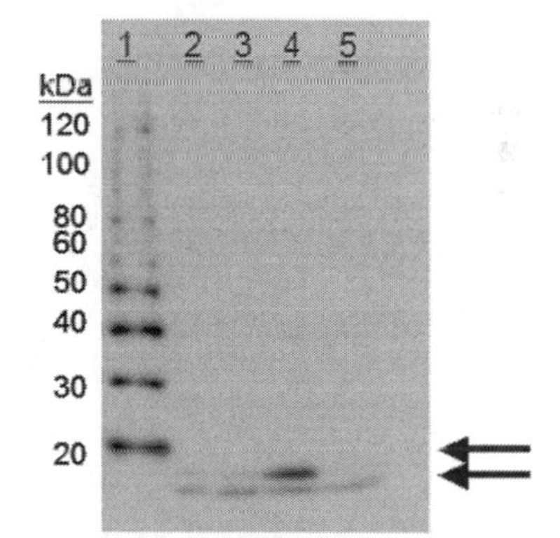

Lane 1: Chemiluminescent/Pre-Stained marker
Lane 2:HeLa cell lysate
Lane 3: A20 cell lysate
Lane 4: WEHI-231 cell lysate
Lane 5: U266 cell lysate

Primary antibody: LC3B

Expected Molecular Weight: 14, 16 kDa

Secondary antibody: Goat anti-rabbit HRP conjugate

Fig. 2. Validated antibodies. (**a**) A blot probed with anti-Beclin shows a band at the expected molecular weight in a variety of commercial cell lysates. *Lanes 3–7* show a faint band at 50 kDa. The band at 50 kDa is less than 20% of the overall signal. (**b**) A blot probed with anti-LC3B show the expected bands at 14 and 16 kDa.

Several organizations provide public access to antibody validation data. These include the Alliance for Cell Signaling (http://www.signaling-gateway.org/data/antibody/cgi-bin/targets.cgi) and antibodypediA (10) (http://www.antibodypedia.org/). This chapter describes antibody validation by classical western blotting techniques with a vacuum-assisted rapid detection method (SNAP i.d.™, Millipore).

2. Materials

Western blotting consists of four steps: sample preparation, 1D gel electrophoresis, immunoblotting, and detection. Western blotting is a process in which denatured complex protein mixtures are separated by sodium dodecyl sulfate-polyacrylamide gel electrophoresis (SDS-PAGE) then electrophoretically transferred to a membrane (blot). The blot is probed with an antibody to the protein of interest, followed by addition of a secondary antibody directed against the primary antibody, with subsequent visual detection of the antigen-antibody complex on the blot. Western blotting provides information concerning the presence of a protein, its molecular weight, relative abundance, and isoforms (11–14).

2.1. Specimen

1. Cell lines, whole tissue sections, or laser capture microdissected cells. The optimal total protein concentration of the sample is 1–2 μg/μL of protein (see Note 2).
2. Serum, plasma or body fluids (see Note 3).
3. Protein extraction buffer (denaturing): 450 μL T-PER Tissue Protein Extraction Reagent (Pierce), 450 μL Novex Tris-Glycine SDS Sample Buffer (2×) (Invitrogen), 100 μL Bond-Breaker TCEP Solution (Pierce). Mix well. Store at room temperature. Make fresh daily.
4. Commercial cell extracts/lysates: for use as positive/negative controls with the appropriate antibody.
5. Recombinant protein of interest (optional) (see Note 4).

2.2. 1D Gel Electrophoresis

1. Invitrogen X-Cell II Mini-Cell electrophoresis chamber with Xcell SureLock lid (or equivalent).
2. Bio-Rad Power Pac 1000 power supply (or equivalent).
3. 4–20% Novex Tris-Glycine Gel (Invitrogen).
4. Tris-Glycine Running Buffer 10× (Invitrogen). Dilute to 1× using Type 1 reagent grade water (see Note 5). Prepare 1 L of 1× running buffer.
5. Benchmark™ Prestained Protein Ladder (Invitrogen).

6. MagicMark™ XP Western Protein Standard (Invitrogen) (see Note 6).
7. Novex 2× SDS Tris-Glycine Loading Buffer (Invitrogen).
8. Whatman Marking pen.
9. Gel loading pipette tips (either flat or round tip).

2.3. Immunnoblot (Transfer)

1. Transfer buffer: 3 g Tris base, 14.4 g glycine, 200 mL methanol, QS to 1 L with Type 1 reagent grade water. Store tightly closed at room temperature. Stable for 2 weeks (see Note 7).
2. Extra thick blot paper.
3. PVDF membrane (Immobilon P, Millipore).
4. Bio-Rad Trans-Blot Semidry transfer apparatus.
5. Wash buffer: 500 mL 1× Phosphate-Buffered Saline (PBS), without calcium and magnesium, and 0.1% Tween 20 (500 μL). Store at 4°C for up to 2 weeks.

2.4. SNAP i.d. ™ Detection System (Millipore)

1. Primary antibody (polyclonal or monoclonal).
2. Secondary antibody (alkaline phosphatase or horseradish peroxidase conjugated).
3. 1,000-mL Vacuum Flask with tubing and a stopper.
4. Millex-FA_{50} filter.
5. Blot holders and Spacers of different types will be used depending on how many primary antibodies are being tested: Single blot, Double blot, or Triple blot.
6. SNAP i.d.™ system (Millipore).
7. Blot Roller.
8. Antibody collection trays.
9. I-Block blocking buffer: 1 g I-Block powder (Applied Biosystems), 500 mL 1× Phosphate Buffered Saline (PBS) without Calcium or Magnesium, 500 μL Tween 20. Add I-Block powder to PBS in a beaker. Heat at low temperature and stir constantly until powder dissolves and solution is slightly cloudy. Do not boil. Cool to room temperature. Add 500 μL of Tween 20. Mix thoroughly. Store at 4°C for up to 2 weeks.
10. Wash Buffer: 500 mL 1× Phosphate-Buffered Saline (PBS), without calcium and magnesium, and 0.1% Tween 20 (500 μL). Store at 4°C for up to 2 weeks.

2.5. Chemiluminescent Detection System

1. Tropix Western *Star*™ Chemiluminescent Detection System (Applied Biosystems) for alkaline phosphatase conjugated secondary antibodies or ECL™ Detection System (Pierce) for horseradish peroxidase conjugated secondary antibodies.
2. Development folder (Applied Biosystems).

3. Chemiluminescent Imager (examples are Carestream Image Station 4000, Cell Biosciences Fluorchem®, Bio-Rad Gel Doc™ or VersaDoc™).

3. Methods

3.1. Sample Preparation

1. Dilute the sample in protein extraction buffer to approximately 2 μg/μL total protein.
2. Heat the sample at 100°C for 5 min. *Do not boil* the molecular weight markers. For commercial cell lysates, follow the manufacturer directions regarding dilutions, extraction buffer components, and heating.
3. Keep the sample at room temperature prior to loading the gel. Proceed immediately to gel electrophoresis.

3.2. 1D SDS-PAGE

SDS-PAGE is a technique to separate protein mixtures in an electric field based on size, molecular weight, and charge. Acrylamide gels are cross-linked polymers that provide a sieving function as the proteins move through the gel. The percentage of acrylamide determines the size of the pores in the gel. Low percentages of acrylamide (i.e. 8%) have large pores that allow large proteins (>150 kDa) to migrate further through the gel. Conversely, high percentage gels (i.e. 14%) have smaller pores that impede the movement of large proteins, allowing smaller proteins (<20 kDa) to migrate through the gel. Gradient gels combine various percentages of acrylamide in one gel (i.e. 4–20%) to facilitate the simultaneous separation of high- and low-molecular-weight proteins.

Prepare the gel for electrophoresis:

1. Rinse the outside of the gel cassette with Type 1 reagent grade water.
2. Mark the bottom of the wells, on the plastic gel casing, using a Whatman marking pen (see Note 8).
3. Remove the white tape from the bottom of the gel case.
4. Rinse the wells with 1× tris-glycine Running Buffer (see Note 9).
5. Place the gel securely in the electrophoresis chamber (see Note 10).
6. Fill the inner chamber with approximately 200 mL 1× tris-glycine running buffer. Observe the outer chambers for leakage. If there is no noticeable leakage, continue by adding approximately 600 mL 1× running buffer to the outer chamber. Ensure that the electrophoresis chamber is half full of buffer.
7. Remove the gel comb using a slow, even motion to avoid tearing the gel.

8. Use gel-loading pipette tips to load the samples into the appropriate wells. Load 10 μL each, in separate lanes, of MagicMark™ and BenchMark™ molecular weight markers. Load 2 μg/μL (10–20 μL) of the test samples (see Note 11).
9. When all the samples have been loaded, place the cover on the electrophoresis chamber and connect the power leads to the power source.
10. Run the gel for about 1.5 h at 125 V constant volts/gel. The expected current at the start is 30–40 and 8–12 mA/gel at the end (see Note 12).
11. The running time should be 1.5 h ± 15 min. As the protein/dye front reaches the bottom of the gel, adjust the time as needed for sufficient separation of the bands, based on the appearance of the prestained MW marker (see Note 13). While the gel is running prepare the transfer apparatus (Subheading 3.3.1).

3.3. Transfer of Proteins

Blotting is a process in which the protein is transferred to a nitrocellulose or PVDF (polyvinylidene fluoride) membrane via electrophoretic transfer. The proteins bind to the membrane via electrostatic interactions and van der Waals forces. This procedure describes a semidry transfer method.

3.3.1. Prepare the Transfer Apparatus

1. Cut two pieces of thick blot paper (or four pieces of thin blot paper) to match the size of the gel. Soak the blotter paper in transfer buffer for at least 10 min.
2. Cut one piece of PVDF membrane for each gel (same size as the blotter paper).
3. Wet the PVDF membrane in methanol for 1 min. Rinse the membrane in several rinses of distilled water to remove the methanol (see Note 14).
4. Place the PVDF membrane in transfer buffer to soak until needed. The PVDF membrane and filter papers can be added to the same container of transfer buffer.

3.3.2. Transfer of Proteins from the Gel to Blot

1. Turn off the power supply after the proteins have migrated the desired distance on the gel. Disconnect the power leads from the power supply.
2. Remove the gel assembly from the chamber and open the hard plastic casing using the gel spatula. Pry off the back of the casing – lay the casing on the counter with the longer side (front) facing up. Cut off the thick bottom of the gel, as well as the comb loading area at the top (see Note 15).
3. Place one thick, or two thin, presoaked blot papers on the Bio-Rad Trans-Blot Semidry transfer apparatus.

4. Place the PVDF membrane on top of the presoaked blot paper(s).
5. Add approximately 500 μL transfer buffer on top of the PVDF membrane to make it easier to adjust the position of the gel on top of the membrane.
6. Place the gel on top of the PVDF membrane. Wearing gloves, adjust the gel as necessary to ensure it is centered on the PVDF membrane. Pick up the gel near the bottom of the gel to avoid tearing.
7. Place the remaining blot paper(s) on top of the gel.
8. Use a clean serological plastic pipette to express any air bubbles in the gel-membrane sandwich. Roll the pipette with firm, even pressure across the surface of the top blot paper. Use a continuous even motion. Repeat the process at a 90° angle to the first surface.
9. Close the Trans-Blot SD transfer assembly. Do not adjust the lid once it has been closed. Turn on the power and set the power supply for constant amps (amperes) at 0.09 A/blot. Transfer for 1.5 h (see Notes 16).
10. At the end of the transfer, disconnect the transfer unit from the power supply. Remove the assembled filter paper, gel, and PVDF membrane from the transfer apparatus (see Note 17).
11. With tweezers, place the PVDF membrane in a clean container. Rinse the blot twice for 5 min each in wash buffer to remove the remaining methanol from the transfer buffer (see Note 18).

3.4. SNAP i.d.™ Detection System

The SNAP i.d. system uses vacuum assisted fluid distribution to rapidly and thoroughly permeate the blot with primary antibody, secondary antibody, and buffers.

1. Prepare the primary and the secondary antibody dilutions (see Note 19) (Table 1).
2. Attach the tube provided with the SNAP i.d. to the back of the unit, and connect the other end to the flask. Attach another tube to the arm on the flask and the vacuum source.
3. Open the blot holder lid and wet the inner white face with Type 1 water until the white face turns gray. Any unused wells must also be wet.
4. Place the prewet PVDF membrane in the center of the blot holder with the protein side down.
5. Roll the blot membrane gently across the PVDF membrane to remove air bubbles.
6. Place the spacer on top of the PVDF membrane making sure it covers the edges.

Table 1
Common antibody dilutions and volume requirements used with the SNAP i.d. detection system

Antibody dilution for traditional western blot	Antibody dilution for SNAP i.d.	Volume required (mL/well)	Total I-Block per well (mL/well)	Antibody required (μL)	I-Block required to dilute antibody (μL)
Single-well blot holder					
1:10,000	1:3,333	3	30	0.9	3000
1:2,000	1:667	3	30	4.5	3000
1:1,000	1:333	3	30	9	2991
1:750	1:250	3	30	12	2988
1:500	1:167	3	30	18	2982
1:250	1:83	3	30	36	2964
1:200	1:67	3	30	45	2955
Double-well blot holder					
1:10,000	1:3,333	1.5	15	0.45	1500
1:2,000	1:667	1.5	15	2.25	1500
1:1,000	1:333	1.5	15	4.5	1495.5
1:750	1:250	1.5	15	6	1494
1:500	1:167	1.5	15	9	1491
1:250	1:83	1.5	15	18	1482
1:200	1:67	1.5	15	22.5	1477.5
Triple-well blot holder					
1:10,000	1:3,333	1	10	0.3	1000
1:2,000	1:667	1	10	1.5	998.5
1:1,000	1:333	1	10	3	997
1:750	1:250	1	10	4	996
1:500	1:167	1	10	6	994
1:250	1:83	1	10	12	988

7. Roll the blot membrane gently across the PVDF membrane to remove air bubbles.
8. Close the blot holder lid.
9. Squeeze firmly at the base of the tab area to secure the lid.
10. Open the lid of the SNAP i.d. system by squeezing latch between thumb and forefingers and lifting upward.
11. Place the blot holder in the system chamber with the well side up, aligning the blot holder tabs with the notches on the chamber.
12. Close and latch the system lid.
13. Add appropriate amount of I-Block to each well being used (Table 1).
14. Immediately turn on vacuum using the vacuum control knob on the system.

15. After the wells have emptied completely (10–20 s) turn the vacuum off using the knobs.
16. If using antibody collection trays, open the lid and remove the blot holder from the system. Antibody Collection Trays can only be used with single- or double-well blot holders. If Antibody Collection Trays are not used, skip to step 20.
17. Place the antibody collection trays in the chamber with the tabs positioned toward the back of the system if using a single-well blot holder. If using a double-well blot holder, position the tabs to face each other.
18. The tabs may be labeled with a Whatman marker for antibody identification.
19. Place the blot holder back into position and close the system lid.
20. Apply the appropriate amount of primary antibody. The antibody solution must evenly cover entire blot holder surface, not just the PVDF membrane.
21. Incubate the primary antibody for 10–20 min at room temperature, with the vacuum off (see Note 20).
22. Turn the vacuum on and wait 10–20 s to make sure that the antibody solution has been completely emptied from the blot holder. There are two vacuum knobs, one on each side of the unit, and they can work independently of each other.
23. Distribute 2.5 mL of wash buffer for a single well or 1 mL of wash buffer for a double-well holder.
24. Turn the vacuum on again and wait 10–20 s. This will drive the remaining primary antibody out of the blot and into the Antibody Collection Tray.
25. Turn the vacuum off and open the lid.
26. Take out the blot holders, and then remove the antibody collection trays being careful to match the collection tray with its corresponding antibody.
27. Place the blot holders back into the SNAP i.d. and close the lid.
28. Turn the vacuum on, and with the vacuum running continuously, wash the blot with appropriate amount of I-Block three times (Table 1). Each wash should take 10–20 s to complete. When the blot holder is empty, turn the vacuum off (see Note 21).
29. With the vacuum off, apply the appropriate amount of secondary antibody evenly across the surface (Table 1).
30. Incubate the secondary antibody for 10 min at room temperature with the vacuum off.
31. Turn the vacuum on and wait for 10–20 s to make sure that the antibody solution has been completely emptied from the blot holder.

32. With the vacuum running continuously, wash the blot three times with appropriate amounts of I-Block (Table 1). Each wash should take 10–20 s to be completed. When the blot holder is empty, turn the vacuum off.
33. Remove the blot holder from the system, place it on the bench with the well side down, and open the lid. Use forceps to remove and discard the spacer.
34. Remove the blot and incubate with the appropriate detection reagent.

3.5. Immunodetection

Polyclonal or monoclonal antibodies directed against the protein of interest are used for immunodetection. Enzymes bound to the secondary antibody react with a substrate that produces light (chemiluminescence) or fluorescence. Chemiluminescence is a common western blot detection system. Immunodetection of antigen-specific antibodies with secondary antibodies conjugated to alkaline phosphatase (AP) is one example of an enzyme-linked immunodetection method (Tropix, Western *Star*™, Applied Biosystems). A chemiluminescent substrate is added to the membrane and subsequently dephosphorylated by the alkaline phosphatase that is linked to the secondary antibody. Once the phosphate group is removed from the substrate an unstable intermediate is generated that emits light while spontaneously decaying to the ground energy level (15).

Enhanced chemiluminescent (ECL™) detection systems use horseradish peroxidase (HRP) conjugated secondary antibodies coupled with enhanced chemiluminescent substrates for the hydrogen-peroxide-catalyzed oxidation of luminol. The detection reagents consist of Detection Reagent Solution 1 (ECL substrate solution containing oxidant) and Detection Reagent Solution 2 (Luminol, enhancer, substrate buffer). In the presence of hydrogen peroxide, the horseradish peroxidase catalyzes the conversion of luminol to 3-aminophthalate* that spontaneously decays to 3-aminophthalate emitting light (16).

Both Alkaline Phosphatase and Horseradish Peroxidase detection chemistries are based on the prolonged emission of light that can be recorded on X-ray film or by a CCD (charge-coupled device) camera.

3.5.1. Alkaline Phosphatase Conjugated Secondary Antibodies

1. Prepare Western*Star* Assay buffer. Add 2 mL of 10× Western*Star* Assay Buffer to 18 mL of Type 1 water. This is sufficient for one PVDF membrane.
2. Wash the membrane twice for 2 min each in 10–20 mL of the 1× Assay Buffer (see Note 22).
3. Place a sheet of plastic wrap on a flat surface, away from drafts.

4. Blot the edge of the PVDF membrane on a paper towel. Place the PVDF membrane on the plastic wrap. Dispense 3 mL of CDP-Star Ready-to-Use substrate solution onto the PVDF membrane.
5. Incubate undisturbed for 5 min.
6. Blot excess substrate solution from the PVDF membrane and place it in a development folder. Smooth out bubbles or wrinkles in the membrane.
7. Use an imager (CCD camera) without an incident light source and emission filter to image the blot (see Notes 23 and 24).
8. Review the image for appropriate bands. Increase or decrease the exposure as needed (see Note 25).

3.5.2. Horseradish Peroxidase Conjugated Secondary Antibodies

1. Combine 2 mL of ECL Reagent 1 with 2 mL ECL Reagent 2.
2. Place a sheet of plastic wrap on a flat surface away from drafts.
3. Blot the edge of the PVDF membrane on a paper towel. Place the PVDF membrane on the plastic wrap.
4. Use a pipette to dispense 4 mL of the ECL Reagent (1+2) onto the PVDF membrane.
5. Incubate for 3 min at room temperature.
6. Blot excess substrate solution from the PVDF membrane and place it in a development folder. Smooth out bubbles or wrinkles in the membrane.
7. Use an imager (CCD camera) without an incident light source and emission filter, or X-ray film and a film developer, to image the blot (see Notes 23 and 24).
8. Review the image for appropriate bands. Increase or decrease the exposure as needed (see Note 25).

3.6. Validation Assessment

Antibodies may be considered to be validated if any of the following conditions are met (8, 10):

1. The presence of a single band at the expected molecular weight.
2. The presence of a single band at the expected molecular weight plus one additional band whose signal intensity is less than 20% of the signal of the band at the expected molecular weight.
3. The presence of multiple bands that can be shown to be cleavage products, splice variants, or isoforms, of the parent protein (3, 10).
4. The presence of a band at the expected molecular weight plus additional bands that can be attributed to be reactive only with the secondary antibody (see Note 26).

4. Notes

1. The presence of specific and nonspecific antibody binding may be difficult to discriminate with techniques such as protein arrays in which many analytes/samples are simultaneously detected. Each antigen–antibody reaction has only one total output signal, regardless if this signal resulted from the sum of many nonspecific binding events or from a single specific antigen–antibody interaction. The specificity of each primary antibody is of utmost importance to ensure specific detection of each different antigen.
2. 5,000–10,000 cells captured by laser capture microdissection are sufficient for 1D-gel electrophoresis and immunoblotting. This correlates to approximately 1,000–2,000 shots using the 30-μm spot size on the Arcturus PixCell IIc or Arcturus^{XT} laser capture microdissection instrument (17). Nondenaturing western blotting can be performed using RIPA buffer (1% IGEPAL, 0.5% sodium deoxycholate, 0.1% SDS, Phosphate Buffered Saline). The expected molecular weight may vary in denaturing conditions compared to native conditions due to the presence of tertiary protein structure. Movement of a large, globular protein in SDS-PAGE will be hindered compared to the denatured form of the same protein.
3. Serum and plasma contain large quantities of immunoglobulins and albumin that can interfere with the detection of proteins at 25 kDa (light chains), 50 kDa (heavy chains), and 65 kDa (albumin). Dilute serum or plasma 1:20 in 0.9% saline (NaCl) prior to denaturation in extraction buffer.
4. Recombinant proteins, or GST-tagged, or HIS-tagged recombinant proteins, may be used as a reference sample to show the expected molecular weight of the protein. It is important to add the molecular weight of the "tag" to the protein when comparing the expected molecular weight of the protein to the observed molecular weight of the tagged recombinant protein. The tagged recombinant protein should have a higher observed molecular weight compared to the expected molecular weight.
5. The type of gel and corresponding running buffer and transfer buffer should be selected based on (1) the molecular weight of the protein and (2) the compatibility of the gels and buffers. In general, high-molecular-weight proteins (>150 kDa) migrate further on 4% gels, whereas low-molecular-weight proteins (<25 kDa) migrate optimally on 12% gels. 4–20% gels are general all-purpose gels for proteins between 15 and 180 kDa. The migration distance of a protein varies with the protein's molecular weight, percentage of acrylamide in the gel, and length of gel running time.

6. The Benchmark™ ladder is a prestained marker for direct visual assessment of the gel/blot, while the MagicMark™ standard is a chemiluminescent marker that provides protein marker assessment on the blot during detection. To combine the markers in one sample/lane, add 6 μL of the Benchmark™ protein ladder with 4 μL of the MagicMark™ Western Protein Standard in a microcentrifuge tube. Mix well. Load entire 10 μL mixture in one lane of the gel. Precaution: The MagicMark™ marker has less sensitivity with mouse monoclonal antibodies compared to rabbit antibodies. This diminished sensitivity is more apparent when the MagicMark™ and BenchMark™ markers are combined because the MagicMark™ marker is more diluted.
7. The methanol in the transfer buffer may evaporate during storage. An alternative method of preparing transfer buffer is to prepare a tris-glycine solution in water without additional methanol. Soak the PVDF membrane in 20 mL methanol for 3 min prior to transfer. Add 80 mL of tris-glycine transfer buffer (minus methanol) to the methanol/PVDF. Allow the PVDF membrane to soak in the methanol/tris-glycine transfer buffer for at least 5 min prior to transfer.
8. It is much easier to visualize the location of the well during sample loading by delineating the bottom of the well with a Whatman marking pen.
9. Rinsing the wells with running buffer can be accomplished in one of two ways. The first way is to fill a squirt bottle with 1× running buffer and then gently squirt the running buffer into the wells, allowing the buffer to overflow each well. The second method is to fill a gel loading pipette with 1× running buffer, insert the tip into each well, and dispense the buffer. The squirt bottle technique is faster than the pipette technique.
10. When running two gels at the same time it is helpful to number the two gels with a Whatman marker to distinguish them from each other. Another method to identify the gels is to load the molecular weight markers in different lanes on each gel. Ensure that both gels face outward in the gel box.
11. The advantage of loading the MagicMark™ and Benchmark™ separately is that the Benchmark™ will not obscure the 60–80 kDa range when the blot is imaged. Loading the MagicMark™ as the first lane on the gel will avoid a common problem of carry-over into the next lane. Load the Benchmark™ marker in the second lane next to the MagicMark™, to prevent carryover signal from the MagicMark™ lane into a sample lane. Load 10 μL of Novex 2× SDS loading buffer in any unused wells. If the wells are left empty, the electrical field will not be uniform and the proteins will migrate at different rates in each lane.

12. If running two gels at constant amps rather than constant volts, double the amperes (0.07 A). In this case, do not adjust the voltage as it will adjust automatically. If using two electrophoresis chambers to run four gels at constant amps, do not quadruple the constant amperes. Quadrupling constant amps causes the proteins to run off the bottom of the gel. Two output jacks need to be used to accommodate two electrophoresis chambers therefore each power supply should be set to 0.07 constant amps.
13. It is important to check the gel after 45–50 min to verify the rate of protein migration. If the proteins migrate more quickly than expected, low-molecular-weight proteins may migrate off the gel.
14. PVDF membranes are highly hydrophobic and prewetting in methanol is necessary to make them polar/hydrophilic.
15. The bottom of a 4–20% gradient gel is slightly thicker than the top of the gel near the wells. Handle the gel near the bottom to prevent tearing or ripping the gel.
16. When transferring two blots at the same time double the constant amperes and set to 0.18A without changing the volts as they will change automatically.
17. One blot membrane may be cut into strips to probe each strip with a different set of primary/secondary antibodies. Prior to removing the PVDF membrane from the transfer apparatus, use a Whatman marking pen to label the PVDF membrane strips. It is also helpful to mark each band of the BenchMark™ prestained molecular weight marker. The marker often fades during blocking. For orientation marks, we draw a short horizontal line for each blue band and a jagged line for the pink band of the BenchMark™ marker. The MagicMark™ marker will not be visible.
18. Commercial antibody manufacturers recommend a variety of wash and blocking buffers. I-Block is a casein-based blocking solution that has a similar protein composition to dry milk based blocking solutions. We strongly recommend using the manufacturer suggestions for blocking and wash buffers prior to using an alternative buffer.
19. Follow manufacturer's recommendations for optimal antibody dilutions if stated in the product description. If low sensitivity is suspected, increase the concentration of the primary antibody, but be aware that this will also increase the background and unspecific signal. The secondary antibody must be raised against the same species as the primary antibody. For example, a mouse primary antibody must be used with an anti-mouse secondary antibody. The secondary antibodies are conjugated to an enzyme, either alkaline phosphatase or horseradish peroxidase,

which catalyze the chemiluminescent reaction. HRP-conjugated secondary antibodies are used with the ECL detection reagents. Alkaline Phosphatase-conjugated antibodies are used with the Tropix Western Star detection kit.

20. All antibody concentrations need to be three times as concentrated using the SNAP i.d. system compared to traditional Western Blot as per the manufacturer's instructions.
21. If the time required to suction the fluid away from the membrane starts to increase try pressing the lid down tightly toward the base to improve the vacuum.
22. This step is fundamental to get rid of the phosphate that is present in the I-Block buffer that would otherwise inhibit the reaction of the alkaline phosphatase with its substrate, resulting in no signal or diminished signal.
23. The kinetics of the chemiluminescence reaction performed with the suggested kits has duration of 6–8 h, but peaks around 20 min to 1 h from the addition of the substrate. It is, therefore, more desirable to perform all the exposures and acquisitions within the first hour from the beginning of the reaction.
24. If low sensitivity is suspected, increase the concentration of the primary antibody and/or secondary antibody. Increasing antibody concentrations will also increase the background and nonspecific signal. An alternative method to increase sensitivity is to increase incubation times or change the reaction temperature, for example incubate at room temperature versus 4°C.
25. High background may be due to (a) too long exposure time, (b) incomplete blocking, or (c) improper washing or incomplete removal of methanol from the membrane prior to blocking with I-Block. Bacterial contamination of one or more of the components of the system can also cause high background.
26. If multiple bands are present, plus a band at the expected molecular weight, the blot should be stripped to remove the primary antibody, washed, blocked, and reprobed with secondary antibody only. The presence of the same extra bands with secondary antibody only, as seen on the original blot, indicates nonspecific staining due to secondary antibody and the antibody could be considered to be validated. If the additional bands are not present when stained with secondary antibody only, the bands are most likely due to reactivity with the primary antibody and the antibody should not be considered to be validated.

Acknowledgments

This work was supported in part by George Mason University and the Istituto Superiore di Sanità, Rome, Italy.

References

1. Elliott, S., Busse, L., McCaffery, I., Rossi, J., Sinclair, A., Spahr, C. et al. (2010) Identification of a sensitive anti-erythropoietin receptor monoclonal antibody allows detection of low levels of EpoR in cells. *J Immunol Methods* 352, 126–39.
2. Elliott, S., Busse, L., Bass, M. B., Lu, H., Sarosi, I., Sinclair, A. M. et al. (2006) Anti-Epo receptor antibodies do not predict Epo receptor expression. *Blood* 107, 1892–5.
3. Bordeaux, J., Welsh, A., Agarwal, S., Killiam, E., Baquero, M., Hanna, J. et al. (2010) Antibody validation. *Biotechniques* 48, 197–209.
4. Knottnerus, J. A., van Weel, C., Muris, J. W. (2002) Evaluation of diagnostic procedures. *Bmj* 324, 477–80.
5. Pozner-Moulis, S., Cregger, M., Camp, R. L., Rimm, D. L. (2007) Antibody validation by quantitative analysis of protein expression using expression of Met in breast cancer as a model. 87, 251–260.
6. Skliris, G. P., Rowan, B. G., Al-Dhaheri, M., Williams, C., Troup, S., Begic, S. et al. (2009) Immunohistochemical validation of multiple phospho-specific epitopes for estrogen receptor alpha (ERalpha) in tissue microarrays of ERalpha positive human breast carcinomas. *Breast Cancer Res Treat* 118, 443–53.
7. Dews, I. C., Mackenzie, K. R. (2007) Transmembrane domains of the syndecan family of growth factor coreceptors display a hierarchy of homotypic and heterotypic interactions. *Proc Natl Acad Sci U S A* 104, 20782–7.
8. VanMeter, A. J., Rodriguez, A. S., Bowman, E. D., Jen, J., Harris, C. C., Deng, J. et al. (2008) Laser capture microdissection and protein microarray analysis of human non-small cell lung cancer: differential epidermal growth factor receptor (EGPR) phosphorylation events associated with mutated EGFR compared with wild type. *Mol Cell Proteomics* 7, 1902–24.
9. Hsi, E. D. (2001) A practical approach for evaluating new antibodies in the clinical immunohistochemistry laboratory. *Arch Pathol Lab Med* 125, 289–94.
10. Bjorling, E., Uhlen, M. (2008) Antibodypedia, a portal for sharing antibody and antigen validation data. *Mol Cell Proteomics* 7, 2028–37.
11. Burnette, W. N. (1981) "Western blotting": electrophoretic transfer of proteins from sodium dodecyl sulfate--polyacrylamide gels to unmodified nitrocellulose and radiographic detection with antibody and radioiodinated protein A. *Anal Biochem* 112, 195–203.
12. Laemmli, U. K. (1970) Cleavage of structural proteins during the assembly of the head of bacteriophage T4. *Nature* 227, 680–5.
13. Towbin, H., Staehelin, T., Gordon, J. (1979) Electrophoretic transfer of proteins from polyacrylamide gels to nitrocellulose sheets: procedure and some applications. *Proc Natl Acad Sci U S A* 76, 4350–4.
14. Towbin, H., Staehelin, T., Gordon, J. (1989) Immunoblotting in the clinical laboratory. *J Clin Chem Clin Biochem* 27, 495–501.
15. Bronstein, I., Voyta, J. C., Murphy, O. J., Bresnick, L., Kricka, L. J. (1992) Improved chemiluminescent western blotting procedure. *Biotechniques* 12, 748–53.
16. Whitehead, T. P., Kricka, L. J., Carter, T. J., Thorpe, G. H. (1979) Analytical luminescence: its potential in the clinical laboratory. *Clin Chem* 25, 1531–46.
17. Espina, V., Wulfkuhle, J. D., Calvert, V. S., VanMeter, A., Zhou, W., Coukos, G. et al. (2006) Laser-capture microdissection. *Nat Protoc* 1, 586–603.

Chapter 11

Laser Capture Microdissection: Arcturus^{XT} Infrared Capture and UV Cutting Methods

Rosa I. Gallagher, Steven R. Blakely, Lance A. Liotta, and Virginia Espina

Abstract

Laser capture microdissection (LCM) is a technique that allows the precise procurement of enriched cell populations from a heterogeneous tissue under direct microscopic visualization. LCM can be used to harvest the cells of interest directly or can be used to isolate specific cells by ablating the unwanted cells, resulting in histologically enriched cell populations. The fundamental components of laser microdissection technology are (a) visualization of the cells of interest via microscopy, (b) transfer of laser energy to a thermolabile polymer with either the formation of a polymer–cell composite (capture method) or transfer of laser energy via an ultraviolet laser to photovolatize a region of tissue (cutting method), and (c) removal of cells of interest from the heterogeneous tissue section. Laser energy supplied by LCM instruments can be infrared (810 nm) or ultraviolet (355 nm). Infrared lasers melt thermolabile polymers for cell capture, whereas ultraviolet lasers ablate cells for either removal of unwanted cells or excision of a defined area of cells. LCM technology is applicable to an array of applications including mass spectrometry, DNA genotyping and loss-of-heterozygosity analysis, RNA transcript profiling, cDNA library generation, proteomics discovery, and signal kinase pathway profiling. This chapter describes the unique features of the Arcturus^{XT} laser capture microdissection instrument, which incorporates both infrared capture and ultraviolet cutting technology in one instrument, using a proteomic downstream assay as a model.

Key words: Cancer, Laser capture microdissection, Tissue, Infrared, Ultraviolet, DNA, Molecular profiling, Protein, RNA, Tissue heterogeneity

1. Introduction

Cellular heterogeneity of tissue is a common problem encountered by both genomic and proteomic researchers during tissue analysis. Molecular analysis of heterogeneous tissue is hindered by extreme variability and inaccuracy because it is impossible to discern which cells contribute which cellular constituents to a given tissue lysate (1). Molecular profiling of pure cell populations, which is reflective of the cell population's in vivo genomic and proteomic state, is

Virginia Espina and Lance A. Liotta (eds.), *Molecular Profiling: Methods and Protocols*, Methods in Molecular Biology, vol. 823, DOI 10.1007/978-1-60327-216-2_11, © Springer Science+Business Media, LLC 2012

essential for correlating molecular signatures in normal and diseased tissue (2–8). Laser capture microdissection (LCM) is a technique that allows the identification, selection, and isolation of pure cell populations from a heterogeneous tissue section or cytological preparation under direct microscopic visualization of the cells (7, 8). LCM enables researchers to isolate normal, premalignant, and malignant cells without contamination from surrounding cells (1–3, 5, 6, 9–14). Xenograft tissue may be isolated from the host via LCM (15). Downstream analysis of the microdissected cells may be performed with any method that has adequate sensitivity. Recently, a method for in situ proteomic analysis of microdissected cells has been described, permitting potential biomarker discovery in small numbers of breast cells (16).

LCM technology encompasses two general technologies, infrared (IR) laser capture systems (7, 8) and ultraviolet (UV) laser cutting systems (17–20). The fundamental components of LCM technology are (a) visualization of the cells of interest via microscopy, (b) transfer of laser energy to a thermolabile polymer with formation of a polymer–cell composite (IR system) or photo volatilization of cells surrounding a selected area (UV system), and (c) removal of the cells of interest from the heterogeneous tissue section (Fig. 1). The Arcturus$^{XT™}$ system (Applied Biosystems/Life Technologies) discussed herein incorporates both laser types in one instrument providing options as to the type of microdissection to be performed.

A stationary near-infrared laser mounted in the optical axis of the microscope stage is used for melting a thermolabile polymer film (see Note 1). The polymer film is manufactured on the bottom surface of an optical-quality plastic support cap. The cap acts as an optic for focusing the laser in the same plane as the tissue section. The polymer melts only in the vicinity of the laser pulse, forming a polymer–cell composite. A dye integrated into the polymer serves two purposes (1) it absorbs laser energy, preventing damage to the cellular constituents, and (2) it aids in visualizing areas of melted polymer. Lifting the polymer from the tissue surface shears the embedded cells of interest away from the heterogeneous tissue section (Fig. 2). The exact cellular morphology, as well as the DNA, RNA, and proteins of the procured cells, remain intact and bound to the polymer.

Ultraviolet lasers photovolatilize cells and/or the mounting medium, typically a polyethylene naphthalate (PEN) membrane. In UV cutting mode, a microdissection cap is placed on the surface of the tissue, encompassing the area of photovolatilization, and the infrared laser melts the thermolabile polymer in several defined areas around the edge of the cut area, thus holding the cut area of tissue/membrane in place. This step is important for preventing accidental removal of the cut tissue area from the slide. Following microdissection, extraction buffer directly applied to the polymer

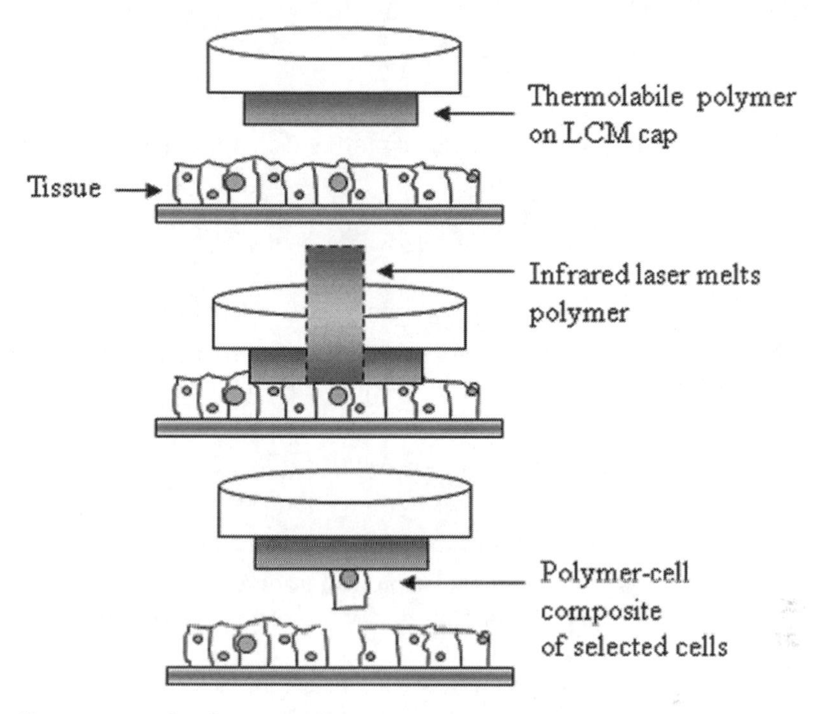

Fig. 1. Principles of laser capture microdissection (LCM). A thermolabile polymer is placed directly on top of a tissue section on a glass slide. A near-infrared laser melts the polymer in the vicinity of the selected cells. A polymer–cell composite is formed. The polymer-embedded cells are removed from the tissue by lifting the cap away from the tissue, which shears the captured cells from the remaining tissue on the slide.

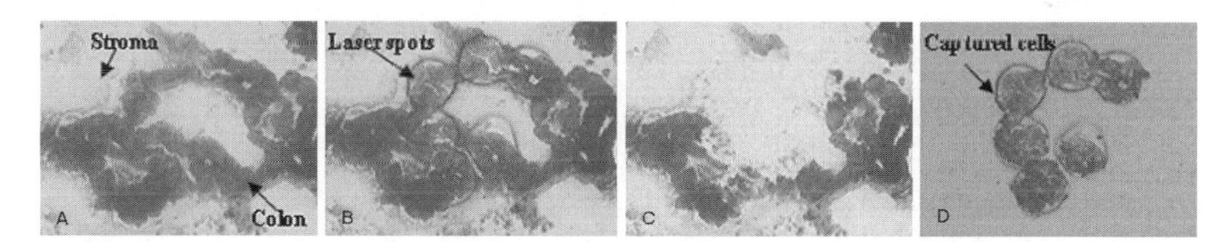

Fig. 2. Infrared LCM process. (**a**) A section of tissue is mounted and stained on a glass slide and viewed under high magnification. (**b**) Cells of interest are identified visually and designated for microdissection using software annotation tools. The infrared laser locally expands a thermoplastic polymer that captures the cell(s) only in the vicinity of the laser pulse. (**c**, **d**) Microdissection occurs when the film is lifted from the tissue section, shearing the selected cells from the tissue section. The presence of stained material inside the melted margins of the polymer indicates effective microdisssection.

film solubilizes the cells, allowing the collection of the molecules of interest for downstream analysis. This chapter presents microdissection of frozen tissue sections for proteomic analysis as an illustration of the operation of the Arcturus^{XT} in both capture and cutting modes.

2. Materials

Tissue for microdissection is usually prepared as a cryosection, or paraffin-embedded section, on a glass slide. The tissue section is stained for histomorphologic identification of cells. The stained section is used immediately for LCM.

2.1. Preparation of Tissue Sections for Infrared Capture Mode

1. Uncharged, precleaned glass microscope slides, 25 mm×75 mm.
2. Specimen for protein analysis: frozen tissue of interest or frozen sections cut at 2–15 µm (thickness of 5–8 µm is optimal for capture methods) (see Note 2).
3. Cryopreservation solution (Tissue-Tek® OCT™) (Sakura Finetek) for embedding frozen tissue sections.
4. Cryomolds.
5. Dry ice; inhalation and contact hazard. Use with appropriate ventilation and personal protective equipment.

2.1.1. Preparation of Tissue Sections for UV Cutting Mode

1. PEN membrane or PEN frame slides, 25 mm×75 mm.
2. Specimen for protein analysis: frozen tissue of interest cut at 2–200 µm (see Note 2).
3. Cryopreservation solution (Tissue-Tek® OCT™) (Sakura Finetek) for embedding frozen tissue sections.
4. Cryomolds.
5. Dry ice.

2.2. Hematoxylin and Eosin Staining

1. Mayer's Hematoxylin Solution (Sigma). Inhalation and contact hazard; wear gloves when handling.
2. Eosin Y Solution, alcoholic (Sigma). Highly flammable; store away from heat, sparks, and open flames. Contact hazard; wear gloves when handling.
3. Scott's Tap Water Substitute (Fisher), also known as Blueing Solution (see Note 3).
4. Ethanol (ethyl alcohol, absolute, 200 proof molecular biology grade). Flammable; store away from heat, sparks, and open flames.
5. Ethanol gradient: 70% (v/v in dH_2O), 95 and 100% ethanol (see Note 4).
6. Type 1 reagent-grade water (dH_2O).
7. Xylene. Xylene vapor is harmful and can be fatal if inhaled. Use in a well-ventilated area and discard in appropriate hazardous waste container. Flammable; contact hazard; wear gloves when handling.

8. Protease inhibitors (Complete Protease Inhibitor Cocktail tablets; Roche). Use one inhibitor tablet for every 10 mL of solution (see Note 5).

2.3. Laser Capture Microdissection Infrared Mode

1. Arcturus^{XT} laser capture microdissection system (Applied Biosystems).
2. CapSure® Macro LCM Caps (Applied Biosystems) are recommended for protein analysis.
3. CapSure® HS LCM Caps (Applied Biosystems) are recommended for DNA or RNA downstream analysis (see Note 6).
4. SealRight tubes 0.5 mL (USA Scientific) or GeneAmp® 0.5-mL Thin-walled Reaction Tubes with Domed Cap (cat. No. N801-0611; Applied Biosystems) (see Note 7).

2.4. Laser Capture Microdissection UV mode

1. Arcturus^{XT} laser capture microdissection system (Applied Biosystems).
2. CapSure® Macro LCM Caps (Applied Biosystems) are recommended for protein analysis.
3. CapSure® HS LCM Caps (Applied Biosystems) are recommended for DNA or RNA downstream analysis (see Note 6).
4. SealRight tubes 0.5-mL (USA Scientific) or GeneAmp® 0.5-mL Thin-walled Reaction Tubes with Domed Cap (cat. No. N801-0611; Applied Biosystems) (see Note 7).

2.5. Extraction Buffer for Cellular Constituent of Interest

Protein extraction buffer: 450 μL T-PER Tissue Protein extraction reagent (Pierce), 450 μL Novex® Tris–glycine 2× SDS loading buffer (Invitrogen), and 100 μL (10% v/v) TCEP [Tris (2-carboxyethyl)phosphine)] Bond Breaker® (Pierce).

3. Methods

The protocols described below illustrate (a) sample preparation of frozen tissue sections, (b) hematoxylin and eosin (H&E) tissue staining for frozen sections and formalin-fixed paraffin-embedded (FFPE) sections, (c) infrared capture mode microdissection, (d) ultraviolet cutting mode microdissection using the Arcturus^{XT}, and (e) cell lysis and protein extraction. The preferred specimen for protein analysis is frozen tissue, whereas ethanol-fixed or formalin-fixed tissue is acceptable for microdissection with RNA and DNA analysis, respectively.

3.1. Frozen Tissue Sectioning

Tissue biopsy samples should be embedded in a cryopreservative solution or snap frozen as soon as the specimen is excised from the patient/animal. Prompt preservation of the sample limits protein and RNA degradation as a result of protease and RNase activity, respectively, and limits reactive changes in phosphorylated proteins (21–24).

1. Embed tissue directly in a cryomold, covering the tissue with cryopreservative solution (OCT). Place OCT-embedded tissue on dry ice, or at –80°C, to freeze the tissue. Store the frozen tissue at –80°C (see Note 8).
2. Cut frozen sections at 2–15 μm thickness (5–8 μm is optimal for IR laser capture) and place sections on labeled, uncharged, precleaned glass microscope slides (see Note 9). Position the tissue section near the center of the slide, avoiding the top and bottom thirds of the slide (see Note 10). Place the slide directly on dry ice or keep it in the cryostat at –20°C or colder, or place the slide in a prechilled slide box on dry ice until the slides can be stored at –80°C. Alternatively, the frozen section can be immediately stained and microdissected (see Note 11).
3. Paraffin-embedded sections for DNA analysis can be stored at room temperature indefinitely prior to microdissection. Ethanol-fixed sections for RNA or protein analysis can be stored at room temperature for up to 3 months. Frozen sections for RNA analysis can be stored at –80°C for 1 month. Frozen sections for protein analysis can be stored at –80°C for up to 3 months.

3.2. Hematoxylin and Eosin Staining

Selection of tissue staining protocols should be based on compatibility with the downstream analysis to be performed with the microdissected tissue (see Note 12). Staining protocols allow visualization and identification of the tissue or cells of interest with a standard inverted light microscope. Incorporation of protease inhibitors in the staining reagents, along with a microdissection session limited to 1 h, minimizes protein degradation during the staining process. Skin tissue, cartilage, and samples prepared on charged slides might be difficult to microdissect and may require additional slide or tissue treatments (see Note 13) (25). The staining procedure employed depends on whether the tissue is frozen or paraffin-embedded.

3.2.1. H&E Staining Procedure for Frozen Tissue Sections

1. Remove the frozen section slide from freezer/dry ice and proceed immediately with the staining protocol. Do not allow the slide to thaw.
2. Dip the slide in each of the following solutions, for the time indicated (see Note 14). Blot the slide on absorbent paper in between the different solutions to prevent carryover from the previous solution:
   (a) 70% Ethanol fixative, 3–10 s
   (b) dH_2O, 10 s
   (c) Mayer's hematoxylin, 15–20 s
   (d) dH_2O, 10 s

(e) Scott's Tap Water Substitute, 10 s
(f) 70% Ethanol, 10 s
(g) Eosin Y (optional), 3–5 s (see Note 15)
(h) 95% Ethanol, 10 s
(i) 95% Ethanol, 10 s
(j) 100% Ethanol, 30–60 s (see Note 16)
(k) 100% Ethanol, 30–60 s
(l) Xylene, 30–60 s (see Note 17)
(m) Xylene, 30–60 s

3. Allow the stained slide to air dry as quickly as possible (see Note 18).
4. Proceed immediately with microdissection. Do not coverslip the slide.

3.2.2. H&E Staining Procedure for Formalin-Fixed or Ethanol-Fixed Paraffin-Embedded Tissue Sections

1. Paraffin-embedded tissue sections must be deparaffinized and rehydrated to allow staining of the tissue elements. Dip the slide in each of the following solutions, for the time indicated. Blot the slide on absorbent paper in between the different solutions to prevent carryover from the previous solution:

(a) Xylene, 5–15 min (see Note 19)
(b) Xylene, 5–15 min
(c) 100% Ethanol, 30 s
(d) 95% Ethanol, 30 s
(e) 70% Ethanol, 30 s
(f) dH_2O, 10 s
(g) Mayer's hematoxylin, 15 s
(h) dH_2O, 10 s
(i) Scott's Tap Water Substitute, 10 s
(j) 70% Ethanol, 10 s
(k) Eosin Y (optional), 3–5 s (see Note 15)
(l) 95% Ethanol, 10 s
(m) 95% Ethanol, 10 s
(n) 100% Ethanol, 30–60 s
(o) 100% Ethanol, 30–60 s
(p) Xylene, 30–60 s
(q) Xylene, 30–60 s

2. Air dry the slide as quickly as possible.
3. Proceed immediately with microdissection. Do not coverslip the slide (see Note 18).

3.3. Laser Capture Microdissection: Infrared Capture Mode

The Arcturus^{XT} system combines IR capture microdissection (LCM) and ultraviolet (UV) laser cutting in one instrument. UV laser cutting microdissection allows the "cut and capture" of cells of interest by first ablating unwanted cells, thus preventing contamination during cell capturing. This system is particularly useful for microdissection of tissue sections up to 200-μm thick, such as plant tissue sections (5, 9).

The Arcturus^{XT} instrument features imaging software for creating stitched images of the tissue, allowing the user to identify differences in cellular morphology during cell selection more accurately. The Graphical User Interface permits the control of all operations in the system, including stage translation, slide selection, focus and light intensity, laser parameters, objective selection, cap transfers, and camera settings (Fig. 3). Images may be captured, annotated, and saved as JPEG or TIFF files; live video can also be taken at any point during the microdissection process. This automated system is

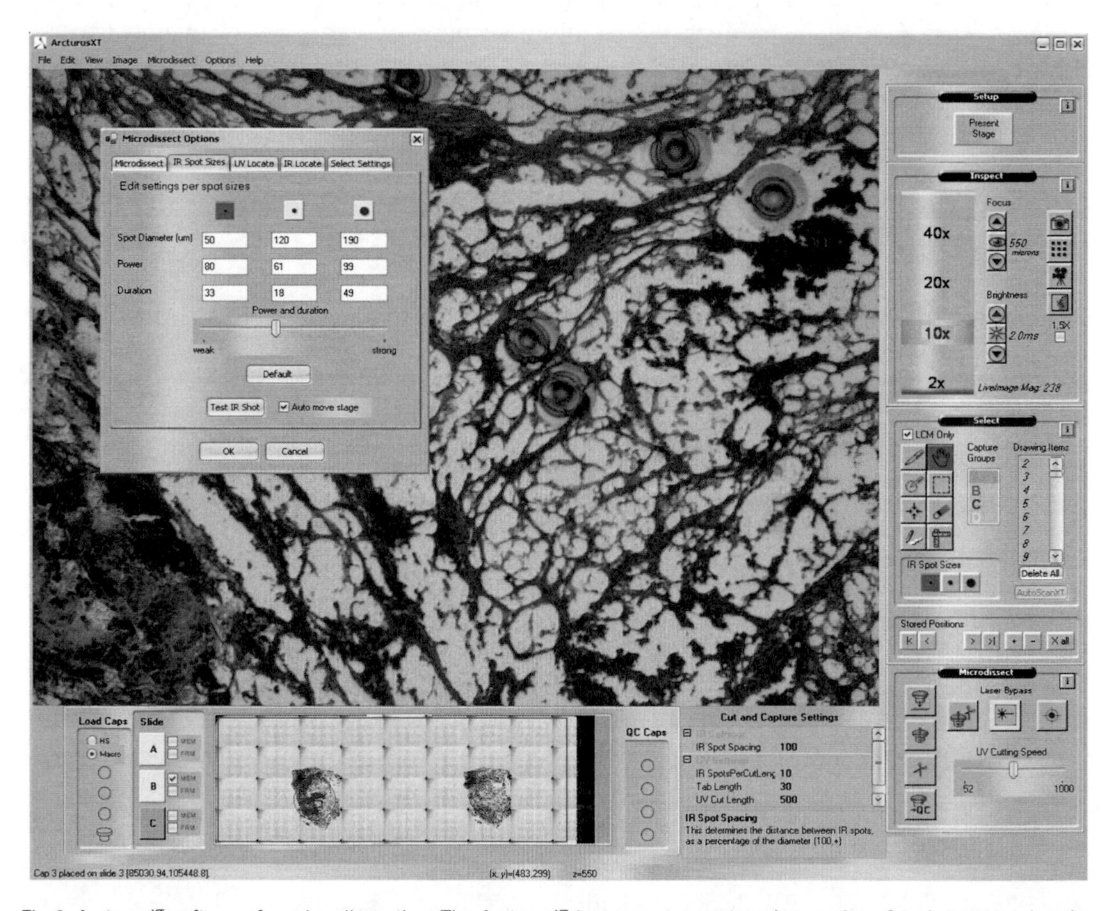

Fig. 3. Arcturus^{XT} software for microdissection. The Arcturus^{XT} instrument operates via a series of tool panes and option buttons. Each step of the microdissection process and the instrument can be accessed from the main software screen. *Inset*: Example of the "Microdissect" options tool pane showing IR laser spot size selection with laser power and duration adjustment options.

equipped with a trackball activated stage and a mouse for navigation across a slide. An interactive touch screen (stylus) monitor permits selection of single cells or groups of cells (26).

A variety of slide types can be used with this automated system including glass, glass membrane, and/or framed membrane slides. This flexibility, along with the IR laser and UV laser capability of the system, allows the user to technically prepare any specimen for automated laser microdissection (see Note 20).

Microdissection is performed without cover slips or immersion oils. Lack of immersion fluids on any of the optics prevents refraction of light from the tissue image. Thus, the color and detail of a given tissue stain is lost as the stained slide dries. Manual LCM methods capitalize on the index refraction of a wet tissue slide for visualizing and reviewing an index-matched image of the tissue (11, 14). The Arcturus^{XT} allows an index-matched image or images to be digitally saved which can be used to directly mark the cells of interest for microdissection.

The microdissection process consists of six steps: (1) loading the slides and LCM caps, (2) locating the cells of interest, (3) LCM cap placement and laser location, (4) marking the cells of interest, (5) capturing the cells, and (6) unloading of the samples and caps containing the captured tissue.

3.3.1. Arcturus^{XT} Instrument Setup

1. Turn on the PC, then the Arcturus instrument, and touch screen monitor.
2. Open the Arcturus^{XT} software program by double clicking on the Arcturus^{XT} software icon.
3. Click on "Present Stage" in the SetUp tool pane. The microscope stage moves forward to allow loading of CapSure® caps and slides.
4. Load the CapSure® caps by sliding a CapSure® cap cartridge into the stage slot. Each cartridge holds four individual LCM caps.
5. Load up to three slides on the stage.
6. Remove any existing caps in the QC/Unload area.
7. Click the "i" options button to open the "Load Options" dialog box. Enter information about your slides and caps:
   (a) Check each slide that is loaded.
   (b) Check "Load with Overview" to create a full slide overview images.
   (c) Select the type of slide: glass, membrane, or frame.
   (d) Enter a slide name/ID in the "SlideName" field.
   (e) Enter any comments for each slide in the "SlideNotes" field.
   (f) Click on the "Caps" tab to enter information about the caps.

(g) Select the type of cap: Macro or HS.

(h) Check the box corresponding to each cap that is loaded.

(i) Click on "File Paths" tab to enter information regarding the location of saved images. Enter your desired file path/ name information.

(j) Click "OK."

3.3.2. Inspect Image and Locate Cells of Interest

1. Use the trackball or mouse to move the stage to an area for microdissection. The live video image is displayed on the monitor. Alternatively, you can tap the stylus on the overview image at the location of interest.
2. To view a different slide, tap the slide button for the slide of interest (see Note 21).
3. In the "Inspect" tool pane, select the desired objective, brightness, and focus by tapping on the corresponding button. The selected objective is red. "Autobrightness" and "Autofocus" settings are the middle icon between the up and down arrows (see Note 22).
4. Locate the cells of interest using the trackball, mouse, or stylus to move the stage to the desired location (see Note 23).

3.3.3. LCM Cap Placement and Laser Location

1. Place a cap on the slide by clicking "Place Cap" icon in the "Microdissect" tool pane. The instrument places a cap at the center of the red box in the slide overview image. The apparent size of this box will change depending on the microscope objective. The cap location is outlined in green.
2. Use the trackball to move to an area without tissue but still under the cap. This can be at the side of the tissue, a luminal area, or any other area without cells.
3. Tap the "i" options button in the "Select" tool pane. Tap the "IR Spot Sizes" tab.
4. Click on the desired spot size button. Click "OK".
5. Double click the mouse in the area without tissue to fire a test IR laser shot. This test spot allows you to verify the polymer wetting, spot size, and IR laser location.
6. Place the mouse cursor directly in the center of the test spot. Right click in the center of the spot and select "Located IR laser" (see Note 24).
7. Move to another area free of cells and fire another test IR laser shot. Assess the quality of the spot for the following parameters (Fig. 4) (see Note 25):

   (a) Clear center.

   (b) Dark ring around spot.

   (c) Appropriate spot size for cell type to be microdissected.

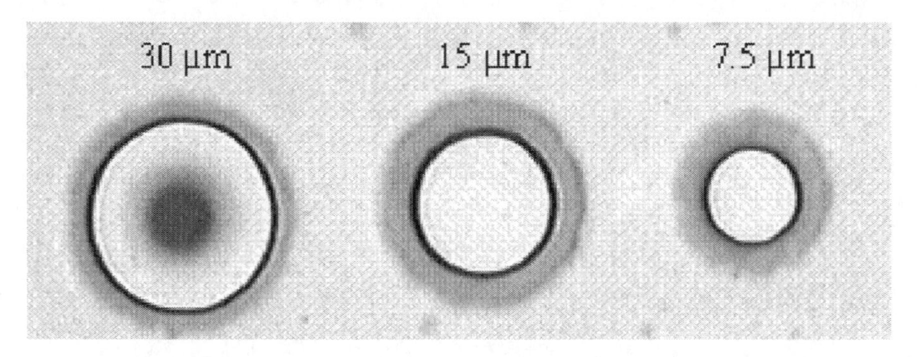

Fig. 4. Properly wetted polymer is essential for microdissection. Properly melted spots have a dark outer ring and a clear center, indicating that the polymer has melted and is in direct contact with the slide. Inadequate power and/or duration settings create spots with a hazy appearance, lacking a distinct black ring. A *dark area* in the center of the *spot* indicates excessive laser power.

Spot adjustments can be made using the options button in the "Select" tool pane. If the spot diameter is larger than desired, reduce the power and/or duration. If the spot diameter is too small, increase the power and/or duration (see Note 26). Fire additional test spots until the desired spot is achieved (see Note 27).

8. Measure the diameter of the test spot. Click on the ruler icon in the "Select" tool pane. Using the mouse or stylus, click and drag from one inner edge of the spot to the opposite inner edge. A line and label will be displayed showing the diameter of the spot. Click on the "Move Stage" (hand) icon to deactivate the ruler tool.
9. Tap the "i" options button in the "Select" tool pane to edit the spot diameter, laser power/duration, and IR laser location.
10. Tap the "IR Spot Sizes" tab. Enter the measured spot diameter for the selected spot size. Click "OK".

3.3.4. Mark the Cells for Microdissection

1. Locate the cells of interest using the trackball, mouse, or stylus to move the stage to the desired location.
2. Click on any tool in the "Select" tool pane. The pencil icon is for freehand drawing. Freehand drawing objects must be a closed figure, i.e., the ends of the drawing must touch to enclose the desired area. The Defined circle tool allows the selection of defined areas of known diameter. The Single IR spot tool is for single cell or single spot microdissection. The IR Spot Line tool permits microdissection of a line, either straight or curved. A drawing tool is active if the icon is gray. Deactivate each drawing tool by clicking again on the icon.
3. Using the stylus or mouse, mark cells on the live image that are to be microdissected.

4. Each drawing object is numbered and appears in the Drawing Items list, located to the right of the Capture Groups List (Fig. 3).
5. Selected areas may be deleted or erased if they were erroneously selected. Right click on the area to be deleted and select "delete" or use the eraser icon to delete partial areas. Alternatively, the drawing object may be deleted by right clicking on the item in the Drawing Items List and selecting "Delete Object." Selected areas may also be copied and pasted to another location or dragged as needed to duplicate or correct the drawing item placement.
6. Different populations of desired cell types can be marked for microdissection and separated out by utilizing Capture Groups. This feature allows cells to be marked from one image but microdissected separately. A Capture Group is a group of drawing items that will be microdissected during one microdissection session, on one cap (see Note 28):
   (a) Prior to marking cells for microdissection, click on the desired Capture Group (A, B, C, D) in the "Select" tool pane.
   (b) Mark the desired cell population for microdissection.
   (c) Select a new Capture Group. Mark the second desired cell population for microdissection.
   (d) Tap the "i" options button in the "Select" tool pane and click on Capture Group. Click the desired group and enter the attributes for that group (name, IR spot color, UV cut color). Click "OK".
7. To view the area of microdissection or the number of laser spots in each area, tap the "i" options button in the "Select" pane. Click on the "Drawing Items" tab to determine the area microdissected (μm^2) or number of laser spots (see Note 29).

3.3.5. Capturing the Cells

1. Tap the "Capture Group" to be microdissected if capture groups were assigned.
2. Click on the IR capture icon in the "Microdissect" tool pane.
3. If necessary, move the cap to new areas of the tissue/slide to microdissect additional cells (of the same cell type) that were not included under the original cap diameter. To move the cap, use the trackball, mouse, or stylus to move the stage to the new location. Right click on the slide overview image and select "Place cap at region center."
4. Fire a test IR pulse in an area lacking cells to check the laser spot morphology (see Subheading 3.3.5, steps 4–8). Adjust the power and/or duration as needed to achieve the desired spot size and quality.
5. Click on the IR capture icon in the "Microdissect" tool pane.

3.3.6. Unload the Samples and Microdissected Tissue

1. After the desired number of cells, or a capture group, has been collected on a cap, move the cap to the QC station. Click on the "Move Cap to QC" icon in the "Microdissect" tool pane. The cap will be moved to the QC position on the stage. The stage will be aligned so the QC station holding the cap is in the optical path of the microscope. The cap image will be displayed in the live image window.
2. Inspect the tissue on the cap for efficiency of microdissection. Adjust the magnification (maximum of 20× in the QC station), focus, and brightness:
   (a) Observe the cap for microdissection of the desired cells and for debris and/or adhesion of nonspecific tissue to the polymer surface.
   (b) Estimate the percentage efficiency of microdissection by observing the polymer for cellular material within the diameter of the melted laser spot. Efficiency of microdissection is a critical factor for estimating the number of cells procured by LCM.
   (c) Debris or nonspecific tissue adhering to the cap may be removed by gently blotting the polymer surface with the tacky side of an adhesive note. Do not use "super sticky" style adhesive notes.
3. The cap can either be removed for storage/cell lysis or placed back on the slide to continue microdissection. To place the cap back on the slide, right click on the appropriate cap icon in the "QC Caps" pane (to the right of the slide overview) (Fig. 3), and select "Replace Cap on Slide."
4. If no further microdissection is to be performed with the cap, remove the cap from the stage by carefully lifting the cap straight up from the QC position. Place the cap in a labeled microcentrifuge tube, label the outer edge of the cap, and place the cap/tube assembly in dry ice or at –80°C (for protein/ RNA analysis) or at room temperature (for DNA analysis).
5. After all dissections are completed, remove all slides and caps from the stage. Click "Present Stage" for access to the slides and caps.
6. Close the Arcturus^{XT} software.
7. Turn off the Arcturus^{XT} instrument.
8. Turn off the PC and the monitor.

3.4. UV Cutting Mode Microdissection Using the Arcturus^{XT}

By default, the instrument will perform IR capture first followed by UV cutting. Depending on the type of slide loaded in the instrument, details of the UV cutting vary (26). In UV cutting mode, glass slides will have a moat cut around the region of interest. For membrane and frame slides, the instrument cuts around the region of interest, leaving tabs, or sections of uncut tissue. Tabs prevent the tissue from curling-up or detaching from the surface of the membrane slide. Initial slide preparation and instrument set-up are similar to the IR capture mode (Subheadings 3.3.1–3.3.3, step 2).

3.4.1. UV Laser Location

1. To locate the UV laser for each objective, tap the “i” options button in the “Microdissection” tool pane. Select the “UV Locate” tab. Follow the steps in the dialog box:
   (a) Tap the locate “UV” button. The UV laser will be fired.
   (b) From the live image, tap the location of the UV laser spot (designated by a circle) then tap the OK button in the dialog box next to the UV button.
   (c) If the UV laser is not visible, tap “Push On” to manually turn on the UV cutting laser.
   (d) Click “OK” to close the dialog box.
2. Click “OK”.

3.4.2. Mark the Cells for UV Cutting

1. Locate the cells of interest using the trackball, mouse, or stylus to move the stage to the desired location.
2. Click on the freehand drawing tool, or the circle tool, in the “Select” tool pane.
3. Using the stylus or mouse, mark the cells on the live image that are to be cut or ablated with the UV laser. Assign capture groups if desired (see Subheading 3.3.4, step 6).
4. To determine the area (μm^2) of tissue to be microdissected tap the “i” options button in the “Select” pane. Click on the “Drawing Items” tab to view the microdissection area (μm^2).

3.4.3. Setting Cut and Capture Properties

1. Tap the “i” options button in the “Microdissect” tool pane. Select the “Select Settings” tab.
2. Tap “IRSpotSpacing” and enter a value of 100%. This allows the spots to touch but not overlap.
3. Tap “UV Settings” and enter values for the following settings. Values will depend on the area to be cut. Smaller areas require fewer tabs and fewer IR spots compared to larger areas:
   (a) IR SpotsPerCutLength – number of IR spots per cut length.
   (b) Tab Length – distance (microns) between each UV cut.
   (c) UV CutLength – length of UV cut before a tab.
   (d) UV CuttingSpeed – speed of the UV laser. Increase the speed to cut more quickly. Alternatively, the UV laser cutting speed can be adjusted using the slide bar in the “Microdissect” tool pane, allowing adjustment in real time during UV cutting if necessary.

3.4.4. Cutting the Cells

1. Tap the "Capture Group" to be microdissected if capture groups were assigned.
2. Click on either the UV cut and capture or UV cut icon in the "Microdissect" tool pane.
3. Move the cap to the QC area and unload caps/slides as described in Subheading 3.3.6.

3.5. Protein Extraction of Microdissected Material for Downstream Analysis

The LCM cap, containing microdissected cells for protein analysis, can be stored at –80°C for extraction at a later date (see Note 30). Extraction of proteins from microdissected cells should be performed just prior to the downstream analysis to prevent aggregation of proteins, degradation of proteins, or binding of protein to the walls of the microcentrifuge tube during prolonged storage. Microdissected samples for western blotting and/or reverse phase protein array analysis can be prepared with the following denaturing cell lysis/protein extraction buffer:

1. Protein extraction buffer: 450 μL T-PER Tissue Protein extraction reagent, 450 μL Novex® Tris–glycine 2× SDS loading buffer, and 100 μL TCEP Bond Breaker®. The final extraction buffer is a 10% (v/v) solution of TECP in T-PER/Tris–glycine 2× SDS buffer (see Note 31).
2. Thaw each LCM cap at room temperature and remove all traces of condensation from the edges and rim of the cap. Place the CapSure® cap containing microdissected cells on a flat-clean surface, film side up (Fig. 5).
3. Using a pipette, dispense the desired quantity of extraction buffer directly on the cap film and incubate for 1 min. The maximum volume of extraction buffer that can be used to cover the surface of a CapSure® cap is 15 μL.
4. Pipette the extraction buffer up and down on the surface of the cap to solubilize the cells. Be careful not to scrape the cap polymer.

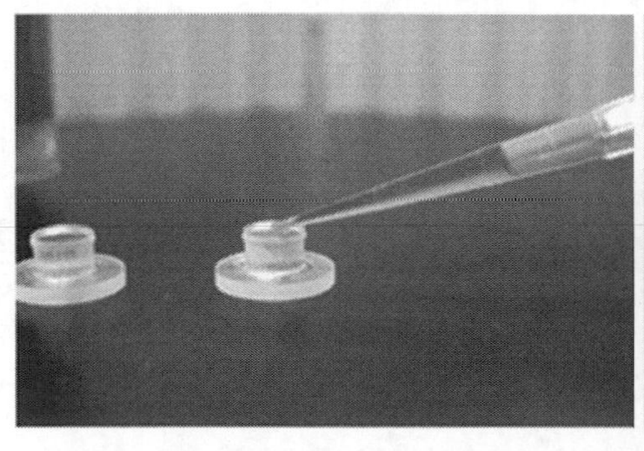
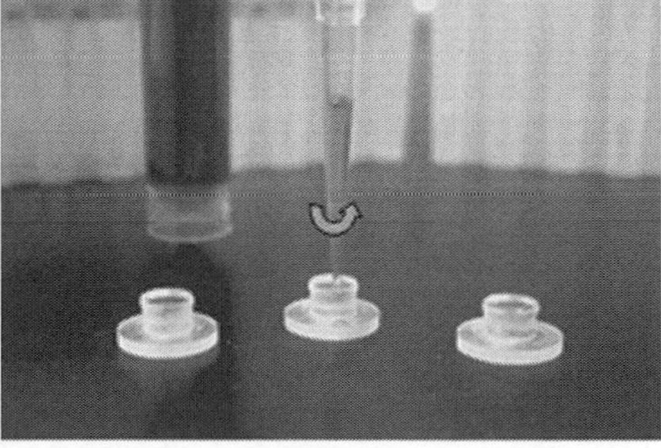

Fig. 5. Scheme for solubilization of microdissected cells. Place the LCM caps on a flat surface. Dispense extraction buffer directly on the cap and incubate for 1 min. Aspirate and dispense delivered extraction buffer up and down several times on the surface of the cap. Collect the whole cell lysates in a 0.5-mL microcentrifuge tube.

5. Collect the extraction buffer containing the solubilized cells in a 0.5-mL microcentrifuge tube. If more than one CapSure® cap was used to microdissect the cells of interest, solubilized cells from these caps can be collected in the same microcentrifuge tube.
6. Denature the proteins by heating the closed microcentrifuge tube at 100°C for 5–8 min prior to downstream proteomic analysis.
7. Briefly spin the microcentrifuge tubes at 950 RCF to pellet any condensation in the tube. Store the samples at –80°C.

4. Notes

1. Significant amounts of heat are not deposited at the tissue surface during capture mode (IR) microdissection. Heat deposition is limited by the following engineering safeguards (a) the short laser pulse durations used, (b) the low laser power levels required (the near-IR laser diode has a maximum output of 100 mW), and (c) absorption of the laser pulse by the dye-impregnated polymer.
2. Optimal tissue thickness for capture mode microdissection is 5–8 μm. Tissue sections cut less than 5 μm may not provide a full cell thickness, requiring microdissection of more cells for a particular downstream assay. Tissue sections thicker than 8 μm may not microdissect completely, leaving essential cellular components adhering to the slide. Recently, a modified LCM technique was described by Iyer and Cox to microdissect neurons from Drosophila embryos that were more than 8 μm thick (27). UV mode microdissection is capable of cutting tissue sections up to 200 μm. Specimen types for DNA or RNA analysis are frozen-tissue sections, ethanol-fixed, or formalin-fixed paraffin-embedded tissue sections cut at 2–15 μm.
3. Scott's Tap Water Substitute is an alkaline solution that is used to develop the blue color of the hematoxylin stain (28). Scott's Tap Water Substitute can be prepared by dissolving 3.5 g sodium carbonate and 20 g magnesium sulfate in 1 L of water.
4. Prepare fresh solutions after staining more than 20–25 slides at a time or if the ambient humidity is greater than 50%. Solutions can be stored at 4°C for 2 days.
5. To limit protein degradation, add protease inhibitors to the 70% ethanol, water, hematoxylin, and Scott's Tap Water Substitute staining solutions. Complete protease inhibitor tablets are soluble in aqueous solutions. Dissolve the tablets in Type 1 reagent grade water (dH_2O) and use this solution to prepare 70% ethanol.

6. CapSure® HS caps can be used successfully for RNA, DNA, or protein extraction. HS caps are designed with a 12-μm rail on the polymer surface, which prevents the polymer from directly touching the tissue except in the vicinity of the laser pulse. An extraction device is designed for use with the CapSure® HS caps, allowing extraction buffer to contact the polymer within a centrally designated area. These features limit any potential RNA contamination from surrounding cells. In contrast, CapSure® Macro caps, placed in direct contact with the tissue, are not equipped with an extraction device, and any cellular material on the polymer surface will be available for extraction.
7. Other brands/types of 0.5-mL microcentrifuge tubes may not form adequate seals with the LCM cap. Leakage may occur if the cap/tube assembly is placed cap-down during incubation with buffer inside the tube.
8. In our experience, it is best to store the block of tissue rather than storing the cut tissue sections. We have had successful protein recovery from frozen rhabdomyosarcoma blocks up to 12 years old when stored at –80°C.
9. Tissue with a thin open architecture, such as lung tissue, can be sectioned and collected on charged or silanized slides to prevent the tissue from nonspecifically adhering to the polymer during microdissection. Coated slides are generally not used for microdissection due to increased adhesive forces between the tissue and the slide. Effective microdissection is a balance between three adhesive forces (a) maximizing downward adhesive forces between the polymer and the tissue, (b) minimizing lateral adhesive forces between the cells, and (c) minimizing upward adhesive forces between the slide and the tissue.
10. The frame on membrane style slides limits the usable region of the microscope slide. In order to prevent potential damage to the instrument, the area available for microdissection on any slide type is limited by the width of the frame. Therefore, it is best to place the tissue sections on the slide near the middle third of the slide. Tissue sections placed at extreme edges of the slide will not be within the usable microdissection region.
11. Do not allow the tissue section to dry on the slide at room temperature. Repeated fluctuations in temperature may cause the tissue to strongly adhere to the slide, preventing procurement of the cells of interest.
12. Stains compatible with LCM include H&E, methylene blue, Wright-Giemsa, or toluidine blue. Staining of the cytoplasm with eosin is not necessary for visualization of cells during microdissection. Fluorescence stains are compatible with fluorescence-equipped systems. Arcturus^{XT} instruments include metal halide lamps with blue (455–495 nm excitation, 510 nm

emission), green (503–548 nm excitation, 565 nm emission), and red filter cubes (570–630 nm excitation, 655 nm emission) that can be used for immuno-LCM protocols (29).

13. Tissue with strong intracellular adhesion might be difficult to microdissect and may require a modified staining protocol. The following protocol incorporates glycerol in the staining procedure for frozen sections as adapted from Agar et al. (30):
    (a) Mayer's hematoxylin, 30 s
    (b) dH_2O, 15 s
    (c) 70% Ethanol fixative, 10 s
    (d) 95% Ethanol, 10 s
    (e) dH_2O, 10 s
    (f) Scott's Tap Water Substitute, 15 s
    (g) 70% Ethanol, 2 min
    (h) 3% Glycerol in PBS, 5–10 min
    (i) 100% Ethanol, 10 s
    (j) 100% Ethanol, 1 min
    (k) Xylene, 30–60 s
    (l) Xylene, 30–60 s
    (m) Air dry the slide as quickly as possible
14. Staining solutions can be prepared in 50-mL conical tubes. Use forceps or tweezers to hold the slide and dip the slide in and out of the solutions for the times indicated.
15. Dilute Eosin Y 1:1 with 100% ethanol to prevent over-staining of cytoplasmic proteins.
16. Complete dehydration of the tissue is necessary for minimizing the upward adhesive forces between the tissue section and the slide. Increasing incubation time to 2 min for the 100% ethanol and xylene rinses may enhance dehydration, resulting in maximized microdissection efficiency.
17. If absolutely necessary, the slide may be left in xylene for a maximum of 5 min before proceeding with microdissection.
18. Xylene dissolves the polymer on CapSure® LCM caps. It is crucial that the tissue slide be completely dry before cap placement for microdissection. When dry, the slide will appear as a grayscale (nonrefractive index-matched) image.
19. Allow formalin-fixed or ethanol-fixed paraffin-embedded slides to soak in Xylene for 5–15 min to dissolve the paraffin.
20. Any specimen preparation may be used with the Arcturus^{XT} system: thin or thick sections, frozen or formalin-fixed tissues, stained, fluorescently stained, or unstained sections, hydrated

or dehydrated specimens, fine needle aspirates, forensic smears, live plant specimens, and cell cultures.

21. If the slide overview is blank or does not update, right click in the slide overview and select "Reacquire Overview Image."

22. Older manual LCM systems such as the PixCell® II/IIe instruments did not allow the user to microdissect directly from an index-matched image of the tissue. As a work around, map images were saved while the tissue was wet (or rewetted with a drop of xylene prior to microdissection), providing a guide for microdissection (11, 14). The Arcturus^{XT} instrument illumination system greatly enhances the image properties during microdissection. For optimization of cell visualization, adjustments to the illumination system can be made by selecting the "i" options button under the Inspect tools pane. Click on the "Illumination" tab to adjust the brightfield lamp settings, white balance, camera gain, and diffuser settings.

23. To capture a static image of the main image window, tap the "Camera" button in the "Inspect" tools pane. The images will be saved in the folder specified in the "File Paths" tab of the "Load Options" dialog box in the "Present Stage" tool pane. The three general types of image are Before image – tissue before microdissection; After image – tissue after microdissection; Cap image – microdissected tissue only.

24. The laser must be located each time the objective is changed for microdissection. It is recommended to fire the test pulse using the objective you intend to use for microdissection. Microdissection may be performed with any suitable laser spot size and a 2×, 10×, 20×, 40×, 60×, or 100× (no oil) objective.

25. The dark ring produced by pulsing the laser is caused by a combination of migration of the dye and changes in the thickness of the polymer wall at the site of the laser pulse, permitting visualization of the melted polymer. The black ring should be sharp in appearance with a clear center (Fig. 4). This pattern indicates proper laser focusing, adequate laser operation, and acceptable performance of the polymer. A "fuzzy" ring could indicate improper focusing of the laser, uneven placement of the cap on the tissue, or inadequate power and/or duration of the laser pulse. The following steps can be followed when troubleshooting a poorly wetted polymer spot:

    (a) Reposition the cap on the tissue; the cap may be crooked or uneven in relation to the tissue.

    (b) Relocate the laser; the laser may be out of alignment.

    (c) Adjust the power and duration of the laser pulse. Increase the laser power in increments of approximately 10 mW and the duration by 2 ms and fire additional laser test pulses after each adjustment. Observe the wetted polymer for the appropriate appearance.

(d) If the above steps fail to resolve the problem, discard the cap and repeat the process with a new cap.

26. A phenomenon termed "polymer depletion" occurs when microdissecting large, polygon-shaped areas from the perimeter toward the center. As the laser melts the polymer downward onto the cells, the polymer is depleted on the edges of the laser fire area. As more and more polymer is melted onto the cells in a localized area, this depletion effect becomes more apparent. This can be prevented by microdissecting large, enclosed areas from the center of the area toward the perimeter, or using the freehand drawing microdissection tool.

27. Single-cell microdissection is possible by adjusting the power and duration settings such that a very narrow area of the polymer is melted with each laser pulse. Select the smallest spot size setting and manually adjust the laser power and duration. Suggested settings for single-cell microdissection are power 45 mW and duration 650 μs using Macro caps.

28. Capture groups are used to mark different types of cells from the same slide such that the cell populations will be microdissected separately, using two different caps. As an example, areas of tumor and stroma on the same slide can be marked for microdissection. Tumor cells can be assigned to Capture Group A, and will be color coded to indicate Group A, while stromal cells could be assigned to Capture Group B, and color coded to indicate Group B.

29. It is possible to estimate the number of captured cells based upon the number of laser pulses counted during microdissection (which is automatically counted on the toolbar), the spot size, and the efficiency of microdissection. The percent efficiency of microdissection can be estimated by observing the polymer for cellular material within the diameter of the melted laser spot:

    30-μm Laser spot size: Number of pulses × 5 × % efficiency = total cells captured.

    15 μm Laser spot size: Number of pulses × 3 × % efficiency = total cells captured.

    7.5-μm Laser spot size: Number of pulses × 1 × % efficiency = total cells captured.

30. Microdissected cells for DNA analysis can be stored desiccated at room temperature up to 1 week prior to extraction. Samples for RNA analysis should be extracted immediately after microdissection because condensation in the microcentrifuge tube during storage may be a potential source of RNase contamination.

31. The optimal protein extraction buffer for electrophoresis or microarray analysis consists of a detergent, a denaturing agent, and a buffer. 10% v/v TCEP in T-PER/Tris–glycine 2× SDS buffer is considered a mild denaturing extraction buffer for the solubilization of cellular proteins. If the microdissected cells are to be analyzed via mass spectrometry, a urea-based buffer is recommended. A mass spectrometry compatible denaturing buffer is 8 M urea in Tris–HCl, pH 7.0–7.5 (25).

References

1. Wulfkuhle, J. D., Speer, R., Pierobon, M., Laird, J., Espina, V., Deng, J. et al. (2008) Multiplexed cell signaling analysis of human breast cancer applications for personalized therapy. *J Proteome Res* 7, 1508–17.
2. Petricoin, E. F., 3rd, Espina, V., Araujo, R. P., Midura, B., Yeung, C., Wan, X. et al. (2007) Phosphoprotein pathway mapping: Akt/mammalian target of rapamycin activation is negatively associated with childhood rhabdomyosarcoma survival. *Cancer Res* 67, 3431–40.
3. Petricoin, E. F., 3rd, Bichsel, V. E., Calvert, V. S., Espina, V., Winters, M., Young, L. et al. (2005) Mapping molecular networks using proteomics: a vision for patient-tailored combination therapy. *J Clin Oncol* 23, 3614–21.
4. Wulfkuhle, J. D., Sgroi, D. C., Krutzsch, H., McLean, K., McGarvey, K., Knowlton, M. et al. (2002) Proteomics of human breast ductal carcinoma in situ. *Cancer Res* 62, 6740–9.
5. Nakazono, M., Qiu, F., Borsuk, L. A., Schnable, P. S. (2003) Laser-capture microdissection, a tool for the global analysis of gene expression in specific plant cell types: identification of genes expressed differentially in epidermal cells or vascular tissues of maize. *Plant Cell* 15, 583–96.
6. Ma, X. J., Dahiya, S., Richardson, E. A., Erlander, M., Sgroi, D. C. (2009) Gene expression profiling of tumor microenvironment during breast cancer progression. *Breast Cancer Res* 11, R7.
7. Emmert-Buck, M. R., Bonner, R. F., Smith, P. D., Chuaqui, R. F., Zhuang, Z., Goldstein, S. R. et al. (1996) Laser capture microdissection. *Science* 274, 998–1001.
8. Bonner, R. F., Emmert-Buck, M., Cole, K., Pohida, T., Chuaqui, R., Goldstein, S. et al. (1997) Laser capture microdissection: molecular analysis of tissue. *Science* 278, 1481,1483.
9. Angeles, G., Berrio-Sierra, J., Joseleau, J. P., Lorimier, P., Lefebvre, A., Ruel, K. (2006) Preparative laser capture microdissection and single-pot cell wall material preparation: a novel method for tissue-specific analysis. *Planta* 224, 228–32.
10. Gallup, J. M., Kawashima, K., Lucero, G., Ackermann, M. R. (2005) New quick method for isolating RNA from laser captured cells stained by immunofluorescent immunohistochemistry; RNA suitable for direct use in fluorogenic TaqMan one-step real-time RT-PCR. *Biol Proced Online* 7, 70–92.
11. Mouledous, L., Hunt, S., Harcourt, R., Harry, J., Williams, K. L., Gutstein, H. B. (2003) Navigated laser capture microdissection as an alternative to direct histological staining for proteomic analysis of brain samples. *Proteomics* 3, 610–5.
12. Nakamura, N., Ruebel, K., Jin, L., Qian, X., Zhang, H., Lloyd, R. V. (2007) Laser capture microdissection for analysis of single cells. *Methods Mol Med* 132, 11–8.
13. VanMeter, A. J., Rodriguez, A. S., Bowman, E. D., Jen, J., Harris, C. C., Deng, J. et al. (2008) Laser capture microdissection and protein microarray analysis of human non-small cell lung cancer: differential epidermal growth factor receptor (EGPR) phosphorylation events associated with mutated EGFR compared with wild type. *Mol Cell Proteomics* 7, 1902–24.
14. Wong, M. H., Saam, J. R., Stappenbeck, T. S., Rexer, C. H., Gordon, J. I. (2000) Genetic mosaic analysis based on Cre recombinase and navigated laser capture microdissection. *Proc Natl Acad Sci U S A* 97, 12601–6.
15. Kennedy, J., Katsuta, H., Jung, M. H., Marselli, L., Goldfine, A. B., Balis, U. J. et al. (2010) Protective unfolded protein response in human pancreatic beta cells transplanted into mice. *PLoS One* 5, e11211.
16. Cha, S., Imielinski, M. B., Rejtar, T., Richardson, E. A., Thakur, D., Sgroi, D. C. et al. (2010) In situ proteomic analysis of human breast cancer epithelial cells using laser capture microdissection: annotation by protein set enrichment analysis and gene ontology. *Mol Cell Proteomics* 9, 2529–44.
17. Kolble, K. (2000) The LEICA microdissection system: design and applications. *J Mol Med* 78, B24-5.

18. Micke, P., Ostman, A., Lundeberg, J., Ponten, F. (2005) Laser-assisted cell microdissection using the PALM system. *Methods Mol Biol* 293, 151–66.
19. Schutze, K., Posl, H., Lahr, G. (1998) Laser micromanipulation systems as universal tools in cellular and molecular biology and in medicine. *Cell Mol Biol (Noisy-le-grand)* 44, 735–46.
20. Schermelleh, L., Thalhammer, S., Heckl, W., Posl, H., Cremer, T., Schutze, K. et al. (1999) Laser microdissection and laser pressure catapulting for the generation of chromosome-specific paint probes. *Biotechniques* 27, 362–7.
21. Espina, V., Edmiston, K. H., Heiby, M., Pierobon, M., Sciro, M., Merritt, B. et al. (2008) A portrait of tissue phosphoprotein stability in the clinical tissue procurement process. *Mol Cell Proteomics* 7, 1998–2018.
22. Botling, J., Edlund, K., Segersten, U., Tahmasebpoor, S., Engstrom, M., Sundstrom, M. et al. (2009) Impact of Thawing on RNA Integrity and Gene Expression Analysis in Fresh Frozen Tissue. *Diagn Mol Pathol*
23. Micke, P., Ohshima, M., Tahmasebpoor, S., Ren, Z. P., Ostman, A., Ponten, F. et al. (2006) Biobanking of fresh frozen tissue: RNA is stable in nonfixed surgical specimens. *Lab Invest* 86, 202–11.
24. Xiang, C. C., Mezey, E., Chen, M., Key, S., Ma, L., Brownstein, M. J. (2004) Using DSP, a reversible cross-linker, to fix tissue sections for immunostaining, microdissection and expression profiling. *Nucleic Acids Res* 32, e185.
25. Espina, V., Wulfkuhle, J. D., Calvert, V. S., VanMeter, A., Zhou, W., Coukos, G. et al. (2006) Laser-capture microdissection. *Nat Protoc* 1, 586–603.
26. Arcturus^{XT} user guide: Flexible and Modular Laser Capture Microdissection, Rev. A. Molecular Devices Corporation, Sunnyvale, CA, 2007.
27. Iyer, E. P., Cox, D. N. (2010) Laser capture microdissection of Drosophila peripheral neurons. *J Vis Exp* 39, e2016, doi: 10.3791/2016.
28. Kiernan, J. (2008) Histological staining in one or two colours, in *Histological and Histochemical Methods* (ed.), Scion, Oxfordshire, pp. 146–147.
29. Buckanovich, R. J., Sasaroli, D., O'Brien-Jenkins, A., Botbyl, J., Conejo-Garcia, J. R., Benencia, F. et al. (2006) Use of immuno-LCM to identify the in situ expression profile of cellular constituents of the tumor microenvironment. *Cancer Biol Ther* 5, 635–42.
30. Agar, N. S., Halliday, G. M., Barnetson, R. S., Jones, A. M. (2003) A novel technique for the examination of skin biopsies by laser capture microdissection. *J Cutan Pathol* 30, 265–70.

Chapter 12

Antibody Microarrays: Analysis of Cystic Fibrosis

Catherine E. Jozwik, Harvey B. Pollard, Meera Srivastava, Ofer Eidelman, QingYuan Fan, Thomas N. Darling, and Pamela L. Zeitlin

Abstract

Cystic fibrosis (CF) is the most common autosomal recessive disease in the USA and Europe, whose life-limiting phenotype is manifest on epithelial cells throughout the body. The principal cause of morbidity and mortality is a massively proinflammatory condition in the lung. The mutation responsible for most cases of CF is [ΔF508]CFTR. However, the penetrance of the disease is quite variable, and adverse events leading to hospitalization cannot be easily predicted. Thus, there is a strong need for prognostic endpoints that might serve to identify impending clinical problems long before they happen. Our approach has been to search for proteomic signatures in easily accessed biological fluids that might identify the molecular basis for adverse events. We describe here a workflow that begins with patient-derived bronchial brush biopsies and progresses to analysis of serum and plasma from patients on antibody microarrays.

Key words: Cystic fibrosis, Laser capture microdissection, Saturation CyDye labeling, Antibody microarrays

1. Introduction

Cystic fibrosis (CF) is a life-limiting, autosomal recessive disease characterized by inflammation in the lung. The proinflammatory phenotype of the CF lung results in the hypersecretion of proinflammatory cytokines, an exaggerated response to bacterial infections and, ultimately, lung obstruction (1). Although therapeutic interventions have increased the average age of survival, approximately one child or young adult with CF still dies each day (2, 3). Identification of CF-specific proteins from lung epithelia (Fig. 1) in CF patients would facilitate the identification of mechanisms underlying the proinflammatory response in CF. However, bronchial brush biopsies from CF patients frequently contain blood,

Virginia Espina and Lance A. Liotta (eds.), *Molecular Profiling: Methods and Protocols*, Methods in Molecular Biology, vol. 823, DOI 10.1007/978-1-60327-216-2_12, © Springer Science+Business Media, LLC 2012

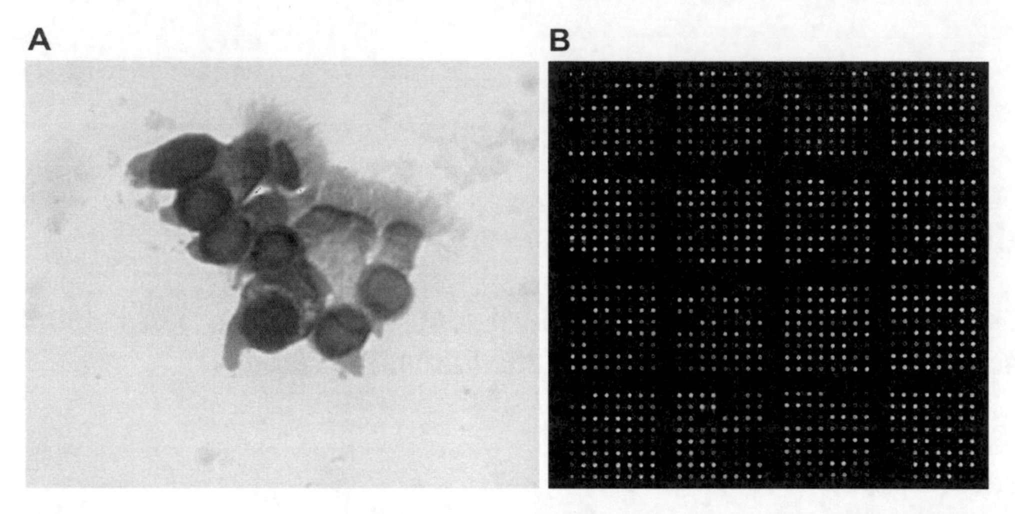

Fig. 1. Epithelial cells obtained from a bronchial brush biopsy. (**a**) Biopsies were prepared on cytospin slides stained with hematoxylin and eosin. (**b**) Antibody microarray image from bronchial epithelial cells. Bronchial epithelial cells were isolated by laser capture microdissection and the resulting Cy5-labeled proteins were used to interrogate the Clontech Ab 500 Microarray.

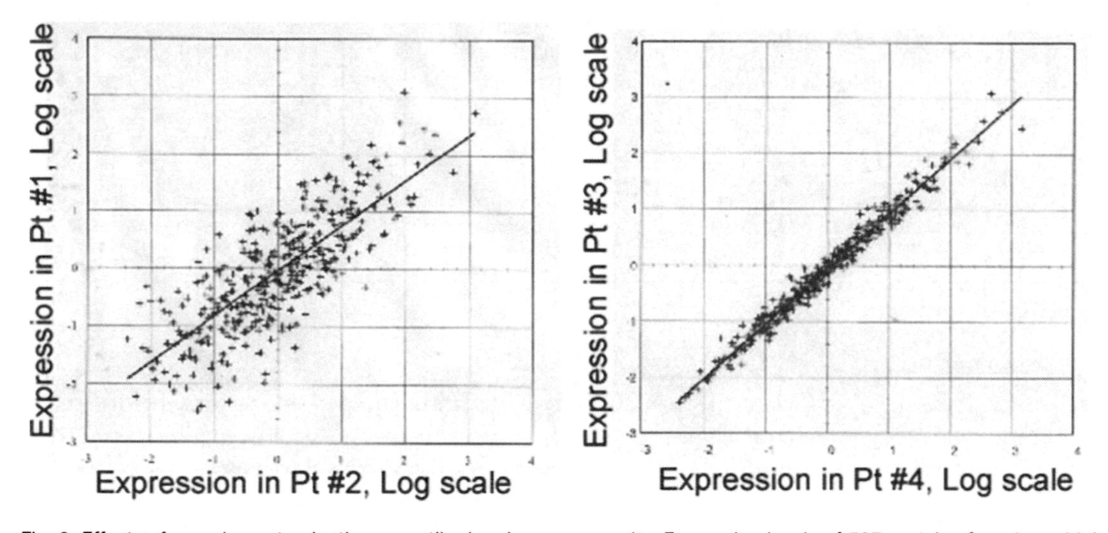

Fig. 2. Effects of sample contamination on antibody microarray results. Expression levels of 507 proteins from bronchial brush biopsies for four different patients were compared based on data from a Clontech Ab 500 Microarray. Samples from patients 1 and 2 had significant contamination from non-epithelial cell types, while samples from patients 3 and 4 had greater than 95% epithelial cells.

microorganisms, and resident inflammatory cells, all of which complicate proteomic analysis of CF epithelial cells (Fig. 2).

Technologies such as laser capture microdissection (LCM) have rapidly become standard for harvesting specific cell types within tissue sections, but LCM is not frequently used for cytological specimens (4, 5). A hindrance to using LCM on cytological specimens is that cells air-dried onto glass slides adhere too strongly for microdissection (6). We have developed a suite of tools for the procurement and proteomic analysis of homogeneous cell populations from bronchial brush biopsies that we term "precision clinical proteomics" (7). Using a combination of specially treated slides for LCM, highly sensitive saturation CyDye labeling, and antibody microarrays, we can determine the expression levels of over 500 lower abundance proteins from 1,000 to 5,000 epithelial cells from bronchial brush biopsies. Diseased epithelial cells may release proteins into the circulation that may constitute candidate CF-specific biomarkers useful for diagnostic screening. Identification of serum biomarkers requires a reliable means for quantifying protein biomarkers in serum or plasma. Therefore, we also describe a protocol for the identification of low abundance proteins in serum or plasma using an antibody microarray (8, 9).

The precision clinical proteomics workflow begins with the acquisition of the sample (Fig. 3). Suitable samples include biopsy tissue or fluids such as serum or plasma. Proteins in the samples are extracted, then fluorescently labeled using CyDyes and differentially expressed proteins are identified using an antibody microarray (Fig. 1) (7, 9, 10). Bioinformatic analyses of disease-specific proteins are then used to identify cellular processes or signaling pathways that are affected by the disease (Fig. 4 and Table 2) (8, 11).

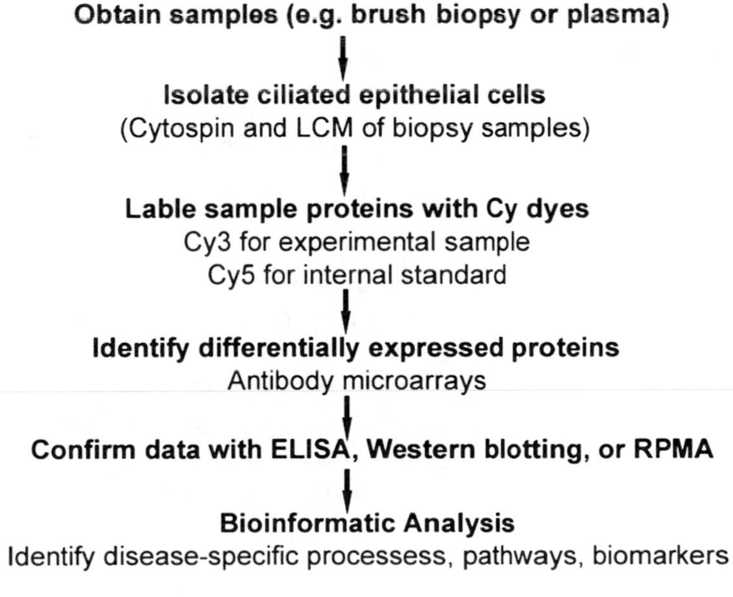

Fig. 3. Precision clinical proteomics workflow.

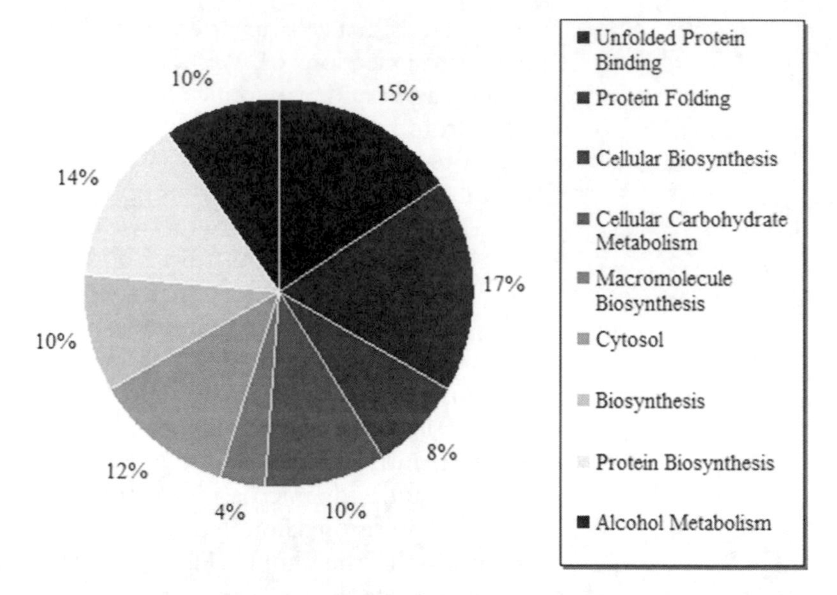

Fig. 4. Differentially expressed proteins from cultured cystic fibrosis cells. Differentially expressed proteins from cultured cystic fibrosis cells were identified using the Ab 500 Microarray and affected protein classes were determined using the GOMiner program (http://discover.nci.nih.gov/gominer/). Gene ontology terms that were most changed in cystic fibrosis cells are shown.

2. Materials

2.1. Sample Acquisition

2.1.1. Bronchial Brush Biopsy

1. Sterile 2 ml cryovials.
2. L15 Media (Invitrogen).
3. Saccomanno Fixative (Fisher Scientific).
4. Phosphate-buffered saline (PBS).
5. Bronchial brush biopsy set (cytology brush; Conmed).

2.1.2. Laser Capture Microdissection of Ciliated Bronchial Cells

1. Rain-X Original Glass Treatment™ water repellant (available at hardware stores).
2. Laser capture microscopy equipment (Arcturus PixCell IIe laser capture microdissection system), ArcturusXT™ Microdissection System (Life Technologies/Applied Biosystems) with CapSure Macro caps.
3. SealRight tubes 0.5 ml (USA Scientific) or GeneAmp™ 0.5 ml Thin-walled Reaction Tubes with Domed Cap (cat. no. N801-0611; Applied Biosystems).
4. Cytocentrifuge (Shandon Cytospin®4; Thermo Scientific) with funnels and clips.
5. HistoGene LCM Frozen Section Staining Kit (Life Technologies/Applied Biosystems).

6. Lysis buffer: 7 M urea, 4% CHAPS, 20 mM Tris-HCl, pH 8.0. Complete-mini Protease Inhibitor tablets (Roche Applied Science) may be included, if desired. Use one tablet in 10 ml of lysis buffer.
7. Micro BCA Protein Assay Kit [Thermo Scientific (formerly Pierce Biotechnology)].

2.1.3. Plasma/Serum

1. Glass red top vacutainer serum collection tubes.
2. 1.5–2.0 ml Microcentrifuge tubes.
3. BCA Protein Assay Kit [Thermo Scientific (Pierce Biotechnology)].

2.2. Protein Labeling

2.2.1. Protein Labeling of LCM Samples

1. pH paper.
2. CyDye DIGE Fluor Labeling Kit for Scarce Samples (GE Healthcare). Reconstituted dyes should be discarded after 8 weeks.
3. Dimethylformamide (99.8% anhydrous DMF). Discard 3 months after opening.
4. TCEP (Tris-(2-carboxyethyl)phosphine; Invitrogen). The 2 mM working solution must be prepared prior to use and used immediately.
5. DTT (1 M dithiothreitol).
6. Lysine (10 mM L-lysine).
7. Internal control extract (see Note 1).

2.2.2. Protein Labeling of Plasma/Serum

1. Cy3 and Cy5 Monoreactive Dye Packs (GE Healthcare).
2. Extraction/labeling and desalting buffer (Ab Microarray Express Buffer Kit; Clontech Laboratories, Inc).
3. Protein Desalting Spin Columns [Thermo Scientific (Pierce Biotechnology)].
4. Internal control extract (see Note 1).

2.3. Antibody Microarray

1. Ab 500 Microarray (Clontech Laboratories, Inc).
2. Ab Microarray Express Buffer Kit (Clontech Laboratories, Inc).
3. SecureSeal Incubation Chambers (Grace Biolabs; the SA200 fit snugly on the Clontech arrays, if you are unsure about your ability to center the incubation chamber over the array, use the larger SA500).
4. Fluorescent scanner with Cy3 and Cy5 filter sets (Axon Genepix 4000).

2.4. Bioinformatic Analysis of Antibody Microarrays

1. Fluorescent scanner spot intensity analysis program (GenePix 6.0; Molecular Devices).
2. Microsoft Excel or other spreadsheet program.

3. Methods

3.1. Sample Acquisition

Suitable samples for antibody microarrays include cells from any tissue type or cell line as well as many biological fluids including serum, cerebrospinal fluid, and urine. Primary human samples must be obtained with prior institutional review board approval. Bronchial brush biopsies are used to obtain cells from the bronchial tree.

Cellular samples may need to be further selected/separated prior to application on an antibody microarray. LCM is an example of a cell separation/selection method utilized in this protocol. Serum samples are prepared for dye labeling with an initial desalting step.

3.1.1. Bronchial Brush Biopsy

1. Prepare sterile cryovials (one per expected brush biopsy; see Note 2). Fill cryovial with 1.5 ml of L15 medium and cap with the blue brush insert followed by sterile cap from cryovial. Transport tubes on ice to bronchoscopy suite.
2. Select biopsy anatomical location. This is determined by the clinical situation – if there is no obvious foci of disease on a chest radiograph, we select the right middle lobe bronchus and the lingula.
3. Appropriately trained medical personnel should perform bronchoalveolar lavage. Use 1 cc/kg to a maximum of 30 cc lavage fluid, three times in rapid succession (see Note 3).
4. Brush two subsegments with a disposable, sterile bronchoscopic brush by inserting the brush through the suction port and twirling it three times while moving it up and down. Pull brush back to the tip of the port.
5. Remove bronchoscope and place brush in L15 medium in cryovial (blue brush insert and cap have been removed). Cut brush and cap tube. Keep brushes on ice during transport and processing. Process brushes as soon as possible.
6. Remove cryovial cap from tube and brush insert (see Note 4). If the clear plastic sheath remains on the brush, draw it off with a sterile hemostat.
7. Grasp the end of the brush wires with the sterile hemostat and pull the brush up and down through opening in the brush insert to dislodge cells from brush (see Note 5). Keep cells and brush submerged in media at all times. Discard brush when cells have been removed.
8. Centrifuge at $1{,}000 \times g$, 4°C. Discard medium.
9. Add 1 ml cold PBS, gently resuspend cell pellet to wash and repeat centrifugation.

10. Discard PBS and gently resuspend cell pellet in 0.4 ml Saccomanno Fixative for each brush.
11. Add 80 μl PBS to 20 μl cells, and prepare a cytospin using the Shandon Single Cytofunnel on untreated slides at 1,000 rpm for 5 min, and stain with hematoxylin and eosin. Use these slides to assess quality and number of cells.
12. Determine an appropriate volume of cells required for coverage of treated slides for LCM. Dilute (in Saccomanno Fixative) or concentrate cells as necessary (see Note 6).

3.1.2. Laser Capture Microdissection (Arcturus PixCell IIe)

LCM is a method to procure pure or enriched cell populations from heterogeneous cellular samples (4, 5, 12). Cell/tissue samples are prepared as a single layer on a glass microscope slide, either as a cytospin (liquid cellular samples) or as tissue sections. A near-infrared laser mounted in the optical axis of a microscope is used to melt a thermolabile polymer in the vicinity of the desired cell(s). The cells in the vicinity of the laser pulse are embedded in the polymer, forming a polymer–cell composite. The cells can be readily solubilized from the inert polymer, yielding a whole cell lysate for downstream analysis. The LCM process consists of six steps: (1) loading the slides and LCM caps, (2) locating the cells of interest, (3) LCM cap placement and laser location, (4) marking the cells of interest, (5) capturing the cells, and (6) unloading of the samples and caps containing the captured tissue (4, 5, 13).

Cytsospin Preparation and Slide Staining

1. Prepare Rain-X-coated slides for LCM. Add one drop of Rain-X to each glass slide and spread across surface with a KimWipe until dry (see Note 7).
2. Prepare a cytospin of each bronchial biopsy sample (onto Rain-X coated slides). Dispense 60 μl of appropriately diluted/concentrated cell mixture into each cytospin funnel/slide set. Spin the cytospin slides at 600 rpm for 3 min in the cytocentrifuge. Make ~20–30 slides per sample.
3. Remove the funnel from the slide. Air dry the slides for 2 min. Place the slides in a slide box, and store immediately at –80°C.
4. For LCM, transfer the slides to be microdissected from –80°C to dry ice. Immediately prior to microdissection, place each slide briefly at room temperature to defrost. Proceed immediately to staining (4, 5, 13).
5. Stain the slide using the HistoGene Staining Kit.
   (a) Place the slide in 70% ethanol for 30 s. Dip the slide up and down several times in the 70% ethanol.
   (b) Place slide in dH_2O for 30 s. Dip the slide up and down several times. Remove from dH_2O and blot the slide on a paper towel to remove excess liquid.

(c) Add 50 μl Staining Solution directly to the cells on the slide. Incubate at room temperature for 30 s.

(d) Remove the Staining Solution with a pipette. Rinse the slide in dH_2O for 30 s. Dip the slide up and down several times.

(e) Dehydrate the slide sequentially in 70% ethanol, 95% ethanol, and 100% ethanol for 30 s. each. Dip the slide up and down several times in each solution.

(f) Rinse the slide in xylene for 5 min. Dip the slide up and down several times in the xylene.

(g) Immediately begin to microdissect the cells of interest.

Laser Capture Microdissection (PixCell IIe)

1. Turn on the power to the LCM instrument and the laser control tower.
2. Load CapSure caps into dovetail assembly (right side of microscope).
3. Ensure that the joystick is perpendicular to the table.
4. Position the slide, using your fingers, not the joystick, until the desired area of cells is located.
5. Turn on the vacuum (push the vacuum button on front of the laser control tower).
6. Pick-up a cap with the cap arm and move it to the work area on the slide. Press the cap firmly but gently on to the slide. Ensure that the cap is in contact with the tissue/slide.
7. Push the "laser enable" button (on front of the laser control tower).
8. Locate laser spot on screen.
9. Push the spot size selection lever (on the left side of the microscope) to the small (7.5 μm) spot size and turn the objective turret to 10× (yellow ring).
10. Focus the laser spot to bright, well-defined spot. The laser focus is a knurled-knob near the spot size selection lever.
11. Fire the laser, in an area without any cells, as a test spot. Observe the melted polymer (spot) for adequate polymer melting, which is evident by a spot with a clear center and a dark outer ring. Adjust the power and/or duration as needed to achieve adequate polymer melting. Typical power and duration settings are listed in Table 1.
12. Locate the cells to be dissected and fire the laser on each cell. Capture the desired cells (preferably ~10,000 cells).
13. After microdissecting the desired cells, view the captured cells on the surface of the cap. Move the cap to an area of the slide

Table 1
Example power and duration settings for laser capture microdissection with an Arcturus PixCell IIe instrument

Laser spot size (μm)	Power (mW)	Duration
Small (7.5)	40–50	750–950 μs
Medium (15)	30–40	1.5–2.0 ms
Large (30)	25–35	5.0–6.0 ms

without cells. Place the cap on the slide. Observe the cap to estimate the efficiency of microdissection.

14. Use the cap arm to move the cap to the capping station. Remove the cap and proceed immediately with cell lysis.
15. Immediately after microdissection, extract protein from the cells directly on the cap. Add 30 μl lysis buffer to the cap. Allow the lysis buffer to remain on the cap for 30 s to 1 min. Transfer the lysate, using a pipette, from the cap surface to a clean microcentrifuge tube. Place the cap on another microcentrifuge tube to collect any residual lysate. Spin the cap/tube assembly at 10,000 × *g* for 10 s. Add any residual lysate collected to the original lysate. Store the lysate at −80°C.
16. Measure the protein concentration of the lysate using the MicroBCA Protein Assay (see Note 8).

3.1.3. Serum Collection

Molecular sieving provides a simple method for removing interfering substances such as salt. Salt can interfere with gel electrophoresis, mass spectrometry and dye labeling. Sieving columns (spin columns) selectively retain salt and small molecules while allowing proteins to pass through the column. Centrifugation facilitates movement of the serum through the column. Serum proteins greater than 7,000 Da are recovered with the desalting columns used in this procedure.

1. Collect blood samples by venipuncture into glass red top (serum) tubes (see Note 9).
2. Allow blood to clot for 30 min (see Note 10).
3. Separate sera by centrifugation at 1,300 × *g* at 4°C for 10 min.
4. Transfer sera to a new centrifuge tube and centrifuge at 2,400 × *g* for 15 min.
5. Remove peptides and amines using a Protein Desalting Spin Column.
   (a) Prepare needed amount of 1× desalting buffer.
   (b) Centrifuge columns (attached to 1.5 ml microcentrifuge tubes) at 1,500 × *g* for 2 min to remove storage buffer.

(c) Add 400 μl 1× desalting buffer to each column and centrifuge at 1,500 × *g* for 2 min. Discard the flow-through.

(d) Repeat step 5 (c).

(e) Attach a clean microcentrifuge tube to each column.

(f) Apply 30–120 μl plasma to the column. Allow the sample to penetrate the column.

(g) Centrifuge at 1,500 × *g* for 2 min.

(h) Detach the column and store the collection tube containing the desalted serum on ice.

6. Measure the protein concentration in the desalted serum using the BCA Assay.
7. Aliquot into and store the serum samples at −80°C.

3.2. Protein Labeling

Cy3 and Cy5 mono-functional maleimide dyes label-free sulphydryl groups, while Cy3 and Cy5 NHS mono-reactive dyes label free lysines. Cy3 and Cy5 dyes provide a green or red fluorescent signal, respectively (14). Cy3 dye has an absorbance maximum at 550 nm with a 570-nm emission. Cy5 maximum absorbance is 649 nm, with a 670-nm emission. Two labeling protocols are described; one for limited volume samples (LCM) using saturation Cy3 and Cy5 dyes to label cysteine residues, and another protocol for labeling more abundant samples (plasma/serum) using Cy dye minimal labeling to label lysine residues.

Antibody microarray results are usually given in terms of ratios between two samples (e.g., "experiment" and "control"). However, given the printing and calibration issues inherent with antibody microarrays, this ratio approach precludes statistically valid interarray comparisons of the individual protein concentrations. We have developed a different strategy to control for interarray variability. We use a benchmark mixture that is applied to every array as an internal standard; this corrects for any differences in antibody activities as well as printing imperfections. The amount of protein bound to each spot is determined relative to the respective protein in the internal standard. Since all samples in the study are compared to the same internal standard, this semiquantitative approach permits the comparison of multiple samples. All data for the same antibody on multiple arrays can then be used to calculate the average and standard deviation for a given population (such as patients or control groups) in a parallel fashion.

3.2.1. Protein Labeling of LCM Samples

1. Confirm that the extract pH is ~8.0 by spotting a small amount on pH paper. If the pH is not 8.0, adjust the pH of the sample (see Note 11).
2. Add the appropriate volume of protein lysate to a sterile microcentrifuge tube (see Note 12). Add the appropriate amount of internal standard protein extract to a second tube (see Note 13).

3. Prepare 2 mM TCEP solution by dissolving 2.8 mg TCEP in 5 ml dH_2O (see Note 14).
4. Add the appropriate volume of 2 mM TCEP (as determined in the labeling optimization experiment; see Note 15).
5. Mix vigorously by pipetting (see Note 16).
6. Spin down the sample in a microcentrifuge and incubate at 37°C for 1 h, in the dark.
7. Add the appropriate volume of resuspended 2 mM CyDye DIGE Fluor saturation dye (as determined in the labeling optimization experiment (see Note 15); typically 1 µl TCEP and 2 µl dye).
   (a) Label the internal control protein extract with the Cy3 saturation dye.
   (b) Label experimental protein extracts (e.g., control, treated, etc.) with the Cy5 saturation dye.
8. Mix vigorously by pipetting.
9. Briefly spin the sample in a microcentrifuge and incubate at 37°C for 30 min, in the dark.
10. To stop the reaction, add DTT and lysine to final concentrations of 65 and 1 mM, respectively.
11. Mix vigorously by pipetting.
12. Spin down the sample in a microcentrifuge tube (see Note 17).
13. Samples should be stored at –80°C.

3.2.2. Protein Labeling of Plasma/Serum

1. Reconstitute the Cy3 and Cy5 dyes by adding 110 µl Clontech Antibody 500 Extraction/labeling buffer directly to the tube in which the dye was supplied.
2. Vortex for 20 s.
3. Centrifuge briefly to collect reconstituted dye in the bottom of the tube.
4. Confirm that the protein extract pH is ~8.5 by spotting a small amount on pH paper. If the pH is not 8.5, adjust the pH of the sample (see Note 18).
5. Add an equal volume of each experimental sample (or pooled sample; see Notes 1 and 19) to microcentrifuge tubes. One hundred micrograms (final protein concentration) is the recommended amount of protein for each array, but this can vary by as much as two- to threefold when using fluid samples.
6. Add enough of the internal standard extract (100 µg per array) to a microcentrifuge tube (see Notes 20 and 21).
7. Add 6 µl Cy3/100 µg of protein to the internal standard tube and the appropriate amount of Cy5 to each experimental sample tube (6 µl Cy5/100 µg of protein – adjust as necessary).

8. Mix vigorously by pipetting.
9. Briefly spin the sample in a microcentrifuge tube and incubate on ice for 90 min, in the dark. Mix by inversion every 20 min.
10. Add 4 μl blocking buffer/100 μg protein to each tube to stop reaction. Incubate on ice for 30 min, in the dark. Mix by vortexing every 10 min.
11. Remove unbound dye using a Protein Desalting Spin Column.
    (a) Prepare needed amount of 1× desalting buffer.
    (b) Centrifuge columns (attached to 1.5 ml microcentrifuge tubes) at 1,500 × *g* for 2 min to remove the storage buffer.
    (c) Add 400 μl 1× desalting buffer to each column and centrifuge at 1,500 × *g* for 2 min. Discard the flow through.
    (d) Repeat step 11 (c).
    (e) Attach a clean microcentrifuge tube to each column.
    (f) Apply the 30–120 μl of the labeled protein extract to the column. Allow the sample to penetrate the column.
    (g) Centrifuge at 1,500 × *g* for 2 min.
    (h) Detach and discard the column and store the collection tubes on ice.
12. Measure the protein concentration using the BCA Assay. Subtract the absorption of Cy3 and Cy5 at 562 nm by preparing a tube that contains the labeled sample in place of the BCA reagent. Calculate $\Delta OD_{562} = [OD_{562\{protein\ sample\}} - OD_{562\{protein\ blank\}}]$. Use this value as the blank when calculating the protein concentration.
13. Estimate the average number of dye molecules covalently attached to each protein.
    (a) Measure Cy3 and Cy5 absorbance at 552 and 650 nm, respectively. Use the Cy3 and Cy5 molar extinction coefficients (Cy3 $\varepsilon_{552} = 150{,}000$/M/cm; Cy5 $\varepsilon_{650} = 250{,}000$/M/cm) to calculate the micromolar concentration of Cy3 and Cy5.
    (b) Absorbance = Extinction coefficient × path length × concentration, therefore, the micromolar concentration equals absorbance/extinction coefficient.
    (c) Determine micromolar concentration of protein. Assume an average molecular weight in the protein extracts of 60 kDa.
    (d) Calculate the average number of dye molecules per protein. The dye/protein ratio should be less than 6, ideally between 2 and 4 (see Note 22).

3.3. Antibody Microarray

Antibody microarrays are composed of hundreds of different antibodies immobilized on a glass surface (15). Antibody microarrays enable protein expression profiling for numerous proteins in a single experiment (16–19). Covalently bound antibodies are used as baits to capture fluorescently labeled proteins (antigens). The antibody microarray workflow comprises several steps: sample collection, protein extraction, protein labeling, sample hybridization to the antibody microarray, fluorescent detection, and analysis. The relative abundance of a protein in one sample is compared to the relative abundance of the same protein in a second sample. Each Clontech Ab 500 consists of 500 distinct monoclonal antibodies covalently bound to a glass slide. Antibodies are arranged in 16 grids; each grid consists of antibodies, plus positive/negative controls, in a six row × eight column format.

3.3.1. Antibody Microarrays with LCM Samples

1. Combine each CyDye labeled sample with the appropriate volume of labeled internal standard in a microcentrifuge tube. Add incubation buffer to a final volume of 120 μl (if using SA500 Incubation Chambers). Incubate at room temperature for 10 min.
2. Decant the storage buffer from the Ab 500 antibody microarrays that are contained in green-capped shipping/storage tubes.
3. Add 30 ml stock incubation buffer. Invert the tube ten times.
4. Decant the stock incubation buffer. Add 20 ml incubation buffer. Invert the tube ten times and decant the buffer.
5. Record slide lot numbers and assign arrays to samples. Place a SecureSeal incubation chamber on each slide, ensuring that a secure seal is formed.
6. Add sample from step 1 (CyDye labeled sample + internal standard) through the port in incubation chamber. Seal the chamber ports.
7. Incubate the microarrays at room temperature for 2 h.
8. Add 5 ml of Wash Buffer A to each wash chambers supplied with arrays. Peel the incubation chambers off the microarray slides and transfer the microarrays to the wash chambers.
9. Incubate the wash chamber containing the microarrays at room temperature for 5 min with gentle rocking.
10. Decant the wash buffer and add 5 ml Wash Buffer B.
11. Incubate the tray at room temperature for 5 min with gentle rocking.
12. Decant the wash buffer and add 5 ml Wash Buffer C.
13. Incubate the tray at room temperature for 5 min with gentle rocking.
14. Transfer each microarray slide (slide label facing down) to a 50-ml conical tube filled with dH_2O.

15. Decant the dH_2O.
16. Transfer the slides to a clean, green-capped vial supplied with microarrays.
17. Centrifuge slides at 1,000 × g for 5 min at room temperature.
18. Remove the slides from the vial and protect the microarrays from light.
19. Scan the microarrays within 24 h of drying. Genepix 4000B recommended settings: Cy3 532 nm, PMT = 550 V, and Laser Power = 33%; Cy5 635 nm, PMT = 670 V, and Laser Power = 33%. Use the CyDye-labeled BSA positive control spots to adjust scanner settings. CyDye-labeled BSA should fluoresce at 2,500–30,000 fluorescence units (FU). If the control spots are greater than 50,000 FU, adjust PMT and laser power to obtain a signal within 2,500–30,000 FU. See Fig. 1b for a representative microarray image.

3.3.2. Antibody Microarrays with Plasma/Serum

1. Set up incubation tray (one four-chambered tray per two arrays with an incubation and wash chamber for each array). Add 5 ml incubation buffer to each incubation chamber.
2. Combine each labeled sample with the appropriate amount of labeled internal standard (usually 104 μl) in a microcentrifuge tube.
3. Transfer the labeled protein from the tubes to the appropriate incubation chamber.
4. Incubate the tray at room temperature for 10 min with gentle rocking.
5. Decant the storage buffer from the antibody microarrays that are contained in green-capped shipping/storage tubes.
6. Add 30 ml stock incubation buffer. Invert the tube ten times.
7. Decant the stock incubation buffer. Add 20 ml incubation buffer. Invert the tube ten times and decant the buffer.
8. Record slide lot numbers and assign arrays to samples.
9. Remove microarrays from the storage tube and place, face-up, in the appropriate incubation chamber that contains the labeled sample and internal standard.
10. Incubate the microarrays at room temperature for 40 min with gentle rocking. Every 10 min, pry up one end of the slide with a pipette tip and rock the tray gently.
11. Add 5 ml Wash Buffer A to each wash chamber. Transfer the microarray slides to wash chambers.
12. Incubate the tray at room temperature for 5 min with gentle rocking.
13. Decant the wash buffer and add 5 ml Wash Buffer B.

14. Incubate the tray at room temperature for 5 min with gentle rocking.
15. Decant the wash buffer and add 5 ml Wash Buffer C.
16. Incubate the tray at room temperature for 5 min with gentle rocking.
17. Transfer each microarray slide (slide label facing down) to a 50-ml conical tube filled with dH_2O.
18. Decant the dH_2O.
19. Transfer the microarray slides to clean, green-capped vial supplied with arrays.
20. Centrifuge slides at 1,000×*g* for 5 min at room temperature.
21. Remove microarray slides from the vial and protect from light.
22. Scan the microarray slides within 24 h of drying. Genepix 4000B recommended settings: Cy3 532 nm, PMT=550 V, and Laser Power=33%; Cy5 635 nm, PMT=670 V, and Laser Power=33%. Use the CyDye-labeled BSA positive control spots to adjust scanner settings. CyDye-labeled BSA should fluoresce at 2,500–30,000 FU. If the control spots are greater than 50,000 FU, adjust PMT and laser power to obtain a signal within 2,500–30,000 FU.

3.4. Bioinformatic Analysis of Antibody Microarrays

The Clontech Ab 500 Microarray workbook contains an analysis worksheet that identifies differentially expressed proteins using the conventional approach of reciprocal CyDye labeling. In this approach, two samples are labeled with either Cy3 or Cy5, and mixed together for multiplex binding to a given antibody feature on the microarray. If there is more Cy3 or Cy5-labeled protein in the mixed sample, the spot will be either green or red, respectively. If the samples are present in equal amounts, the mixture will show as yellow. However, because of protein-specific differences in quantum yields and labeling efficiencies for Cy 3 and Cy5, a parallel set of two samples must be labeled reciprocally with the opposite dye. The geometric average of the two relative levels of Cy3 and Cy5 occurring at a given feature under the two conditions are then averaged, and the ratio is then used to express the *relative* abundance of the given protein in the two conditions. This strategy, while logical on its face, yields data that is intrinsically noisy. This approach also carries with it the limitation that the data are always ratios. In addition to statistical issues with ratios, this also means that one cannot easily compare one assay with another assay performed the next day or the next year. This is especially critical when trying to establish statistical differences between patient and control groups in clinical samples where each patient and each control may be different from the group as a whole. We believe that in using one dye for an internal standard, we circumvent the issues with ratios while also correcting for printing irregularities. Therefore, we only use the Ab 500 Microarray workbook to obtain the coordinates/names for the antibodies on the array.

A complimentary approach, which we believe has significant advantages, is to treat each dye signal as an independent variable (Rank Order Method) (see Note 23). Rank order methods, either rank-sum or rank-product tests, are frequently used in clinical studies, including serum proteomics of cystic fibrosis patients (20).

1. Use array analysis software program to determine the intensity of each spot.
2. Download the Ab 500 Microarray workbook (an Excel file) that corresponds to the microarray lot number from the "Online Tools" at www.clontech.com. Please note that this number differs from the lot number of the kit; the slide lot number is found on the label affixed to the slide. This workbook contains the names and coordinates for the antibodies on the microarray.
3. Cut and paste the fluorescence data intensity for each antibody into the workbook.
4. Flag all spots with a signal-to-noise ratio (SNR) <2 or those that are below background (see Note 24).
5. Determine a cohort of good spots (i.e., those spots not flagged in all of the arrays to be compared; see Note 25).
6. Calculate the median for the cohort on each array.
7. Normalize all spots on the array regardless of cohort inclusion by dividing the spot intensity by the cohort median for the respective array.
8. Correct for interarray variability by dividing the Cy5 (experimental samples) values obtained for each spot in step 7 by the Cy3 (internal standard) values obtained in step 7.
9. Calculate the average ratio for each antibody and experimental group using the ratios obtained in step 7.
10. Use these average ratios for each antibody to identify differentially expressed proteins in the experimental groups (see Note 23).
11. Test for statistical significance (e.g., Student's *t*-test, SAM, etc.) (see Note 26).
12. Identify affected pathways and cellular processes using gene ontology and pathway analysis programs (see Note 27; Fig. 4 and Table 2).
13. Validate differentially expressed proteins (see Note 28).

Table 2
Example gene ontology (GO) terms most changed in bronchial cells from patients with cystic fibrosis (CF)

No. of proteins in term	No. of term proteins changed in cystic fibrosis	*P*-value, depletion	*P*-value, enrichment	Term
20	10	0.999997436	**2.2477E-05**	Unfolded protein binding
25	9	0.999767826	**0.001171257**	Protein folding
52	13	0.998420061	**0.004617451**	Cellular biosynthesis
11	5	0.999401489	**0.004849652**	Cellular carbohydrate metabolism
31	9	0.998340856	**0.006151711**	Macromolecule biosynthesis
37	10	0.997925974	**0.006900409**	Cytosol
57	13	0.996012328	**0.010393376**	Biosynthesis
28	8	0.997113416	**0.010642812**	Protein biosynthesis
13	5	0.998183998	**0.011119135**	Alcohol metabolism
23	7	0.997184123	**0.01153684**	Translation
14	5	0.997126475	**0.01570003**	Carbohydrate metabolism
15	5	0.995668437	**0.021384264**	Ribonucleoprotein complex
12	4	0.992458993	**0.039018162**	Cofactor metabolism
24	6	0.985280376	**0.047808276**	Mitochondrion
191	29	0.957508768	0.065132437	Cellular macromolecule metabolism
14	4	0.98429997	0.06570481	Nucleotide metabolism
195	29	0.946524684	0.080226639	Macromolecule metabolism
21	5	0.974136042	0.081883871	Generation of precursor metabolites and energy
28	6	0.966318675	0.090837479	Endoplasmic reticulum
120	3	**0.000373545**	0.999931167	Membrane
205	9	**0.000377034**	0.99987298	Cell communication
58	0	**0.00091291**	1	Plasma membrane
109	3	**0.001075818**	0.999783164	Signal transducer activity
176	9	**0.003335047**	0.998700235	Signal transduction
100	4	**0.008603372**	0.997492663	Protein modification
72	2	**0.008811524**	0.998274988	Intrinsic to membrane
72	2	**0.008811524**	0.998274988	Integral to membrane
37	0	**0.011507014**	1	Transcription factor activity
37	0	**0.011507014**	1	Receptor activity

4. Notes

1. It is important to avoid primary amines and reducing agents in preparation of the samples. Serum and other fluid samples should be compared according to volume, not mass (i.e., equal volumes of fluid, not equivalent masses of protein should be used for each array).
2. Bronchial brush biopsies should only be performed by qualified, trained medical personnel. The number of brushes obtained per patient varies according to age – we typically have three (children less than 2 years), four (children greater than 2 years), or five (teens and adults). One brush is generally sufficient for the procurement of ~10,000 cells.
3. All biosafety precautions must be followed when using human specimens.
4. Dislodge cells from brush and resuspend them during subsequent steps as gently as possible.
5. Cells do not stick as tightly to the Rain-X treated slides. Calculate the volume required for slide coverage assuming a loss of ~20% of cells on the treated slides.
6. Rain-X glass coating is necessary to prevent the cells from adhering tightly to the glass slide. The cells will be microdissected and thus the cells must adhere to the slide with minimal adhesive forces. If the cells adhere tightly to the slide, the efficiency of microdissection will be compromised.
7. We use a NanoDrop 1000 (NanoDrop Technologies) to limit the amount of sample required for the protein assay. Theoretically, only 2 μl are needed for the assay, but we use 30 μl for the MicroBCA assay (sample and working reagent). Cap the tube tightly and seal with Parafilm to reduce evaporation during incubation. Any other highly sensitive protein assay that is not sensitive to the detergents in the lysis buffer can also be used.
8. Plasma was not used in this study, but it would be preferred to avoid the extra manipulations necessary to obtain serum. These extra manipulations can result in unacceptable sample variation. Please see the US HUPO guidelines for more information on plasma preparation (20).
9. Keep all collection and processing procedures as consistent as possible to reduce pre-analytical variability.
10. It is critical for labeling efficiency that the sample pH is between 7.8 and 8.2. Adjust the sample pH as necessary. Sample pH can be increased through the careful addition of lysis buffer (pH 9.5 instead of 8.0) or 50 mM NaOH.

11. We have used this technique for 1,000–10,000 cells (up to 5 μg protein). Higher amounts of protein will yield a better signal on the microarray. It is important to use the same amount of sample for each array in a given experiment.
12. The composition of the internal standard is not important. Ideally, it would be a mixture of all proteins on the array at a known concentration. However, this is not practical, so the internal standard should consist of a protein sample that contains all of the proteins represented on the microarray. We use the same combination of cell culture and tissue extracts for every array run in our laboratory. The amount of internal standard is also not critical (1–5 μg is sufficient).
13. The TCEP solution is unstable in phosphate buffer and should be used immediately.
14. Typically, 2 nmol TCEP and 4 nmol dye work well for labeling. However, samples containing more cysteines than average (2%) or more glutathione may require more TCEP and dye. Additionally, sample/buffer components may affect labeling. It is useful to perform an initial labeling optimization experiment whenever new sample types or buffers are employed. In the optimization test, the new samples are incubated with different amounts of TCEP and dye (always at a 1:2 TCEP to dye ratio, however) and the samples run on a 2D polyacrylamide gel. Under-labeled samples will display MW "trains" or vertical streaking and over-labeled samples will show pI charge "trains" and horizontal streaking. Please see the instruction booklet for the CyDye DIGE Fluor Labeling Kit for Scarce Samples for more information.
15. Vortexing does not mix the samples sufficiently; pipetting up and down is preferred.
16. It is not necessary to remove the unbound dye before using the labeled samples for the microarray.
17. Sample pH can be increased through the careful addition of lysis buffer (pH 9.5 instead of 8.0) or 50 mM NaOH.
18. Serum and other fluid samples should be compared according to volume, not mass (i.e., equal volumes of fluid, not equivalent masses of protein used for each array) as analytes in such fluids are generally reported per volumetric unit, not per mass unit. However, it is important to maintain the dye:protein ratio in the labeling reaction; adjust the amount of dye if necessary.
19. The composition of the internal standard is not important. Ideally, it would be a mixture of all proteins on the array at a known concentration. However, this is not practical, so the internal standard should consist of a protein sample that contains all of the proteins represented on the microarray. We use

the same combination of cell culture and tissue extracts for every array run in our laboratory.

20. This tube should contain enough internal standard extract for all of the arrays.
21. A ratio higher than six can interfere with antigen binding.
22. Please *see* Pollard et al. (7) for more information regarding rank order sum analysis.
23. Genepix software calculates a SNR, otherwise you can estimate the signal-to-noise ration by the following formula: SNR = net counts/SD of the background.
24. Bioinformatic analysis should be done in consultation with a biostatistician after all the data has been collected.
25. Reliance on the classical *t*-test for determining significance of a difference between two groups can be difficult when dealing with microarray data. This is because the number of experimental replicates (*n* value in the *t*-test) is low due to the high cost of each array. Tusher et al. (21), building on the concept of a false discovery rate (FDR), have suggested an alternative strategy in the Statistical Analysis of Microarrays (SAM) Algorithm. Their strategy has been to take data associated with each condition, and to calculate statistical parameters based on many permutations of the data. They generate an algorithm to calculate the statistical validity of the difference based on a local FDR, and an associated *q*-statistic. The *q*-value gives an estimate of the probability of falsely identifying a protein as "significant" within the group of proteins that have *q*-values lower than that for the protein in question. The local FDR gives a measure of the probability that a given feature is identified as significant by random chance. The apparent advantage of combining nonparametric inference from rank orders, and the random permutation steps in the SAM algorithm, have been interpreted as an example of "borrowed power," because the resultant study seems to be more locally powered than might have been calculated from a standard statistical approach.
26. There are many excellent online resources for gene ontology analysis and the identification of functional protein classes. These include NAIAD DAVID (http://niaid.abcc.ncifcrf.gov/home.jsp), GATHER (http://gather.genome.duke.edu/) and NCI GOMiner (http://discover.nci.nih.gov/gominer/). Pathway analysis software must be purchased (Ingenuity, GeneGo).
27. It is important to validate microarray results by an independent method such as western blotting, immunohistochemistry, or reverse phase capture protein microarrays. Protein-based

microarrays require quality control parameters and reduction of pre-analytical/analytical variability particularly as the technology moves from research to clinical use. Microarray quality control and technical challenges are reviewed in several recent publications (22–26).

References

1. Welsh, M. J., Ramsey, B. W., Accurso, F., Cutting, G. R. (2001) Cystic Fibrosis, in *The Metabolic Basis of Inherited Disease* (Scriver, C. L., Beaudet, A. L., Valle, D., and Sly, W. S., ed.), McGraw-Hill, New York, pp. 5121–5188.
2. Cystic Fibrosis Foundation (2008) Patient Registry 2008 Annual Report. Bethesda, MD.
3. Liou, T. G., Adler, F. R., Fitzsimmons, S. C., Cahill, B. C., Hibbs, J. R., Marshall, B. C. (2001) Predictive 5-year survivorship model of cystic fibrosis. *Am J Epidemiol* **153**, 345–52.
4. Espina, V., Wulfkuhle, J. D., Calvert, V. S., VanMeter, A., Zhou, W., Coukos, G. et al. (2006) Laser-capture microdissection. *Nat Protoc* **1**, 586–603.
5. Emmert-Buck, M. R., Bonner, R. F., Smith, P. D., Chuaqui, R. F., Zhuang, Z., Goldstein, S. R. et al. (1996) Laser capture microdissection. *Science* **274**, 998–1001.
6. Orba, Y., Tanaka, S., Nishihara, H., Kawamura, N., Itoh, T., Shimizu, M. et al. (2003) Application of laser capture microdissection to cytologic specimens for the detection of immunoglobulin heavy chain gene rearrangement in patients with malignant lymphoma. *Cancer* **99**, 198–204.
7. Pollard, H. B., Srivastava, M., Eidelman, O., Jozwik, C., Rothwell, S. W., Mueller, G. P. et al. (2007) Protein microarray platforms for clinical proteomics. *Proteomics Clin Appl* **1**, 934–52.
8. Srivastava, M., Eidelman, O., Jozwik, C., Paweletz, C., Huang, W., Zeitlin, P. L. et al. (2006) Serum proteomic signature for cystic fibrosis using an antibody microarray platform. *Mol Genet Metab* **87**, 303–10.
9. Eidelman, O., Jozwik, C., Huang, W., Srivastava, M., Rothwell, S. W., Jacobowitz, D. M. et al. (2010) Gender dependence for a subset of the low-abundance signaling proteome in human platelets. *Hum Genomics Proteomics* **2010**, 164906.
10. Miller, J. C., Zhou, H., Kwekel, J., Cavallo, R., Burke, J., Butler, E. B. et al. (2003) Antibody microarray profiling of human prostate cancer sera: antibody screening and identification of potential biomarkers. *Proteomics* **3**, 56–63.
11. Hamelinck, D., Zhou, H., Li, L., Verweij, C., Dillon, D., Feng, Z. et al. (2005) Optimized normalization for antibody microarrays and application to serum-protein profiling. *Mol Cell Proteomics* **4**, 773–84.
12. Espina, V., Wulfkuhle, J., Liotta, L. A. (2009) Application of laser microdissection and reverse-phase protein microarrays to the molecular profiling of cancer signal pathway networks in the tissue microenvironment. *Clin Lab Med* **29**, 1–13.
13. Iyer, E. P., Cox, D. N. (2010) Laser capture microdissection of Drosophila peripheral neurons. *J Vis Exp* **24**(39). pii: 2016. doi: 10.3791/2016.
14. Neblock, D. S., Chang, C. H., Mascelli, M. A., Fleek, M., Stumpo, L., Cullen, M. M. et al. (1992) Conjugation and evaluation of 7E3 x P4B6, a chemically cross-linked bispecific F(ab')2 antibody which inhibits platelet aggregation and localizes tissue plasminogen activator to the platelet surface. *Bioconjug Chem* **3**, 126–31.
15. Haab, B. B. (2001) Advances in protein microarray technology for protein expression and interaction profiling. *Curr Opin Drug Discov Devel* **4**, 116–23.
16. Yu, X., Schneiderhan-Marra, N., Joos, T. O. (2010) Protein microarrays for personalized medicine. *Clin Chem* **56**, 376–87.
17. Yu, X., Schneiderhan-Marra, N., Hsu, H. Y., Bachmann, J., Joos, T. O. (2009) Protein microarrays: effective tools for the study of inflammatory diseases. *Methods Mol Biol* **577**, 199–214.
18. Qin, S., Qiu, W., Ehrlich, J. R., Ferdinand, A. S., Richie, J. P., O'Leary M, P. et al. (2006) Development of a "reverse capture" autoantibody microarray for studies of antigen-autoantibody profiling. *Proteomics* **6**, 3199–209.
19. Ehrlich, J. R., Qin, S., Liu, B. C. (2006) The 'reverse capture' autoantibody microarray: a native antigen-based platform for autoantibody profiling. *Nat Protoc* **1**, 452–60.

20. Rai, A. J., Gelfand, C. A., Haywood, B. C., Warunek, D. J., Yi, J., Schuchard, M. D. et al. (2005) HUPO Plasma Proteome Project specimen collection and handling: towards the standardization of parameters for plasma proteome samples. *Proteomics* **5**, 3262–77.
21. Tusher, V. G., Tibshirani, R., Chu, G. (2001) Significance analysis of microarrays applied to the ionizing radiation response. *Proc Natl Acad Sci USA* **98**, 5116–21.
22. Hartmann, M., Roeraade, J., Stoll, D., Templin, M. F., Joos, T. O. (2009) Protein microarrays for diagnostic assays. *Anal Bioanal Chem* **393**, 1407–16.
23. Ellington, A. A., Kullo, I. J., Bailey, K. R., Klee, G. G. (2009) Measurement and quality control issues in multiplex protein assays: a case study. *Clin Chem* **55**, 1092–9.
24. Ellington, A. A., Kullo, I. J., Bailey, K. R., Klee, G. G. (2010) Antibody-based protein multiplex platforms: technical and operational challenges. *Clin Chem* **56**, 186–93.
25. Kricka, L. J., Master, S. R. (2008) Validation and quality control of protein microarray-based analytical methods. *Mol Biotechnol* **38**, 19–31.
26. Kricka, L. J., Master, S. R. (2009) Quality control and protein microarrays. *Clin Chem* **55**, 1053–5.

Chapter 13

Tissue Microarrays as a Tool in the Discovery and Validation of Predictive Biomarkers

Stephen M. Hewitt

Abstract

The tissue microarray (TMA) is the embodiment of high-throughput pathology. The platform combines tens to hundreds of tissue samples on a single microscope slide for interrogation with routine molecular pathology tools. TMAs have enabled the rapid and cost-effective screening of biomarkers for diagnostic, prognostic, and predictive utility. Most commonly applied to the field of oncology, the TMA has accelerated the development of new biomarkers, and is emerging as an essential tool in the discovery and validation of tissue biomarkers for use in personalized medicine. This chapter provides an overview of TMA technology and highlights the advantages of using TMAs as tools toward rapid introduction of new biomarkers for clinical use.

Key words: Biomarker, FFPE, Immunohistochemistry, Pathology, Protein, RNA, Tissue microarray

1. Introduction

The concept of a multisample platform for pathology traces its origins to Hector Battifore, who developed approaches to present multiple samples of tissue on a single section (1). His methods were laborious and low throughput in nature, limiting the utility to primarily control slides for the development of immunohistochemical assays. In 1998, Juha Kononnen (2), working in the National Human Genome Research Institute (National Institutes of Health), developed the first tissue microarray (TMA) and functional tissue arrayer, which allowed the production of high-throughput TMAs with a relatively simple instrument. A key feature of these arrays was the precise and orderly arrangement of the tissue cores in the recipient block so that the individual tissue cores could be traced back to the individual patients and their clinical information (3). Although

Virginia Espina and Lance A. Liotta (eds.), *Molecular Profiling: Methods and Protocols*, Methods in Molecular Biology, vol. 823,
DOI 10.1007/978-1-60327-216-2_13, © Springer Science+Business Media, LLC 2012

primarily conceived as a research tool for understanding protein expression, it was not long before the TMA was being used to confirm the nature of prognostic and predictive biomarkers (4, 5).

The most common assay used in conjunction with TMAs is immunohistochemistry (IHC). This combination provides a widely used tool for the rapid translation of biomarkers into clinical benefit. TMAs are constructed of archival tissue samples obtained as a part of treatment and diagnosis. IHC applied to TMAs is now a common routine method for validation of gene and protein array data (1, 2, 5), image analysis technology development (6), and molecular research (4). By utilizing antibodies, a common tool of biomarker discovery and research, it requires only a fraction of the effort to move from the research setting to evaluate hundreds of tissue samples at one time with an already clinically accepted assay. The remaining challenge is collection and annotation of the samples (7). Fundamentally, the TMA shifts the balance of how tissue is interpreted with reference to defining disease. Previously, histomorphologic features of a tumor have been used to characterize the behavior of a tumor. Increasingly, it is the expression of specific biomarkers, as detected by IHC, which provides molecular predictors of tumor behavior.

2. Materials

2.1. Samples

1. Within the field of cancer research, TMAs have focused on malignant lesions or screening cohorts for examining normal and dysplastic lesions, although any tissue type can be used to construct TMAs (8, 9).
2. Translational research is founded on principles of *translating* findings from molecular biology to clinical utility; therefore, use of clinically derived tissue, handled according to the same precepts of surgical pathology, is preferred (7).
3. Formalin-fixed, paraffin-embedded tissue is the most commonly applied tissue for TMA construction, but this is a reflection of its predominance in pathology archives and practice (10). Alternative fixatives can be utilized; however, knowledge of the fixative composition is essential, as assays are developed based on these preanalytical features. Ethanol-fixed tissue is also encountered in research settings (9, 11) (see Note 1).

2.2. Tissue Microarray Construction

1. A histopathologist, via microscopic examination, identifies the regions of interest on corresponding tissue section slides.
2. Areas corresponding to these regions of interest are cored from the original donor blocks. A TMA block consists of a recipient block of paraffin into which cores of tissue have been placed with needles after extraction from paraffin donor blocks.

3. A number of instruments now exist for the construction of TMAs, offering a variety of diameters of tissue core size, typically from 0.6 mm (the field of view of a 40× objective on a standard pathology microscope) to 2 mm or greater (see Note 2).
4. Diameter of the tissue cores determines the density of the array. With 0.6-mm needles, arrays on the order of 500 cores are routine while 2-mm cores result in arrays of 40 or so samples.
5. The recipient (TMA) block is sectioned with a microtome onto microscope slides, which are subsequently used for specific assays.

2.3. TMA Assays

1. The assays performed on TMAs are as diverse as the arrays pursued on all tissues. The most common assay is IHC, accounting for approximately 95% of all stains (10) (see Note 3).
2. The routine hematoxylin and eosin (H&E) stain accounts for ~3% of assays, and the last 1% are in situ assays for either RNA or DNA. The H&E stain is a routine stain for diagnostic histo- and cytomorphologic examination. As such, it is typically used on sections from every TMA block to confirm the diagnosis of the tissue present, usually on every 50th section.

3. Methods

3.1. General Process

3.1.1. Identification of Material

Identification of appropriate tissue material and patient cohorts for development of clinical predictive assays is essential. Key points that must be addressed include ethical approval and access to sufficient numbers of cases, which have sufficient diagnostic material for assay and reassay during the development process (10). Lastly, the material must be well-annotated both at the diagnostic and the clinical outcomes levels. Depending on the phase of predictive biomarker development, an epidemiologic approach must be employed to determine the impact of different populations and the molecular differences of the disease between populations. This process is typically the most time consuming, and the time and effort required are often underestimated.

3.1.2. Tissue Fixation and Embedding

The process of preparing formalin-fixed, paraffin-embedded tissue includes (1) acquisition of tissue from clinicians and surgeons; (2) preparation of the tissue specimen such that it is appropriate for fixation and processing (e.g., cutting/dissecting the tissue into pieces approximately 10×10×4 mm); (3) fixation, most commonly in formalin; and (4) tissue processing, which is the process of replacing the aqueous environment of tissue with paraffin using a series of alcohol dehydration steps. Ultimately, the paraffin-impregnated tissue is embedded in a paraffin block, and sectioned onto glass slides for staining. The block is archived, and may eventually be used in a TMA.

3.2. Tissue Microarray Construction

A number of excellent reviews and technical sources are available on the construction of TMAs (12, 13). In so far as construction and validation of TMAs for predictive biomarkers, there are two essential elements: redundant arrays and representativeness of the arrays against the anticipated clinical standard – analysis on whole sections (14). These issues are intertwined. Given the efforts of obtaining a trial or cohort for analysis for a predictive marker, it would be foolish to produce only a single array. Replicate arrays are invaluable for assay development, reproducibility assays, and comparison of new reagents and methodologies, let alone comparison of other biomarkers. The number of replicates that can be produced is a function of the tumor type, size of specimens available for analysis, and array design. Simply, "more is better" to ensure a larger sample size for adequate tumor/stroma area and multiple pieces of tissue from the same or similar blocks. The issue of tissue core size can be a function of preference and number of samples. Although significantly more samples can be applied to a section with 0.6-mm needles, 1.0-mm needles may be chosen for a number of reasons. Larger cores should not be assumed as a simple solution to the issue of representativeness (14), and may reduce the number of replicates; however, in some instances, such as biopsies, they are a preferred solution (9).

Representativeness of a TMA is very challenging and the literature is confusing at best. The goal is to ensure that a TMA accurately represents what would be seen on whole tissue sections. In the context of a predictive assay, this is demonstration that the results of the assay on a TMA would give the same prediction of response or outcome if applied to whole sections cut from the same cases (14). A strict definition is that the TMA demonstrates exactly the same biomarker profile as determined by examination of the whole sections, not just a statically equally proportion. Fundamentally, the only means to accomplish this is to test this comparison, based on the proven utility – a catch-22. It is inappropriate to assume that the representativeness of one marker predicts the representativeness of the next marker. They are independent events. Ultimately, the only solution is a best effort and, then when a potential marker has been found, validation on whole sections (10). This is where replicate arrays are important, as staining of replicate arrays allows determination of how many cores are required to approximate a whole section. Many investigators perform pilot tests on core size and replicate numbers to solve these problems before they construct a large TMA.

3.3. Input Sample Quality

It cannot be emphasized enough that tissue quality is a first-order concern (15). Despite the widely used phrase "per standard protocol," there are functionally no standard protocols between hospitals, and differences between countries of origin are significant (16). To date, no factor in specimen handling has been identified that does not impact quality when measured by the most stringent assays. However, useful tissue biomarkers do exist, demonstrating the intersection of robust biomarker assays and quality tissue preparation. Two factors most commonly associated with specimen preanalytical variability are tissue collection and tissue preparation.

3.3.1. Tissue Collection and Warm Ischemia Time

Acquisition of tissue from the surgeon and clinician, before fixation, is referred to as the warm ischemia time, during which time the tissue is devitalized but not preserved. Prolonged warm ischemia times are clearly associated with suboptimal specimens, which demonstrate altered gene and protein profiles (17–20). However, it is impractical to collect many tissues with extremely short warm ischemia times, making many markers of tissue hypoxia impractical for clinical implementation. Reasonable goals for warm ischemia time are under 30 min if one wishes to maintain the fidelity of the in vivo state of the nucleic acids and proteins (18, 19).

3.3.2. Tissue Preparation

Preparation of the tissue in the pathology laboratory is called "grossing" and includes the process of inking, dissecting, and sectioning of the specimen (16). Depending on the specimen, it may be submitted in toto, in total without the aforementioned steps, such as a small biopsy, but excisional biopsies and expatriations of masses or organs require extensive grossing. Depending on specimen type and size, all of the tissue or a small sampling of the tissue is selected for tissue processing and microscopic examination. The key issue is that this is done in a timely manner and the sections are of appropriate size for tissue processing.

3.4. Biomarker Validation

When an assay developer constructs an IHC assay and correlates it with some utility, the process is called *validation*. When an end user purchases and applies the assay, the user goes through a process of *verification*, in which the end user demonstrates the assay performance as intended (21–24). In so far as development of a validated assay, it is essential to test as many potential variables as possible to ensure that the assay is accurate – reporting a true result. For material obtained from a single center, this is often straightforward; however, in multi-institutional trials, the effects of preanalytic variables, typically fixation and processing, must be incorporated into the validation strategy. In a very simple fashion, this can be determination that the distribution of results from the assay is the same between sites and that any variance in this distribution can be accounted by other methods.

3.5. Proficiency Testing

Another simple element of validation is appropriate crossover testing, also known as proficiency testing, With TMAs, this approach is far simpler than that was previously possible (22, 24). In a crossover validation, TMAs containing tissue of the same diagnostic specifications are shared between two laboratories, which perform the same assay on the specimens. When an assay is robust, the results in lab A on TMA A are identical to the results in lab A on TMA B while at the same time lab B gets identical results on TMAs A and B.

3.6. Total Test Approach to Assays

One element of the development of a predictive biomarker is the concept of the *total test assay*. An assay is calibrated to give a result and this calibration is dependent on the elements of the assay – specimen, analyte, as well as other reagents, and interpretation. For many biomarkers, especially those used in diagnosis (often referred to as markers of lineage), calibration of the assay is simple as the results are typically binary – presence or absence of the marker. In contrast, predictive biomarkers are often based on a specific cutoff or predetermined level of the biomarker of interest. As a result, the calibration of the assay is more specific. To obtain this specificity of the assay, the parameters of the assay require stringent specification.

To determine the predictive value of a biomarker, the assay is defined and performed on a series of specimens for which the response to therapy is determined (25). From this test, a correlation of the assay result with response is developed. Previously, a biomarker was used for selection of patients, and only those patients with a positive result were treated. Although this approach does work, it fails to generate a positive and negative predictive values of the biomarker as it relates to response, as patients who were biomarker negative were never challenged with the treatment, and their correlation with outcome was not determined (26).

The correlation of the biomarker with response is that of a *total test*. Alterations in the assay directly result in alterations in correlation with response. Most commonly, the alteration is introduction of new/alternative antibodies or detection systems, with the belief that these new reagents, typically with higher molecular specificities, result in a superior assay. It is inappropriate to make these assumptions, and the correlation requires formal testing. Another common failure is specification of the specimen, where specimens not handled as specified in the original assay are tested and give inappropriate results. This is an all too common problem, and stems from both poor specimen preparation as well as inadequate specification of the assay (27).

3.7. Interpretation

For over a century, pathology has been defined by interpretation of histomorphology and cytomorphology, generating a description, leading, within the context of the patient, to a diagnosis. This process was mediated by the gross examination of tissue and microscopic examination of slides stained with compounds of differing specificities to different classes of biomolecules.

A paradigm shift occurred with the advent of IHC in which an antibody is utilized to identify a specific protein (or other antigen) in tissue. Interpretation of IHC requires a description of where the antibody–antigen complex is detected – both at the histomorphologic and cytomorphologic levels, combined with qualification of its abundance.

With this molecular approach, the definition of disease began to shift from pattern and appearance of cells to definition by the presence of individual proteins that define the disease. Cellular pattern and appearance continue to be the lynch pin of diagnostic pathology while the use of IHC is an adjunct used for confirmation of the diagnosis, along with the clinical features of the patient. As the breadth of antibodies widens, a panel of antibodies may be more prognostic than histomorphologic-based grading, not to mention the utility of predictive biomarkers that may be applicable to individualized therapy.

In practice, the majority of immunohistochemical stains are interpreted based on the anticipated localization of the antibody–antigen complex (nuclear, cytoplasmic, membranous, or a combination thereof) in a binary (present or absent) fashion. In general, the binary interpretation is based on the staining of *any* of the appropriate cell types and subcellular compartments. Unfortunately, the basis on which these relationships are founded are too frequently made on small case reports of selected tumors and never tested appropriately to define both false-positive and false-negative rates. Many markers probably lack the specificity ascribed to them (28).

3.7.1. Prognostic Markers

The application of IHC to prognostic markers requires establishing a “cutoff” value that is applied to the interpretation to develop a prognostic relationship. Common cutoff values are 1 and 10%, meaning that at a minimum this predefined number of cells of interest should express the marker of study to constitute a positive reaction and a clinically relevant prognostic correlation. In this context, the application of a total test is especially important. Should the assay, specimen, or interpretation of the assay be altered, the correlation of immunohistochemical reaction with a prognosis is lost.

3.7.2. Predictive Biomarkers

Predictive biomarkers typically adhere to the same principles as prognostic biomarkers. In fact, most predictive biomarkers are prognostic as well. In the same way, a graded approach to the result of an immunohistochemical assay would theoretically result in a probability of response to an intervention or drug. If a sufficiently large cohort of patients is both tested for a predictive biomarker and are treated with the agent/intervention to which the biomarker predicts response, it would be possible to define, for every interpretative step in the immunohistochemical assay, a probability of response to the agent/intervention. In practice, this is not done; rather a cutoff is set and patients above the cutoff are considered positive and treated. This approach results in potential undertreatment of patients who have a lower but definable probability of response while those who are positive have a probability of not responding. In practice, this is one rationale for treating a woman who has *any* estrogen receptor-positive breast cancer tumor cells (29). The probability of response is sufficiently high compared to the risk and cost of the therapy to warrant treatment.

It should be noted that theory is theory, and not practice. To determine these relationships, three elements are essential: (1) a total test assay, incorporating interpretation; (2) an interpretative schema that allows multiple categories in a fixed progression that correlates with increased expression; (3) a clinical trial design that treats patients who are biomarker positive and negative and is of sufficient size to meet the power calculations of the assay (26, 30). Within the realm of trials, it is essential to demonstrate that the effect of the biomarker is selection of the patients, with appropriate change in outcome, and not the underlying prognostic impact of the biomarker (25, 30).

3.8. Image Analysis

To answer the complexity of immunohistochemical interpretation, image analysis/image quantification has been developed as a means of standardization and quantification of IHC. It must be emphasized that image analysis is but the last step of the process of a test, and that if the test is not appropriately specified, calibrated, and carried out image analysis cannot improve results. Tools are available to define the cell type of interest (who and where); however, none of these tools are foolproof and always require human verification/supervision. Image analysis is clearly superior at the quantification elements of IHC – how much and how many. By generating continuous data, more sophisticated analysis is feasible, especially for prognostic and predictive biomarkers (6). Incorporating image analysis as an element of interpretation is a crucial facet of personalized medicine.

As these issues apply to TMAs, it is the development of the TMA that has driven the development of these assays (31). Although prognostic and predictive biomarkers were previously identified, the advent of the TMA has opened up the pipeline to rapid testing of potential biomarkers. alpha-Methylacyl coenzyme A racemase (32) is an excellent example of a biomarker identified by these means. It is anticipated that TMAs will likewise leverage image analysis to result in a predictive assay that cannot be accurately reproduced by manual interpretation, but is robust with a well-designed image analysis algorithm, as exemplified in the utility of survivin as a prognostic biomarker in breast cancer (31).

3.9. Multiplexed Markers

Despite the obvious high-throughput nature of TMAs, in general, there is a lack of publications that evaluate biomarkers as panels for personalized medicine. The majority of papers examine a single protein or examine multiple proteins individually. Rare papers have examined panels of biomarkers; however, they have failed to define the utility of the individual biomarkers in a panel and by their nature are difficult to reduce to clinical utility (33). Shou et al. demonstrate the potential to utilize two markers concurrently, although only within a complex model of tumor progression and survival (9).

3.10. Discovery

Although TMAs are routinely used in testing and validation of biomarkers for personalized medicine, their application is not limited to human pathology samples taken directly from patients. The general platform of an array of cellular material can be applied to cell lines grown in vitro or xenograft samples. Cell line microarrays (CMAs) and xenograft microarrays (XMAs) are used at the discovery level as well as tools toward assay development (13).

In their role as discovery tools, CMAs and XMAs are potent tools for transition from experimental platforms, such as western blotting, to a clinically relevant immunohistochemical assay (13). Direct information regarding cell types affected by drugs can be derived, as well the capacity to interrogate pathways, to define the best biomarker for the intended therapy. Often, new agents target specific proteins or classes of proteins, and construction of directed assays toward these targets is useful prior to clinical trials to prove the mechanism of action. However, it should not be assumed that this approach would result in the most clinically relevant biomarker. In some instances, downstream markers may function as more robust markers of predictive medicine. Two primary factors are responsible for alternative (typically, downstream) targets providing a better correlation with response. One is the limitation of antibodies; some epitopes are appropriate for IHC in FFPE tissue, but differences in antibody affinities may render a particular epitope/protein as a suboptimal biomarker. In some instances, the solution may be to select a neighboring epitope within the same antigen. For example, with HER2/neu, the therapeutic monoclonal antibody Trastuzumab (Herceptin®) binds the extracellular domain of HER2 while the antibodies used in the HercepTest® target the intracellular domain. The second issue is molecular cross talk. Alternative members of a signaling pathway may be better reporters of signaling due to differences in phosphorylation or magnitude of expression. Aspects of these relationships may be exhibited by agents targeting mTOR and AKT in the PTEN-PI3 Kinase pathway (34, 35).

3.11. New Methodologies

3.11.1. Multiplex Assays

Rimm et al. have developed a fluorescence-based, multiplex-based immunohistochemical assay approach (36) that has been very successful at identifying relationships of biomarkers with clinical outcomes that were not previously appreciated (37). Another approach is a multiplex immunoblot method, by which proteins are transferred from an FFPE tissue section to specially treated membranes which are then probed like an immunoblot (38). This method, a bit of a hybrid of western blot and IHC, provides histo-geographic spatial resolution. The primary advantages of this approach are multiplex (up to five antibodies), normalization against total protein content, and quantitative data, allowing development of ratio-based biomarkers. The data generated by this multiplex immunoblot technology has been replicated by IHC; however, the quantification and normalization methodologies appear to offer some benefit (34, 38, 39).

3.11.2. Fourier Transform Infrared Spectroscopy Coupled with TMA

In a more exotic application, Ira Levin's group has applied Fourier transform infrared spectroscopy (FTIR) to TMAs (40). In the first demonstration of the capacity of infrared spectroscopy of tissue, Fernandez et al. were able to generate a set of metrics that could segregate prostate tissue into its individual tissue components and with reasonable success identify cancer vs. benign prostatic epithelium (41). Without the high-throughput nature of TMAs, the baseline metric would not have been feasible. This approach is sufficiently robust such that the original metrics can be applied to other tissue types and sources (including frozen tissue) and function as classifiers.

3.11.3. CMA Coupled with Fourier Transform Infrared Spectroscopy

In an application more focused on personalized medicine, Chen et al. (42) utilized FTIR to define metrics of molecular responses to a drug. In their experiments, peripheral blood leukocytes (PBLs), arrayed in a CMA-like format, were exposed to histone deacetylase (HDAC) inhibitors in vitro, and a series of metrics that defined and quantified the changes in lysine acetylation were developed. The assay requires no specimen processing or affinity reagents, depending solely on changes in the vibrations of peptides based on the presence or absence of an acetyl group. In a Phase 0 clinical trial, patients were treated with HDAC inhibitors, PBLs were collected, and the alterations in acetylation were measured in response to drug treatment.

3.12. Troubleshooting: Tissue Preservation is the Major Source of Variability

Unfortunately, it is impossible at this time to provide strict guidance on the best practices, and the current effort is to educate the community on this issue as well as continue research to better characterize these differences. Defining "best" is very challenging, as the factors impacting FFPE tissue quality are interlocking. The two salient issues are: (1) producing an archival-quality FFPE tissue and (2) treating these preanalytic variables as elements of the specification of the specimen within the concept of a total test (see Note 4).

Fixation is a key, and too often mismanaged step, in the tissue preservation chain. Although there are a number of historic and recently described fixatives, the number of tissue fixatives in daily use has diminished over the last decade.

1. Ten percent neutral buffered formalin is the overwhelming choice of fixative worldwide (16). Buffer formulations are varied and have some impact on both RNA and protein in tissue, although this remains poorly characterized (20) (see Note 5).

2. Adequate fixation is a function of time, but requires approximately 24 h, depending on specimen size (18). Despite the assumption that overfixation is worse than underfixation, data demonstrates that underfixation leads to significant quality problems and unreliable assays at both the RNA and protein levels (18, 20). Current recommendations vary greatly; however, 16–32 h are a relatively safe window, as well as obtainable in clinical practice.
3. Fixative volume should be a minimum of ten times the volume of the tissue.
4. Tissue processing protocols impact specimen quality and extent of cross-linking (16, 18, 19). Limited data supports that a slower process, with vacuum and heat, results in high-quality specimens (20) (see Note 6).

3.13. Summary

The TMA has become an essential platform for validation of predictive biomarkers. This platform brings high-throughput technologies together with routine clinical specimens to create validated assays. The TMA is frequently the platform in which a final clinical assay is developed, having been discovered via other methodologies, but in some instances TMA is the platform for discovery of novel predictive biomarkers.

4. Notes

1. Other fixatives, most notably Bouins (formaldehyde, picric acid, and glacial acetic acid) and B5 (formaldehyde and mercuric chloride), are suboptimal, as the tissue tends to be brittle and difficult to array.
2. For a detailed description of TMA production, see Hewitt et al. in Protein Microarrays: Methods and Protocols, 2004 (10).
3. The predominance of IHC as an assay on TMA mirrors the use of whole tissue sections in general research. IHC is the application of an antibody to a tissue section to determine the histo- and cytologic localization of an antigen. Although generally not conceived as quantitative, the benefits of identifying the cell of interest and subcellular localization overcome the limitations of quantification. Performance of an IHC assay on TMAs is not different from IHC on whole tissue sections, but the approach has uncovered deficiencies that have been difficult to study previously (10, 14, 43).
4. Predictive biomarkers, especially those indexed on survival, may be assayed (or reassayed) at the conclusion of the trials. Additionally, it is routine to test patient specimens at relapse, not at diagnosis, for the use of predictive markers, not to mention

the application of new markers to antecedent patients' samples as these assays are developed and approved. In so far as the *total test* paradigm, the tissue specimen has a set of specified preanalytic elements appropriate for the assay. Fixative type is essential, and fixation time is an element of specification for some assays. There is no doubt that alternative tissue processing methods alter the specimen and may render the specimen inappropriate.

5. Formalin is an aqueous form of formaldehyde, the simplest aldehyde, which acts to cross-link proteins, as well as nick and cross-link (to proteins most commonly) nucleic acids.
6. It is believed that this higher quality is a result of more complete tissue dehydration. Alternative tissue processor methodologies have not been rigorously validated (16).

References

1. Battifora, H. (1986) The multitumor (sausage) tissue block: novel method for immunohistochemical antibody testing. *Lab Invest* **55**, 244–8.
2. Kononen, J., Bubendorf, L., Kallioniemi, A., Barlund, M., Schraml, P., *et al.* (1998) Tissue microarrays for high-throughput molecular profiling of tumor specimens. *Nat Med* **4**, 844–7.
3. Fridman, E., Daya, D., Srigley, J., Whelan, K. F., Lu, J. P., *et al.* (2011) Construction of tissue micro array from prostate needle biopsies using the vertical clustering re-arrangement technique. *Prostate.*
4. Ke, A. W., Shi, G. M., Zhou, J., Huang, X. Y., Shi, Y. H., *et al.* (2011) CD151 Amplifies Signaling by Integrin alpha6beta1 to PI3K and Induces the Epithelial-Mesenchymal Transition in HCC Cells. *Gastroenterology.*
5. Simon, R., Nocito, A., Hubscher, T., Bucher, C., Torhorst, J., *et al.* (2001) Patterns of her-2/neu amplification and overexpression in primary and metastatic breast cancer. *J Natl Cancer Inst* **93**, 1141–6.
6. Kwak, J. T., Hewitt, S. M., Sinha, S., and Bhargava, R. (2011) Multimodal microscopy for automated histologic analysis of prostate cancer. *BMC Cancer* **11**, 62.
7. Hewitt, S., Takikita, M., Braunschweig, T., and Chung, J. Y. (2007) Promises and challenges of predictive tissue biomarkers. *Biomarkers in Medicine* **1**, 313–18.
8. Dissanayake, S. K., Olkhanud, P. B., O'Connell, M. P., Carter, A., French, A. D., *et al.* (2008) Wnt5A regulates expression of tumor-associated antigens in melanoma via changes in signal transducers and activators of transcription 3 phosphorylation. *Cancer Res* **68**, 10205–14.
9. Shou, J. Z., Hu, N., Takikita, M., Roth, M. J., Johnson, L. L., *et al.* (2008) Overexpression of CDC25B and LAMC2 mRNA and protein in esophageal squamous cell carcinomas and premalignant lesions in subjects from a high-risk population in China. *Cancer Epidemiol Biomarkers Prev* **17**, 1424–35.
10. Hewitt, S. M. (2004) Design, Construction, and Use of Tissue Microarrays, *in* Protein Arrays: Methods and Protocols (Fung, E. T., ed.), Springer, New York, NY, pp. 61–72.
11. Whiteford, C. C., Bilke, S., Greer, B. T., Chen, Q., Braunschweig, T. A., *et al.* (2007) Credentialing preclinical pediatric xenograft models using gene expression and tissue microarray analysis. *Cancer Res* **67**, 32–40.
12. Sauter, G., Simon, R., and Hillan, K. (2003) Tissue microarrays in drug discovery. *Nat Rev Drug Discov* **2**, 962–72.
13. Braunschweig, T., Chung, J. Y., and Hewitt, S. M. (2005) Tissue microarrays: bridging the gap between research and the clinic. *Expert Rev Proteomics* **2**, 325–36.
14. Fergenbaum, J. H., Garcia-Closas, M., Hewitt, S. M., Lissowska, J., Sakoda, L. C., *et al.* (2004) Loss of antigenicity in stored sections of breast cancer tissue microarrays. *Cancer Epidemiol Biomarkers Prev* **13**, 667–72.
15. Lim, M. D., Dickherber, A., and Compton, C. C. (2011) Before you analyze a human specimen, think quality, variability, and bias. *Anal Chem* **83**, 8–13.

16. Hewitt, S. M., Lewis, F. A., Cao, Y., Conrad, R. C., Cronin, M., *et al.* (2008) Tissue handling and specimen preparation in surgical pathology: issues concerning the recovery of nucleic acids from formalin-fixed, paraffin-embedded tissue. *Arch Pathol Lab Med* **132**, 1929–35.
17. Webb, D., Hamilton, M. A., Harkin, G. J., Lawrence, S., Camper, A. K., *et al.* (2003) Assessing technician effects when extracting quantities from microscope images. *J Microbiol Methods* **53**, 97–106.
18. Espina, V., Edmiston, K. H., Heiby, M., Pierobon, M., Sciro, M., *et al.* (2008) A portrait of tissue phosphoprotein stability in the clinical tissue procurement process. *Mol Cell Proteomics* 7, 1998–2018.
19. Espina, V., Mueller, C., Edmiston, K., Sciro, M., Petricoin, E. F., *et al.* (2009) Tissue is alive: New technologies are needed to address the problems of protein biomarker pre-analytical variability. *PROTEOMICS - CLINICAL APPLICATIONS* **3**, 874–82.
20. Chung, J. Y., Braunschweig, T., Williams, R., Guerrero, N., Hoffmann, K. M., *et al.* (2008) Factors in tissue handling and processing that impact RNA obtained from formalin-fixed, paraffin-embedded tissue. *J Histochem Cytochem* **56**, 1033–42.
21. Wolff, A. C., Hammond, M. E., Schwartz, J. N., Hagerty, K. L., Allred, D. C., *et al.* (2007) American Society of Clinical Oncology/College of American Pathologists guideline recommendations for human epidermal growth factor receptor 2 testing in breast cancer. *Arch Pathol Lab Med* **131**, 18–43.
22. Wasielewski, R., Hasselmann, S., Ruschoff, J., Fisseler-Eckhoff, A., and Kreipe, H. (2008) Proficiency testing of immunohistochemical biomarker assays in breast cancer. *Virchows Arch* **453**, 537–43.
23. Gown, A. M. (2008) Current issues in ER and HER2 testing by IHC in breast cancer. *Mod Pathol* **21 Suppl 2**, S8-S15.
24. Bogen, S. A., Vani, K., McGraw, B., Federico, V., Habib, I., *et al.* (2009) Experimental validation of peptide immunohistochemistry controls. *Appl Immunohistochem Mol Morphol* **17**, 239–46.
25. Oldenhuis, C. N., Oosting, S. F., Gietema, J. A., and de Vries, E. G. (2008) Prognostic versus predictive value of biomarkers in oncology. *Eur J Cancer* **44**, 946–53.
26. Hewitt, S. M., Takikita, M., Abedi-Ardekani, B., Kris, Y., Bexfield, K., *et al.* (2008) Validation of proteomic-based discovery with tissue microarrays. *PROTEOMICS - CLINICAL APPLICATIONS* **2**, 1460–66.
27. Leong, A. S. (2004) Pitfalls in diagnostic immunohistology. *Adv Anat Pathol* **11**, 86–93.
28. Takikita, M., Altekruse, S., Lynch, C. F., Goodman, M. T., Hernandez, B. Y., *et al.* (2009) Associations between selected biomarkers and prognosis in a population-based pancreatic cancer tissue microarray. *Cancer Res* **69**, 2950–5.
29. Fisher, B., Costantino, J. P., Wickerham, D. L., Cecchini, R. S., Cronin, W. M., *et al.* (2005) Tamoxifen for the prevention of breast cancer: current status of the National Surgical Adjuvant Breast and Bowel Project P-1 study. *J Natl Cancer Inst* **97**, 1652–62.
30. Amado, R. G., Wolf, M., Peeters, M., Van Cutsem, E., Siena, S., *et al.* (2008) Wild-type KRAS is required for panitumumab efficacy in patients with metastatic colorectal cancer. *J Clin Oncol* **26**, 1626–34.
31. Brennan, D. J., Rexhepaj, E., O'Brien, S. L., McSherry, E., O'Connor, D. P., *et al.* (2008) Altered cytoplasmic-to-nuclear ratio of survivin is a prognostic indicator in breast cancer. *Clin Cancer Res* **14**, 2681–9.
32. Rubin, M. A., Zhou, M., Dhanasekaran, S. M., Varambally, S., Barrette, T. R., *et al.* (2002) alpha-Methylacyl coenzyme A racemase as a tissue biomarker for prostate cancer. *Jama* **287**, 1662–70.
33. Braunschweig, T., Kaserer, K., Chung, J.-Y., Bilke, S., Krizman, D., *et al.* (2007) Proteomic expression profiling of thyroid neoplasms. *PROTEOMICS - Clinical Applications* **1**, 264–71.
34. Chung, J. Y., Hong, S. M., Choi, B. Y., Cho, H., Yu, E., *et al.* (2009) The expression of phospho-AKT, phospho-mTOR, and PTEN in extrahepatic cholangiocarcinoma. *Clin Cancer Res* **15**, 660–7.
35. Tsurutani, J., Fukuoka, J., Tsurutani, H., Shih, J. H., Hewitt, S. M., *et al.* (2006) Evaluation of two phosphorylation sites improves the prognostic significance of Akt activation in non-small-cell lung cancer tumors. *J Clin Oncol* **24**, 306–14.
36. Camp, R. L., Chung, G. G., and Rimm, D. L. (2002) Automated subcellular localization and quantification of protein expression in tissue microarrays. *Nat Med* **8**, 1323–7.
37. Nadler, Y., Camp, R. L., Giltnane, J. M., Moeder, C., Rimm, D. L., *et al.* (2008) Expression patterns and prognostic value of Bag-1 and Bcl-2 in breast cancer. *Breast Cancer Res* **10**, R35.
38. Chung, J. Y., Braunschweig, T., Baibakov, G., Galperin, M., Ramesh, A., *et al.* (2006) Transfer

and multiplex immunoblotting of a paraffin embedded tissue. *Proteomics* **6**, 767–74.

39. Chung, J. Y., Braunschweig, T., Hu, N., Roth, M., Traicoff, J. L., *et al.* (2006) A multiplex tissue immunoblotting assay for proteomic profiling: a pilot study of the normal to tumor transition of esophageal squamous cell carcinoma. *Cancer Epidemiol Biomarkers Prev* **15**, 1403–8.
40. Bhargava, R., Fernandez, D. C., Hewitt, S. M., and Levin, I. W. (2006) High throughput assessment of cells and tissues: Bayesian classification of spectral metrics from infrared vibrational spectroscopic imaging data. *Biochim Biophys Acta* **1758**, 830–45.
41. Fernandez, D. C., Bhargava, R., Hewitt, S. M., and Levin, I. W. (2005) Infrared spectroscopic imaging for histopathologic recognition. *Nat Biotechnol* **23**, 469–74.
42. Chen, T., Lee, M. J., Kim, Y. S., Lee, S., Kummar, S., *et al.* (2008) Pharmacodynamic assessment of histone deacetylase inhibitors: infrared vibrational spectroscopic imaging of protein acetylation. *Anal Chem* **80**, 6390–6.
43. DiVito, K. A., Charette, L. A., Rimm, D. L., and Camp, R. L. (2004) Long-term preservation of antigenicity on tissue microarrays. *Lab Invest* **84**, 1071–8.

Chapter 14

Reverse-Phase Protein Microarrays

Mariaelena Pierobon, Amy J. VanMeter, Noemi Moroni, Francesca Galdi, and Emanuel F. Petricoin III

Abstract

Cancer is the consequence of intra and extracellular signaling network deregulation that derives from alteration of genetic and proteomic cellular homeostasis. Mapping the individual molecular circuitry of a patient's tumor cells is the starting point for rational personalized therapy.

While genes and RNA encode information about cellular status, proteins are considered the engine of the cellular machine, as they are the effective elements that drive cellular functions, such as proliferation, migration, differentiation, and apoptosis. Consequently, investigations of the cellular protein network are considered a fundamental tool to understand cellular functions. In the last decades, increasing interest has been focused on the improvement of new technologies for proteomic analysis. In this context, reverse-phase protein microarrays (RPMAs) have been developed to study and analyze posttranslational modifications that are responsible for principal cell functions and activities. This innovative technology allows the investigation of protein activation as a consequence of protein–protein interaction or biochemical reactions, such as phosphorylation, glycosylation, ubiquitination, protein cleavage, and conformational alterations.

Intracellular balance is carefully conserved by constant rearrangements of proteins through the activity of a series of kinases and phosphatases. Therefore, knowledge of the key cellular signaling cascades reveal information regarding the cellular processes driving a tumor's growth (such as cellular survival, proliferation, invasion, and cell death) and response to treatment.

Alteration to cellular homeostasis, driven by elaborate intra- and extracellular interactions, has become one of the most studied fields in the era of personalized medicine and targeted therapy. RPMA technology is a valid tool that can be applied to protein analysis of several diseases for the potential to generate protein interaction and activation maps that lead to the identification of critical nodes for individualized or combinatorial target therapy.

Key words: Array, Lysates, Protein, Proteomics, Reverse-phase protein microarray, Tissue

Virginia Espina and Lance A. Liotta (eds.), *Molecular Profiling: Methods and Protocols*, Methods in Molecular Biology, vol. 823, DOI 10.1007/978-1-60327-216-2_14, © Springer Science+Business Media, LLC 2012

1. Introduction

Reverse-phase protein microarray (RPMA) technology allows simultaneous analysis of phosphorylated, glycosylated, cleaved, or total cellular proteins on multiple samples (1, 2). Advantages of this technology for individualized patient molecular profiling are use of automated, high-throughput robotic instruments, minimal total cellular volume of clinical specimens, and high sensitivity (1–7).

Denatured cellular lysates, obtained by laser capture microdissected material, cell culture, serum, protein fractions, peptides, native proteins, or other body fluids, such as serum, CSF, synovial fluid, and vitreous, can be used as analyte sources (2–4, 8–10). Fresh or fixed tissues (ethanol- or formalin-fixed, paraffin-embedded material) can also be processed and applied to RPMAs (2, 11).

Protein array analysis is conventionally based on immobilized bait molecules (antibodies, cells, phage, proteins, or peptides) printed onto nitrocellulose-coated glass slides, which are subsequently probed with specific antibodies. The unique aspect of RPMA technology is the ability to immobilize multiple samples onto a single solid support and simultaneously probe multiple samples with a single primary antibody, thus allowing comparison of one analyte across many samples on the array (1).

Protein detection is based on a single, specific antibody–epitope interaction (Fig. 1) from a few picograms of cellular lysate or protein. The high sensitivity that RPMA provides is due to signal amplification system that acts independently from the immobilized analytes (2, 12). Commercially or noncommercially available antibodies

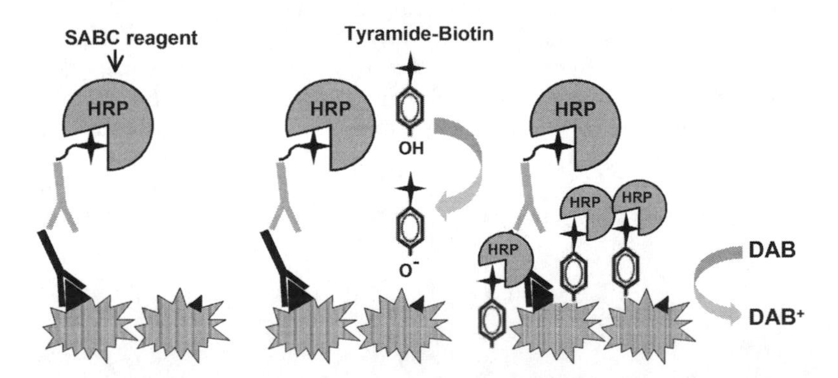

Fig. 1. Catalyzed signal amplification for protein detection. Amplification is achieved through a three-step process that involves a primary antibody, a secondary biotinylated antibody, streptavidin-conjugated horseradish peroxidase (HRP), and diaminobenzidine (DAB) as a chromogenic substrate (CSA kit, Dako). The HRP-catalyzed deposition of an amplification reagent is essential for signal detection. Amplification reagent is a biotin-coupled phenolic compound (tyramide) that is converted to a highly reactive radical species by peroxidase activity and subsequently binds electron-rich molecules (e.g., proteins) in the vicinity of the antigen–antibody complex. Further addition of streptavidin-conjugated peroxidase allows for the in situ precipitation of DAB.

targeting the protein of interest are used as a probe and coupled with signal amplification systems, such as third-generation enzymatic reactions, such as streptavidin-mediated horseradish peroxidase deposition of biotinyl tyramide.

Fluorescent, chemiluminescent, or colorimetric procedures can be employed as signal detection systems. These detection methods are capable of producing a signal that is proportional to the concentration of the analyzed molecules. Electronic images of the arrays are captured with standard image analysis software; spot intensities are calculated, with subsequent comparison of spot intensities by common statistical methods. Thus, the intensity values of each spot can be quantitatively compared to differentiate high/low-abundance analytes (1, 2, 7).

This multiplexed assay generates multitudes of data from single experiments, providing molecular information regarding potential drug targets, prospective biomarker identification, as well as diagnostic and/or prognostic information useful for both researchers and clinicians (2–6, 13–15). We describe a complete RPMA project for preparing cellular lysates and low-molecular-weight (LMW) serum fractions, printing RPMAs, determining total protein using Sypro Ruby Blot stain, array immunostaining with a colorimetric method, and array detection/spot quantification. The arrayer described herein is an Aushon Biosystems 2470 arrayer, but the principles can be applied to any microarray robotic printer.

2. Materials

Sample preparation for RPMAs:

A variety of cellular and noncellular protein-containing samples may be printed on an RPMA. The sample type (cell culture, tissue, or body fluid) dictates the preparation, printing, and immunostaining protocols to be used. Three different protein extraction/lysis buffers are described below because the buffer is dependent on (a) the sample type and (b) whether or not a total protein concentration needs to be determined for the sample prior to printing the microarray. In general, it is desirable to print all samples on a microarray at similar protein concentrations because the total protein value per spot is commonly used to normalize data between different samples on the microarray. Consistency in total protein concentration reduces analytical variability between the samples.

Cell lysis buffer is generally used for cell culture samples, peripheral blood mononuclear cells, or any other nonfixed cellular sample. The cell lysis buffer contains phosphatase and protease inhibitors that inhibit protein and phosphoprotein degradation during sample preparation. Cell lysis buffer is compatible with spectrophotometric (Coomassie) protein assays, but it does not contain any reducing agents.

Cell extraction buffer is typically used with fixed tissue samples, such as cells procured by laser capture microdissection (LCM) (5, 16–19). Extraction buffer is a Laemmli-based buffer containing a reducing agent (20). Extraction buffer is incompatible with spectrophotometric analysis because it contains sodium dodecyl sulfate (SDS) and bromophenol blue.

Extraction buffer for serum is a modified Laemmli-based buffer, which contains a higher percentage of glycerol than cell extraction buffer. The higher glycerol content provides a higher viscosity, which limits spot diffusion on the printed microarray.

2.1. Samples

1. Cell culture pellet or purified peripheral blood cell suspension. Store at –80°C.
2. Phosphate-buffered saline (PBS) without calcium or magnesium.
3. LCM (19, 21) caps containing microdissected cells from tissue that has previously been frozen, ethanol fixed, or formalin fixed. Caps are stored at –80°C in 0.5-mL Safe-Lock Eppendorf tubes.
4. Plasma/serum/body fluids. Store at –80°C.
5. LMW serum samples in 1× Novex® Tris–Glycine SDS Running Buffer (Invitrogen).

2.1.1. Cell Lysis and Protein Extraction Buffers

Cell lysis buffer: Combine reagents in a polypropylene tube.

1. 915 µL Tissue Protein Extraction Reagent (T-PER, Pierce).
2. 60 µL 5 M sodium chloride.
3. 10 µL 100 mM Orthovanadate (Sigma–Aldrich). Add 0.0184 g orthovanadate to 1 mL dH_2O in a screw-cap tube and heat at 100°C for a minimum of 10 min. Prepare fresh daily (see Note 1).
4. 10 µL 200 mM PEFABLOC (Roche). Resuspend PEFABLOC to 200 mM in dH_2O. Stable at –20°C for 2 months.
5. 1 µL 5 mg/mL Aprotonin: Prepare a 5 mg/mL solution of Aprotonin (Sigma–Aldrich) in dH_2O. Stable at –20°C for 6 months.
6. 5 µL 1 mg/mL Pepstatin A: Dissolve Pepstatin A (Sigma–Aldrich) at 1 mg/mL in a solution of 10% (v/v) acetic acid and 90% (v/v) methanol. Stable at –20°C for 6 months.
7. 1 µL 5 mg/mL Leupeptin: Resuspend Leupeptin (Sigma–Aldrich) at 5 mg/mL in dH_2O. Stable at –20°C for 6 months.

Extraction buffer: 950 µL Novex® Tris–Glycine SDS Sample Buffer 2× (Invitrogen). Store at 4°C. Bring to room temperature before use. 50 µL 2-mercaptoethanol (final concentration 2.5%

v/v) (use with adequate ventilation). 1 mL T-PER (Pierce). Prepare fresh immediately prior to use.

Extraction buffer for LMW serum fractions: 775 μL Novex® Tris–Glycine SDS Sample Buffer 2X, 25 μL 2-mercaptoethanol, 200 μL glycerol.

2.1.2. Spectrophotometric Protein Concentration

Bradford Coomassie spectrophotometric method (Pierce) or NanoDrop microvolume instrument (Thermo Fisher).

2.2. Printing Reverse-Phase Protein Microarrays

Although there are many robotic arrayer devices capable of printing reverse-phase protein arrays, the arrayer described in this text is the Aushon 2470 arrayer (Aushon Biosystems) equipped with 350-μm-diameter pins.

1. Aushon 2470 arrayer (Aushon Biosystems).
2. Nitrocellulose coated slides (FAST™ slides, KeraFAST; Nexterion® Slide NC-C, Schott Nexterion; ONCYTE® Nitrocellulose Film Slides, Grace Bio-Labs).
3. 384-Well microtiter plates (Genetix).
4. Commercial cell lysates and/or recombinant peptides containing the analytes of interest.
5. Desiccant (Drierite, anhydrous calcium sulfate).
6. Ziploc-style plastic storage bags.
7. Extraction buffer: 950 μL Novex® Tris–Glycine SDS Sample Buffer 2× (Invitrogen), 50 μL 2-mercaptoethanol, and 1 mL T PER (Pierce).

2.3. Immunostaining

1. Reblot™ Mild Antigen Stripping solution 10× (Millipore/Chemicon).
2. Blocking solution: I-Block™ Protein Blocking Solution (Applied Biosystems). Pour 500 mL PBS, calcium and magnesium free, in a beaker and add a magnetic stir bar. Place the beaker and the magnetic stir bar on a hot plate and adjust the temperature to low/mid heat with the magnetic stir bar rotation intensity set to mid/high range. Add 1 g of I-Block powder. Heat solution until the I-Block power is completely solubilized (see Note 2). Remove the beaker containing the I-Block solution from the hot plate. Leave the beaker on the bench or place it in a container of wet ice until the solution cools to room temperature. Add 500 μL of Tween-20 (final concentration 0.1% v/v) and mix well. Store I-Block at 4°C for a maximum of 14 days.
3. Validated primary antibodies (see Note 3).
4. Biotinylated species-specific secondary antibody.
5. Dako Autostainer (Dako) – optional, staining can be done manually.

6. Catalyzed Signal Amplification (CSA) kit (Dako).
7. Biotin blocking system (Dako).
8. Antibody diluent with background reducing components (Dako).
9. Tris Buffered Saline with Tween (TBST) (Dako).
10. Diaminobenzidine (DAB)+solution (Dako) is carcinogenic, so wear gloves while handling. Dispose DAB solutions following your institution's policies for chemical waste.

2.4. Reverse-Phase Protein Microarray Total Protein Assay

1. Sypro Ruby Protein Blot Stain (Invitrogen).
2. Fixative solution: 41.5 mL dH_2O, 3.5 mL acetic acid (final concentration 7% v/v), 5 mL methanol (final concentration 10% v/v). Store tightly closed at room temperature. Stable for 2 months.
3. Fluorescent image capturing system, such as a UV transilluminator or laser scanner (excitation 280 nm, emission 618 nm) (see Note 4).

2.5. Microarray Spot Analysis

1. High-resolution flatbed scanner, provided with grayscale option (Epson® Perfection Scanner series 1640 or UMAX PowerLook 2100XL).
2. Adobe® Photoshop software.
3. Spot analysis software, such as ImageQuant® (GE Healthcare) or MicroVigene™ (Vigene Tech).

3. Methods

3.1. Protein Lysate Preparation

Human, animal, or bacterial cells from cell culture (4, 5, 17, 22–25), body fluids (8), or peripheral blood mononuclear cell preparations (26) can be lysed and printed onto RPMAs. Cell culture samples should be pelleted by centrifugation and washed in PBS to remove contaminating immunoglobulins or serum present in the medium, prior to cell lysis. Cell lysis buffer for unfixed cells does not contain reducing agents. After initial cell lysis, a Laemmli buffer (Novex® Tris–Glycine SDS Sample Buffer 2× plus 2-mercaptoethanol) is added to the cell lysate to make the final lysate that is printed on the array.

Fixed cells are lysed directly in Laemmli buffer plus a reducing agent (extraction buffer). Serum fractionated by continuous elution electrophoresis is diluted in a modified Laemmli-based buffer, which contains a higher percentage of glycerol than cell extraction buffer.

3.1.1. Cell Pellet Lysis

1. Prepare cell lysis buffer (see Subheading 2.1.1): 915 μL T-PER, 60 μL 5 M NaCl, 10 μL 100 mM orthovanadate, 10 μL 200 mM PEFABLOC, 1 μL 5 mg/mL Aprotonin, 5 μL 1 mg/mL Pepstatin A, and 1 μL 5 mg/mL Leupeptin.
2. Thaw cell pellets on ice for 20 min to 1 h.
3. Wash cells 3× in ice-cold PBS without calcium or magnesium. Centrifuge pellet at 4°C for 10 min at 900 × *g*. Remove and discard supernatant from each wash.
4. After last wash, centrifuge pellet at 4°C for 10 min at 900 × *g*, and remove and discard supernatant. Add cell lysis buffer (100 μL lysis buffer/1 × 10^6 cells).
5. Vortex sample for 15 s and store on ice for 20 min.
6. Centrifuge sample at 3,000 × *g* for 5 min, remove and save the supernatant, which contains the cellular proteins, and place on ice in labeled screw-cap tubes.

3.1.2. Determine Protein Concentration of Sample

Any method with adequate sensitivity may be used for measuring the protein concentration in the sample lysate. Example methods are Bradford Coomassie spectrophotometric method (Pierce) or a NanoDrop microvolume instrument (Thermo Fisher). Only lysates in cell lysis buffer are compatible with spectrophotometric analysis because bromophenol blue and SDS are known interfering substances.

1. Measure the total protein concentration in the cell lysate.
2. Adjust the total protein concentration of the lysates to 0.5–1 mg/mL. This is a two-step process. In order to maintain the SDS at a final concentration of 1× and 2-mercaptoethanol at a final concentration of 2.5%, the lysate is first diluted in an equal volume of Novex® Tris–Glycine SDS Sample Buffer (2×) plus 5% 2-mercaptoethanol (950 μL Novex® Tris–Glycine SDS Sample Buffer (2×) plus 50 μL 2-mercaptoethanol). Then, this diluted lysate is further diluted to the final desired concentration using extraction buffer (*see* Subheading 2.1.1, 950 μL Novex® Tris–Glycine SDS Sample Buffer 2×, 50 μL 2-mercaptoethanol, and 1 mL T-PER).
   (a) Determine the volume of lysate needed to print the RPMA such that the undiluted spot has a total protein concentration of 0.5 μg/μL. 0.5 μg/μL total protein concentration produces robust signal:noise ratios. 40 μL of lysate is sufficient to print over 100 microarrays, in duplicate, using a series of four spots in a twofold dilution curve:

   Example calculation: $C_1 V_1 = C_2 V_2$

   C_1 = Measured protein concentration = 2 μg/μL

   V_1 = Volume of lysate to use to make dilution = unknown

C_2 = Desired protein concentration = 0.5 μg/μL

V_2 = Desired volume = 40 μL

Calculation: 2 μg/μL × V_1 = 0.5 μg/μL × (40 μL)

V_1 = 0.5 μg/μL × (40 μL)/2 μg/μL

V_1 = 20 μg/2 μg/μL

V_1 = 10 μL of lysate

(b) The lysate, which is currently in T-PER, requires further dilution in Novex® Tris–Glycine SDS Sample Buffer (2×) and 2-mercaptoethanol. Based on the calculation in step (a), add equal volumes of lysate and a solution of Novex® Tris–Glycine SDS Sample Buffer (2×) supplemented with 5% (v/v) 2-mercaptoethanol (950 μL Novex® Tris–Glycine SDS Sample Buffer 2× plus 50 μL 2-mercaptoethanol). This dilution step uses 5% 2-mercaptoethanol, so the final concentration of 2-mercaptoethanol is 2.5% (v/v). Following the above example, one would add 10 μL lysate to 10 μL Novex® Tris–Glycine SDS Sample Buffer (2×) supplemented with 5% 2-mercaptoethanol for a total volume of 20 μL lysate.

(c) Dilute this new volume of lysate in extraction buffer (950 μL Novex® Tris–Glycine SDS Sample Buffer (2×), 50 μL 2-mercaptoethanol, and 1 mL T-PER) to the desired final volume. Calculation for this step: *Volume of extraction buffer to add to lysate = [desired final volume of lysate] − [calculated volume of lysate based on protein concentrations × 2]*. Again following the above example, one would add 20 μL of extraction buffer to 20 μL of diluted lysate for a final total volume of 40 μL at a concentration of 0.5 μg/μL.

3. Heat the lysates in a dry heat block at 100°C or a boiling water bath for 10 min. Cool to room temperature.
4. Use lysates immediately to print arrays (recommended) or store lysates at −80°C until the time of array printing.

3.1.3. Laser Capture-Microdissected Tissue Lysate Preparation

This protocol is designed for extraction of proteins from laser capture-microdissected cells (2, 4, 5, 17, 19). LCM of tissue is generally performed with fixed, stained cells; therefore, protein extraction and cell lysis are performed in one step using extraction buffer (Novex® Tris–Glycine SDS Sample Buffer 2×, 50 μL 2-mercaptoethanol, and 1 mL T-PER).

The extraction buffer volume is calculated by assuming 1 μL of extraction buffer/1,000 cells. Often times, multiple LCM caps containing the cells of interest are collected for one sample (Fig. 2). The lysate from these multiple caps per sample are pooled together to increase the total volume of the lysate and to ensure a sufficient

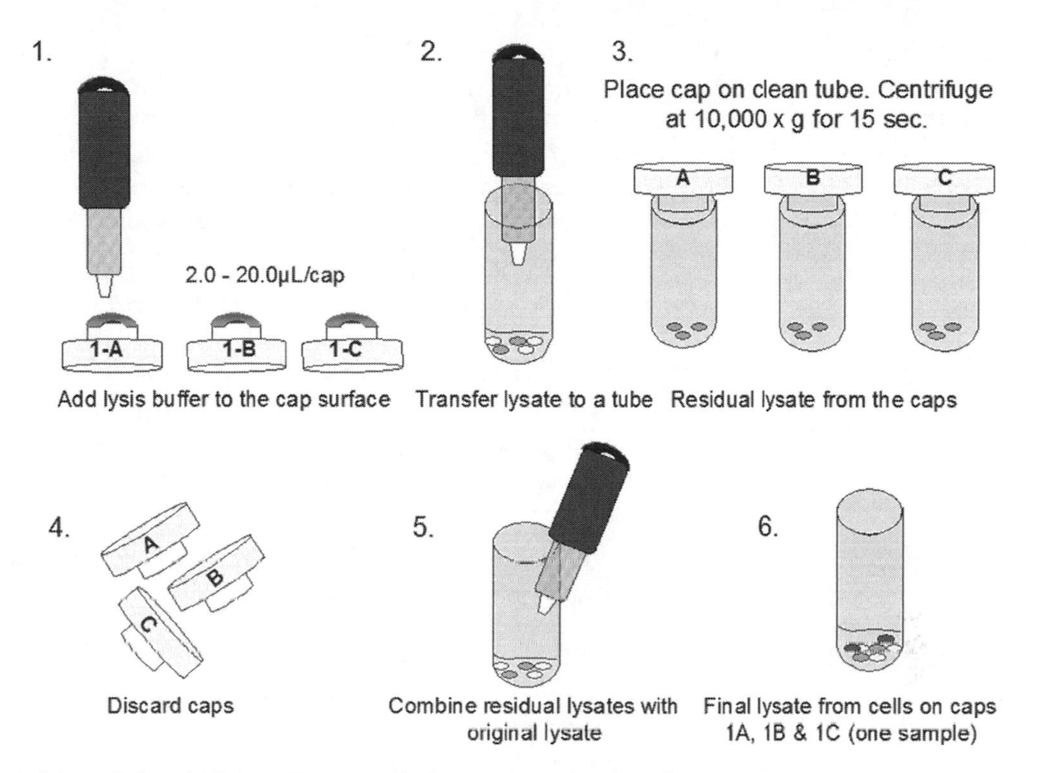
1.
2.0 - 20.0µL/cap
1-A
1-B
1-C
Add lysis buffer to the cap surface
2.
Transfer lysate to a tube
3.
Place cap on clean tube. Centrifuge at 10,000 x g for 15 sec.
A
B
C
Residual lysate from the caps
4.
A
B
C
Discard caps
5.
Combine residual lysates with original lysate
6.
Final lysate from cells on caps 1A, 1B & 1C (one sample)

Fig. 2. Schematic for solubilizing cells procured by laser capture microdissection.

number of cells for adequate total protein concentration (0.5 μg/μL).

1. Prepare extraction buffer: 950 μL Novex® Tris–Glycine SDS Sample Buffer 2×, 50 μL 2-mercaptoethanol, and 1 mL T-PER.
2. Determine the extraction buffer volume for each sample based on 1 μL extraction buffer/1,000 cells.
3. Place all caps from the same sample on a flat, dry, draft-free surface, with the polymer facing up. Using a pipette, distribute the entire extraction buffer volume onto the multiple caps and incubate for 1 min at room temperature (Fig. 2).
4. Gently pipette the lysate on each cap up and down, being careful to avoid generating bubbles. Transfer the protein lysates to a labeled screw-cap tube (see Note 5).
5. Place the cap on a microcentrifuge tube. Spin the cap/tube assembly at 10,000 × *g* for 15 s.
6. Remove and discard the caps. Add the residual lysate in the tubes to the original lysate in the screw-cap tube.
7. Heat lysates in a dry heat block or boiling water bath for 10 min at 100°C. Cool to room temperature.
8. Use the lysates to print arrays (recommended) or store lysates at –80°C until the time of array printing.

3.1.4. Low-Molecular-Weight Serum Fractions Sample Preparation

Printing LMW proteins that were previously recovered by continuous elution electrophoresis allows the identification of proteins or protein fragments less than 25 kDa. An advantage of using this sample type compared to unfractionated serum is the elimination of heavy- and light-chain immunoglobulins, thereby preventing excessive background staining by secondary antibodies.

1. Prepare 1:2 dilutions of the concentrated LMW serum fractions (one part LMW serum:1 LMW protein extraction buffer). LMW protein extraction buffer: 775 μL Novex® Tris–Glycine SDS Sample Buffer 2×, 25 μL 2-mercaptoethanol, 200 μL glycerol.
2. Heat the diluted LMW serum samples in a dry heat block at 100°C or boiling water bath for 10 min. Cool to room temperature.
3. Use the samples to print arrays (recommended) or store lysates at −80°C until the time of array printing.

3.2. Reverse-Phase Protein Microarray Printing

RPMAs are a multiplexed proteomic platform used to evaluate cell-signaling protein levels or phosphoprotein profiles in many samples printed on one array for one specific end point per array.

In addition to printing sample lysates, it is also essential to print control lysates, such as commercial cell lysates, recombinant peptides, or peptide mixtures, that are known to contain the antigens being investigated. The exact protein concentration for any given analyte is unknown in each sample and each antibody has a unique affinity constant. In order to match the protein concentration with the antibody affinity, all samples are printed in a dilution curve on the array. The dilution format allows one to select the protein concentration (dilution) that optimally matches the antibody affinity, thus giving a signal:noise ratio within the linear dynamic range of the dilution curve.

In this chapter, we describe how to print reverse-phase protein arrays with an Aushon 2470 arrayer equipped with twenty 350-μm pins, in a 4×5 format (Fig. 3) (4, 27). Multiple robotic printing platforms can be used to print RPMA. Printing layouts vary with each instrument and printhead configuration. Readers should refer to the vendor instrument manuals for specific instructions.

1. Fill water wash container with dH_2O and empty waste container. Care should be taken to verify that tubing is sufficiently inserted into each container.
2. Fill the humidifier with dH_2O water (see Note 6).
3. Load nitrocellulose-coated slides onto platens with the nitrocellulose pad facing up and the label edge to the right. Insure that the slides are securely held by the platen clips. Load platens into the Aushon 2470 arrayer.

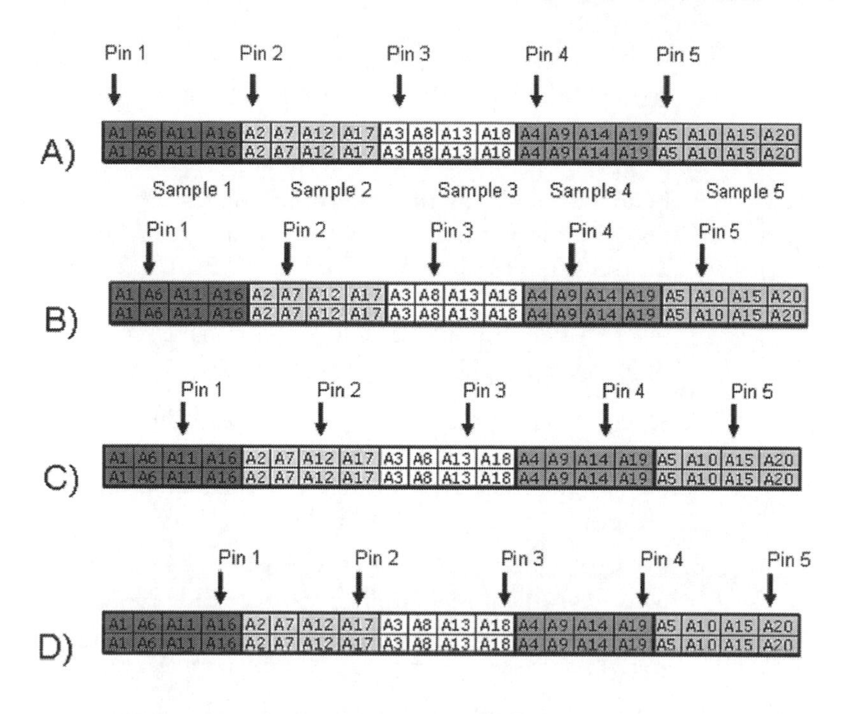

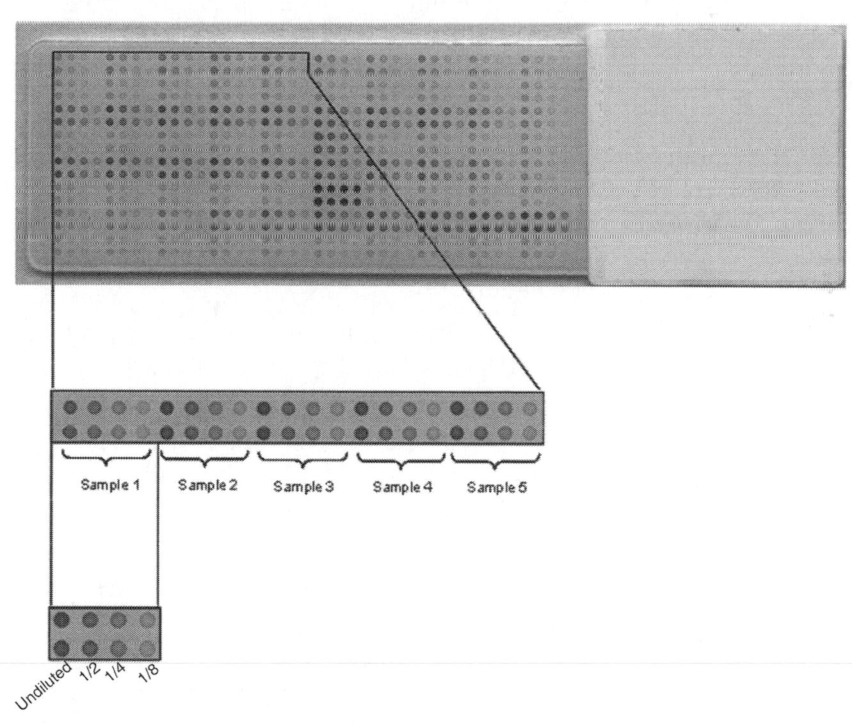

Fig. 3. Example microtiter plate layout and array map. *Top*: Rows A–D represent rows of a microtiter plate. Samples are placed in nonconsecutive wells (e.g., A1, A6, A11, A16) in order to be printed in adjacent spots on the array. An Aushon Biosystems 2470 arrayer equipped with a 20-pin printhead transfers samples from the indicated wells during the printing process. *Bottom*: The printed slide depicts a typical array map for samples printed in duplicate, in a twofold dilution series (undiluted, 1:2, 1:4, and 1:8).

4. If the experimental lysates have been stored frozen, thaw the lysates and heat the lysates (do not heat commercial cell lysates samples) in a dry heat block or boiling water bath for 7 min at 100°C. Cool to room temperature.
5. Load samples into 384-well microplates, creating a 4-point, twofold dilution curve. Use extraction buffer (950 μL Novex® Tris–Glycine SDS Sample Buffer 2×, 50 μL 2-mercaptoethanol, and 1 mL T-PER) to make the sample dilutions. Refer to Fig. 2 for an example sample loading map (see Note 7).
6. Place the lid on top of the 384-well microplates (see Note 8).
7. Open the two metal clips on the plate holder. Position A1 should be in the upper left-hand corner. Slide the 384-well plate to the back of the holder. Flip the metal clips to the closed position.
8. Place the plate holder in the Aushon 2470 arrayer elevator with microplate well A1 facing the outside of the instrument.
9. Turn the power switch on and open the Aushon 2470 arrayer software.
10. Select the number of microplates by double clicking on "Genetix Polystyrene." Use the selected well plate grid to determine which samples are loaded into each plate. Click "continue."
11. On the next screen, define the top offset as 4.5 mm (see Note 9).
12. Designate 5 mm for the left offset and 1,125 μM for the feature-to-feature spacing in the *x* and *y* axes. To print duplicate spots, select one replicate to be printed in a linear vertical position (Fig. 3).
13. Select three depositions per feature. Approximately 30 nL of lysate is deposited on the nitrocellulose per spot (see Note 10) with three depositions/feature.
14. Select two "super arrays" if samples are occupying rows I–P in the microtiter plate or if more than 40 samples (prepared in a 4-point dilution curve) will be printed on one array. Click "continue."
15. On the following screen, choose 4 s for adequate pin washing (see Note 11). Click "continue."
16. On the final screen, select the number of slides loaded on each platen. Ten slides can be loaded per platen.
17. Set the humidity to 50% to prevent sample evaporation and to aid in spot drying between depositions.
18. Use the Aushon 2470 arrayer software checklist to ensure that the instrument/slides are ready to be printed and click start.
19. When printing is complete, the software prompts the user to either continue or quit. If additional slides are not being

printed, select quit, save the run file, and close the software (see Note 12). Remove the microtiter plates and slides from the arrayer.

20. Turn the Aushon 2470 arrayer power off, and place the printed microarray slides in a slide box. Store the slide box in a plastic storage bag with desiccant at –20°C (see Notes 13 and 14).

3.3. Immunostaining

3.3.1. RPMA Slide Pretreatment

RPMA slides printed with cellular lysates, but not serum, should be pretreated and blocked prior to the staining procedure. Pretreatment with Reblot™ solution is as an antigen retrieval method, providing enhanced signal:noise ratios. Blocking reduces background signal due to nonspecific protein binding. The RPMA slides are incubated in a protein-based solution to bind all available protein-binding sites on the nitrocellulose. Blocking is required prior to staining the arrays with antibodies. Two important exceptions to this method are slides designated by the user for total protein staining with Sypro Ruby Blot stain and RPMA slides printed with serum. RPMA slides for staining with Sypro Ruby Blot stain are neither pretreated with Reblot™ nor blocked. RPMA slides comprised with serum samples are blocked but are not pretreated with Reblot™ prior to staining.

1. Allow frozen RPMA slides to warm at room temperature for approximately 5–10 min. Do not remove slides from the plastic bag containing desiccant until the slides have warmed slightly.
2. Prepare a 1× solution of Mild Reblot™ (stock is 10×) in dH_2O. 50 mL is sufficient for blocking ten RPMA arrays in a 20 × 15-cm dish.
3. Incubate the RPMA slides in 1× Mild Reblot™ solution for 15 min on a rocker/shaker (see Notes 15 and 16). Do not use Reblot™ for arrays printed with serum or LMW serum fractions. For serum/LMW serum fraction arrays, place the slides directly in I-Block solution. Do not wash serum arrays with PBS.
4. Decant the Reblot™ solution and wash the RPMA slides with PBS (calcium and magnesium free) twice for 5 min each.
5. Decant the last PBS wash and immediately add blocking solution (I-Block solution) for a minimum of 60 min (see Note 17).

3.3.2. Microarray Immunostaining

The sensitivity of the RPMA platform is due in part to the use of catalyzed signal amplification (CSA) chemistry, effectively amplifying the signal compared to nonamplified methods (12, 28–31). CSA methods are based on streptavidin–biotin-mediated deposition of biotinylated tyramide as an amplification reagent. Horseradish peroxidase catalyzes the precipitation of DAB around the site of the tyramide–avidin–biotin complex, resulting in a visible brown precipitate (12, 28–31). This method requires a single antibody–epitope interaction with the protein of interest.

The number of RPMA slides to be stained is chosen in relation to the number of end points of interest. The Dako autostainer allows simultaneous staining of 48 slides (see Note 18). Antibodies from different animal species can be used during the same autostainer run. However, to quantify the nonspecific background signal generated from the interaction between the secondary antibody and samples, it is essential in each staining run to include one slide that is stained with secondary antibody only for each species of secondary antibody used. If in a single run, antibodies belonging to different species are used, it is necessary to stain a negative-control slide for each species. The signal intensity of the slide probed with secondary antibody only is subtracted from the signal intensity of the primary/secondary antibody-stained slide during data analysis.

1. Select the unconjugated primary antibodies of interest.
2. Select biotinylated secondary antibodies correlating to the species of the primary antibodies (see Note 19).
3. Program the Dako autostainer or perform manual staining of the microarrays (Table 1).
4. Prepare CSA solutions according to the manufacturer's directions.
5. Fill the buffer reservoir with 1× TBST and the water carboy with dH_2O. Empty the waste container if necessary.
6. Load the reagents and slides on the autostainer. Prevent the nitrocellulose from drying during slide loading. If necessary, rinse the slides with 1× TBST during the slide loading process (see Note 18).
7. Prime the water first and then the buffer before starting the run.
8. At the end of the autostainer run, remove the slides, rinse them with dH_2O, and allow them to air dry.
9. Label the microarray slides specifying the date, study, and antibody that has been used in the staining procedure.
10. Store the slides in the dark at room temperature. See Subheading 3.5 for RPMA slide scanning.

3.4. Sypro Ruby Blot Stain for Total Protein

Sypro Ruby is a fluorescent dye detection method to quantify the amount of protein present in each individual array spot. Sypro Ruby has an excitation wavelength of 280 nm and an emission of 618 nm. Sypro Ruby staining has a sensitivity of 1 ng to 1 µg protein (32, 33).

Table 1
Reverse-phase protein microarray staining protocol

Category	Reagent	Time in minutes
Buffer	TBST buffer	Rinse
Endogenous enzyme block	Hydrogen peroxidase blocking	5
Buffer	TBST buffer	Rinse
Auxiliary	Avidin blocking	10
Buffer	TBST buffer	Rinse
Auxiliary	Biotin blocking	10
Buffer	TBST buffer	Rinse
Protein block	Protein blocking	5
Buffer	Blow air	Rinse
Primary antibody	Primary antibody	30
Buffer	TBST buffer	Rinse
Buffer	TBST buffer	Rinse
Auxiliary	TBST buffer	3
Buffer	TBST buffer	Rinse
Buffer	TBST buffer	Rinse
Secondary reagent	Biotinylated secondary antibody	15
Buffer	TBST buffer	Rinse
Auxiliary	TBST buffer	3
Buffer	TBST buffer	Rinse
Auxiliary	Streptavidin–biotin complex	15
Buffer	TBST buffer	Rinse
Auxiliary	TBST buffer	3
Buffer	TBST buffer	Rinse
Auxiliary	Amplification reagent	15
Buffer	TBST buffer	Rinse
Tertiary reagent	Streptavidin–HRP	15
Buffer	TBST buffer	Rinse
Auxiliary	TBST buffer	3
Buffer	TBST buffer	Rinse
Switch to toxic waste	–	–
Chromogen	Diaminobenzidine	4
Buffer	TBST buffer	Rinse
Overnight water	Deionized water	840

1. Select the number of RPMA slides to be stained with Sypro Ruby (see Note 20).
2. Allow the RPMA slide(s) to reach room temperature if the slides were stored at –20°C. This usually takes 5–10 min.
3. Wash the slide(s) in dH_2O for 5 min with constant rocking.
4. Incubate the slide(s), on a rotator at low speed, for 15 min with fixative solution (Subheading 2.4, item 2; 41.5 mL dH_2O, 3.5 mL acetic acid, 5 mL methanol). Use an adequate volume of fixative to cover the surface of the slides.
5. Discard the fixative solution and wash the slides with dH_2O four times for 5 min each.
6. Incubate the slide(s) with Sypro Ruby Blot stain solution for 30 min. Protect the slides from light during staining (see Note 21). Rinse the slides with dH_2O.
7. Allow the slide to air dry in an area protected from light, such as a drawer or cupboard.
8. Proceed to image detection with a fluorescent imaging system of choice (see Note 4).

3.5. RPMA Slide Scanning: Colorimetric Image Acquisition and Data Analysis

Any high-resolution scanner, provided with grayscale option, can be employed for image acquisition of DAB-stained microarrays, providing it generates 14- or 16-bit scanned images. TIFF images can be imported to a variety of data analysis software programs, including MicroVigene (Vigene Tech) or ImageQuant (GE Healthcare). Data analysis produces a single-pixel intensity value for each array spot that is proportional to the amount of measured analyte per spot.

Final intensity values generated by the spot analysis software are obtained after subtraction of the negative control (secondary antibody alone) intensity value/spot and normalization to a suitable analyte, such as the total protein value/spot. Once the corrected, normalized intensity values of each sample have been calculated, statistical analyses can be applied and the data can be displayed using graphing programs or hieratical clustering procedures.

1. Place the RMPA slides nitrocellulose side down on the flat bed scanner.
2. Scan each slide at 600–1,200 dpi as a 14- or 16-bit image in grayscale mode. Save the image as a TIFF file.
3. Adjust the image appearance (inverted/not inverted) as required by the image analysis software. Save the adjusted image as a TIFF file (see Note 22).

4. Notes

1. Throughout the text, we use the abbreviation "dH_2O" to indicate type I (reagent grade) water. Numerous terms are used colloquially to describe type I water, such as "deionized," "distilled," or "purified." Type I water as defined herein is deionized, with a greater than or equal to 18 MΩ cm resistivity, a total organic carbon (TOC) content less than 20 ppb, and has been filtered through a 0.2-μm filter such that the final product contains less than 10 CFU/mL (34).
2. Avoid boiling the I-Block solution. Heating the solution at low/mid-heat levels for 10–15 min is usually adequate to completely solubilize the I-Block powder. I-Block is a casein-based protein solution. Boiling causes protein degradation and potential alterations in blocking efficiency.
3. Each primary antibody must be validated by Western blotting to confirm specific interaction between the protein of interest and the antibody using complex samples similar to those which are used on the array.
4. A UV transilluminator (~300 nm), a blue-light transilluminator, or a laser scanner that emits at 450, 473, 488, or 532 nm is appropriate for imaging a Sypro Ruby-stained array. Example imaging systems are Kodak 4000 MM imager; Alpha Innotech NovaRay; Tecan Reloaded LS; PerkinElmer ProScanArray HT; and Molecular Devices GenePix 4000B.
5. Use a microscope to inspect at least one LCM cap from every sample to ensure complete cell lysis. Place the LCM cap on a glass microscope slide with the polymer facing up. Focus on the polymer surface. Cellular material, if present, will be visible as heterogeneous dark areas on the polymer. Occasionally, the polymer may contain an imprint of the cells that should not be mistaken for actual cellular material. One way to distinguish a polymer imprint from cellular material is to look for nuclear staining of the cells, often noted as very dark, purplish areas due to the hematoxylin stain commonly used in LCM.
6. Use dH_2O in the humidifier to prevent mold/bacterial growth in the humidifier. Empty the water, and clean and air dry the humidifier every 2 weeks.
7. Samples that fill an entire array (640 spots maximum from 350-μm pins) can be loaded into four individual 384-well plates to prevent significant evaporation during the printing process. Samples in rows A–D can be loaded in plate 1, rows E–H in plate 2, rows I–L in plate 3, and rows M–P in plate 4.
8. The Aushon 2470 arrayer is equipped with suction cups to remove the lid from the microtiter plates. A lid MUST be

placed on every microtiter plate that is loaded in the arrayer. The lid should be clean and dry, free of dust, adhesive, and liquid.

9. Top offset settings may vary slightly with each arrayer and/or slide manufacturer. Food dyes, used for baking or egg coloring, can be diluted in PBS or water to substitute as "samples" for evaluating spot placement by the robotic arraying device. Clean the arrayer pins thoroughly following printing of food dyes. 70% (v/v) ethanol dispensed into a microtiter plate can be used to effectively clean the pins. Dispense 20 μL of 70% ethanol into wells 1–20 in rows A, B, C, and D of a 384-well microtiter plate. Load one nitrocellulose-coated slide into the arrayer. Program and execute a print run for one slide, from one microtiter plate with four unique extractions, at three depositions/feature.

10. Nitrocellulose film coatings have a finite protein binding capacity based on their porosity and depth (35, 36). Printing more than five depositions/spot is generally not recommended as the nitrocellulose becomes saturated at approximately five depositions/spot and the spots become more diffuse. Samples with protein concentrations less than 0.5 μg/μL can be effectively concentrated on an array by printing more than three depositions per feature (spot). If more than three depositions/feature are necessary, it is helpful to print a test microarray in which a sample is printed at three, four, and five depositions/feature to determine the effect on spot quality.

11. Carryover experiments should be conducted with each instrument to determine the optimal pin washing time for various sample matrices.

12. A gal file is generated with each print run and is saved in C:\Documents and Settings\user\My Documents\user\Array Data Files. It can be uploaded into software analysis packages for array layout spot identification.

13. Proteins immobilized on nitrocellulose slides are stable at –20°C for up to 3 years (personal experience) if stored in dry (with desiccant) conditions. If the remaining lysate in the microtiter plates needs to be saved for further analysis or printing additional arrays at a later date, wrap the 384-well microtiter plates in plastic wrap or parafilm. Store microtiter plates containing lysates at –80°C.

14. To reprint arrays from 384-well microtiter plates containing frozen lysates, unwrap the microtiter plate and thaw the plate quickly on a heat block set at 37°C. Do not allow condensation to form on the inside of the lid. Remove condensation with a paper towel. Mix the samples well prior to array printing. Replace the original microtiter plate lid with a clean lid.

Any dust, adhesive, or moisture on the original lid causes failure of the suction cups lid remover device.

15. Do not use Reblot for arrays composed of serum or LMW serum samples. Reblot causes diffusion of the serum sample beyond the printed spot resulting in a blurry, poorly defined spot.
16. The Reblot solution further denatures the protein immobilized onto the nitrocellulose slides improving the antibody–epitope recognition. Do not exceed the suggested incubation time (15 min). Reblot is a very basic solution (pH 14). Overexposure to Reblot solutions may cause nitrocellulose alterations or nitrocellulose detachment (delamination) from the glass slide.
17. I-Block solution is a protein-based blocking reagent that is useful for blocking the nitrocellulose prior to immunostaining the array. A minimum blocking time of 1 h at room temp, with gently rocking, is recommended while longer blocking times are not detrimental. If blocking must be performed overnight, block the slides at 4°C.
18. TBST contains a high concentration of salt. If the buffer is not rinsed from the stained nitrocellulose arrays before the slides, dry salt crystals may form on the nitrocellulose surface. Consequently, it is suggested to program the autostainer for a water rinse after DAB deposition. Moreover, it is possible to add a further water rinse (auxiliary step) after the colorimetric reaction in order to rinse the slides with deionized water and pause the instrument for 840 min. By doing this, the autostainer is programmed to be in an idle status for 14 h, at which time the slides are rinsed again with deionized water.
19. The total protein concentration in a large set of printed array slides may vary from the initial slides printed to the last slides printed due in part to sample evaporation. It is recommended to stain 1 of every 25 slides with Sypro Ruby Protein Blot stain. For example, if 120 microarrays have been printed, it is suggested to use slides 25, 50, 75, and 100 for total protein quantification.
20. The secondary antibody should be of the same species as the primary antibody. Historically, monoclonal antibodies were produced in mice and polyclonal antibodies were produced in other species. However, monoclonal antibodies have recently been commercially produced in rabbits. Therefore, it is important to consult the antibody vendor specifications regarding species rather than relying on the terms monoclonal and polyclonal. Monoclonal rabbit primary antibodies should be used with anti-rabbit secondary antibodies.

21. Sypro Ruby is a photosensitive solution; in order to avoid alteration and reduction of the signal, it is suggested to protect the slides from light by covering the staining container with aluminum foil.

22. Image adjustments should only include those adjustments that change all pixel intensities in an image in a linear, consistent manner. All image adjustments must be performed prior to spot analysis and must be consistent for all array slides. It is important to note that some image manipulation programs are capable of changing the actual pixels in the image and should not be applied to microarray images (37).

References

1. Liotta, L. A., Espina, V., Mehta, A. I., Calvert, V., Rosenblatt, K., Geho, D. et al. (2003) Protein microarrays: meeting analytical challenges for clinical applications. *Cancer Cell* **3**, 317–25.
2. Paweletz, C. P., Charboneau, L., Bichsel, V. E., Simone, N. L., Chen, T., Gillespie, J. W. et al. (2001) Reverse phase protein microarrays which capture disease progression show activation of pro-survival pathways at the cancer invasion front. *Oncogene* **20**, 1981–9.
3. Grote, T., Siwak, D. R., Fritsche, H. A., Joy, C., Mills, G. B., Simeone, D. et al. (2008) Validation of reverse phase protein array for practical screening of potential biomarkers in serum and plasma: accurate detection of CA19-9 levels in pancreatic cancer. *Proteomics* **8**, 3051–60.
4. VanMeter, A. J., Rodriguez, A. S., Bowman, E. D., Jen, J., Harris, C. C., Deng, J. et al. (2008) Laser capture microdissection and protein microarray analysis of human non-small cell lung cancer: differential epidermal growth factor receptor (EGPR) phosphorylation events associated with mutated EGFR compared with wild type. *Mol Cell Proteomics* **7**, 1902–24.
5. Wulfkuhle, J. D., Speer, R., Pierobon, M., Laird, J., Espina, V., Deng, J. et al. (2008) Multiplexed cell signaling analysis of human breast cancer applications for personalized therapy. *J Proteome Res* **7**, 1508–17.
6. Rapkiewicz, A., Espina, V., Zujewski, J. A., Lebowitz, P. F., Filie, A., Wulfkuhle, J. et al. (2007) The needle in the haystack: application of breast fine-needle aspirate samples to quantitative protein microarray technology. *Cancer* **111**, 173–84.
7. Espina, V., Mehta, A. I., Winters, M. E., Calvert, V., Wulfkuhle, J., Petricoin, E. F., 3rd et al. (2003) Protein microarrays: molecular profiling technologies for clinical specimens. *Proteomics* **3**, 2091–100.
8. Davuluri, G., Espina, V., Petricoin, E. F., 3rd, Ross, M., Deng, J., Liotta, L. A. et al. (2009) Activated VEGF receptor shed into the vitreous in eyes with wet AMD: a new class of biomarkers in the vitreous with potential for predicting the treatment timing and monitoring response. *Arch Ophthalmol* **127**, 613–21.
9. Longo, C., Patanarut, A., George, T., Bishop, B., Zhou, W., Fredolini, C. et al. (2009) Core-shell hydrogel particles harvest, concentrate and preserve labile low abundance biomarkers. *PLoS One* **4**, e4763.
10. Mueller, C., Zhou, W., Vanmeter, A., Heiby, M., Magaki, S., Ross, M. M. et al. (2009) The Heme Degradation Pathway is a Promising Serum Biomarker Source for the Early Detection of Alzheimer's Disease. *J Alzheimers Dis*
11. Becker, K. F., Schott, C., Hipp, S., Metzger, V., Porschewski, P., Beck, R. et al. (2007) Quantitative protein analysis from formalin-fixed tissues: implications for translational clinical research and nanoscale molecular diagnosis. *J Pathol* **211**, 370–8.
12. Bobrow, M. N., Harris, T. D., Shaughnessy, K. J., Litt, G. J. (1989) Catalyzed reporter deposition, a novel method of signal amplification. Application to immunoassays. *J Immunol Methods* **125**, 279–85.
13. Kornblau, S. M., Tibes, R., Qiu, Y., Chen, W., Kantarjian, H. M., Andreeff, M. et al. (2008) Functional proteomic profiling of AML predicts response and survival. *Blood*
14. Hennessy, B. T., Lu, Y., Poradosu, E., Yu, Q., Yu, S., Hall, H. et al. (2007) Pharmacodynamic markers of perifosine efficacy. *Clin Cancer Res* **13**, 7421–31.

15. Belluco, C., Mammano, E., Petricoin, E., Prevedello, L., Calvert, V., Liotta, L. et al. (2005) Kinase substrate protein microarray analysis of human colon cancer and hepatic metastasis. *Clin Chim Acta* **357**, 180–3.
16. Silvestri, A., Colombatti, A., Calvert, V. S., Deng, J., Mammano, E., Belluco, C. et al. (2010) Protein pathway biomarker analysis of human cancer reveals requirement for upfront cellular-enrichment processing. *Lab Invest* **90**, 787–96.
17. Petricoin, E. F., 3rd, Espina, V., Araujo, R. P., Midura, B., Yeung, C., Wan, X. et al. (2007) Phosphoprotein pathway mapping: Akt/mammalian target of rapamycin activation is negatively associated with childhood rhabdomy osarcoma survival. *Cancer Res* **67**, 3431–40.
18. Espina, V., Wulfkuhle, J. D., Calvert, V. S., VanMeter, A., Zhou, W., Coukos, G. et al. (2006) Laser-capture microdissection. *Nat Protoc* **1**, 586–603.
19. Emmert-Buck, M. R., Bonner, R. F., Smith, P. D., Chuaqui, R. F., Zhuang, Z., Goldstein, S. R. et al. (1996) Laser capture microdissection. *Science* **274**, 998–1001.
20. Laemmli, U. K. (1970) Cleavage of structural proteins during the assembly of the head of bacteriophage T4. *Nature* **227**, 680–5.
21. Bonner, R. F., Emmert-Buck, M., Cole, K., Pohida, T., Chuaqui, R., Goldstein, S. et al. (1997) Laser capture microdissection: molecular analysis of tissue. *Science* **278**, 1481,1483.
22. Popova, T. G., Turell, M. J., Espina, V., Kehn-Hall, K., Kidd, J., Narayanan, A. et al. (2010) Reverse-phase phosphoproteome analysis of signaling pathways induced by Rift valley fever virus in human small airway epithelial cells. *PLoS One* **5**, e13805.
23. Popova, T., Espina, V., Bailey, C., Liotta, L., Petricoin, E., Popov, S. (2009) Anthrax infection inhibits the AKT signaling involved in the E-cadherin-mediated adhesion of lung epithelial cells. *FEMS Immunol Med Microbiol* **56**, 129–42.
24. Agarwal, R., Gonzalez-Angulo, A. M., Myhre, S., Carey, M., Lee, J. S., Overgaard, J. et al. (2009) Integrative analysis of cyclin protein levels identifies cyclin b1 as a classifier and predictor of outcomes in breast cancer. *Clin Cancer Res* **15**, 3654–62.
25. Nishizuka, S., Charboneau, L., Young, L., Major, S., Reinhold, W. C., Waltham, M. et al. (2003) Proteomic profiling of the NCI-60 cancer cell lines using new high-density reverse-phase lysate microarrays. *Proc Natl Acad Sci U S A* **100**, 14229–34.
26. Accordi, B., Espina, V., Giordan, M., VanMeter, A., Milani, G., Galla, L. et al. (2010) Functional protein network activation mapping reveals new potential molecular drug targets for poor prognosis pediatric BCP-ALL. *PLoS One* **5**, e13552.
27. Mueller, C., Liotta, L. A., Espina, V. (2010) Reverse phase protein microarrays advance to use in clinical trials. *Mol Oncol* **4**, 461–81.
28. Bobrow, M. N., Shaughnessy, K. J., Litt, G. J. (1991) Catalyzed reporter deposition, a novel method of signal amplification. II. Application to membrane immunoassays. *J Immunol Methods* **137**, 103–12.
29. Bobrow, M. N., Litt, G. J., Shaughnessy, K. J., Mayer, P. C., Conlon, J. (1992) The use of catalyzed reporter deposition as a means of signal amplification in a variety of formats. *J Immunol Methods* **150**, 145–9.
30. Hunyady, B., Krempels, K., Harta, G., Mezey, E. (1996) Immunohistochemical signal amplification by catalyzed reporter deposition and its application in double immunostaining. *J Histochem Cytochem* **44**, 1353–62.
31. King, G., Payne, S., Walker, F., Murray, G. I. (1997) A highly sensitive detection method for immunohistochemistry using biotinylated tyramine. *J Pathol* **183**, 237–41.
32. Berggren, K., Steinberg, T. H., Lauber, W. M., Carroll, J. A., Lopez, M. F., Chernokalskaya, E. et al. (1999) A luminescent ruthenium complex for ultrasensitive detection of proteins immobilized on membrane supports. *Anal Biochem* **276**, 129–43.
33. Berggren, K. N., Schulenberg, B., Lopez, M. F., Steinberg, T. H., Bogdanova, A., Smejkal, G. et al. (2002) An improved formulation of SYPRO Ruby protein gel stain: comparison with the original formulation and with a ruthenium II tris (bathophenanthroline disulfonate) formulation. *Proteomics* **2**, 486–98.
34. Miller, W. G., Gibbs, E. L., Jay, D. W., Pratt, K. W., Rossi, B., Vojt, C. M. et al. (2006) Preparation and testing of reagent water in the clinical laboratory; Approved Guideline-Fourth Edition. **26**. Clinical Laboratory Standards Institute: Wayne, PA,
35. Stillman, B. A., Tonkinson, J. L. (2000) FAST slides: a novel surface for microarrays. *Biotechniques* **29**, 630–5.
36. Tonkinson, J. L., Stillman, B. A. (2002) Nitrocellulose: a tried and true polymer finds utility as a post-genomic substrate. *Front Biosci* 7, c1-12.
37. Rossner, M., Yamada, K. M. (2004) What's in a picture? The temptation of image manipulation. *J Cell Biol* **166**, 11–5.

Chapter 15

Serum Low-Molecular-Weight Protein Fractionation for Biomarker Discovery

Amy J. VanMeter, Serena Camerini, Maria Letizia Polci, Alessandra Tessitore, Nishant Trivedi, Michael Heiby, Yasmin Kamal, Jonathan Hansen, and Weidong Zhou

Abstract

Protein biomarkers provide the key diagnostic information for the detection of disease, risk of disease progression, and a patient's likely response to drug therapy. Potential biomarkers exist in biofluids, such as serum, urine, and cerebrospinal fluid. Unfortunately, discovering and validating protein biomarkers are hindered by the presence of high-molecular-weight proteins, such as serum albumin and immunoglobulins, which comprise 90% of the proteins present in these samples. High-abundance, high-molecular-weight proteins mask the low-molecular-weight (LMW) proteins and peptides using conventional protein detection methods. Candidate biomarkers are believed to exist in very low concentrations and comprise less than 1% of serum proteins, and may be highly labile as well. Therefore, it is imperative to isolate and enrich LMW proteins from complex mixtures for biomarker discovery. This chapter describes a continuous elution electrophoresis method, based on molecular weight sieving, to isolate specific molecular weight fractions for mass spectrometric, western blotting, or protein array analysis.

Key words: Biomarker, Continuous elution electrophoresis, Fractionation, Prep cell, Low-molecular-weight protein, Mass spectrometry, Protein, Serum

1. Introduction

Potential biomarkers exist in biofluids, such as serum, urine, and cerebrospinal fluid (1). Protein biomarker discovery/validation are hindered by the presence of high-molecular-weight proteins, such as serum albumin and immunoglobulins, which comprise 90% of the proteins present in these samples and therefore makes it difficult to detect low-molecular-weight (LMW) proteins and peptides (2, 3). Candidate biomarkers are believed to exist in very low concentrations and comprise less than 1% of serum proteins, and may

Virginia Espina and Lance A. Liotta (eds.), *Molecular Profiling: Methods and Protocols*, Methods in Molecular Biology, vol. 823, DOI 10.1007/978-1-60327-216-2_15, © Springer Science+Business Media, LLC 2012

be highly labile as well (4). Many methods exist to reduce sample complexity by eliminating serum albumin and immunoglobulins (5, 6). Proteins can be extracted following 2-dimensional electrophoresis; however, a disadvantage of this system is that a large amount of starting material is required to detect low-abundance proteins (7, 8). Other methods include membrane-size filtration and size-exclusion chromatography (1, 9–13). A disadvantage of using these methods is the loss of LMW biomarkers that are bound to carrier proteins, such as albumin. Continuous elution electrophoresis is an attractive alternative because it provides predictable size separation using sodium dodecyl sulfate-polyacrylamide gel electrophoresis (SDS-PAGE), avoids the excision and digestion of bands in a gel, and provides a way to isolate and enrich LMW proteins and peptides under denaturing conditions while eliminating high-abundance proteins (14–16). This platform also has the capability to collect proteins ranging from microgram to milligram concentrations (14–16).

Continuous elution electrophoresis is performed by running a single heterogeneous sample on a cylindrical SDS-PAGE gel. Protein molecules migrate through the gel based on molecular weight and charge. Proteins of specific molecular weight classes pass into an elution chamber and are then collected as aliquots based on time of elution. Multiple sample collections are made and can subsequently be used for mass spectrometry (17), reverse-phase protein microarrays (RPMAs) (18), western blotting, and other proteomic techniques (Fig. 1). Using these downstream analysis technologies, posttranslational modifications can be observed, relative abundance of proteins can be quantified, and sample populations can be compared to discover differences in cell-signaling pathways/proteins.

2. Materials

2.1. Sample Preparation

1. 25 μL of serum.
2. Sample buffer: 75 μL Novex® Tris–glycine SDS sample buffer 2×, 15 μL 1 M DTT, 3 μL bromophenol blue, and 35 μL dH_2O (see Note 1). Store Novex sample buffer at 4°C and bring to room temperature before use. Prepare concentrated bromophenol blue by adding dH_2O to powdered bromophenol blue, mix, centrifuge, and collect supernatant. Store at room temperature.

2.2. Resolving and Stacking Gel Preparation

1. 1.5 M Tris–HCl, pH 8.8. Store at room temperature.
2. 0.5 M Tris–HCl, pH 6.8. Store at room temperature.
3. *N*,*N*,*N*′,*N*′-tetramethylethylenediamine (TEMED) (Bio-Rad).

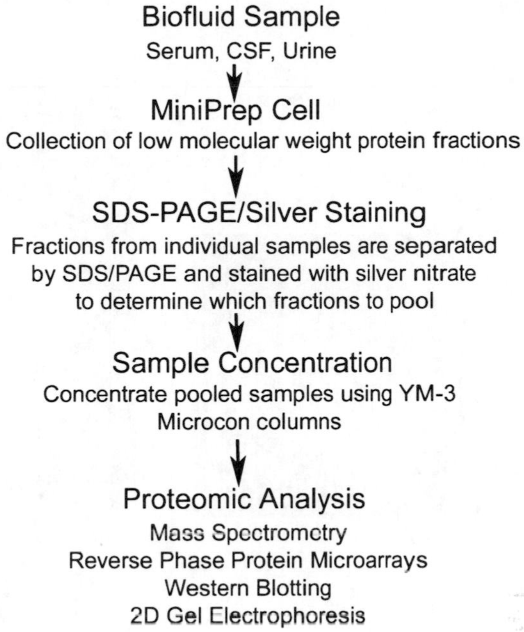

Biofluid Sample
Serum, CSF, Urine

MiniPrep Cell
Collection of low molecular weight protein fractions

SDS-PAGE/Silver Staining
Fractions from individual samples are separated by SDS/PAGE and stained with silver nitrate to determine which fractions to pool

Sample Concentration
Concentrate pooled samples using YM-3 Microcon columns

Proteomic Analysis
Mass Spectrometry
Reverse Phase Protein Microarrays
Western Blotting
2D Gel Electrophoresis

Fig. 1. Workflow for continuous elution electrophoresis. Biological fluid samples are separated into protein-containing fractions based on molecular weight, an aliquot of the fraction is resolved by electrophoresis on a 1D gel, and the protein bands are visualized with a silver stain. The corresponding eluted protein fractions are pooled, concentrated, and prepared for downstream proteomic analysis.

4. Ammonium persulfate (APS): Prepare a 10% solution in dH_2O. Store at 4°C.
5. Ultra Pure Protogel 30% Acrylamide 0.8% Bis Acrylamide Stock (37:5:1) (National Diagnostics). Wear personal protective gear while handling acrylamide. Acrylamide is a potent neurotoxin.
6. 9″ glass Pasteur pipette.
7. Novex® Tris–Glycine SDS Running Buffer 10× (Invitrogen). Prepare 700 mL of a 1× solution (70 mL of the 10× stock plus 630 mL dH_2O). Mix well and store at room temperature.
8. 10% resolving gel: 2.075 mL dH_2O, 1.66 mL Ultra Pure Protogel 30% Acrylamide 0.8% Bis Acrylamide Stock (37:5:1), 1.25 mL 1.5 M Tris–HCl, pH 8.8, 10% APS (see Note 2), 2.5 μL TEMED.
9. 4% stacking gel: 1.23 mL dH_2O, 270 μL 30% Acrylamide 0.8% Bis Acrylamide Stock (37:5:1), 500 μL 0.5 M Tris–HCl, pH 6.8, 10 μL 10% APS (see Note 2), 2.0 μL TEMED.

2.3. Mini Prep Cell Apparatus

1. Mini Prep Cell Apparatus (Bio-Rad) (Fig. 2). Elution manifold base with attached support frit, harvest ring, and sealing gasket is stored at 4°C submerged in ddH_2O. Before first use, wet the

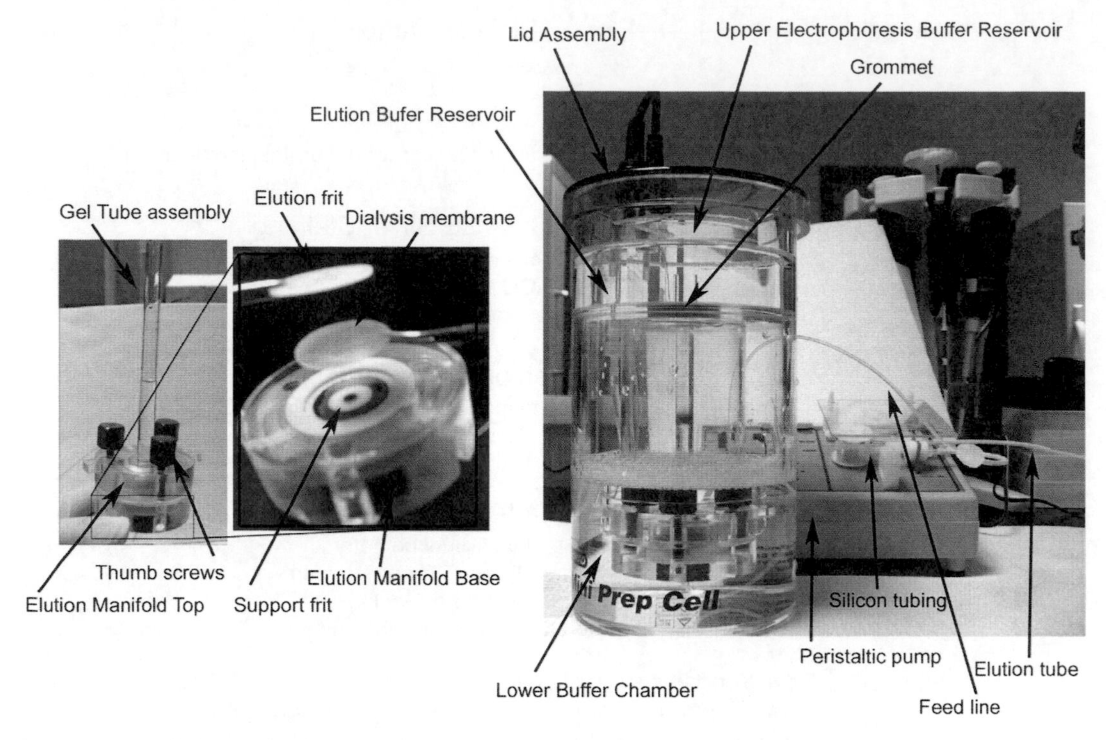

Fig. 2. Mini Prep Cell apparatus for continuous elution electrophoresis

dialysis membrane and elution frit by incubating in 1× running buffer overnight. Thereafter, store the membrane and frit in dH_2O at 4°C. Both the dialysis membrane and elution frit can be used repeatedly assuming that no holes or tears are found in the membranes.

2. Determine the optimal setting on the variable flow peristaltic pump to achieve an elution buffer flow rate of 75–100 μL/min.
3. 20-mL plastic syringe. Connect an approximately 5-cm-long piece of Teflon tubing to the end of the syringe.

2.4. Silver Staining 1-Dimensional SDS-PAGE

1. 4–20% Tris–glycine gel with 1.0 mm × 12 wells.
2. Electrophoresis apparatus: Gel box and power supply.
3. Benchmark Protein Molecular Weight Marker (Invitrogen).
4. Novex® Tris–glycine SDS sample buffer 2×.
5. 2-mercaptoethanol.
6. Novex® Tris–glycine SDS running buffer 10×.
7. Fixing solution: 50% methanol and 10% acetic acid (glacial) in dH_2O.
8. 6% acetic acid (glacial).

9. Prepare 500 mL 0.02% sodium thiosulfate pentahydrate solution in dH_2O. Prepare fresh daily.
10. Prepare 100 mL 0.2% silver nitrate solution.
11. Developing solution: Dissolve 3.0 g sodium carbonate in 98 mL dH_2O. Immediately before use, add 50 μL 37% formaldehyde and 2.0 mL 0.02% sodium thiosulfate.

2.5. Low-Molecular-Weight Serum Fractions: Preparation for Reverse-Phase Protein Microarray and Western Blotting

1. Microcon Ultracel YM-3 centrifugal filter device (Millipore).
2. Glycerol 20% (v/v).
3. 2-mercaptoethanol.
4. Sample buffer for reverse-phase protein arrays and western blotting: 775 μL Novex® Tris–glycine SDS sample buffer 2×, 25 μL 2-mercaptoethanol, 200 μL glycerol.

2.6. Low-Molecular-Weight Serum Fractions: Preparation for Mass Spectrometry

1. ProteoSpin™ Detergent Clean-Up Micro Kit (Norgen Biotek).
2. Modified column activation and wash buffer (acidic): For 25 samples, prepare 0.5 mL pH binding buffer (acidic) (provided in ProteoSpin™ kit), 12.5 mL isopropanol, 12.0 mL dH_2O.
3. Modified column activation and wash buffer (basic): For 25 samples, prepare 0.5 mL pH buffer (basic) (provided in ProteoSpin™ kit), 12.5 mL isopropanol, 12.0 mL dH_2O.

3. Methods

3.1. Preparation of Serum Samples

Sample preservation is crucial for obtaining reliable proteomic results. Proteases degrade proteins and phosphatases cleave phosphorylated residues necessitating strict adherence to processing time lines. The specimen of choice is serum obtained from whole blood collected in red top vacutainer tubes with no additives or clot activators. The serum should be free of hemolysis and clots. Samples subjected to repeated freezing and thawing are not acceptable due to the detrimental effects on the proteins.

1. The whole-blood specimen should be allowed to clot for 15–30 min.
2. Centrifuge the whole blood sample at 1,500 × g for 10 min.
3. Immediately after centrifugation, transfer the serum into smaller working aliquots (50–100 μL each) in plastic cryovials.
4. Store serum aliquots at –80°C to avoid repeated freeze/thaw cycles.
5. Thaw serum samples on ice to prevent protein degradation.
6. In a labeled, screw-cap tube (1.5–2.0 mL volume), dilute 25 μL of serum into 128 μL sample buffer (75 μL Novex®

Tris–glycine SDS sample buffer 2×, 15 μL 1 M DTT, 3.0 μL bromophenol blue, and 35 μL dH_2O) (see Note 3).

7. Heat samples at 100°C in a heat block for 10 min to denature tertiary protein structure.

3.2. Gel Preparation and Mini Prep Cell Setup

The apparatus used for continuous elution electrophoresis described in this chapter is Bio-Rad's Mini Prep Cell (Fig. 2). Assembly and cleanup are performed according to manufacturer's instructions.

1. Use the leveling bubble attached to the casting stand to ensure a level surface. Adjust the plastic screws located on the corners if needed. Place gel tube through the center of the elution manifold top and securely attach the assembly to the casting stand by tightening the three rubber-thumb screws.
2. Prepare a 10% resolving gel in a 15-mL conical tube, mix well, and use a glass pipette to add resolving gel up to the 5.2-cm mark on the gel tube (Subheading 2.2, item 8). Use a new glass pipette to add three drops of dH_2O to the top of the gel. Allow 15–20 min for solidification (see Note 4).
3. After the resolving gel has solidified and immediately before adding the stacking gel, carefully use a glass pipette to remove the three drops of water without puncturing the gel.
4. Prepare a 4% stacking gel in a 15-mL conical tube, mix well, and use a glass pipette to add the stacking on top of the resolving gel, bringing the total volume up to 5.7 cm (Subheading 2.2,v item 9). Allow 15–20 min for solidification (see Note 4).
5. While the gels are solidifying, attach a 3-mL syringe to the luer fitting of the Teflon tube which is attached to the elution manifold base and purge the system with dH_2O several times to effectively remove all air bubbles.
6. After complete solidification, the mini prep cell can be assembled. Loosen the rubber-thumb screws on the elution manifold top. Place a clean dialysis membrane on top of the support frit of the elution manifold base, and a clean elution frit on top of the dialysis membrane. Place the gel tube/elution manifold top assembly on top of the elution manifold base by aligning the screws. Tighten the thumb screws for a secure fit.
7. Push the gel tube through the grommet located in the center of the upper buffer chamber assembly. Place the protruding plastic wings of the elution manifold top into the groves of the upper buffer chamber assembly and twist to lock into place. Securely connect the elution buffer feedline to the luer fitting located on the chamber.
8. Place the apparatus into the lower buffer chamber, with the elution collection tube resting in the carved slot of the lower chamber. Use a 3-mL syringe to flush 1× running buffer

through the feedline and into the elution buffer reservoir to prime the apparatus.

9. Use the luer fitting to attach the feedline to the peristaltic pump's silicone tubing/elution tube and then place it around the peristaltic pump (Fig. 2).
10. Add 400 mL 1× running buffer to the lower buffer chamber, 100 mL to the elution buffer reservoir, and approximately 125 mL running buffer to upper electrophoresis buffer reservoir (see Note 5).
11. Use a glass pipette to carefully add the entire sample volume (153 μL) to the gel tube.
12. Connect the lid to the mini prep apparatus and plug the electrodes into a power source. Run the gel at 200 constant volts for unlimited time.
13. Prime the collection tubing by turning the pump on forward and prime for 1 min. Turn the pump off and run gel for approximately 55 min.

3.3. Mini Prep Cell Operation

3.3.1. Low-Molecular-Weight Sample Collection

1. After approximately 55 min, when the tracking dye reaches the 1-cm mark on the gel tube, switch the pump that has been set to achieve 75–100 μL elution buffer flow rate to forward and slow. After approximately 20–25 min, the blue dye front begins running through the tubing. Once it reaches the first clamp, start sample collection (see Note 6).
2. Collect samples at 5-min intervals into five different labeled collection tubes to obtain samples that are equal to or less than approximately 25 kDa.
3. Place samples on ice after collection or store at –80°C.
4. After sample collection is complete, turn off the pump and power source.

3.3.2. Disassembly of Mini Prep Cell and Cleanup

1. Ensure that the power source has been switched off and remove the lid assembly from the apparatus. Detach the feedline from the tubing attached to the peristaltic pump. Pour buffer out of the upper and lower chambers and unscrew the elution buffer feedline from the luer attached to the upper buffer chamber.
2. Carefully, twist the elution manifold so that it is no longer locked into the upper chamber and pull the gel tube out of the grommet.
3. Unscrew the rubber screws on the elution manifold top that connects it to the elution manifold base. While separating the two pieces, locate the dialysis membrane and frit and place them into a 50-mL conical tube containing ddH_2O. Additionally, place the elution manifold into a beaker containing dH_2O (see Note 7).

4. Remove gel tube from elution manifold top. Use a 2-mL pipette to force the polymerized gel from the tube and then rinse with dH_2O. Use a 20-mL syringe with an attached piece of Teflon tubing to repeatedly purge water through the feedline.
5. Clean the elution manifold base by rinsing with dH_2O and repeatedly flush the tubing with dH_2O using a 3-mL syringe. After washing is complete, store in a beaker of dH_2O at 4°C.
6. Repeatedly rinse the dialysis membrane and elution frit with dH_2O (see Note 8). Store in a 50-mL conical tube with dH_2O at 4°C.
7. Connect a 20-mL syringe filled with ddH_2O to the silicone tubing/elution tubing and flush repeatedly.

3.4. 1-D Gel Silver Staining

Silver staining is performed for visualization of proteins present at low concentrations. This step is critical for determining the range of the molecular weights of proteins in each prep cell fraction (Fig. 3). If samples are collected by precisely following the same procedure each time, it is necessary to only silver stain a subset of the samples to determine the molecular weight range of each fraction. Specific molecular weight fractions can be combined for further concentration.

1. Prepare 1:2 dilutions of the LMW serum fractions in a sample buffer containing Novex® Tris–glycine SDS sample buffer 2×, 5% 2-mercaptoethanol. Heat dilute samples in a 100°C heat block for 10 min.
2. Remove white tape from the bottom of a 4–20% Tris–glycine 1.0-mm × 12-well gel and place into the gel apparatus. Remove

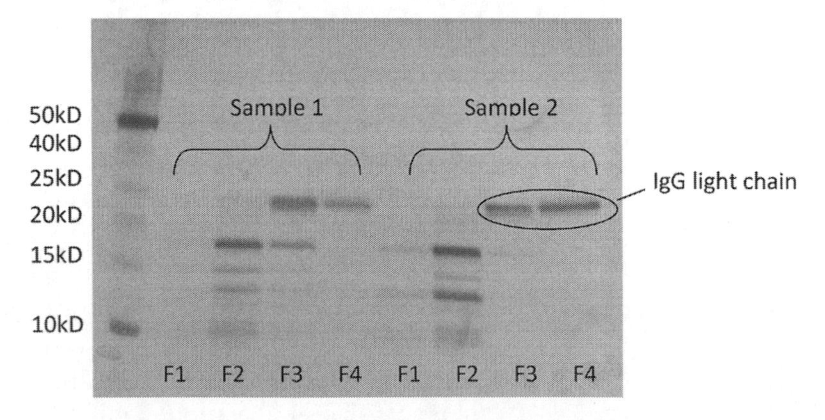

Fig. 3. 1D SDS-PAGE with silver stain for low-molecular-weight fractions eluted with the Mini Prep Cell. Two different samples, each representing four different protein fractions, were resolved on a 4–20% gel and stained with silver nitrate. Light chain immunoglobulins (25 kDa) show prominent bands in fractions 3 (F3) and 4 (F4) which can be used as a marker for determining which fractions contain the protein molecular weights of interest.

comb and fill the inner and outer chamber with 1× Tris–glycine running buffer.

3. Load 20 μL of each sample and 10 μL protein molecular weight ladder into separate wells.
4. Complete assembly of the apparatus and run gel for 1 h at 35 constant milliamps.
5. Turn off the power supply and remove gel from plastic casing.
6. In a glass dish, submerge gel in fixing solution (50% methanol and 10% acetic acid in dH_2O). Incubate for 15 min on a shaker in order to diffuse the SDS out of the gel, leaving the proteins fixed in the gel.
7. Rinse the gel quickly with dH_2O and then wash in excess dH_2O for 15 min. Repeat two times for 15 min and then again for 45–60 min (see Note 9).
8. Incubate gel for 90 s in freshly prepared sodium thiosulfate and then rinse in dH_2O, three times for 30 s each.
9. Cover dish with tin foil and stain the gel in 0.2% silver nitrate solution for 45 min. After incubation, wash gel in dH_2O, three times for 1 min each (see Note 10).
10. Develop the gel for 5–8 min in freshly prepared developing solution (Subheading 2.4, item 11). Stop color development by adding 6% acetic acid. Wash gel in dH_2O twice for 1 min each.

3.5. LMW Sample Preparation for Reverse-Phase Protein Microarray and Western Blotting

LMW fractions can be printed onto RPMAs or resolved on western blots for analysis of specific proteins, such as posttranslationally modified proteins, present in each fraction. While western blotting is the conventional technology for determining the presence or activation status of proteins, RPMA is an alternative multiplexed proteomic assay in which denatured proteins from individual patient samples or cell lysates are immobilized onto nitrocellulose-coated microscope slides (18). The arrays are stained with validated antibodies on a Dako Autostainer using a horseradish peroxidase-mediated, biotinyl tyramide amplification chemistry (19). Spot analysis software is used to measure spot intensities and background intensity. Net intensities are normalized to total protein per spot and mean values are obtained. This technology allows the evaluation of relative intensities of up to 150 proteins (total and post-translationally modified proteins) across hundreds of samples with minimal sample volume.

1. After examining the LMW fractions resolved on SDS-PAGE gels and stained with silver nitrate, combine all fractions containing proteins ≤25 kDa.
2. Place a Microcon Ultracel YM-3 filter cartridge into a labeled 1.5-mL tube provided with the filters.

3. Add 500 μL of sample to the filter cartridge. Spin the sample at 14,000 × *g* for 60 min. Proteins greater than 3 kDa remain on the filter and excess volume flows through.
4. Discard the flow through and add another 400 μL of sample to the filter cartridge. Spin at room temperature at 14,000 × *g* for 60 min. Repeat this process for the entire volume (see Note 11).
5. Approximately 100 μL remains on the filter. To reduce the volume, spin the filter column for an additional 30 min at room temperature at 14,000 × *g*.
6. Invert the filter cartridge containing the collected proteins into a new, sterile, labeled 1.5-mL tube provided with the filters (see Note 12). Centrifuge at 1,000 × *g* for 3 min. Discard the filter and save the protein lysate.
7. Add 20% glycerol (8 μL) and 2.5% 2-mercaptoethanol (1.0 μL) to each sample. Bring each sample volume up to 40 μL with 1× Tris–glycine SDS running buffer.
8. To use the samples for reverse-phase protein arrays and western blotting, prepare 1:2 dilutions of the concentrated LMW fractions in sample buffer (775 μL Novex® Tris–glycine SDS sample buffer 2×, 25 μL 2-mercaptoethanol, 200 μL glycerol (see Subheading 2.5, item 4)).

3.6. LMW Serum Fractions: Preparation for Mass Spectrometry

Tandem mass spectrometry coupled with liquid chromatography (LC-MS/MS) is currently one of the most powerful proteomic tools for protein sequencing. SDS must be removed from samples prior to mass spectrometry. The method described below removes SDS by ion-exchange matrix sequestration and concentrates the fractions isolated from the Mini Prep Cell for MS analysis.

1. After examining the LMW fractions run on SDS-PAGE gels and stained with silver nitrate, combine all fractions with proteins ≤25kDa. Split the volume into two different vials with one labeled basic and the other acidic.
2. Adjust the pH of the vial-labeled acidic to 4.5 by adding acidic pH-binding buffer and adjust the pH of the vial-labeled basic to 7.0 using the basic pH-binding buffer.
3. Add equal volume isopropanol to both tubes and invert repeatedly to ensure adequate mixing.
4. Place spin column into collection tube labeled with the sample name and acidic or basic. Apply 250 μL acidic or basic modified column activation and wash buffer into the labeled acidic and basic columns, respectively. Centrifuge for 1 min at 14,000 × *g*. Discard flow through and repeat washing step.
5. Add 650 μL protein solution to the column and centrifuge at 14,000 × *g*. Discard flow through and repeat process until all protein volume has been applied to the column.

6. Pipette 250 μL of acidic or basic modified column activation and wash buffer to the acidic and basic tubes, respectively, and centrifuge at 14,000 × *g* for 1 min. Discard flow through and repeat column washing once again. Repeat an additional 1-min centrifugation step if any volume remains on filter.
7. Label a clean set of elution tubes, apply 5 μL neutralizer solution to each, and place the washed column into the appropriate tube. Pipette 25 μL elution buffer to the center of the column and centrifuge at 14,000 × *g* for 1 min to collect proteins in the elution tube. Repeat with another 25 μL.
8. Remove column. Samples are ready for mass spectrometric analysis.

4. Notes

1. Prepare fresh 1 M DTT daily.
2. 10% APS and TEMED should be added immediately before use.
3. Use screw-cap tubes to prevent the lid from popping off or sample loss during heating.
4. Save the remaining gel in the 15-mL tube to determine when the acrylamide gel has solidified. Periodically tip/rotate the tube containing the residual gel. Lack of movement/flow of the gel indicates that it has solidified.
5. 1× running buffer should cover the electrode in the upper electrophoresis buffer reservoir for proper conductivity.
6. The blue dye front can be difficult to see. Place a KimWipe or other white absorbent towel underneath the tubing for enhanced contrast.
7. Elution manifold base with attached support frit, the dialysis membrane, and the elution frit must always stay hydrated. Do not allow these components of the Mini Prep Cell to dry.
8. Washing of the dialysis membrane and elution frit should be performed over a beaker so that they are not lost in the sink drain or counter.
9. The last washing step can be extended overnight.
10. Silver staining is a very sensitive technique for protein detection. It is possible to overstain a gel. To prevent saturation of signal, monitor the color development of the gel during the 10-min incubation with the development solution. After adequate signal is observed, stop development by adding 6% acetic acid.
11. The o-ring can detach from the filter, resulting in sample loss. If this occurs, collect the entire sample at the bottom of the

tube, place the filter upside down in a clean microcentrifuge tube, and spin at 1,000 × *g* force for 3 min. Combine this with the flow through and place the entire volume into a clean Microcon filter and repeat the concentration process.

12. When tubes are centrifuged, lids may detach from tubes. Ensure that the tubes are labeled on the side as well as the lid.

References

1. Merrell, K., Southwick, K., Graves, S. W., Esplin, M. S., Lewis, N. E., Thulin, C. D. (2004) Analysis of low-abundance, low-molecular-weight serum proteins using mass spectrometry. *J Biomol Tech* **15**, 238–48.
2. Hortin, G. L., Sviridov, D., Anderson, N. L. (2008) High-abundance polypeptides of the human plasma proteome comprising the top 4 logs of polypeptide abundance. *Clin Chem* **54**, 1608–16.
3. Anderson, N. L., Anderson, N. G. (2002) The human plasma proteome: history, character, and diagnostic prospects. *Mol Cell Proteomics* **1**, 845–67.
4. Petricoin, E. F., Belluco, C., Araujo, R. P., Liotta, L. A. (2006) The blood peptidome: a higher dimension of information content for cancer biomarker discovery. *Nat Rev Cancer* **6**, 961–7.
5. Lowenthal, M. S., Mehta, A. I., Frogale, K., Bandle, R. W., Araujo, R. P., Hood, B. L. et al. (2005) Analysis of albumin-associated peptides and proteins from ovarian cancer patients. *Clin Chem* **51**, 1933–45.
6. Mehta, A. I., Ross, S., Lowenthal, M. S., Fusaro, V., Fishman, D. A., Petricoin, E. F., 3rd et al. (2003) Biomarker amplification by serum carrier protein binding. *Dis Markers* **19**, 1–10.
7. Gygi, S. P., Corthals, G. L., Zhang, Y., Rochon, Y., Aebersold, R. (2000) Evaluation of two-dimensional gel electrophoresis-based proteome analysis technology. *Proc Natl Acad Sci U S A* **97**, 9390–5.
8. O'Farrell, P. H. (1975) High resolution two-dimensional electrophoresis of proteins. *J Biol Chem* **250**, 4007–21.
9. Zolotarjova, N., Martosella, J., Nicol, G., Bailey, J., Boyes, B. E., Barrett, W. C. (2005) Differences among techniques for high-abundant protein depletion. *Proteomics* **5**, 3304–13.
10. Wang, Y. Y., Cheng, P., Chan, D. W. (2003) A simple affinity spin tube filter method for removing high-abundant common proteins or enriching low-abundant biomarkers for serum proteomic analysis. *Proteomics* **3**, 243–8.
11. Sato, A. K., Sexton, D. J., Morganelli, L. A., Cohen, E. H., Wu, Q. L., Conley, G. P. et al. (2002) Development of mammalian serum albumin affinity purification media by peptide phage display. *Biotechnol Prog* **18**, 182–92.
12. Rothemund, D. L., Locke, V. L., Liew, A., Thomas, T. M., Wasinger, V., Rylatt, D. B. (2003) Depletion of the highly abundant protein albumin from human plasma using the Gradiflow. *Proteomics* **3**, 279–87.
13. Adkins, J. N., Varnum, S. M., Auberry, K. J., Moore, R. J., Angell, N. H., Smith, R. D. et al. (2002) Toward a human blood serum proteome: analysis by multidimensional separation coupled with mass spectrometry. *Mol Cell Proteomics* **1**, 947–55.
14. Camerini, S., Polci, M. L., Bachi, A. (2005) Proteomics approaches to study the redox state of cysteine-containing proteins. *Ann Ist Super Sanita* **41**, 451–7.
15. Camerini, S., Polci, M. L., Liotta, L. A., Petricoin, E. F., Zhou, W. (2007) A method for the selective isolation and enrichment of carrier protein-bound low-molecular weight proteins and peptides in the blood. *Proteomics Clin Appl* **1**, 176–84.
16. Tran, J. C., Doucette, A. A. (2008) Gel-eluted liquid fraction entrapment electrophoresis: an electrophoretic method for broad molecular weight range proteome separation. *Anal Chem* **80**, 1568–73.
17. Mueller, C., Zhou, W., Vanmeter, A., Heiby, M., Magaki, S., Ross, M. M. et al. (2010) The Heme Degradation Pathway is a Promising Serum Biomarker Source for the Early Detection of Alzheimer's Disease. *J Alzheimers Dis* **19**, 1081–91.
18. Paweletz, C. P., Charboneau, L., Bichsel, V. E., Simone, N. L., Chen, T., Gillespie, J. W. et al. (2001) Reverse phase protein microarrays which capture disease progression show activation of pro-survival pathways at the cancer invasion front. *Oncogene* **20**, 1981–9.

19. Bobrow, M. N., Harris, T. D., Shaughnessy, K. J., Litt, G. J. (1989) Catalyzed reporter deposition, a novel method of signal amplification. Application to immunoassays. *J Immunol Methods* **125**, 279–85.
20. Bobrow, M. N., Shaughnessy, K. J., Litt, G. J. (1991) Catalyzed reporter deposition, a novel method of signal amplification. II. Application to membrane immunoassays. *J Immunol Methods* **137**, 103–12.
21. Bobrow, M. N., Litt, G. J., Shaughnessy, K. J., Mayer, P. C., Conlon, J. (1992) The use of catalyzed reporter deposition as a means of signal amplification in a variety of formats. *J Immunol Methods* **150**, 145–9.
22. Hunyady, B., Krempels, K., Harta, G., Mezey, E. (1996) Immunohistochemical signal amplification by catalyzed reporter deposition and its application in double immunostaining. *J Histochem Cytochem* **44**, 1353–62.
23. King, G., Payne, S., Walker, F., Murray, G. I. (1997) A highly sensitive detection method for immunohistochemistry using biotinylated tyramine. *J Pathol* **183**, 237–41.

Chapter 16

Mass Spectrometry-Based Biomarker Discovery

Weidong Zhou, Emanuel F. Petricoin III, and Caterina Longo

Abstract

Discovery of candidate biomarkers within the entire proteome is one of the most important and challenging goals in proteomic research. Mass spectrometry-based proteomic is a modern and promising technology for semiquantitative and qualitative assessment of proteins, enabling protein sequencing and identification with exquisite accuracy and sensitivity. For mass spectrometry analysis, protein extractions from tissues of interest or body fluids with subsequent protein fractionation represent an important and unavoidable step in the workflow for biomarker discovery. The aim of our chapter is to provide practical lab procedures for sample digestion and protein fractionation for subsequent mass spectrometry analysis.

Key words: Biomarker discovery, Mass spectrometry, In-gel digestion of proteins, In-solution digestion of proteins, LC-MS/MS, LTQ-Orbitrap, SEQUEST

1. Introduction

The emerging field of tissue proteomics stated the importance of the myriads of proteins and fragments generated by tissues which are correlated with disease outcomes (1, 2) that can be targeted for more efficacious and appropriate therapy. In fact, the discovery and characterization of valuable candidate biomarkers have the benefit to select patients who are the most inclined to respond and, subsequently, develop a patient-tailored therapy (3, 4).

In these efforts to identify and measure informative biomarkers from patient body fluids and tissue samples, sensitive mass spectrometry instruments coupled to bioinformatics analysis play a central role. Mass spectrometers are powerful, versatile, and analytical instruments with the ability to sequence and characterize disease-related candidate biomarkers, both qualitatively and quantitatively (5, 6) (Table 1). Independently of the mass spectrometer employed, protein extractions from tissues of interest or body fluids

Virginia Espina and Lance A. Liotta (eds.), *Molecular Profiling: Methods and Protocols*, Methods in Molecular Biology, vol. 823,
DOI 10.1007/978-1-60327-216-2_16, © Springer Science+Business Media, LLC 2012

Table 1
Comparison of analytical multiplex platforms for measuring the peptidome

	Mass spectrometry based		Protein array based		
Attribute	**Immuno-MS**	**MS profiling**	**Suspension beads**	**Antibody array**	**Reverse-phase array**
Multiplex measurement	Yes	Yes	Yes	Yes	Yes
High throughput	Yes	Yes	Yes	Yes	Yes
Highly sensitive	No	No	Yes	Yes	
Highly quantitative	No	No	Yes	Yes	Yes
Readout of posttranslational modification	Yes	Yes	Limited	Limited	Yes
Readout of size/mass	Yes	Yes	No	No	No
Two-site antibody capture required	No	No	Yes	Yes	No
Works with single antibody	Yes	No	No	No	Yes

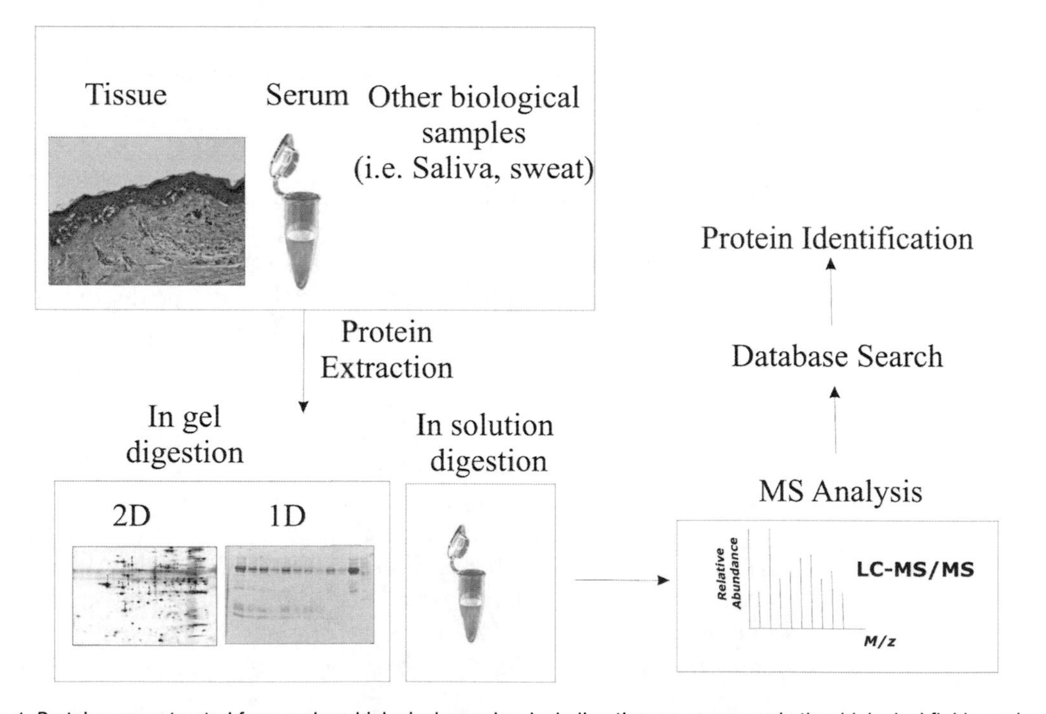

Fig. 1. Proteins are extracted from various biological samples, including tissues, serum and other biological fluids, and are separated either by 1-dimensional (1D) or 2D gel electrophoresis. Proteins are digested with proteolytic enzyme (i.e., trypsin) before mass spectrometric analysis. Mass spectrometry identifies masses of the peptides and their primary structures. The mass data obtained are then used in database searches, enabling identification of proteins.

and subsequent protein fractionation represent an important and unavoidable step in the workflow for biomarker isolation and sequencing (Fig. 1). Samples obtained from polyacrylamide gels, tissues, or directly from solution are digested with a proteolytic enzyme (i.e., trypsin digestion) into peptide fragments prior to mass spectrometer analysis (Table 2).

Herein, we describe the materials and procedures necessary to isolate and sequence proteins from multiple biological samples by using nanoelectrospray ionization mass spectrometry. In detail, we describe the following procedures: acetone precipitation of proteins, Coomassie and silver staining of SDS-PAGE, in-gel digestion of proteins, in-solution digestion of protein, desalting digestion, purification of phosphopeptides, LC-MS/MS analysis of the sample using LTQ-Orbitrap, and SEQUEST analysis of LTQ-Orbitrap raw MS data.

2. Materials

2.1. Acetone Precipitation of Proteins

80% (v/v) acetone solution. Store at −20°C.

2.2. Coomassie Staining of SDS-PAGE Gel

1. Stain solution: 40% (v/v) methanol, 10% (v/v) acetic acid, 50% (v/v) water, 0.1% (w/v) Coomassie brilliant blue R250.
2. Destain solution: 40% (v/v) methanol, 10% (v/v) acetic acid, 50% (v/v) water.

2.3. Silver Staining of SDS-PAGE Gel

1. Fixative solution: 50% (v/v) methanol, 10% (v/v) acetic acid.
2. Sensitizer solution: 0.02% (w/v) sodium thiosulfate. The solution is prepared by dissolving 0.1 g sodium thiosulfate ($Na_2S_2O_3 \cdot 5H_2O$) in 500 mL water.
3. Stain solution: 0.2% (w/v) silver nitrate. The solution is prepared by dissolving 0.1 g silver nitrate ($AgNO_3$) in 50 mL water. Protect solution from light.
4. Developer solution: 3% (w/v) sodium carbonate, 0.185% (w/v) formaldehyde, 0.0004% (w/v) sodium thiosulfate. The solution is prepared by dissolving 3 g sodium carbonate (Na_2CO_3) in 98 mL water, and adding 50uL 37% formaldehyde (HCOH) and 2 mL of sensitizer solution.
5. Stopper solution: 6% (v/v) acetic acid.

2.4. In-Gel Digestion of Proteins

1. 500 mM ammonium bicarbonate solution is prepared by dissolving 3.95 g ammonium bicarbonate (NH_4HCO_3) in 100 mL water. 50 mM ammonium bicarbonate solution is prepared by mixing 10 mL 500 mM solution with 90 mL water.

Table 2
Comparison of the available methods used to purify and study the peptidome

Method	Sensitivity	Peptide identification	Throughput	Enrichment	Input volume capacity	Complexity resolution
SELDI-TOF	Low	No	High	Low–medium	Low	Low
Direct LC-ESI-MS	Low	Yes	High	Low	Low	Medium
Hydrophobic bead capture MS	Low	Yes	Medium	High	Low	Low–medium
Particle capture MS	Low–medium	Yes	Medium	High	Low	Medium
Native carrier-protein-harvesting MS	High	Yes	Low	High	Low	High
Albumin depletion MS	Low	Yes	Low	Low	Low	Low
2D PAGE MS	Low	Yes	Low	Low	Low	Low
Centrifugation MS	Medium	Yes	Low	High	High	Medium
Preparative 1D PAGE MS	High	Yes	Low	High	High	High
ICAT/labeling MS	Medium	Yes	Low	Low	High	High

SELDI-TOF surface-enhanced laser desorption/ionization time of flight, *LC-ESI* liquid chromatography-electrospray ionization, *MS* mass spectrometry, *2D* two-dimensional, *PAGE* polyacrylamide gel electrophoresis, *ICAT* isotope-coded affinity tagging

2. 1 M dithiothreitol (DTT) solution is prepared by dissolving 15.4 mg DTT in 90 μL water.
3. Reducing solution: 10 mM DTT, 50 mM ammonium bicarbonate. The solution is prepared by mixing 495 μL 50 mM ammonium bicarbonate and 5 μL of 1 M DTT.
4. 500 mM iodoacetamide solution is prepared by dissolving 18.5 mg iodoacetamide in 200 μL 500 mM ammonium bicarbonate.
5. Alkylating solution: 20 mM iodoacetamide, 50 mM ammonium bicarbonate. The solution is prepared by mixing 480 μL 50 mM ammonium bicarbonate and 20 μL 500 mM iodoacetamide.
6. Dry solution: 80% (v/v) acetonitrile, 50 mM ammonium bicarbonate. The solution is prepared by mixing 80 mL acetonitrile, 10 mL 500 mM ammonium bicarbonate, and 10 mL water.
7. Trypsin (0.5 μg/μL) (Promega).
8. Digestion solution: 10 ng/μL trypsin, 50 mM ammonium bicarbonate. The solution is prepared by mixing 1 μL trypsin (0.5 μg/μL) with 49 μL 50 mM ammonium bicarbonate.
9. Extraction buffer: 50% (v/v) acetonitrile, 2% (v/v) acetic acid.

2.5. In-Solution Digestion of Proteins

1. 1 M DTT solution is prepared by dissolving 15.4 mg DTT in 90 μL water.
2. 8 M urea solution is prepared by dissolving 48 g urea in ~60 mL water. Add additional water to bring final volume to 100 mL.
3. 500 mM ammonium bicarbonate (NH_4HCO_3) solution is prepared by dissolving 3.95 g ammonium bicarbonate in 100 mL water.
4. 500 mM iodoacetamide solution is prepared by dissolving 18.5 mg iodoacetamide in 200 μL 500 mM ammonium bicarbonate.
5. Trypsin (0.5 μg/μL) (Promega).

2.6. Desalting Digestion Mixture by Sep-Pak Column

1. Sep-Pak column from Waters.
2. Buffer A: 0.1% (v/v) trifluoroacetic acid (TFA).
3. Buffer B: 0.1% (v/v) TFA, 80% (v/v) acetonitrile.

2.7. Desalting Digestion Mixture by ZipTip

1. ZipTip from Millipore.
2. Buffer A: 0.1% (v/v) TFA.
3. Buffer B: 0.1% (v/v) TFA, 80% (v/v) acetonitrile.

2.8. Purification of Phosphopeptides

1. Bovine β-casein (Sigma).
2. Human angiotensin I (Ang I) and tyrosine phosphorylated angiotensin II (Ang II-Phos) (Calbiochem).
3. TiO_2 resin (5 μm, loose media; GL Sciences, Inc.).

4. Inline MicroFilter Assembly (Upchurch Scientific).
5. Fused silica tubing (ID 100 µm and ID 200 µm; Polymicro Technologies).
6. Pressure Cell from Brechbühler Inc.
7. TiO_2 loading buffer: 200 mg/mL 2,5-dihydroxybenzoic acid (DHB), 5% (v/v) TFA, 80% (v/v) acetonitrile.
8. TiO_2 washing buffer 1: 40 mg/mL DHB, 2% (v/v) TFA, 80% (v/v) acetonitrile.
9. TiO_2 washing buffer 2: 2% (v/v) TFA, 50% (v/v) acetonitrile.
10. TiO_2 elution buffer: 5% (w/v) ammonia solution.

2.9. LC-MS/MS Analysis of the Sample Using LTQ-Orbitrap

1. Laser puller (Model P-2000; Sutter Instrument Co.).
2. C18 resin (5 µm, loose media; Michrom Bioresources, Inc.).
3. Water with 0.1% formic acid and acetonitrile with 0.1% formic acid.
4. Mobile phase A: 0.1% (w/v) formic acid.
5. Mobile phase B: 0.1% (w/v) formic acid, 80% (v/v) acetonitrile.

2.10. SEQUEST Analysis of LTQ-Orbitrap Raw MS Data

1. SEQUEST Bioworks Browser (ThermoFisher).
2. Scaffold software from Proteome Software.

3. Methods

Always wear powder-free nitrile gloves when performing gel staining, protein digestion, and LC-MS analyses.

3.1. Acetone Precipitation of Proteins

The method removes salts and many lipid-soluble contaminants and concentrates proteins (7).

1. Chill one vial of 100% acetone and one vial of 80% acetone at –20°C.
2. Add four volumes cold 100% acetone to the sample and mix well. Place sample at –20°C for 1 h.
3. Spin at 16,000 × *g* for 10 min at 4°C. Remove supernatant using a pipette.
4. Wash pellet with cold 80% acetone. Spin at 16,000 × *g* for 30 s at 4°C. Repeat wash once.
5. Remove supernatant and evaporate residual acetone in a SpeedVac for 2 min.
6. Dissolve pellet in 8 M urea in desired volume.
7. Sonicate the sample for 1 min. Repeat sonication several times to resuspend the proteins.

3.2. Coomassie Staining of SDS-PAGE Gel

Coomassie staining is a classic method of detecting proteins after SDS-PAGE (8). The Coomassie dye binds directly to the basic and aromatic side chains of the amino acids of the protein. The detection limit of this staining is 50–100 ng compared to 1–10 ng limit of silver staining (9).

1. Put the gel (1 mm thick) in a plastic staining container, add enough stain solution to cover gel well, and stain while rocking for 10 min (see Note 1).
2. Pour out stain solution, add destain solution to cover gel, and destain while rocking for 5 min. Repeat two more times.
3. Pour out destain solution, add a large amount of water, and leave the container on rocker overnight.

3.3. Silver Staining of SDS-PAGE Gel

Silver staining is preferable for detection of low-abundant proteins in the 1–10-ng range. The basic mechanisms underlying silver staining of proteins in gels are relatively well-understood (10). Basically, protein detection depends on the binding of silver ions to the amino acid side chains, primarily the sulfhydryl and carboxyl groups of proteins (11), followed by reduction to free metallic silver (12). The protein bands are visualized as spots, where the reduction occurs.

1. Fix gel (1 mm thick). In a glass container, submerge gel in fixative solution and leave the container on rocker for 15 min. Pour off fixative solution and rinse briefly with water (see Note 1).
2. Wash gel. Submerge gel in a large amount of water, and leave the container on rocker for 15 min. Pour off water, and repeat two more times.
3. Sensitize gel. Pour off water, and submerge gel in freshly made sensitizer solution for 90 s.
4. Rinse with water for 30 s. Repeat two more times for a total of three rinses.
5. Stain gel. Pour off water, and submerge gel in freshly made stain solution for at least 25 min. Protect gel from light during staining.
6. Pour off stain solution, and rinse gel with water for 60 s. Repeat two more times for a total of three rinses.
7. Develop gel. Pour off water, and submerge gel in freshly made developer solution. When desired contrast is attained (this may take less than a minute or up to several minutes), pour off developer solution, and submerge gel in stopper solution for 10 min.
8. Pour off stopper solution, and submerge gel in water.

3.4. In-Gel Digestion of Proteins

The following procedure describes the trypsin digestion of proteins from gel bands (1D) or spots (2D) and their subsequent extraction, thereby preparing samples for mass spectrometric analysis (13).

1. Cut the band/spot from SDS-PAGE gel to 1 × 1-mm pieces using razor blade and put them into 0.6-mL centrifuge tube.
2. For silver-stained gel, go to step 3. For Coomassie-stained gel, add 500 μL destain solution to the gel pieces and destain while rocking for 15 min. Repeat one more time if necessary until gel piece is clear. Discard destain solution.
3. Reduce the sample proteins. Add 500 μL reducing solution into centrifuge tube. Incubate the sample on rocker at 37°C for 30 min.
4. Alkylate the sample proteins. Discard reducing solution, and add 500 μL alkylating solution into centrifuge tube. Incubate the sample on rocker at room temperature for 20 min in darkness by wrapping the tube with aluminum foil.
5. Discard alkylating solution, and add 500 μL 50 mM ammonium bicarbonate into centrifuge tube. Incubate the sample on rocker at room temperature for 5 min.
6. Discard the solution, add 500 μL of a dry solution, and incubate for 15 min.
7. Remove dry solution, and dry the gel pieces for 20 min with SpeedVac. Then, put the tube on ice.
8. Add 30 μL (or more if the volume of gel pieces is larger) of digestion solution to the tube to rehydrate the gel pieces, and incubate on ice for 30 min.
9. Put the tube in 37°C water bath for 4 h or overnight digestion.
10. Transfer the supernatant to a clean 0.6-mL centrifuge tube.
11. Add 30 μL of extraction buffer to the gel pieces, and incubate at room temperature for 15 min.
12. Transfer the supernatant to the tube in step 15 and combine with previous supernatant.
13. Dry the supernatant by SpeedVac.
14. Resuspend the peptides in 1% acetic acid for LC-MS analysis.

3.5. In-Solution Digestion of Proteins

This protocol can be used for the digestion of purified proteins in solvent, precipitated proteins, or body fluids, such as serum and vitreous. Here, we take the digestion of serum as an example.

1. In a 0.6-mL centrifuge tube, add 10 μL serum (~500 μg).
2. Reduce the sample proteins. Add 80 μL 8 M urea, 1 μL 1 M DTT (final concentration of ~10 mM), and mix well. Incubate the sample at 37°C water bath for 30 min.

3. Alkylate the sample proteins. Add 6 µL 500 mM iodoacetamide (final concentration of ~30 mM) into tube and mix well. Incubate the sample at room temperature for 20 min in darkness.
4. Add 260 µL water, 40 µL 500 mM ammonium bicarbonate, and mix well.
5. Add 5 µL trypsin (0.5 µg/µL) and mix well. Put the tube in 37°C water bath for overnight digestion.
6. The next day, add 10 µL of glacial acetic acid (100%) to stop digestion and acidify the sample (see Note 2).
7. Desalt the sample using Sep-Pak column or ZipTip.

3.6. Desalting Digestion Mixture by Sep-Pak Column

The binding capacity of the Sep-Pak columns is dependent on the amount of C18 resin packed in column. Here, we take the 1 cc/50 mg (catalogue number WAT054955) as an example (the estimated binding capacity is ~5 mg peptides).

1. Acidify digestion mixture by adding glacial acetic acid (final concentration of ~2%); see step 6, Subheading 3.5.
2. Put the column into a clamp attached with iron stand.
3. Wash the column with 1 mL buffer B, and repeat once.
4. Wash the column with 1 mL buffer A, and repeat two more times.
5. Load sample to the column.
6. Collect the flow through, and load to the column. Repeat one more time.
7. Discard flow through, wash the column with 1 mL buffer A. Repeat wash once.
8. Apply 0.5 mL buffer B to elute the column, and collect the eluted peptides into 0.6-mL centrifuge tube.
9. (Optional) Concentrate the eluted sample by SpeedVac if you need to remove the acetonitrile from the sample for LC-MS analysis.

3.7. Desalting Digestion Mixture by ZipTip

ZipTip desalting is recommended for small amounts of proteins/peptides in the sample. The binding capacity of the ZipTip is dependent on the amount of C18 resin packed in tip. Here, we take the catalogue number ZTC18S960 ZipTip as an example (the estimated binding capacity is ~50 µg peptides).

1. Acidify digestion mixture by adding glacial acetic acid (final concentration of ~2%).
2. Pick up one ZipTip using a 20 µL-capacity pipette.
3. Wash the ZipTip with 20 µL buffer B, and repeat once.

4. Wash the ZipTip with 20 μL buffer A, and repeat two more times.
5. Pipette sample through resin several times for efficient binding.
6. Discard flow through, and wash the ZipTip with 20 μL buffer A. Repeat wash once.
7. Pipette 20 μL buffer B to elute peptides into 0.6-mL centrifuge tube. Repeat once and combine with previous elution.
8. (Optional) Concentrate the eluted sample by SpeedVac if you need to remove the acetonitrile from the sample for LC-MS analysis.

3.8. Purification of Phosphopeptides

1. Digest 100 μg (up to 1 mg) cell lysate or tissue homogenate in solution using trypsin as described above. Add 100 ng bovine β-casein (10 μL of a 10 ng/μL stock solution) into the sample to serve as a protein standard.
2. Clean up the digestion mixture using a Sep-Pak column or ZipTip as described above.
3. Add 200 fmol AngII-Phos (2 μL of a 100 fmol/μL stock) into the elution to serve as a peptide standard.
4. Concentrate the sample in a SpeedVac (no heat) to a volume of ~100 μL. Add 100 μL TiO_2 loading buffer and mix well.
5. Pack 200 μm TiO_2 column. Resuspend TiO_2 resin with 1.5 mL of TiO_2 washing buffer 1 in a 2-mL tube with a tiny magnetic stir bar, and place the tube in a pressure cell on a stir plate. Cut fused silica capillary using a ceramic blade, attach ~30-cm tubing (OD 360 μm, ID 200 μm) to the frit end of Inline MicroFilter Assembly, and attach ~30-cm tubing (OD 360 μm, ID 100 μm) to the other end of the Assembly to make a blank column. Insert the ID 200 μm tubing into pressure cell, apply ~400 Psi of N_2, and observe the packing of the resin using light microscopy. Halt the packing when the column is ~2 cm in length, and wash the column with 100 μL TiO_2 washing buffer 1.
6. Put the sample tube into pressure cell and load the sample to TiO_2 column by ~600 Psi N_2 with flow rate of 3 μL per minute.
7. Stop loading when the volume of flow through is ~200 μL, put a new tube containing 500 μL TiO_2 washing buffer 1 into pressure cell, and wash the column by 200 μL.
8. Put another tube containing 500 μL TiO_2 washing buffer 2 into pressure cell, and wash the column by 200 μL. Repeat wash once.
9. Put one tube containing 500 μL TiO_2 elution buffer into pressure cell, and elute phosphopeptides off column into 0.6-mL

siliconized, low-retention microcentrifuge tube. Collect ~50 μL elution.

10. Evaporate ammonia in the elution by SpeedVac quickly (~3 min), and acidify the sample by adding glacial acetic acid to a final concentration of 2%.
11. Use a ZipTip to clean up the sample as described above.
12. Concentrate the sample in a SpeedVac. Add 100 fmol AngI (1 μl of a 100fmol/μL stock solution) to serve as a peptide standard.

3.9. LC-MS/MS Analysis of the Sample Using LTQ-Orbitrap

Tryptic peptides or TiO_2 (14)-enriched phosphopepitdes are analyzed by liquid chromatography nanospray tandem mass spectrometry using a Thermo Fisher LTQ-Orbitrap. Separations were performed using 100 μm i.d. × 10-cm-long fused silica capillary column packed in-house with 5 μm C18 resin.

1. Pack 100-μm fused silica capillary C18 column.
   (a) Cut a 30-cm length of tubing (OD 360 μm, ID 100 μm) using a ceramic blade, and pull the tubing to produce a ~10-μm id tip at one end using laser puller.
   (b) Resuspend C18 resin with 1.5 mL of 70% ethanol in a 2-mL tube with a tiny magnetic stir bar, and place the tube in a pressure cell on a stir plate. Insert the blunt end of column in pressure cell, and apply 1,000 Psi of N_2 to pack the resin into column. If the liquid phase is not flowing through the tip, very gently rub the tip with a diamond-tipped engraving pen under microscope.
   (c) Halt the resin packing when the column is ~10 cm in length.
2. Test the packed capillary C18 column. Attach the column to Packed Tip Probe and Nano Electrospray Ionization Source from Thermo Fisher.
   (a) Connect the column to the HPLC system and verify that satisfactory electrospray can be obtained at nanoflow (~200 nL/min) on column when high voltage (~2 kV) is turned on.
   (b) Wash the column with HPLC mobile phase B for 15 min at flow rate of 1 μL/min, and wash the column by mobile phase A for 15 min at flow rate of 1 μl/min.
   (c) Next, load standard (100 fmol yeast enolase digest and 100 fmol AngI) onto the column using a pressure cell at 1,000 Psi N_2. Operate LTQ-Orbitrap to acquire raw data and verify that chromatographic peak shape is symmetrical and peak intensity is satisfactory.

3. Load the sample onto the column using a pressure cell at 1,000 Psi N_2 (see Note 3).
   (a) Load the desired amount of sample by measuring the volume of the flow through using a volumetric capillary.
   (b) Attach the column to LTQ-Orbitrap, wash the column with mobile phase A at ~1 μL/min flow for 3 min, then reduce the flow rate on column to ~200 nL/min, and operate the mass spectrometer to acquire raw data.
4. The LTQ-Orbitrap is calibrated and tuned with positive polarity mode according to manufacturer's instruction.
   (a) The "Capillary Temperature" is set to 200°C.
   (b) "Source Voltage" is 2 kV, and "Injection Waveforms" is turned on.
   (c) The "Ion Trap Full AGC Target" is set to 30,000, "Ion Trap MSn AGC Target" is 10,000, and "FTMS Full AGC Target" is 1,000,000.
   (d) "Microscan" is set to 1 for both the Ion Trap and Orbitrap.
   (e) The "Ion Trap Full Max Ion Time (ms)" is set to 50, "Ion Trap MSn Max Ion Time (ms)" is 100, and "FTMS Full Max Ion Time (ms)" is 1,000.
5. The mass spectrometer is operated in a data-dependent MS/MS mode in which each full MS scan by Orbitrap (60,000 resolution, 400–1,600 mass range) is followed by eight MS/MS CID scans with dynamic exclusion.
   (a) "Min. Signal Required" for CID is set to 500.
   (b) "Isolation Width" is 2.
   (c) "Normalized Coll. Energy" is 30.
   (d) "Default Charge State" is 3.
   (e) "Activation Q" is 0.25, and "Activation Time" is 30.
   (f) "FT master scan preview mode," "Charge state screening," and "Monoisotopic precursor selection" are enabled.
   (g) "Charge state rejection" is enabled, and "Unassigned charge state ions" and "Charge states 4+ ions" are rejected.
   (h) For the enabled "Dynamic exclusion" parameters, the "Repeat Count" is set to 1, "Repeat Duration" is 20, "Exclusion List Size" is 300, and "Exclusion Duration" is 60.
6. The LTQ-Orbitrap HPLC is set up to load samples using either autosampler or pressure cell.
   (a) The HPLC is operated with flow rate 40–100 μL/min, and nano flow rate (~200 nL/min) on C18 capillary column is achieved by using a splitter (see Note 4).

(b) After sample injection, the column is washed for 3 min with mobile phase A and peptides are eluted using a linear gradient of 0% mobile phase B to 10% B in 5 min, then to 50% B in 90 min, and lastly to 100% B in an additional 10 min at flow rate 200 nL/min on column.

(c) After the gradient, regenerate the column by washing it 15 min with mobile phase B and then equilibrating it 15 min with mobile phase A at higher flow rate on column (~1 μL/min).

3.10. SEQUEST Analysis of LTQ-Orbitrap Raw MS Data

1. Tandem mass spectra were matched against human database downloaded from the National Center for Biotechnology Information (NCBI) through the SEQUEST Bioworks Browser using full tryptic cleavage constraints and static cysteine alkylation by iodoacetamide (15).

   (a) The result is filtered by proper criteria to get a list of candidates with reasonable low false-positive identification. Mostly, for a peptide to be considered identified, it should be Top #1 matched, and has to achieve cross-correlation scores of 1.7, 2.2, 3.5 for 1+, 2+, 3+ ions, $\Delta Cn > 0.1$, probabilities of randomized identification <0.01, and 10 ppm mass accuracy.

   (b) Tandem mass spectrometry of phosphopeptide ions using collision-induced dissociation (CID) often produces product ions dominated by the neutral loss of phosphoric acid, and the quality of the resulted MS2 spectrum is low. For a phosphopeptide to be considered identified, it should be Top #1 matched, and has to achieve cross-correlation scores of 1.7, 1.9, 3.0 for 1+, 2+, 3+ ions, and 10 ppm mass accuracy.

   (c) The corresponding MS and MS2 spectra of identified peptides are manually inspected to verify correct identification.

2. The SEQUEST result (*.srf) can be exported to Excel for further comparison and statistical analysis or loaded to Scaffold software to visualize and validate complex MS/MS proteomics experiments.

4. Notes

1. Coomassie stain and silver stain should be discarded in appropriate waste container. Coomassie stain should be discarded in an organic solvent waste container. Silver stain should be discarded in a heavy-metal waste container.

2. Do not digest samples longer than 14 h prior to addition of glacial acetic acid.

3. Because the sample is manually loaded in this manner, the phosphopeptides are never exposed to metal (e.g., iron in metal tubing, needles, or valves common to autosamplers – phosphopeptides are known to bind to metal).
4. If autosampler is used, a splitter is put before the sample loop; if manual loading using pressure cell is used, a splitter is put before the C18 capillary column.

Acknowledgments

The authors thank Prof. Enrico Garaci from the Italian Istituto Superiore di Sanità for his financial and academic support in the framework of the Italy/USA cooperation agreement between the US Department of Health and Human Services, George Mason University, and the Italian Ministry of Public Health. Caterina Longo was supported by Prof. Giovanni Pellacani, Department of Dermatology, University of Modena and Reggio Emilia, Italy.

References

1. Anderson, N. L., and Anderson, N. G. (2002) The human plasma proteome: history, character, and diagnostic prospects. *Mol Cell Proteomics* **1**, 845–67.
2. Liotta, L. A., Ferrari, M., and Petricoin, E. (2003) Clinical proteomics: written in blood. *Nature* **425**, 905.
3. Petricoin, E. F., 3 rd, Bichsel, V. E., Calvert, V. S., *et al.* (2005) Mapping molecular networks using proteomics: a vision for patient-tailored combination therapy. *J Clin Oncol* **23,** 3614–21.
4. Espina, V., Wulfkuhle, J., Calvert, V. S., *et al.* (2008) Reverse phase protein microarrays for theranostics and patient-tailored therapy. *Methods Mol Biol* **441**, 113–28.
5. Domon, B., and Aebersold, R. (2006) Mass spectrometry and protein analysis. *Science* **312**, 212–7.
6. Cravatt, B. F., Simon, G. M., and Yates, J. R., 3rd (2007) The biological impact of mass-spectrometry-based proteomics. *Nature* **450**, 991–1000.
7. Pechar, J., and Hrabane, J. (1951) [Electrophoretic analysis of serum proteins by cold precipitation with acetone.]. *Cas Lek Cesk* **90**, 225–7.
8. Bennett, J., and Scott, K. J. (1971) Quantitative staining of fraction I protein in polyacrylamide gels using Coomassie brillant blue. *Anal Biochem* **43**, 173–82.
9. Winkler, C., Denker, K., Wortelkamp, S., and Sickmann, A. (2007) Silver- and Coomassie-staining protocols: detection limits and compatibility with ESI MS. *Electrophoresis* **28**, 2095–9.
10. Shevchenko, A., Jensen, O. N., Podtelejnikov, A. V., *et al.* (1996) Linking genome and proteome by mass spectrometry: large-scale identification of yeast proteins from two dimensional gels. *Proc Natl Acad Sci U S A* **93**, 14440–5.
11. Merril, C. R., and Pratt, M. E. (1986) A silver stain for the rapid quantitative detection of proteins or nucleic acids on membranes or thin layer plates. *Anal Biochem* **156**, 96–110.
12. Rabilloud, T. (1990) Mechanisms of protein silver staining in polyacrylamide gels: a 10-year synthesis. *Electrophoresis* **11**, 785–94.
13. Castellanos-Serra, L., Ramos, Y., and Huerta, V. (2005) An in-gel digestion procedure that facilitates the identification of highly hydrophobic proteins by electrospray ionization-mass spectrometry analysis. *Proteomics* **5**, 2729–38.
14. Thingholm, T. E., Jorgensen, T. J., Jensen, O. N., and Larsen, M. R. (2006) Highly selective enrichment of phosphorylated peptides using titanium dioxide. *Nat Protoc* **1**, 1929–35.
15. Yates, J. R., 3rd, Eng, J. K., McCormack, A. L., and Schieltz, D. (1995) Method to correlate tandem mass spectra of modified peptides to amino acid sequences in the protein database. *Anal Chem* **67**, 1426–36.

Chapter 17

Mitochondrial Proteome: Toward the Detection and Profiling of Disease Associated Alterations

Paul C. Herrmann and E. Clifford Herrmann

Abstract

Existing at the heart of cellular energy metabolism, the mitochondrion is uniquely positioned to have a major impact on human disease processes. Examples of mitochondrial impact on human pathology abound and include etiologies ranging from inborn errors of metabolism to the site of activity of a variety of toxic compounds. In this review, the unique aspects of the mechanisms related to the mitochondrial proteome are discussed along with an overview of the literature related to mitochondrial proteomic exploration. The review includes discussion of potential areas for exploration and advantages of applying proteomic techniques to the study of mitochondria.

Key words: Cancer, Human disease, Mitochondria, Pathology, Proteomics

1. Introduction

1.1. Mitochondrial Associated Pathology

At the cellular level, the mitochondrion provides the scaffold for oxidative phosphorylation. Consequently, the mitochondrion has been described as the cell's energy powerhouse. The mitochondrion is at the cross-roads of cellular physiology and by extension uniquely positioned to have a profound impact on cellular pathophysiology. The extreme toxicity of poisons such as arsenic, which uncouples cellular oxidative phosphorylation from stable ATP production, and cyanide, which blocks the active site of the mitochondrial respiratory enzyme cytochrome *c* oxidase, underscores the important, dynamic role the mitochondrion plays in cellular life (1, 2).

A variety of human pathologic processes appear to relate directly to the chemical environment of the mitochondrion. These

Virginia Espina and Lance A. Liotta (eds.), *Molecular Profiling: Methods and Protocols*, Methods in Molecular Biology, vol. 823,
DOI 10.1007/978-1-60327-216-2_17,

processes include changes associated with hypoxia and reperfusion injury experienced in myocardial infarction (3). The predominant cellular damage appears to result from reactive oxygen species (ROS) that are mediated at least partially by superoxide dismutase found within mitochondria (4). Additionally, mitochondrial associations have been made with alterations in liver metabolism commensurate with ethanol toxicity and steatosis (5, 6). Chemical toxicities of compounds such as bile acids, acetaminophen (7), and pathologies associated with aspirin and Reye's syndrome, along with doxorubicin-induced mitochondrial cardiomyopathy (8) all appear to be linked to mitochondrial function.

1.2. Gene-Based Diseases

There are a number of inborn errors of metabolism, which are effected by gene mutations in mitochondria-associated proteins. These are known collectively as mitochondrionopathies and are some of the most commonly diagnosed metabolic disorders in humans, affecting an estimated 1/5,000 live births (9). The genetics of these diseases are fascinating because of the dual source of genetic coding for the mitochondrial proteins. The mitochondrial genome as well as the cell's nuclear genome can serve as sites of pathogenic mutations. An example disorder resulting from mitochondrial DNA mutations is MELAS syndrome, an acronym for myopathy, encephalopathy, lactic acidosis, and stroke-like episodes. The disorder is characterized by stroke-like episodes occurring within the first 20 years of life, which are associated with migraine headaches. MELAS syndrome results from mutations within one of the mitochondrially encoded transfer RNA molecules. Another example is Neurogenic weakness accompanied by Ataxia and Retinitis Pigmentosa (NARP syndrome) characterized by paraplegia, neuropathy, myopathy, and seizures (10).

Mitochondrionopathies, resulting from nuclear DNA mutations of the respiratory chain genes, present with symptoms of hypotonia, lactic acidosis, and failure to thrive. A variety of organs may be affected. Most such disorders are inherited in an autosomal recessive fashion consistent with mutations within the nuclear genome (11). Additionally, there appear to be a number of neurodegenerative disorders with incompletely understood etiology linked to the mitochondrion, such as Alzheimer's and Parkinson's disease (12).

1.3. Cancer

Cancer has long been associated with metabolic alterations. Over 80 years ago, Warburg noted a shift in tumor metabolism favoring fermentative over oxidative pathways (13). Interest in the association of cancer and metabolism waned somewhat in the intervening years, but is currently becoming fashionable again. Changes in mitochondrial metabolism due to nononcogenetic mutations have recently been identified as possible mechanisms for modulating oxygen consumption in tumors (14). Recent work has focused on alterations in glycolytic metabolism and cytochrome *c* oxidase pathways in mitochondria as well as studies of mutations within the tumor mitochondrial genome (15–18). Elucidation of specific mitochondrial functions may be exploited as potential therapeutic targets as recently described for multiple myeloma (19).

2. The Promise of Proteomics

With so many disease processes being linked to the mitochondrion, a renewed interest in its structure and function has emerged. This "mitochondrial renaissance" has generated an interest in focusing high-throughput proteomic technologies to study mitochondria (20). Proteomic techniques allow small differences in proteins to be detected quantitatively as well as qualitatively. Quantitative and qualitative differences are revealing alterations in subcellular organelle function that correspond to alterations at the protein level (17, 18). Mitochondria can be thought of as an interface-rich region within the cell. In addition to unraveling disease-associated changes in mitochondria, the study of this interface-rich cellular organelle should also lead to a new and better understanding of compartmentalized subcellular physiology and patterns of protein translocation within cells.

2.1. Genetic Considerations

A mitochondrion contains proteins encoded in the mitochondrion itself combined with proteins encoded in the parent cell nucleus. This dual source of molecules creates a broad and rather intricate scenario of inherited disorders. The mitochondrial genome is composed of a double-stranded circular DNA molecule composed of 16,569 nucleotide pairs. This genetic information leads to the formation of two ribosomal RNA, 22 transfer RNA, and 13 protein molecules. These 13 proteins, along with nuclear encoded proteins, comprise complexes I, III, IV, and V of the mitochondrial respiratory chain. The set of 22 encoded transfer RNA molecules affects the unique intramitochondrial translation code, which is different from that in extramitochondrial space. Multiple copies of each mitochondrial DNA molecule occur in individual mitochondria and there are multiple mitochondria per cell. The mutation rate of mitochondrial DNA is greater than that of nuclear DNA, presumably due to increased levels of damaging oxidizers and diminished repair machinery relative to the nucleus (10). The mitochondrion is inherited overwhelmingly from the mother with some debate as to a possible diluting role of a few sperm-contained mitochondria present at fertilization. Additionally, the extramitochondrially encoded proteins are products of genes located in a variety of chromosomal locations on multiple chromosomes. Protein phenotypic dysfunction localized very specifically to the mitochondria most likely results from a multitude of disparate genotypic etiologies (20).

Mitochondria within the fertilized embryo are inherited from the sperm and egg in a ratio on the order of 1:1,000. Each mitochondrion contains two to ten DNA molecules, and each cell contains multiple mitochondria. Thus, genotypically variable mitochondrial DNA can co-exist within the same cell. This condition, known as heteroplasmy, allows an otherwise lethal mutation to persist. An interesting discussion of heteroplasmy is found in the context of human ooplasm donation, which is performed occasionally in fertility clinics (21). Through the course of cell division, mitochondrial populations within cells can drift toward homoplasmy following the laws of population dynamics (22). The proportion of mutant mitochondrial DNA required for the occurrence of a deleterious phenotype is known as the threshold effect and varies among cells, organs, and individuals.

Despite fundamental differences between nuclear and mitochondrial genomes, there are sufficient similarities to allow inferences to be drawn about how the mutated messages in these two information storage compartments might find their way consistently into the proteome such that a highly sensitive disease signature might be generated.

3. The Challenge of Defining the Mitochondrial Proteome

3.1. DNA Coding Considerations

Defining the mitochondrial proteome in itself is a difficult task. A large number and variety of proteins are present within the mitochondria and the mitochondrial composition varies between tissues. Estimates of the total number of unique proteins comprising this proteome are quite variable with reported estimates ranging from 600 to 4,000 with very little certainty (23–26). Part of the problem rests simply in defining what criteria to use in assigning a particular protein to the proteome. Some of the proteins present within the mitochondria are encoded in mitochondrial DNA and synthesized in the mitochondrion using the transcription code unique to the mitochondrion. Examples include some of the subunits of the electron transport system (27–29). These types of proteins are easily assigned to the mitochondrial proteome. Other proteins are encoded in the nucleus with the mRNA transcript being sent to the mitochondria where it is apparently translated within that organelle making assignment to the mitochondrial proteome reasonably accurate (30, 31). Examples of nuclear encoded proteins that are functionally altered by mutations of mitochondrial transfer RNA include some nuclear encoded respiratory chain subunits and metabolic enzymes (32, 33). Some mitochondrial proteins are encoded in the nuclear genome and are transcribed and translated in the cellular cytoplasm before being imported into the mitochondrion. These proteins are by no means amenable to simple organeller classification.

3.2. Protein Localization

Some proteins synthesized from nuclear encoded DNA reside predominantly within the mitochondrion, such as certain members of the electron transport system complex subunits. Other proteins such as carnitine–acylcarnitine translocase appear to be specifically targeted to at least one subcellular location in addition to the mitochondrion (34), where promoter sequences appear to play a role in targeting (35). Other mitochondrial proteins are found in multiple subcellular locations with similar properties in each location, such as superoxide dismutase (4). Further complexity exists because some mitochondrial proteins found in other subcellular locations exhibit different properties. For instance, isoforms of glycerol phosphate acyltransferase are present in the mitochondrial and cellular microsomal fractions of rat liver. The isoenzymes from the two subcellular locations exhibit different enzyme kinetic values (enzyme affinity and binding reactions), different resistance to heat inactivation and different catalytic activity in the presence of various inhibitors (36). Thus, intracellular compartmentalization can provide separate distinct nanoenvironments with profoundly different protein function. Furthermore, location-specific post-translational modification also alters protein behavior. Knockout of a particular enzyme can have completely different effects in different organs as observed in double creatine kinase knockout mice apparently completely lacking creatine kinase. The knockout mice demonstrated very large changes in skeletal muscle function with very little change in cardiac function (3). Ceramide is synthesized within the mitochondria as well as other subcellular locations but through different pathways (37). Additionally, Acyl-CoA synthetase which is found in a variety of isoforms and subcellular locations including, but not limited to, mitochondria can be locally inhibited at certain sites independently of the other locations (38). The ability to inhibit enzymes in different locations of the mitochondrial respiratory chain has been suggested to underlie the differential sensitivity of different regions of the brain to hypoxia (12).

3.3. Mitochondrial Compartments

Within the mitochondrion, deciphering the proteome is further complicated by a variety of suborganelle compartments (Fig. 1). Proteins are localized specifically to the outer mitochondrial membrane, inner mitochondrial membrane, matrix, and intermembrane regions. The proteins localized to membranes have totally different properties from those found in the cytosol, and hence have very disparate behaviors requiring vastly different separation methods.

A final difficulty in addressing the proteomic problem is the number and variety of mitochondria within a given cell. The number of individual mitochondria within a cell ranges over two orders

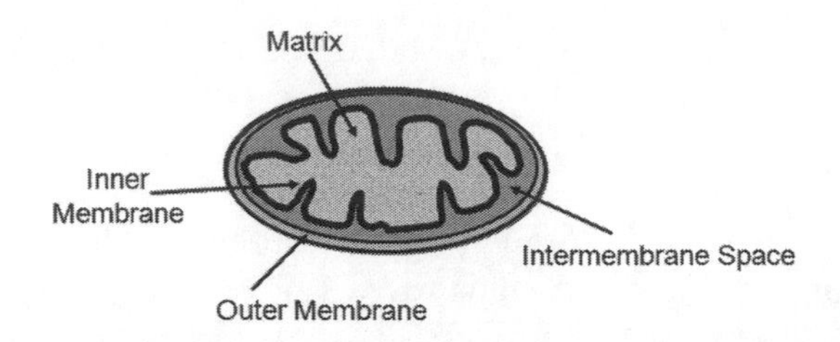

Fig. 1. Simplified schematic of the mitochondrial architecture. The mitochondrion is composed of at least four suborganellar spaces/compartments including the matrix, inner mitochondrial membrane, outer mitochondrial membrane, and intermembrane space. The machinery associated with oxidative phosphorylation is associated with the inner mitochondrial membrane. The intermembrane space is key to the production of ATP by storing electromotive force generated from the electron transport system. The strict compartmentalization of the organelle is important for organeller function and adds an additional level of complexity to the study of the mitochondrial proteome.

of magnitude. Evidence exists demonstrating that mitochondria in different subcellular regions may be unique in regard to their proteome as well as their genome. All this suggests that a rather complicated nomenclature including subcellular localization will probably be required for annotation of a mitochondrial proteome since inadequate cataloguing will leave the proteome ill defined and poorly characterized.

3.4. Proteomic Tools, Techniques, and Processes

In addition to the difficulties in proteome cataloguing, a variety of novel proteomic techniques must be utilized to ensure a truly mitochondrial-specific proteome rather than one deduced from mitochondrial-enriched cellular fractions (Fig. 2).

3.4.1. Mitochondrial Identification

The first hurdle consists of unambiguous identification of the mitochondria. Mitochondrial identification methods must not drastically alter the structural protein relationships in order to have utility in proteomics (see Note 1). Methods of mitochondrial identification include antibody staining of mitochondrial-specific proteins, mitochondrial specific dyes, and electron microscopy.

3.4.2. Antibodies

A variety of apparently specific antibodies exist to individual proteins from the mitochondrion. These include antibodies against the subunits of the electron transport system (Molecular Probes/Invitrogen). These are fairly robust if utilized for proteins transcribed from the mitochondrial genome, although the possibility that such proteins may be present in other undiscovered locations, and hence lead to nonspecific staining, must not be dismissed lightly.

3.4.3. Dyes

A variety of mitochondria specific dyes exist that are generally small molecules with fluorescent properties (MitoTracker probes, Molecular Probes/Invitrogen). Some dyes seem specific to the unique membrane composition of the organelle, while others are

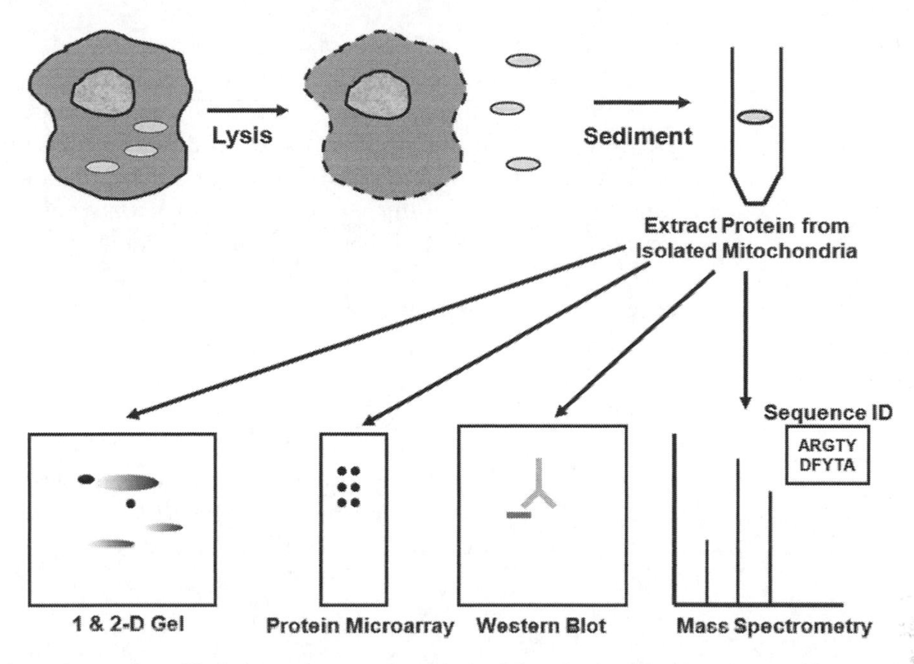

Fig. 2. Schematic overview of isolation and proteomic analysis of the mitochondrion. In order to achieve a mitochondrial specific proteome the cells must first be lysed and the mitochondria extracted. The method shown here employs sedimentation. After mitochondrial purification, various proteomic techniques complimentary to each other may be applied. These include electrophoretic gel methods, protein microarray, western blotting, and mass spectrometry. Mass spectrometry can be coupled to gel electrophoresis methodologies or liquid chromatography nanospray to focus on specific proteins leading to amino acid sequence identification.

made to fluoresce by the intermembrane electromotive potential of the organelle (see Note 2). A few dyes can be used in living cells without apparent interference with vital metabolic processes. These have profound potential provided they do not deleteriously alter the cellular proteome over time.

Techniques based on the organelle architecture such as electron microscopy give unequivocal assessment. Unfortunately, these techniques are very likely to irreversibly alter mitochondrial protein structure to a degree incompatible with further proteomic studies.

3.4.4. Mitochondrial Isolation

The next pragmatic hurdle in the pursuit of the mitochondrial proteome is the isolation of the organelle. Mitochondria isolation has traditionally been performed by ultracentrifugation with a variety of gradients such as glucose and cesium chloride (39) (see Note 3). While subcellular fractionation with such methods greatly enriches for mitochondria there is still a purification dilemma because lysosomes have a very similar size profile to mitochondria. Various schemes to circumvent this include novel subfractionation schemes using metrizamide gradients to remove subcellular contaminant proteins and combining sucrose gradient centrifugation with 2D gel electrophoresis (8, 40–42).

3.4.5. Submitochondrial Compartmentalization

Once the organelle has been isolated, the proteome can begin to be disassembled. A variety of suborganeller regions exist within the mitochondrion, including the matrix, inner mitochondrial membrane, outer mitochondrial membrane, and intermembrane space. The definition of the mitochondrial proteome has profound implications based on excluding or including proteins found on the outer surface of the outer mitochondrial membrane.

3.5. Early Proteomic Ventures into the Mitochondrion

The number of studies of the mitochondrial proteome is still quite low, so a review of a broad set of categories is appropriate and will hopefully inspire the reader to more studies directly targeting human disease. Most of the studies to date are applications of mass spectral techniques and gel electrophoresis (43). The studies can be grouped into three broad categories: (1) shotgun approach to discover and define the proteome, (2) comparisons between normal and tumor proteomes, (3) diseases specifically linked to metabolism.

3.5.1. Shotgun Approach

Studies utilizing the shotgun approach have been targeted to a variety of organisms and in some cases specific organs within particular organisms. These have included a broad range of subjects including plant (44), Arabidopsis thaliana (45), yeast (46), rat liver (47), human (23, 48), and even an assessment of the thermoreceptive pit membrane of the western diamondback rattle snake (49). Some of these studies utilize proteomics to analyze gross organ structure at the mitochondrial level. The particular study on the Western Diamondback rattlesnake revealed heavy neural involvement of the pit organ suggesting a molecular function correlation (49). Roxo-Rosa et al. (50) analyzed proteomic differences between normal human nasal cells and those derived from cystic fibrosis patients homozygous for deletion of phenylalanine residue 508. The majority of the proteins with differing concentrations were ubiquitously expressed proteins. However, lower levels of some mitochondrial proteins were also noted. Basso et al. (51) performed proteomic analysis of human substantia nigra specimens from Parkinson's disease patients. Increased mitochondrial protein expression was observed along with increased expression of ROS scavenging proteins. Studies such as these uncover previously unrecognized mitochondrial-based pathophysiology.

3.5.2. Tumor and Normal Comparisons

Studies comparing normal and tumor proteomes have been performed on tumors such as carcinomas and neuroblastomas (52). Some of these are carried out specifically tracking particular proteins while others make global comparisons looking for any observable differences between normal and cancer. While global analyses are of value in new diagnostic paradigms and finding new pathways, following disease pathogenesis through analysis of specific proteins brings the wisdom of past biochemistry to current problems. One example is the observation of differences in the subunit levels of mitochondrial metabolic machinery with correlation of those differences to measurable metabolic alterations within tumor and normal tissue derived cell lines (17). Such studies begin to give insight into the structure–function relationship of tumor-altered mitochondrial proteomes.

3.5.3. Metabolically Altered Diseases

The third category of mitochondrial proteomic study follows metabolically altered diseases. Studies comparing two groups that had diverged independently of each other include hypoxia/reoxygenation in rat liver mitochondria (6), cardiotoxicity studies (53), and variations in protein ratios (54). In such experiments, a group of cells or organisms are split into two groups, or two separate groups of cells or organisms are procured. One of the groups is exposed to toxic compounds while the other is not. The proteomic profiles obtained from each group are then compared and differences are attributed to the various conditions or disease state differences between the groups.

3.6. The Issue of Chronology

Very few studies to date have effectively utilized proteomics over a time course. Instead most studies appear to be like vacation pictures, which have been haphazardly snapped at particular times and conditions with little analysis given to the events leading up to the snapshot. Considering the early status of the proteomics field, this is inevitable. One very interesting study of plants utilized mitochondrial assessment of pea roots and leaves in an attempt to gain insight into developmental changes by exploiting the chronological issue (55). Most studies to date seem to rest on the implicit assumption that the mitochondrial proteome is static. The mitochondrial proteome is probably quite dynamic, however. Studies of a small number of proteins demonstrated drastically altered concentration ratios over apparent disease progression from benign cancer precursors to full blown carcinoma (17). Consequently, it is imperative that the chronological aspects of proteomic studies not be ignored.

3.7. A Glimpse Forward

The future for mitochondrial proteomics looks very bright. A number of mitochondrial proteomic databases are being set up related to the important task of cataloguing and archiving mitochondrial proteomic data. These include MITOP and MITOP2 in Germany (56, 57) and another in Grenoble, France (58). Additionally, there is a MitoProteome database run by the supercomputing center at The University of California at San Diego (59). New techniques allowing high throughput and analysis of multiple parameters should lead to a much enhanced view of mitochondrial basic science. The difficulties associated with mitochondrial analyses such as large number of mitochondria per cell, the overall organelle-specific variability within and between cells, difficulty in understanding the complexities of mitochondrial reproduction and compartmentalization, should be overcome as better separation tools are designed and analysis and integration concepts progress. These new tools and concepts should prove useful to the general field of proteomics as well as specifically to mitochondrial studies since they open up a path to general organelle studies. From a clinical perspective, these new findings should result in a medical revolution. Techniques of mitochondrial proteomics should put us well along the path to specific protein-based diagnosis, therapy, and monitoring of human disease and treatment.

3.8. Concluding Remarks

As history has illustrated repeatedly, new technology leads to new discovery and additional technological advancement. While the first canned food products were placed in glass bottles and sealed with pitch, within a few short years the process had evolved into a method very similar to the one responsible for a large amount of the food storage of today. Along the way, practitioners found that the preservation of food came not only from the exclusion of oxygen, but also by the sterilization of the container and contents. Concurrent with the technological advances came a scientific understanding of the biological processes involved, opening the way to ever-improved consumer products. By analogy, the new technologies of proteomics when applied to the classical biochemical enigmas of the mitochondrion should also lead to new understanding of disease, and therapeutic benefit to the ultimate medical consumer, the patient.

4. Notes

1. Mitochondria isolation methods use differential centrifugation, density gradient centrifugation, or immunomagnetic bead separation. Density gradient separation is the most labor-intensive process, but results in a more purified mitochondria fraction compared to differential centrifugation. Each of these processes requires permeabilization of cell membranes in order to liberate the mitochondria.
2. While mitochondrial dyes appear to be specific, the precise mechanism of their action remains somewhat elusive; hence care must be taken regarding the assumption that the dyes are specific for mitochondrial staining.
3. Differential centrifugation mitochondria isolation kits are available from Pierce (catalog # 89801 (tissue) or 89874 (cells) (60)), or Qiagen (Qproteome kit catalog #37612). Immunomagnetic separation kits are available from Miltenyi Biotec (catalog # 130-094-532).

References

1. el Bahri, L., and Ben Romdane, S. (1991) Arsenic poisoning in livestock. *Vet Hum Toxicol* **33**, 259–64.
2. Pearce, L. L., Bominaar, E. L., Hill, B. C., and Peterson, J. (2003) Reversal of cyanide inhibition of cytochrome c oxidase by the auxiliary substrate nitric oxide: an endogenous antidote to cyanide poisoning? *J Biol Chem* **278**, 52139–45.
3. Kernec, F., Unlu, M., Labeikovsky, W., Minden, J. S., and Koretsky, A. P. (2001) Changes in the mitochondrial proteome from mouse hearts deficient in creatine kinase. *Physiol Genomics* **6**, 117–28.
4. Yang, J., Marden, J. J., Fan, C., Sanlioglu, S., Weiss, R. M., *et al.* (2003) Genetic redox preconditioning differentially modulates AP-1 and NF kappa B responses following cardiac ischemia/reperfusion injury and protects against necrosis and apoptosis. *Mol Ther* **7**, 341–53.
5. Venkatraman, A., Landar, A., Davis, A. J., Ulasova, E., Page, G., *et al.* (2004) Oxidative modification of hepatic mitochondria protein thiols: effect of chronic alcohol consumption. *Am J Physiol Gastrointest Liver Physiol* **286**, G521-7.
6. Reinheckel, T., Korn, S., Mohring, S., Augustin, W., Halangk, W., *et al.* (2000) Adaptation of protein carbonyl detection to the requirements of proteome analysis demonstrated for hypoxia/reoxygenation in isolated rat liver mitochondria. *Arch Biochem Biophys* **376**, 59–65.
7. Ruepp, S. U., Tonge, R. P., Shaw, J., Wallis, N., and Pognan, F. (2002) Genomics and proteomics analysis of acetaminophen toxicity in mouse liver. *Toxicol Sci* **65**, 135–50.
8. McDonald, T. G., and Van Eyk, J. E. (2003) Mitochondrial proteomics. Undercover in the lipid bilayer. *Basic Res Cardiol* **98**, 219–27.
9. Thorburn, D. R. (2004) Diverse powerhouses. *Nat Genet* **36**, 13–4.
10. Blass, J., and McDowell, F. H. (1999) Oxidative/energy metabolism in neurodegenerative disorders, *in* "Annals of the New York Academy of Sciences", New York Academy of Sciences, New York, NY,
11. Stocker, J., and Dehner, L. (2002) Pediatric Pathology 2nd edition, Lippincott William and Wilkins.
12. Hinerfeld, D., Traini, M. D., Weinberger, R. P., Cochran, B., Doctrow, S. R., *et al.* (2004) Endogenous mitochondrial oxidative stress: neurodegeneration, proteomic analysis, specific respiratory chain defects, and efficacious antioxidant therapy in superoxide dismutase 2 null mice. *J Neurochem* **88**, 657–67.
13. Warburg, O., Posener, K., and Negelein, E. (1924) Uber den Stoffwechsel der Carcinomzelle. *Biochem Zeitschrift* **152**, 309–44.
14. Chen, Y., Cairns, R., Papandreou, I., Koong, A., and Denko, N. C. (2009) Oxygen consumption can regulate the growth of tumors, a new perspective on the Warburg effect. *PLoS One* **4**, e7033.
15. Mathupala, S. P., Rempel, A., and Pedersen, P. L. (1997) Aberrant glycolytic metabolism of cancer cells: a remarkable coordination of genetic, transcriptional, post-translational, and mutational events that lead to a critical role for type II hexokinase. *J Bioenerg Biomembr* **29**, 339–43.
16. Maitra, A., Cohen, Y., Gillespie, S. E., Mambo, E., Fukushima, N., *et al.* (2004) The Human MitoChip: a high-throughput sequencing microarray for mitochondrial mutation detection. *Genome Res* **14**, 812–9.
17. Herrmann, P. C., Gillespie, J. W., Charboneau, L., Bichsel, V. E., Paweletz, C. P., *et al.* (2003) Mitochondrial proteome: altered cytochrome c oxidase subunit levels in prostate cancer. *Proteomics* **3**, 1801–10.
18. Lopez, M. F., Kristal, B. S., Chernokalskaya, E., Lazarev, A., Shestopalov, A. I., *et al.* (2000) High-throughput profiling of the mitochondrial proteome using affinity fractionation and automation. *Electrophoresis* **21**, 3427–40.
19. Kurtoglu, M., Philips, K., Liu, H., Boise, L. H., and Lampidis, T. J. (2009) High endoplasmic reticulum activity renders multiple myeloma cells hypersensitive to mitochondrial inhibitors. *Cancer Chemother Pharmacol* 66, 129–40.
20. Scheffler, I. E. (1999) Mitochondria, Wiley-Liss, New York, NY.
21. Cui, H., Cruz-Correa, M., Giardiello, F. M., Hutcheon, D. F., Kafonek, D. R., *et al.* (2003) Loss of IGF2 imprinting: a potential marker of colorectal cancer risk. *Science* **299**, 1753–5.
22. Kopsidas, G., Kovalenko, S. A., Heffernan, D. R., Yarovaya, N., Kramarova, L., *et al.* (2000) Tissue mitochondrial DNA changes A stochastic system. *Ann N Y Acad Sci* **908**, 226–43.
23. Taylor, S. W., Fahy, E., Zhang, B., Glenn, G. M., Warnock, D. E., *et al.* (2003) Characterization of the human heart mitochondrial proteome. *Nat Biotechnol* **21**, 281–6.
24. Richly, E., Chinnery, P. F., and Leister, D. (2003) Evolutionary diversification of mitochondrial proteomes: implications for human disease. *Trends Genet* **19**, 356–62.
25. Kumar, A., Agarwal, S., Heyman, J. A., Matson, S., Heidtman, M., *et al.* (2002) Subcellular

localization of the yeast proteome. *Genes Dev* **16**, 707–19.

26. Steinmetz, L. M., Scharfe, C., Deutschbauer, A. M., Mokranjac, D., Herman, Z. S., *et al.* (2002) Systematic screen for human disease genes in yeast. *Nat Genet* **31**, 400–4.
27. Tsukihara, T., Aoyama, H., Yamashita, E., Tomizaki, T., Yamaguchi, H., *et al.* (1996) The whole structure of the 13-subunit oxidized cytochrome c oxidase at 2.8 A. *Science* **272**, 1136–44.
28. Hofmann, S., Lichtner, P., Schuffenhauer, S., Gerbitz, K. D., and Meitinger, T. (1998) Assignment of the human genes coding for cytochrome c oxidase subunits Va (COX5A), VIc (COX6C) and VIIc (COX7C) to chromosome bands 15q25, 8q22→q23 and 5q14 and of three pseudogenes (COX5AP1, COX6CP1, COX7CP1) to 14q22, 16p12 and 13q14→q21 by FISH and radiation hybrid mapping. *Cytogenet Cell Genet* **83**, 226–7.
29. Hey, Y., Hoggard, N., Burt, E., James, L. A., and Varley, J. M. (1997) Assignment of COX6A1 to 6p21 and a pseudogene (COX6A1P) to 1p31.1 by in situ hybridization and somatic cell hybrids. *Cytogenet Cell Genet* **77**, 167–8.
30. Ozawa, T., Sako, Y., Sato, M., Kitamura, T., and Umezawa, Y. (2003) A genetic approach to identifying mitochondrial proteins. *Nat Biotechnol* **21**, 287–93.
31. Tryoen-Toth, P., Richert, S., Sohm, B., Mine, M., Marsac, C., *et al.* (2003) Proteomic consequences of a human mitochondrial tRNA mutation beyond the frame of mitochondrial translation. *J Biol Chem* **278**, 24314–23.
32. Florentz, C. (2002) Molecular investigations on tRNAs involved in human mitochondrial disorders. *Biosci Rep* **22**, 81–98.
33. Rabilloud, T., Strub, J. M., Carte, N., Luche, S., Van Dorsselaer, A., *et al.* (2002) Comparative proteomics as a new tool for exploring human mitochondrial tRNA disorders. *Biochemistry* **41**, 144–50.
34. Fraser, F., and Zammit, V. A. (1999) Submitochondrial and subcellular distributions of the carnitine-acylcarnitine carrier. *FEBS Lett* **445**, 41–4.
35. Kato, A., Sugiura, N., Saruta, Y., Hosoiri, T., Yasue, H., *et al.* (1997) Targeting of endopeptidase 24.16 to different subcellular compartments by alternative promoter usage. *J Biol Chem* **272**, 15313–22.
36. Nimmo, H. G. (1979) Evidence for the existence of isoenzymes of glycerol phosphate acyltransferase. *Biochem J* **177**, 283–8.
37. Bionda, C., Portoukalian, J., Schmitt, D., Rodriguez-Lafrasse, C., and Ardail, D. (2004) Subcellular compartmentalization of ceramide metabolism: MAM and/or mitochondria? *Biochem J* **382**, 527–33.
38. Lewin, T. M., Kim, J. H., Granger, D. A., Vance, J. E., and Coleman, R. A. (2001) Acyl-CoA synthetase isoforms 1, 4, and 5 are present in different subcellular membranes in rat liver and can be inhibited independently. *J Biol Chem* **276**, 24674–9.
39. Welter, C., Meese, E., and Blin, N. (1988) Rapid step-gradient purification of mitochondrial DNA. *Mol Biol Rep* **13**, 117–20.
40. Taylor, S. W., Warnock, D. E., Glenn, G. M., Zhang, B., Fahy, E., *et al.* (2002) An alternative strategy to determine the mitochondrial proteome using sucrose gradient fractionation and 1D PAGE on highly purified human heart mitochondria. *J Proteome Res* **1**, 451–8.
41. Hanson, B. J., Schulenberg, B., Patton, W. F., and Capaldi, R. A. (2001) A novel subfractionation approach for mitochondrial proteins: a three-dimensional mitochondrial proteome map. *Electrophoresis* **22**, 950–9.
42. Rabilloud, T., Kieffer, S., Procaccio, V., Louwagie, M., Courchesne, P. L., *et al.* (1998) Two-dimensional electrophoresis of human placental mitochondria and protein identification by mass spectrometry: toward a human mitochondrial proteome. *Electrophoresis* **19**, 1006–14.
43. Pflieger, D., Le Caer, J. P., Lemaire, C., Bernard, B. A., Dujardin, G., *et al.* (2002) Systematic identification of mitochondrial proteins by LC-MS/MS. *Anal Chem* **74**, 2400–6.
44. Bykova, N. V., Egsgaard, H., and Moller, I. M. (2003) Identification of 14 new phosphoproteins involved in important plant mitochondrial processes. *FEBS Lett* **540**, 141–6.
45. Kruft, V., Eubel, H., Jansch, L., Werhahn, W., and Braun, H. P. (2001) Proteomic approach to identify novel mitochondrial proteins in Arabidopsis. *Plant Physiol* **127**, 1694–710.
46. Marc, P., Margeot, A., Devaux, F., Blugeon, C., Corral-Debrinski, M., *et al.* (2002) Genome-wide analysis of mRNAs targeted to yeast mitochondria. *EMBO Rep* **3**, 159–64.
47. Fountoulakis, M., Berndt, P., Langen, H., and Suter, L. (2002) The rat liver mitochondrial proteins. *Electrophoresis* **23**, 311–28.
48. Langen, H., Berndt, P., Roder, D., Cairns, N., Lubec, G., *et al.* (1999) Two-dimensional map of human brain proteins. *Electrophoresis* **20**, 907–16.

49. Zischka, H., Keller, H., Kellermann, J., Eckerskorn, C., and Schuster, S. C. (2003) Proteome analysis of the thermoreceptive pit membrane of the western diamondback rattlesnake Crotalus atrox. *Proteomics* **3**, 78–86.
50. Roxo-Rosa, M., da Costa, G., Luider, T. M., Scholte, B. J., Coelho, A. V., *et al.* (2006) Proteomic analysis of nasal cells from cystic fibrosis patients and non-cystic fibrosis control individuals: search for novel biomarkers of cystic fibrosis lung disease. *Proteomics* **6**, 2314–25.
51. Basso, M., Giraudo, S., Corpillo, D., Bergamasco, B., Lopiano, L., *et al.* (2004) Proteome analysis of human substantia nigra in Parkinson's disease. *Proteomics* **4**, 3943–52.
52. Fountoulakis, M., and Schlaeger, E. J. (2003) The mitochondrial proteins of the neuroblastoma cell line IMR-32. *Electrophoresis* **24**, 260–75.
53. Kang, Y. J. (2003) New understanding in cardiotoxicity. *Curr Opin Drug Discov Devel* **6**, 110–6.
54. Murray, J., Gilkerson, R., and Capaldi, R. A. (2002) Quantitative proteomics: the copy number of pyruvate dehydrogenase is more than 10(2)-fold lower than that of complex III in human mitochondria. *FEBS Lett* **529**, 173–8.
55. Bardel, J., Louwagie, M., Jaquinod, M., Jourdain, A., Luche, S., *et al.* (2002) A survey of the plant mitochondrial proteome in relation to development. *Proteomics* **2**, 880–98.
56. Scharfe, C., Zaccaria, P., Hoertnagel, K., Jaksch, M., Klopstock, T., *et al.* (2000) MITOP, the mitochondrial proteome database: 2000 update. *Nucleic Acids Res* **28**, 155–8.
57. Andreoli, C., Prokisch, H., Hortnagel, K., Mueller, J. C., Munsterkotter, M., *et al.* (2004) MitoP2, an integrated database on mitochondrial proteins in yeast and man. *Nucleic Acids Res* **32 Database issue,** D459-62.
58. Lescuyer, P., Strub, J. M., Luche, S., Diemer, H., Martinez, P., *et al.* (2003) Progress in the definition of a reference human mitochondrial proteome. *Proteomics* **3**, 157–67.
59. Cotter, D., Guda, P., Fahy, E., and Subramaniam, S. (2004) MitoProteome: mitochondrial protein sequence database and annotation system. *Nucleic Acids Res* **32 Database issue**, D463-7.
60. Zhang, Q., Raoof, M., Chen, Y., Sumi, Y., Sursal, T., Junger, W. et al. (2010) Circulating mitochondrial DAMPs cause inflammatory responses to injury. *Nature* 464, 104–7.

Chapter 18

Adult Neural Stem Cells: Isolation and Propagation

Jamin M. Letcher and Daniel N. Cox

Abstract

Individualized therapy using adult stem cells constitutes a revolutionary vision for molecular medicine of the future. The field of stem cell biology has accelerated dramatically such that it now appears feasible to treat an individual patient's disease with native or modified stem cells collected from the same patient. Neurodegenerative disease is a high-priority goal for stem cell therapy due to the tremendous clinical urgency to reduce the worldwide suffering associated with this class of diseases. This chapter focuses on adult neural stem cells as a prototype for the general field of adult stem cell therapy. Studies of the origin and function of neural stem cells reveals that the adult brain can generate new neurons. This finding provides the rationale for the therapeutic application of adult neural stem cells to treat neuronal damage or loss. Experimental progress in treating Parkinson's disease is discussed in some detail as an example of one of the most promising areas for adult neural stem cell therapy. Methods for neural stem cell isolation and propagation are included.

Key words: Brain, Stem cell, Neurodegenerative disease, Neuron, Parkinson's disease, Embryonic stem cells

1. Introduction

1.1. Neural Stem Cells: Origins and Functions

Neural stem cells comprise the subset of multipotent cells which gives rise to neurons and glial cells (1). Through asymmetric divisions coordinated by their microenvironment niches, these cells give rise to the most complex of all structures – the human brain. Neural stem cells are first identifiable in the developing embryo and are thought to function in differing capacities throughout the life of the individual (2). All areas of neurogenesis and neural repair are of interest to researchers, but an area of particular medical importance is how these phenomena function in the adult. Many neurodegenerative conditions, such as Parkinson's disease, amyotrophic lateral sclerosis (ALS), stroke, or spinal cord injury, are

Virginia Espina and Lance A. Liotta (eds.), *Molecular Profiling: Methods and Protocols*, Methods in Molecular Biology, vol. 823, DOI 10.1007/978-1-60327-216-2_18, © Springer Science+Business Media, LLC 2012

especially devastating because we currently have no way of replacing the brain or cord cells they destroy. Recent research, though, suggests that stem cell therapy might contribute significantly to future treatment of these disorders (3, 4).

Experiments using embryonic stem cells have provided the scientific community with valuable knowledge concerning the body's innate capacity to rejuvenate (5–7). However, there are several drawbacks to the use of human embryonic stem cells for therapeutic purposes. Many individuals have ethical qualms about the use of cells derived from human embryos. For this reason, the US government has imposed serious limitations on federal financial aid to such research using embryonic stem cells. Furthermore, implantation of embryonic stem cells has frequently led to the formation of teratomas, cancers containing cells of all three germ layers. Because of the concerns about embryonic stem cells, many investigators are evaluating adult neural stem cells. Because we know that adult brains are naturally capable of cellular regeneration, scientists pursuing cures to neurodegenerative diseases might be able to sidestep many of ethical, legal, and practical quandaries that accompany the use of embryonic or fetal stem cells (8).

During neurogenesis, the multipotent stem cells of the brain give rise to lineage-restricted cells which themselves are progenitors. They in turn give rise to neuronal and neuroglial cells (1). It has long been known that stem cells are important in brain formation and development. However, it was once thought that neurogenesis in the human reaches a finite endpoint, preventing adult neurons from regenerating. Early clues regarding axonal regenerative capacity were revealed by David and Aguayo in 1981 in which they showed that transplanted peripheral nerves could act as bridges between injured spinal cord nerves and the medulla (9, 10). Discoveries of adult neural stem cells in other organisms spurred stem cell therapy research (11). Further advances in our understanding of cell fate are being gleaned from cancer stem cell (CSC) research. Proponents of the cancer stem cell theory propose that a CSC may be capable of generating new cancer cells (12–14). Glioma stem cell differentiation was thought to be restricted to the neural lineage. However, using orthotopic versus heterotopic xenograft models and in vitro differentiation assays, Ricci-Vitiani et al. found that a subset of glioblastomas contained CSCs with both neural and mesenchymal differentiation potential, thus challenging the restricted lineage paradigm of neurogenesis (15). Recently, two separate groups have identified glioblastoma stem-like cells that give rise to endothelial cells (16, 17), which potentially provide increased vasculature for the growing tumor. Thus, central nervous system-derived stem cells, whether they are embryonic, adult, or cancer cells, are providing cellular resources for deciphering cell fate and rejuvenation.

1.2. Adult Neural Stem Cells: Adult Brains Can Generate New Neurons

The first indication that neural stem cells might persist in adult mammals came in the 1960s with a series of investigations conducted by Joseph Altman and others. In the first of these experiments, Altman and Das injected ^{3}H-labeled thymidine into adult rats (18). The tritiated nucleoside is incorporated into the daughter strands of DNA during the S phase of cell division, and can thus be used to identify newly generated nuclei. Autoradiography revealed cell division occurring in the rat hippocampus. However, the interpretation that this suggested the development of neurons was challenged by some biologists on the basis that the labeled cells were indistinguishable from glial cells using light microscopy (19, 20).

A later study by Kaplan and Hinds sought to establish with certainty the fates of cells that resulted from divisions in the post-adolescent brain. Adult rats were injected with ^{3}H-thymidine and allowed to live for 30 days before being sacrificed. Various regions of the brain were then sectioned at 1-μm thickness and mounted to slides. If a labeled cell displayed the morphological characteristics of a neuron, specifically dentritic processes, investigators further cut and mounted a thin section of this slide. The new slides were then observed using electron microscopy, and they revealed labeled newborn cells in the hippocampus, the olfactory bulb, and the neocortex. The greater resolution of the electron microscope allowed for a more accurate identification of morphological features; these cells were seen to possess uniquely neuronal characteristics, including dendrites and Gray's type 2 synapses (19).

These studies demonstrated that some mammalian brains are capable of generating new neurons even in adulthood. However, initial experiments using animals more closely related to humans, such as monkeys, yielded discouraging results. Pasko Rakic was the first to investigate primates by Altman's method of radiolabeling DNA (21). In his study, postadolescent rhesus monkeys were injected with tritiated thymidine and sacrificed at varying intervals after treatment. Some animals were killed only after 3 days to prevent the possibility that neurons might divide numerous times and essentially wash out the ^{3}H-thymidine concentration in their DNA. Other monkeys, though, were permitted to live an additional 6 years after injection; the stated purpose of this long interval was to give maximal time for the appearance of any slow-developing neuronal characteristics. Rakic analyzed all major structures of the brain by dark field illumination and Nomarski differential interference contrast illumination, yet he found no indication of newly formed cells with neuronal traits (21). Additionally, while oligodendrocytes were observed in younger animals, they were largely absent from older individuals (21). These findings had the effect of greatly discouraging further investigation into adult neurogenesis for the next several years (20).

During this time, researchers continued to investigate a broad range of stem cells, but the idea that neurogenesis could occur in normal human adults was virtually without support until the late 1990s. Adult primate neurogenesis was first observed in vivo by Gould et al. in 1998 (22). Adult marmoset monkeys were administered bromodeoxyuridine (BrdU) intravenously. BrdU is a thymidine analog and so is incorporated into DNA during replication. The animals were killed at either 2 h or 3 weeks, and their brains were sectioned and analyzed immunohistochemically. Researchers screened for both BrdU and neuron-specific enolase (NSE), a glycolytic enzyme not expressed in non-neuronal cells in primates. Many labeled cells were observed in the dentate gyri of monkeys killed at 2 h, and more than 80% of these possessed the morphological traits of granule cell precursors. Likewise, many BrdU-labeled nuclei were found in monkeys killed at 3 weeks. More than 80% of these had morphological characteristics of terminally differentiated granule neurons. This experiment presented the first direct evidence of in vivo adult neurogenesis in primates (22).

In the same study, Gould's team investigated whether primate neurogenesis shares a particular trait with neurogenesis in other mammals, namely, whether stress diminishes the proliferation of cells in the dentate gyrus (22). By introducing male monkeys into foreign cages belonging to other, dominant male monkeys, researchers created an environment where the so-called intruder monkey would fight the resident monkey, lose, and spend the next hour attempting not to provoke the resident monkey's aggression. This experience increased measurable indicators of stress, as was confirmed by elevated mean arterial blood pressure. The stressed monkeys were then injected with BrdU and sacrificed at 2 h. Investigators found far fewer BrdU-labeled nuclei in these monkeys than in those which had experienced no recent stress, indicating a dramatic reduction in the brain's ability to renew neurons under certain conditions (22).

The discovery of postadolescent neurogenesis in various mammals, including primates, offered strong evidence that similar phenomena probably occur in humans as well. Although the existence of neural stem cells in adult humans was long suspected, the first direct evidence came from a study performed later the same year, when BrdU was again used to identify new neuronal growth (23). In this study, five cancer patients and one nondiseased control were treated with a BrdU-saline solution, and their brains were examined upon death. All the subjects in this study suffered from squamous cell carcinomas of the tongue; therefore, the investigators examined the hippocampal samples to ensure that any observed cell division was normal and not a result of the cancer. No evidence of metastasis was found in any of the patients. Brain tissue samples were sectioned and fixed and compared to specimens from BrdU-injected adult rats. Immunohistochemical staining was used to quantify BrdU incorporation into new nuclei and to identify specific localization of BrdU positive cells. Every patient who had received BrdU exhibited BrdU-positive cells in the granule cell layer and subgranular zone of the dentate gyrus. Furthermore, researchers triple-labeled tissue sections for BrdU and cell-specific markers (e.g., astroglial and neuronal markers). A cell possessing BrdU in its nucleus and also expressing a marker unique to a differentiated neuron or neuroglia, offers strong evidence of proliferating progenitor cells in the adult brain. This experiment revealed newly generated cells in the dentate gyrus expressing NeuN (i.e., granule cell neurons) and GFAP (i.e., glial cells) and provided compelling evidence that the neurogenesis previously observed in animals has a corollary in humans (23). The finding of BrdU-positive cells in relatively high concentration in the dentate gyrus, a region known to be neurogenic in many other animals, strongly suggests that neural stem cells are found in this region of human brains. The observation that these new cells display phenotypes of differentiated neurons and neuroglia has produced a broad consensus in support of this view (23).

1.3. Adult Neural Stem Cells: Implications for Therapy

Studies such as the stress experiments performed by Gould and her team provide ample data to draw inferences from animal models regarding human neurogenesis. In the last decade, evidence has been gleaned from models that neurogenesis occurs predominantly in two regions of the adult human brain: the subgranular zone (SGZ) of the hippocampal dentate gyrus (24) and the subventricular zone (SVZ)/olfactory bulb (25, 26). Despite having relatively few and small neurogenic sites, the human central nervous system is now believed to possess neural precursor cells throughout the brain and spinal cord. Corresponding cells in rodents and higher mammals can be translocated to neurogenic areas and incited to differentiate (8, 11). Additionally, direct studies of humans, though less frequent, have provided even more certain data, eliminating the possible disconnect between model and human physiologies (8).

If certain areas of the human brain are neurogenic, it is at least theoretically possible for us to artificially stimulate the genesis of new neurons. Our knowledge that the adult human brain is constitutively neurogenic has bolstered efforts to combat neurological pathologies through manipulation of the brain's own machinery (9, 10). If neural tissue is able to incorporate newly formed or newly introduced cells, we may be able to minimize or even eradicate the deleterious effects of maladies such as Parkinson's disease (27–31), stroke (32, 33), ALS (34, 35), and spinal cord injury (36–40). For example, neural cell therapy may be used in several ways to promote axonal regeneration after spinal cord injury: (a) as a chemoattractant at the site of injury allowing trophic factors to accumulate thus enhancing regenerative capacity, (b) as a scaffold for regeneration of axonal tissue, and (c) to replace damaged/dead cells (38). However, therapies relying on embryonic and fetal stem cells have been hindered by the associated ethical, political, and availability concerns. Although many of these studies were not concerned with adult stem cells, they have laid groundwork for the effective pursuit of regenerative therapy, in addition to their own direct contributions to medicine.

1.4. Parkinson's Disease: An Urgent Target for Neural Stem Cell Therapy

One of the pathologies in which stem cell therapy is expected to have a high potential for success is Parkinson's disease. This neurodegenerative disorder is characterized by trembling, loss of balance, and slowness of coordinated movements. It occurs in the substantia nigra of the midbrain, caused by a progressive loss of dopaminergic neurons (i.e., the neurons which release the neurotransmitter dopamine into the presynaptic cleft). The disease is currently irreversible, leading to complete paralysis and ultimately to death. At present, the primary method of treatment is chemical replacement in the form of synthetic dopamine. However, this approach merely addresses a symptom of the disease rather than providing a cure for the loss of neurons; patients receiving this treatment eventually develop dyskinesia (41). Because Parkinson's affects millions of patients in the USA alone, and because it primarily affects only one type of neuron, it has been the subject of interest for many in the stem cell community since at least the late 1980s (42). Beginning in 1992 several papers were published reporting the transplantation of human fetal neurones into patients with Parkinson's disease. These initially elicited hope, because the treatments were reported to improve the patients' motor abilities. However, subsequent studies found previously unnoticed effects, such as persistent dyskinesia. Furthermore, because these approaches required neural material from up to four fetuses per patient, they seem especially untenable as broadly applied therapeutic practices (43).

The possibility of embryonic stem cell therapy for Parkinson's has also been investigated. However, as with fetal stem cell therapy, there are ethical and (in this and several other nations) possible legal concerns over this type of procedure. Additionally, embryonic stem cell transplantation frequently leads to the formation of teratomas (43). For these and other reasons, many researchers have focused on utilizing adult stem cells in the treatment of Parkinson's. An important pair of studies details the stimulation of endogenous stem cells in the brains of rats as a Parkinson's disease model (41, 44). These studies are of particular interest because researchers took into consideration the role of the stem cells' microenvironment, which provides chemo-attractant, adhesion factors, and other cell signaling molecules that dictate cell behavior within the microenvironment niche. Microenvironmental signals are essential to the proper maintenance and function of stem cell populations, and so it is important that they be considered when attempting to modify the process of neurogenesis in the brain.

In the first experiment, investigators sought to determine whether administration of a particular neuronal substrate would stimulate the genesis of new neurons (44). They introduced a dopamine D_3 receptor agonist, 7-hydroxy-*N*,*N*-di-*n*-propyl-2-aminotetralin (7-OH-DPAT), into the brains of rats for a period of 2 weeks. Neurotransmitters, such as dopamine, are known to contribute to prenatal neural development, and D_3 receptors are expressed only in neurogenic regions of the adult brain, therefore this experiment was designed to test the role of D_3 stimulation in adults. At the end of treatment, the rats' brains were sectioned and new nuclei were distinguished by the presence of BrdU. Samples were double labeled and analyzed using immunohistochemical techniques. Data showed that rats which had received 7-OH-DPAT displayed significantly more neurogenesis in the SVZ and neostriatum than their negative control counterparts (44).

The second of these experiments utilized this knowledge about dopamine/D_3 receptor to study its neurogenic and behavioral effects on rats with simulated Parkinson's disease (41). As in the previous experiment, 7-OH-DPAT was administered over the course of several weeks to rats in which a Parkinson-like neural state had been created. Neurogenesis was again indicated by integration of BrdU into neural nuclei and an increase in the production of new cells, roughly half of which bore morphological characteristics of neurons, while the other half resembled neuroglia. Further, animals improved in locomotor skills after 4 weeks of 7-OH-DPAT treatment. The same rats improved even more when treated for 8 weeks, while no improvement was seen in those which had received treatment for only 2 weeks, nor in those which had received only saline (41). Together these data provide support for the hope that an individual patient's own stem cells can be used in the treatment of such diseases as Parkinson's disease.

The field of stem cell biology has revolutionized our ability to study tissue progenitor cells, stem cell differentiation mechanisms, and the role of stem cells in the repair of injured or diseased tissue. Neural stem cells are an important research tool that constitutes a bridge between basic research and translational neuroscience, we describe a general procedure for the isolation and propagation of rodent neural stem cells based on the neurosphere assay (45, 46).

2. Materials

2.1. Brain Dissection

1. CO_2 asphyxiation chamber for rodents.
2. Dissecting instruments: Scapel blades (No. 10 curved) and scissors, cleaned with Cidex antiseptic (*ortho*-Phthalaldehyde solution).
3. 95% Ethanol.

4. Sterile water.
5. Dissecting microscope.
6. Betadine antiseptic solution (10% povidone-iodine).
7. Balanced salt solution (TV = 250 mL): 15.5 mL of 2.0 M sodium chloride, 1.25 mL of 1.0 M potassium chloride, 0.80 mL of 1.0 M magnesium chloride, 41.9 mL of 155 mM sodium bicarbonate, 2.5 mL of 1.0 M glucose, 0.23 mL of 108 mM calcium chloride, 188 mL of Type I reagent grade water. Filter using a 0.22-μm filter. Store at 4°C. Stable for 6 months. Discard if solution becomes cloudy (see Note 1).

2.2. Tissue Dissociation

1. Phosphate-buffered saline (PBS) without calcium, magnesium, or phenol red containing 0.1% bovine serum albumin.
2. 15% Sodium bicarbonate (w/v) in Type 1 reagent grade water (dH_2O). Add 15.0 g $NaHCO_3$ to 100 mL dH_2O. Filter with a 0.22-μm pore size filter. Store at room temperature. Stable for 2 weeks.
3. 30% w/v glucose in Type 1 reagent grade water (dH_2O). Store at 4°C. Stable for 1 week.
4. 1.0 M Hepes Buffer. Store at 4°C.
5. Type 1 reagent grade water (dH_2O), sterilized with a 0.22-μm pore size filter.
6. Putrescine (1,4-Diaminobutane dihydrochloride) (100× stock): dissolve 0.096 g Putrescine in 100 mL sterile, filtered Type 1 water (dH_2O). Store at 4°C. Stable for 1 week.
7. Progesterone (1,000× stock): dissolve 0.0063 g in 100 mL sterile dH_2O. Store at 4°C. Stable for 1 week.
8. B-27® serum-free supplement (Invitrogen) (50× stock). Store at –20°C.
9. Insulin–transferrin–sodium selenite (ITSS; Roche): dissolve in 5.0 mL sterile dH_2O (1,000× stock). Store at –20°C.
10. Heparin – dissolve 0.05 g in 2.0 mL sterile, filtered Type 1 water. Store at 4°C. Stable for 1 week.
11. Epidermal growth factor (EGF): dissolve 0.1 mg in 1.0 mL of PBS containing 0.1% bovine serum albumin. Store at –80°C. Stable for 6 months.
12. Fibroblast growth factor-basic (FGF2): dissolve 25 μg in 1.0 mL of 5 mM Tris-HCl (pH 7.6). Store at –80°C. Stable for 6 months.
13. DMEM:F12 medium (500 mL) plus 8.0 mL 15% sodium bicarbonate (w/v). Store at 4°C.
14. Serum-Free Culture Medium (supplemented with nutrients) (TV = 100 mL) (47): 94.3 mL DMEM:F12/15% $NaHCO_3$

medium, 2.0 mL 30% glucose, 0.5 mL 1 M Hepes, 1.0 mL Putrescine (100× stock), 0.1 mL Progesterone (1,000× stock), 2.0 mL B-27® supplement (50× stock), 100 μL ITSS, 7.32 μL Heparin, 20 μL EGF, and 20 μL FGF2 (see Note 2).

15. 0.25% Trypsin diluted in balanced salt solution: 0.04 g Trypsin and 0.02 g Type 1-S Hyaluronidase in 30 mL balanced salt solution. Filter using a 0.22-μm filter. Store in aliquots at –20°C.
16. DMEM:F12 medium. Store at 4°C.
17. Trypsin inhibitor (Egg white trypsin inhibitor, Roche): add 4.0 mg/4.0 mL in DMEM:F12 medium. Prepare fresh daily.

2.3. Neural Stem Cell Culture

1. Trypan blue dye (0.4% w/v).
2. PBS without calcium, magnesium, or phenol red.
3. Hemocytometer with calibrated glass coverslip.
4. Serum-Free Culture Medium (supplemented with nutrients) (47): 94.3 mL DMEM:F12/15% $NaHCO_3$ medium, 2.0 mL 30% glucose, 0.5 mL 1 M Hepes, 1.0 mL Putrescine (100× stock), 0.1 mL Progesterone (1,000× stock), 2.0 mL B-27® supplement (50× stock), 100 μL ITSS, 7.32 μL Heparin, 20 μL EGF, and 20 μL FGF2.
5. 24-Well microplate.

2.4. Neurosphere Propagation

1. 24-Well microplate.
2. TrypLE™ Express Stable Trypsin Replacement Enzyme without Phenol Red (Invitrogen).
3. PBS without calcium, magnesium, or phenol red.
4. 1 mM EDTA: Add 4.162 g EDTA to 100 mL dH_2O.
5. Serum-Free Culture Media (supplemented with nutrients) (47): 94.3 mL DMEM:F12/15% $NaHCO_3$ medium, 2.0 mL 30% glucose, 0.5 mL 1 M Hepes, 1.0 mL Putrescine (100× stock), 0.1 mL Progesterone (1,000× stock), 2.0 mL B-27® supplement (50× stock), 100 μL ITSS, 7.32 μL Heparin, 20 μL EGF, and 20 μL FGF2.

3. Methods

The neurosphere assay uses a serum-free culture system for isolation and propagation of CNS dervied stem cells, while preventing the differentiation of the cells. Neural tissue is dissociated and plated at low density in growth medium containing EGF and basic FGF2. Neurospheres develop as distinct cell clusters which float in the liquid medium. Propagation of the neural stem cells is achieved by dissociating the neurospheres with trypsin, and replating in fresh medium. The neuroshpheres can also be differentiated into specific neural cell types by withdrawal of the growth factors and/or addition of serum to the medium (1, 45–47) (see Note 3).

3.1. Harvesting and Dissection of the Brain

Tissue harvesting using sterile technique is critical for limiting bacterial/fungal contamination of the neural cell culture. Antibiotics are not used in the culture medium.

1. Sterilize the scissors and scalpels to be used for the dissection.
2. Prewarm the balanced salt solution and serum-free medium to 37°C.
3. Clean the work area with 95% ethanol.
4. Euthanize the animal using CO_2 asphyxiation.
5. Clean the fur around the skull with sterile gauze soaked in Betadine antiseptic solution (10% povidone-iodine).
6. Expose the skull using a midline incision. Using scissors, cut through the hard palate and skull. Remove the brain using sterile technique (see Note 4). Cut the spinal cord and blood vessels at the base of the brain.
7. Place the brain tissue into a sterile dish. Add an adequate volume of the warm balanced salt solution to barely cover the tissue.
8. Remove the meninges from the surface of the brain. Rinse the brain tissue gently with fresh, warm balanced salt solution.
9. Dissect the desired regions of the brain (see Note 5). Place the dissected brain in sterile dishes with a small amount of warm serum-free culture medium.

3.2. Neural Tissue Dissociation

Dissociation of neural cells from their associated matrix provides single cells capable of generating neurospheres. Trypsin serves as a protease to digest the matrix with the caveat that it also cleaves membrane receptors from the surface of the cells.

1. Mince the brain tissue into pieces no larger than 3–4 mm^2.
2. Immediately place the tissue in a sterile round bottom tube. Weigh the tissue/tube to determine how much trypsin should be used to dissociate the tissue (in step 6).
3. Add warm balanced salt solution to wash off any contaminating blood/cells.
4. Centrifuge at $250 \times g$ for 1 min. Remove the supernatant. Add fresh warm balanced salt solution. Repeat 1×.
5. Remove the balanced salt solution. Add 0.25% trypsin diluted in warm balanced salt solution. Use 1.0 mL of trypsin solution for each 100 mg of tissue.

6. Incubate at 37°C for 30–90 min. Periodically, triturate the samples using a sterile Pasteur pipette (see Note 6).
7. Prewarm DMEM:F12 medium at 37°C. Add 4.0 mg of trypsin inhibitor to 4.0 mL warm DMEM:F12.
8. Centrifuge the dissociated tissue at 250×*g* for 5 min.
9. Decant and discard the trypsin solution.
10. Add 4 mL of warm trypsin inhibitor/medium solution (see Note 7).
11. Triturate the samples using a sterile Pasteur pipette.
12. Incubate at 37°C for 10 min.
13. Centrifuge the dissociated tissue at 500×*g* for 5 min.
14. Resuspend in 500 μL Serum-Free Culture Media (supplemented with nutrients). Triturate to produce a single cell suspension.

3.3. Neural Stem Cell Culture

Addition of growth factors, hormones, and nutritional factors to the basal medium reduces the need to add serum. B-27® serum-free supplement contains antioxidants to reduce free radical damage.

3.3.1. Viable Cell Count

1. Dilute 20 μL Trypan Blue dye to 20 μL PBS. Mix well.
2. Prepare a hemocytometer for cell counting. Clean the counting surface of the hemocytometer and the cover glass with 70% ethanol. Dry the surface thoroughly with a lintless cloth (Kimwipe). Place a calibrated cover glass on the hemocytometer.
3. Add 10 μL of the neural cell suspension to 10 μL Trypan Blue dye. Mix gently.
4. Add 10 μL of the cell/dye suspension to each side of the hemocytometer. Allow the cells to settle for 1–2 min.
5. Count the number of blue (dead) cells and viable cells (undyed) in the four large corner squares on each side of the hemocytometer.
6. Calculate the number of viable cells/mL: average number of viable (unstained) cells in 4 squares × 2,500 × 2.

3.3.2. Plating the Cells

1. Plate the cells at a density of 50–100 cells/μL in Serum-Free Culture Media (supplemented with nutrients) (see Note 8).
2. Incubate the cells for 7 days at 37°C in a humidified atmosphere with 5% CO_2.
3. Observe each microwell daily using phase contrast to (a) observe number and size of neurospheres, and (b) look for bacterial or fungal contamination. Neurospheres less than 100 μm diameter are an optimal size for propagation.

3.4. Propagation of Neural Stem Cell Spheroids (47)

1. Prepare 1× solution of TrypLE™. Add 4.162 g EDTA to 100 mL dH_2O (final concentration 1 mM EDTA). Filter with a 0.22-µm filter. Add 1.0 mL of the 1 mM EDTA solution to 99 mL PBS without calcium, magnesium, or phenol red. Add 200 µL of TrypLE™ to 1.8 mL of the EDTA/PBS solution. Mix well. Prepare fresh daily.
2. Transfer the neurospheres and medium to a 15-mL conical tube (see Note 9).
3. Centrifuge at 200 × *g* for 5 min.
4. Remove and discard the supernatant.
5. Add 2.0 mL of 1× TrypLE™ solution to the tube. Mix gently. Do not vortex.
6. Incubate the tube at 37°C for 20 min.
7. Centrifuge at 500 × *g* for 5 min.
8. Remove the supernatant and resuspend the cells in 500 µL of Serum-Free Culture Media (supplemented with nutrients).
9. Stir gently with a sterile Pasteur pipette.
10. Determine the number of viable cells using the Trypan Blue dye exclusion described in Subheading 3.3.1.
11. Plate the cells at a density of 50–100 cells/µL in a new 24-well plate.
12. Incubate the cells at 37°C in a humidified atmosphere with 5% CO_2.
13. Observe each microwell daily using phase contrast to (a) observe number and size of neurospheres, and (b) look for bacterial or fungal contamination. Neurospheres less than 100 µm diameter are an optimal size for propagation.

3.5. Conclusions: The Future Potential of Adult Neural Stem Cell Therapy

Neural stem cell therapy is an active field, and data collected to date in human and animal models have contributed to our growing understanding of the role of adult stem cells in constitutive neurogenesis and as possible treatments for neurodegenerative disorders. Although embryonic stem cell research has provided incalculable knowledge in our pursuit to understand the human organism and to provide treatments for disease, such endeavors face significant limitations. Adult neural stem cells appear to be a means of sidestepping many of the undesired elements of other fields. Through the use of a patient's own stem cells, genetically induced from adult progenitor cells (48), or harvested and enriched from neural or bone marrow biopsies, we might be able to develop treatments which are ethically acceptable and unencumbered by the current legal environment. Autologous stem cells will not require immune suppression and do not offer the concern for donor matching or the risk of neoplasm development. The principles of stem cell therapy are being elucidated in labs around the world. With the confirmation that human adults are capable of generating new neural cells, there is now hope that stem cell therapy can be successfully applied to many degenerative diseases affecting millions worldwide.

4. Notes

1. Alternatively, Hanks' Balanced Salt Solution without calcium or magnesium can be used.
2. Prepare fresh serum-free medium weekly. The biologic activity of the growth factors and the enzymes decreases drastically over time.
3. Variables affecting the cellular components of the neurosphere include age or developmental stage of the animal tissue source, growth and differentiation factors, plating density, and passage number (47). The age of the animal markedly affects the yield of neurospheres. The younger the animal the higher the yield.
4. Use one set of sterile scalpels and scissors for cutting the skin and opening the skull. Use a clean set of instruments for removing and dissecting the brain.
5. The subventricular zone, hippocampus, and rostral migratory stream will yield abundant mammospheres (47).
6. The length of time used for enzymatic digestion can significantly affect cell viability. Carefully monitor the dissociation during digestion and utilize the minimum digestion time necessary. It is better to reduce the yield somewhat but retain a higher viability.
7. Trypsin activity can be inhibited by adding a trypsin inhibitor such as soybean or egg white trypsin inhibitor, or aprotinin. It is important to monitor the action of trypsin because trypsin cleaves protein receptors from the cell surface. Over-trypsinization can delay the propagation of neurospheres or kill the cells.
8. Plating density for clonal assays should be done at 10–15 cells/μL. Higher density cell plating produces nanospheres more rapidly compared to low density plating methods.
9. Ensure that none of the wells are contaminated with bacteria, yeast, or fungus. If any well is found to be contaminated, do not add the contents of the contaminated wells into the pooled wells.

References

1. Reynolds, B. A., Weiss, S. (1996) Clonal and population analyses demonstrate that an EGF-responsive mammalian embryonic CNS precursor is a stem cell. Dev Biol 175, 1–13.
2. Gage, F. H., Ray, J., Fisher, L. J. (1995) Isolation, characterization, and use of stem cells from the CNS. Annu Rev Neurosci 18, 159–92.
3. Lindvall, O., Kokaia, Z. (2010) Stem cells in human neurodegenerative disorders – time for clinical translation? J Clin Invest 120, 29–40.
4. Zeng, X., Rao, M. S. (2007) Human embryonic stem cells: long term stability, absence of senescence and a potential cell source for neural replacement. Neuroscience 145, 1348–58.
5. Carlson, M. E., Suetta, C., Conboy, M. J., Aagaard, P., Mackey, A., Kjaer, M. et al. (2009) Molecular aging and rejuvenation of human muscle stem cells. EMBO Mol Med 1, 381–91.
6. Hanson, S. E., Gutowski, K. A., Hematti, P. (2010) Clinical applications of mesenchymal stem cells in soft tissue augmentation. Aesthet Surg J 30, 838–42.
7. Shin, D. M., Liu, R., Klich, I., Wu, W., Ratajczak, J., Kucia, M. et al. (2010) Molecular signature of adult bone marrow-purified very small embryonic-like stem cells supports their developmental epiblast/germ line origin. Leukemia 24, 1450–61.
8. Emsley, J. G., Mitchell, B. D., Kempermann, G., Macklis, J. D. (2005) Adult neurogenesis and repair of the adult CNS with neural progenitors, precursors, and stem cells. Prog Neurobiol 75, 321–41.
9. Aguayo, A. J., David, S., Bray, G. M. (1981) Influences of the glial environment on the elongation of axons after injury: transplantation studies in adult rodents. J Exp Biol 95, 231–40.
10. David, S., Aguayo, A. J. (1981) Axonal elongation into peripheral nervous system "bridges" after central nervous system injury in adult rats. Science 214, 931–3.
11. Sohur, U. S., Emsley, J. G., Mitchell, B. D., Macklis, J. D. (2006) Adult neurogenesis and cellular brain repair with neural progenitors, precursors and stem cells. Philos Trans R Soc Lond B Biol Sci 361, 1477–97.
12. Dirks, P. B. (2008) Brain tumor stem cells: bringing order to the chaos of brain cancer. J Clin Oncol 26, 2916–24.
13. Dirks, P. B. (2008) Brain tumour stem cells: the undercurrents of human brain cancer and their relationship to neural stem cells. Philos Trans R Soc Lond B Biol Sci 363, 139–52.
14. Vescovi, A. L., Galli, R., Reynolds, B. A. (2006) Brain tumour stem cells. Nat Rev Cancer 6, 425–36.
15. Ricci-Vitiani, L., Pallini, R., Larocca, L. M., Lombardi, D. G., Signore, M., Pierconti, F. et al. (2008) Mesenchymal differentiation of glioblastoma stem cells. Cell Death Differ 15, 1491–8.
16. Wang, R., Chadalavada, K., Wilshire, J., Kowalik, U., Hovinga, K. E., Geber, A. et al. (2010) Glioblastoma stem-like cells give rise to tumour endothelium. Nature 468, 829–33.
17. Ricci-Vitiani, L., Pallini, R., Biffoni, M., Todaro, M., Invernici, G., Cenci, T. et al. (2010) Tumour vascularization via endothelial differentiation of glioblastoma stem-like cells. Nature 468, 824–8.
18. Altman, J., Das, G. D. (1966) Autoradiographic and histological studies of postnatal neurogenesis. I. A longitudinal investigation of the kinetics, migration and transformation of cells incorporating tritiated thymidine in neonate rats, with special reference to postnatal neurogenesis in some brain regions. J Comp Neurol 126, 337–89.
19. Kaplan, M. S., Hinds, J. W. (1977) Neurogenesis in the adult rat: electron microscopic analysis of light radioautographs. Science 197, 1092–4.
20. Gould, E. (2007) How widespread is adult neurogenesis in mammals? Nat Rev Neurosci 8, 481–8.
21. Rakic, P. (1985) Limits of neurogenesis in primates. Science 227, 1054–6.
22. Gould, E., Tanapat, P., McEwen, B. S., Flugge, G., Fuchs, E. (1998) Proliferation of granule cell precursors in the dentate gyrus of adult monkeys is diminished by stress. Proc Natl Acad Sci U S A 95, 3168–71.
23. Eriksson, P. S., Perfilieva, E., Bjork-Eriksson, T., Alborn, A. M., Nordborg, C., Peterson, D. A. et al. (1998) Neurogenesis in the adult human hippocampus. Nat Med 4, 1313–7.
24. Kempermann, G., Gast, D., Kronenberg, G., Yamaguchi, M., Gage, F. H. (2003) Early determination and long-term persistence of adult-generated new neurons in the hippocampus of mice. Development 130, 391–9.
25. Lois, C., Alvarez-Buylla, A. (1993) Proliferating subventricular zone cells in the adult mammalian forebrain can differentiate into neurons and glia. Proc Natl Acad Sci U S A 90, 2074–7.
26. Gritti, A., Bonfanti, L., Doetsch, F., Caille, I., Alvarez-Buylla, A., Lim, D. A. et al. (2002) Multipotent neural stem cells reside into the

rostral extension and olfactory bulb of adult rodents. J Neurosci 22, 437–45.

27. Yang, D., Zhang, Z. J., Oldenburg, M., Ayala, M., Zhang, S. C. (2008) Human embryonic stem cell-derived dopaminergic neurons reverse functional deficit in parkinsonian rats. Stem Cells 26, 55–63.
28. Roy, N. S., Cleren, C., Singh, S. K., Yang, L., Beal, M. F., Goldman, S. A. (2006) Functional engraftment of human ES cell-derived dopaminergic neurons enriched by coculture with telomerase-immortalized midbrain astrocytes. Nat Med 12, 1259–68.
29. Freed, C. R., Breeze, R. E., Rosenberg, N. L., Schneck, S. A., Wells, T. H., Barrett, J. N. et al. (1990) Transplantation of human fetal dopamine cells for Parkinson's disease. Results at 1 year. Arch Neurol 47, 505–12.
30. Freed, C. R., Breeze, R. E., Rosenberg, N. L., Schneck, S. A., Kriek, E., Qi, J. X. et al. (1992) Survival of implanted fetal dopamine cells and neurologic improvement 12 to 46 months after transplantation for Parkinson's disease. N Engl J Med 327, 1549–55.
31. Freed, C. R., Breeze, R. E., Rosenberg, N. L., Schneck, S. A. (1993) Embryonic dopamine cell implants as a treatment for the second phase of Parkinson's disease. Replacing failed nerve terminals. Adv Neurol 60, 721–8.
32. Daadi, M. M., Maag, A. L., Steinberg, G. K. (2008) Adherent self-renewable human embryonic stem cell-derived neural stem cell line: functional engraftment in experimental stroke model. PLoS One 3, e1644.
33. Daadi, M. M., Davis, A. S., Arac, A., Li, Z., Maag, A. L., Bhatnagar, R. et al. (2010) Human neural stem cell grafts modify microglial response and enhance axonal sprouting in neonatal hypoxic-ischemic brain injury. Stroke 41, 516–23.
34. Benkler, C., Offen, D., Melamed, E., Kupershmidt, L., Amit, T., Mandel, S. et al. (2010) Recent advances in amyotrophic lateral sclerosis research: perspectives for personalized clinical application. 1, in The EPMA Journal. Springer Netherlands, 343–361.
35. Suzuki, M., McHugh, J., Tork, C., Shelley, B., Klein, S. M., Aebischer, P. et al. (2007) GDNF secreting human neural progenitor cells protect dying motor neurons, but not their projection to muscle, in a rat model of familial ALS. PLoS One 2, e689.
36. Sahni, V., Kessler, J. A. (2010) Stem cell therapies for spinal cord injury. Nat Rev Neurol 6, 363–72.
37. Salazar, D. L., Uchida, N., Hamers, F. P., Cummings, B. J., Anderson, A. J. (2010) Human neural stem cells differentiate and promote locomotor recovery in an early chronic spinal cord injury NOD-scid mouse model. PLoS One 5, e12272.
38. Perrin, F. E., Boniface, G., Serguera, C., Lonjon, N., Serre, A., Prieto, M. et al. (2010) Grafted human embryonic progenitors expressing neurogenin-2 stimulate axonal sprouting and improve motor recovery after severe spinal cord injury. PLoS One 5, e15914.
39. Barnabe-Heider, F., Frisen, J. (2008) Stem cells for spinal cord repair. Cell Stem Cell 3, 16–24.
40. Abematsu, M., Tsujimura, K., Yamano, M., Saito, M., Kohno, K., Kohyama, J. et al. (2010) Neurons derived from transplanted neural stem cells restore disrupted neuronal circuitry in a mouse model of spinal cord injury. J Clin Invest 120, 3255–66.
41. Van Kampen, J. M., Eckman, C. B. (2006) Dopamine D3 receptor agonist delivery to a model of Parkinson's disease restores the nigrostriatal pathway and improves locomotor behavior. J Neurosci 26, 7272–80.
42. Einstein, O., Ben-Hur, T. (2008) The changing face of neural stem cell therapy in neurologic diseases. Arch Neurol 65, 452–6.
43. Hess, D. C., Borlongan, C. V. (2008) Stem cells and neurological diseases. Cell Prolif 41 Suppl 1, 94–114.
44. Van Kampen, J. M., Hagg, T., Robertson, H. A. (2004) Induction of neurogenesis in the adult rat subventricular zone and neostriatum following dopamine D3 receptor stimulation. Eur J Neurosci 19, 2377–87.
45. Reynolds, B. A., Tetzlaff, W., Weiss, S. (1992) A multipotent EGF-responsive striatal embryonic progenitor cell produces neurons and astrocytes. J Neurosci 12, 4565–74.
46. Reynolds, B. A., Weiss, S. (1992) Generation of neurons and astrocytes from isolated cells of the adult mammalian central nervous system. Science 255, 1707–10.
47. Pacey, L. K. K., Stead, S., Gleave, J.A., Tomczyk, K., Doering, L. C. (2006) Neural stem cell culture: neurosphere generation, microscopical analysis and cyropreservation. Protocol Exchange: doi:10.1038/nprot.2006.215.
48. Zhou, H., Wu, S., Joo, J. Y., Zhu, S., Han, D. W., Lin, T. et al. (2009) Generation of induced pluripotent stem cells using recombinant proteins. Cell Stem Cell 4, 381–4.

Chapter 19

Evanescent-Wave Field Imaging: An Introduction to Total Internal Reflection Fluorescence Microscopy

Bryan A. Millis

Abstract

Advancements in technology and computational power in recent years have directly impacted modern microscopy through improvements in light detection, imaging software platforms, as well as integration of complex hardware systems. These successes have allowed for mainstream utilization of previously complex microscopic techniques such as total internal reflection fluorescence (TIRF) microscopy, revealing many aspects of cell biology not previously appreciated. Through the restriction of illumination to areas of cell-coverslip interfaces in combination with modern detectors, TIRF microscopy allows researchers in the life sciences a glimpse of dynamic cellular phenomena with resolutions never before achieved.

This chapter provides a basic overview to the concept of TIRF microscopy and some considerations to setting up this technique in the lab.

Key words: CCD, Evanescence, Fluorescence, Microscopy, Total internal reflection

1. Introduction

Traditional approaches in the research lab have given way to demands for faster, more sensitive, and undoubtedly higher throughput methods than once thought possible. It could be argued that this trend is certainly not a new one, but the recent acceleration that is now being felt at many levels of the life science community is a result of an exciting merge between two previously distinct disciplines: physical and life sciences.

Now, through the aid of physical and computational sciences, biologists are able to detect and analyze increasingly complex and dynamic events. One approach to understand these dynamics is through the use of advanced microscopy and digital imaging techniques.

Virginia Espina and Lance A. Liotta (eds.), *Molecular Profiling: Methods and Protocols*, Methods in Molecular Biology, vol. 823, DOI 10.1007/978-1-60327-216-2_19, © Springer Science+Business Media, LLC 2012

When Gordon Moore, postulated his "Moore's Law" in 1965 (estimating that the number of transistors per chip should approximately double every 2 years), the number was in the tens of thousands (1). As of 2008, the number of transistors per chip has hit 2 billion (http://www.intel.com/technology/architecture-silicon/2billion.htm). These advancements have directly impacted modern microscopy. Notable examples are improvements in charge-coupled devices (CCDs), imaging software platforms, and management of the ever-increasing demand for larger datasets. Dramatic benefits to three primary attributes have been speed, sensitivity, and resolution. The efficiency and simplicity of operating these advanced microscope systems have also been enhanced by faster and more intelligent integration, turning complicated operations into something that can be accomplished with the single push of a button. When combined with a parallel increase in the processing power of today's workstations, these integrated systems result in the ability to present extremely complex, multidimensional microscopic data in simpler and more intuitive ways. This data rendering reveals interactions, morphologies, and subcellular dynamics that have been previously unseen. Each advanced microscopy technique is designed to answer specific questions about cellular behavior, and therefore each presents both benefits and challenges or limitations to the researcher during the experimental design process. Obviously, microscopy is not the only field that has benefited from such leaps in digital technologies. As this book illustrates, there are many more ways to unearth the dynamics of networks, pathways, and specific cellular interactions. Microscopy can add to this picture in ways that cannot be easily addressed by other methods. One of these aspects is the extremely dynamic interaction of a cell and its extracellular environment, including cell–cell and cell–matrix interactions. Cellular interaction within a microenvironment has been an ubiquitous subject of investigation across many life science disciplines and has become a significant area of interest in cancer research as it relates to invasion, metastasis, and signal transduction (2–6). One technique which has grown in popularity in recent years to meet the experimental demands of modern cell biologists is total internal reflection fluorescence (TIRF) microscopy (7) (see Note 1).

2. Challenges of Basic Fluorescence Microscopy

Since the introduction of fluorescence microscopy, a prevailing problem facing scientists in the field has been the proper resolution of fluorescent probes within the focal plane of a given sample. Resolution of a fluorescent probe's specific localization in three-dimensional space is nearly impossible using traditional wide-field fluorescence techniques. In wide-field fluorescence, light from an arc lamp or other source is filtered and focused on a particular focal plane of a sample to excite probes of various compositions. These excited probes, which may be derivatives of proteins (such as green fluorescent protein (GFP)), antibody conjugates (such as IgG–AlexaFlour® conjugates), stains (such as eosin), dyes (such as hoechst), or nanoparticles (such as Q-dots®), as well as many others, emit energy in the form of light resulting from electrons in excited states relaxing back down to their ground states. The entire column is inundated by normally very intense and harmful light. Not only does this light causes problems for the integrity of the probes themselves (via photobleaching and quenching), but it can also be extremely harmful to live cells, which form thymine dimers in DNA as well as accumulate reactive oxygen species leading to unnecessary oxidative stress and adverse effects (8–10). Aside from physical damage of the cell, three-dimensional localization becomes a problem as all of the probes present in that column become illuminated. This means that the focal plane is compromised by emission of light from many probes at many focal planes which results in potentially high background and low signal-to-noise ratio, even if the probe itself has low cross-reactivity/background characteristics. Many times, the result is the inability to resolve small cellular constituents, especially ones of a dynamic nature. Of course, multiple techniques have been applied to circumvent this problem. Deconvolution, confocal, and TIRF microscopy represent a few of the ways to address problems with out of focus light degrading image quality. To discuss all methods is beyond the scope of this article, however TIRF microscopy in particular is able to resolve a cell's microenvironment interactions at the level of the cell surface through exquisite restriction of illumination and will be explored further here.

2.1. Physical Basis of TIRF

At the core of optical physics are the fundamental properties of reflection and refraction of light. As light passes through media of one refractive index to another, there are varying degrees of reflection and refraction which may occur. As the angle of incident light is varied, the proportions of each of these phenomena change. For example, if a light source is aimed close to perpendicular on a mostly transparent medium, the majority of the light will be transmitted, though refracted, through the medium while a small portion of the light is reflected (11). As the light source is increasingly angled at the surface, depending on the materials that the light is propagating through, Fresnel's equations show that the incident light may become less *refracted* and more *reflected* (Fig. 1) (11). This can have different consequences as the light approaches a greater angle. If light is passing through a material of higher refractive index and into a material of lower refractive index, it may approach what is known as the *critical angle*. At critical angle, refraction into the lower index media does not occur, but instead

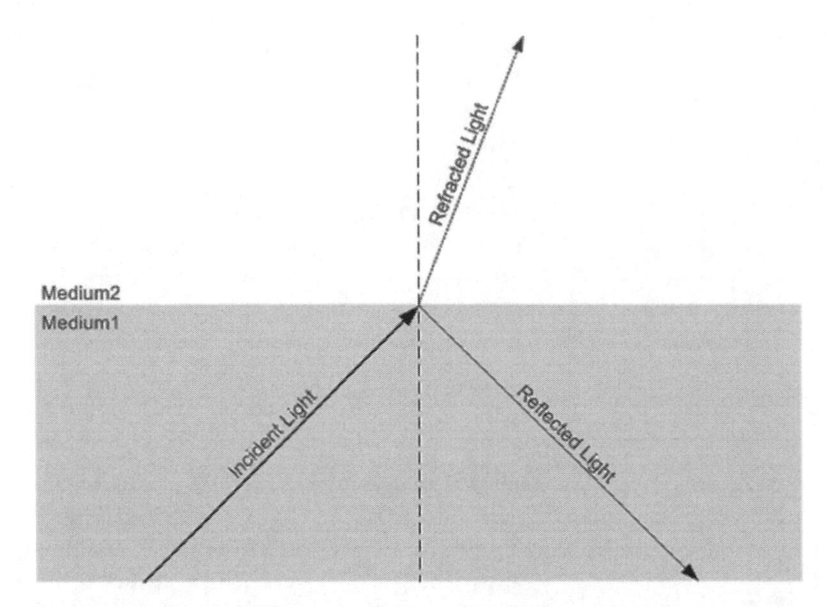

Fig. 1. Basic refraction and reflection. As light passes through media of one refractive index to another, there are varying degrees of reflection and refraction. As the angle of incident light is varied, the proportions of each of these phenomena change. If a light source is aimed close to perpendicular on a mostly transparent medium, the majority of the light will be transmitted, though refracted, through the medium while a small portion of the light is reflected. As the light source is increasingly angled at the surface, depending on the materials that the light is propagating through, Fresnel's equations show that the incident light may become less *refracted* and more *reflected* as shown here.

is transmitted parallel to the boundary between the two mediums (Fig. 2). Critical angle is dependent on several variables; the most important are the refractive indices themselves. The critical angle of a system can be calculated easily and is defined as:

$$\mathrm{Sin}(\theta c) = n2 \,/\, n1, \quad \text{for} \quad n1 > n2,$$

where θc is the critical angle and n is refractive index (11). Once the critical angle is passed, all of the light is being completely reflected internally; in other words, *total internal reflection* has been achieved. In TIRF microscopy, this light, which may come from a wide-field fluorescent source or more commonly from a laser, is angled toward the coverslip of a sample. In this case, the interface between the glass coverslip (higher refractive index, ~1.5) and the media of the sample (lower refractive index, ~1.3 for tissue culture media) serves as the boundary upon which this reflection occurs. So if all of the light is being reflected back into the coverslip, then how is the light source exciting fluorophores within the sample? An important rule of electromagnetic radiation states that electrical and magnetic fields (such as present in light) cannot

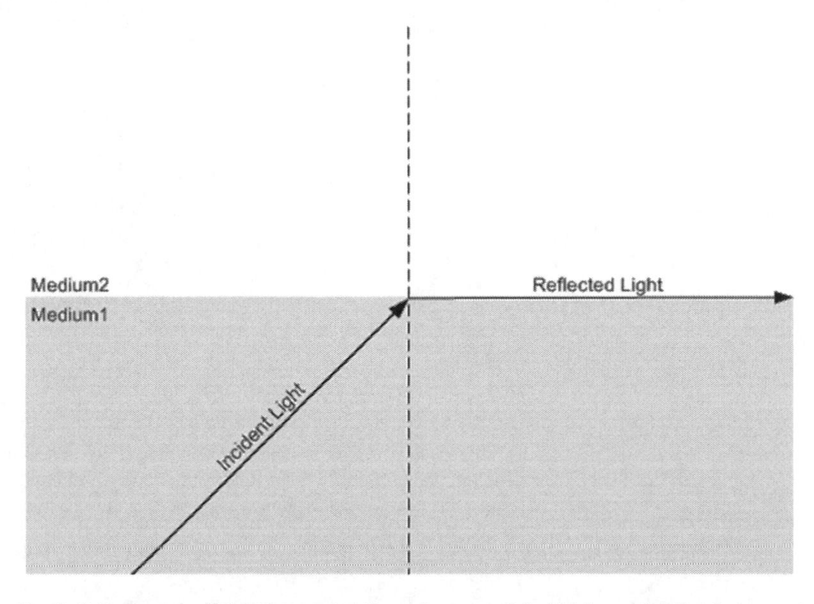

Fig. 2. Critical angle. If light is passing through a material of higher refractive index and into a material of lower refractive index, it may approach what is known as the *critical angle*. At the critical angle, the light hits a point at which none of the light is refracted through the media, but instead it is reflected completely parallel to the surface, on the boundary of the two materials.

become discontinuous at a boundary, but instead produce what is referred to as an *evanescent wave* that dissipates exponentially with distance away from the point of reflection (12) (see Note 2). In TIRF, this is the point of reflection between the cover glass and the sample medium (Fig. 3). The word *evanescence* itself means "tends to vanish" and it is the extremely small distance in which this wave is present (normally, an evanescent wave is most intense within one-third of its wavelength) that is the key to the axial resolution of TIRF (12). With common lasers for microscopic imaging, this theoretical distance is on the order of 100–200 nm, with some systems capable of exciting on the order of tens of nanometers (13). Very few optical sectioning techniques can come close to this kind of axial resolution.

2.2. Numerical Aperture and TIRF

TIRF microscopy has not been widely popular with mainstream labs until recently. This has been mainly due to limitations in the ability to introduce light at angles approaching critical angles in biological specimens. Previously, this problem was solved by using a prism on top of the specimen (the side opposite the objective), which ensured that the beam could be totally internally reflected and that the user could control the angle by altering the beam onto the prism. Several inconveniences with this approach included the fact that the prism was on the other side of the specimen such that

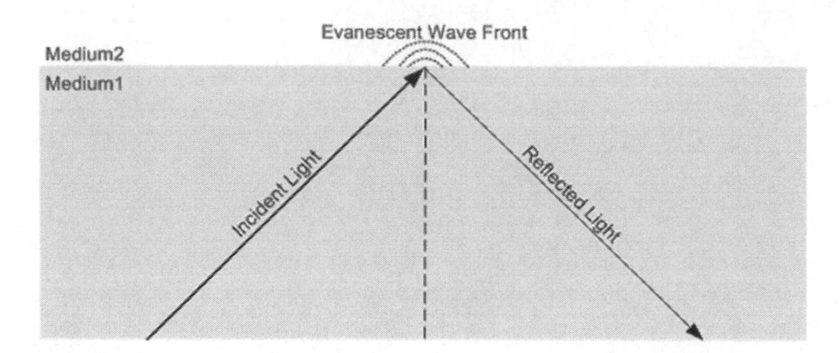

Fig. 3. Evanescent wave front. Once the critical angle is passed, all of the light is completely reflected internally: *total internal reflection* has been achieved. In TIRF microscopy, an evanescent wave excites fluorophores within the sample. Electromagnetic radiation, such as light, cannot become discontinuous at a boundary, but instead produce what is referred to as an *evanescent wave* that dissipates exponentially with distance away from the point of reflection.

the objective would have to gather light from this very small area of emission through the thickness of the tissue itself, which could be significant depending on the sample. For this reason, among others that included frustrations with configuring the prisms on top of certain specimens, it was not the preferred TIRF method, especially in the biological world.

The more popular method utilizes the ability to introduce the TIRF illumination source through the objective itself, similar to wide-field epi-fluorescence, which uses the objective as the condenser of the emitted light in addition to delivering the excitation energy (Fig. 4). Until recently, objectives with high enough numerical apertures (NAs) were not readily available from major microscopic manufacturers to achieve the high incidence angles required to reach the critical angle and beyond. By classical definition, numerical aperture defines how well an objective lens (or other optic) gathers light and resolves fine detail of a specimen (12). The numerical aperture of an objective is mathematically defined as the refractive index multiplied by the sine of one half of the angular aperture of the objective:

$$\mathrm{NA} = n\sin(\alpha).$$

Figure 5 illustrates how the angular aperture or angle α can be greatly affected by the NA of the objective. For example, if air is considered as the projecting medium, when the angular aperture is limited 10° due to focal distance, numerical apertures can be in the range of 0.17 toward the lower end. However, when oil is used as the immersion medium and the objective's focal distance is greatly decreased, angular apertures rise proportionally and numerical apertures greater than 1.4 can be achieved. In fact, as manufacturing techniques have become more refined, objectives with

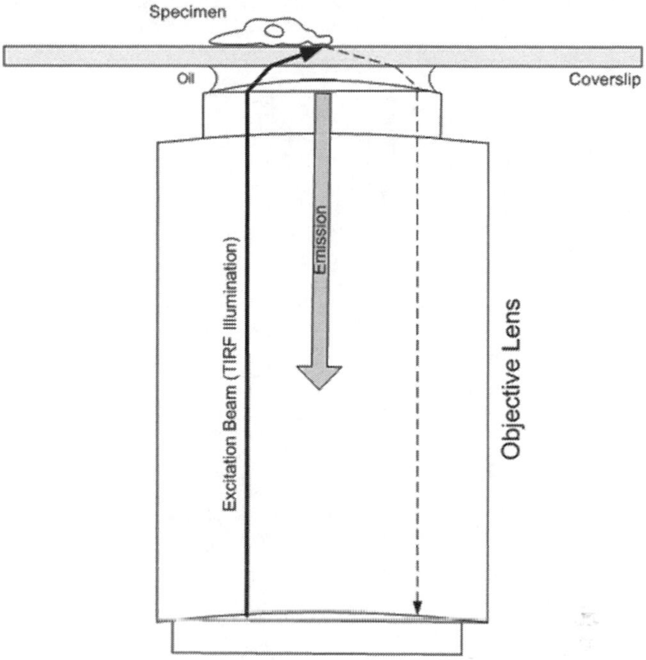

Fig. 4. Objective with total internal reflection fluorescence (TIRF). A common TIRF microscopy method introduces the TIRF illumination source through the objective itself, similar to wide-field epi-fluorescence, which uses the objective as the condenser of the emitted light in addition to delivering the excitation beam. The emitted light is generated by excitation caused by the evanescent wave at the interface between the coverglass and the liquid media.

numerical apertures of 1.49 and above can be readily purchased today. Despite the ability to reach critical angles with slightly lower NA objectives, the demand for higher NA objectives for TIRF applications continues. Besides light gathering, resolution, and the ability to reach critical angles, another major benefit of high NA objectives is the ability to push the light source further past critical angle (at higher angles of total internal reflection) and thus have the ability to control the evanescent field wave depth due to higher angular apertures. This refined ability to control focal plane illumination and gather more light via higher NA has resulted in increases in signal-to-noise ratios and stunning images of cellular dynamics has resulted (14).

2.3. Detection Technologies

The ability to reach total internal reflection in biological applications and achieve highly resolved axial resolutions in TIRF may sometimes be offset if the ability to detect emission energies of probes in that space is limited. Since TIRF pushes the limit of resolution in illumination, it also has the effect of pushing the limit of detection technologies and thus a firm understanding of the basic concepts of detectors, specifically CCDs, is required. The proper choice of detector can many times be a daunting task for investigators if

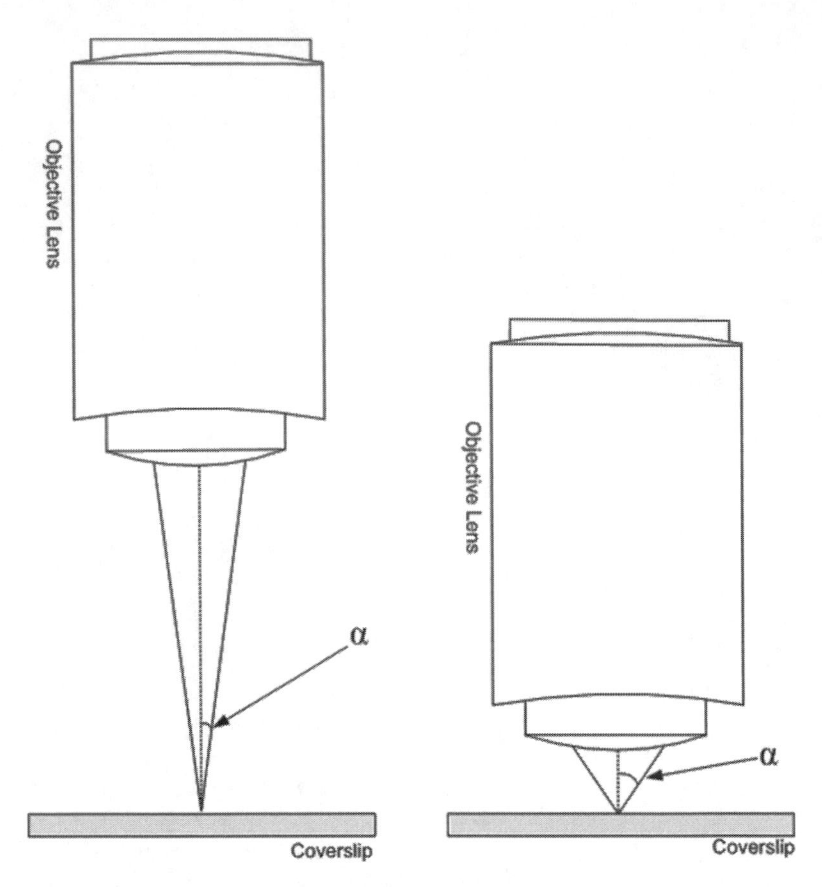

Fig. 5. Half-aperture angles. This diagram illustrates how the half-angular aperture or angle α relates to NA of the objective. If air is considered as the projecting medium, when the angular aperture is limited 10° due to focal distance, numerical apertures can be in the range of 0.17, a low degree of resolving power (*left*). However, when oil is used as the immersion medium and the objective's focal distance is greatly decreased (*right*), angular apertures rise proportionally and numerical apertures higher than 1.4 can be achieved. This results in the ability to achieve the higher incendent angles necessary for TIRF.

adequate time is not spent considering the options. A full discussion on detector architecture and the intricacies that are inherent in the different designs is beyond the scope of this review as each investigator has his/her own requirements. However, this discussion is meant to give the reader an idea of some basic concepts of detection as it relates to TIRF microscopy and important options for instrumentation.

Two general detection methods are employed in high-end microscopy: CCDs and PMTs. The overwhelming majority of biological TIRF applications use CCDs in the form of cameras as their main mode of detection and thus the majority of the discussion here is devoted to CCD considerations. Monochromatic CCDs are discussed due to their superiority in quantitative fluorescence applications over color-based CCDs. While speed, sensitivity, and resolution have increased in certain detectors, it would be inaccurate to assume that every camera has benefited from all three. First, it is important to accurately define what is meant by these terms as not all sources define each the same way. The biggest source of confusion among newcomers to digital imaging and quantification, other than monochromatic superiority over color cameras, is the concept of resolution, as the meaning can vary depending on the context. In microscopy, resolution is most commonly used to describe either *spatial resolution*, referring to the minimum distance that two objects can be distinguished as separate entities, or *temporal resolution*, which is the ability to differentiate separate events in time. This definition of spatial resolution is to be understood as separate from what many consider as display resolution that can be used to quantify the dimensions or total number of pixels on a monitor display, for example. Not all cameras have benefited from increases in each of these characteristics because they are all related to one another. It is best to visualize the relationship among all three as a triangle with each characteristic on the tips of each point. As you move toward one, you have to move away and sacrifice some of another. The reason for this relationship is due to both the physics of the chips themselves as well as the way they are designed.

3. Basic CCD Architecture and Operation

A CCD is simply a semiconductor material (normally, a silicon wafer) that is divided up into thousands and sometimes millions of pixels depending on the chip. It is the goal of the optics in the system to project the image onto the surface of the chip (or to the retina of our eye) in the same way as it is represented in the sample. CCDs take advantage of a physical phenomenon known as the photoelectric effect. This phenomenon refers to the conversion of energy from a photon of light to an electrical charge. In this process, a photon of light (resulting from emission of an excited probe in the case of fluorescence microscopy) strikes the semiconductor's surface and is absorbed (11). As a result of the incident photons, electrons are released from the surface and are stored in what is called a full well or potential well that is associated with each pixel in the array of the chip. During an exposure, light is continually collected from a sample, and as photons strike the surface of a pixel this energy is continuously converted to electrons which contribute to the building charge that is stored in that pixel's full well. A helpful analogy to understand this process is thinking of the array of pixels and potential wells underneath them as an array of rain buckets that collect rain for certain intervals of time (exposure time). This process occurs simultaneously across an array of possibly millions of pixels per exposure. What happens next is dependent on the type of chip that is being used. In one of the most common motifs, when the duration of an exposure is completed, the chip moves the charge down in parallel shifts, one entire row at a time. The row that is closest to the bottom of the chip is shifted into what is known as a serial register at the base of the chip. The serial register is simply a linear strip of pixels at the base of the array into which the bottom row of charges can shift before being read out by the analog-to-digital converter (ADC). In this second process, the serial register moves the charges across the linear array one pixel at a time into the ADC so that the analog voltage that has accumulated in the potential wells can be converted into a digital signal that is then interpreted by the computer as intensity levels or gray levels between black and white. The bit depth of the ADC defines the number of gray levels and thus intensities that can be measured. Here is where the concepts of ***dynamic range*** and ***bit depth*** are extremely important in quantitative digital imaging and thus detector selection when setting up a system for a particular type of microscopy, which are both described in more detail below.

3.1. CCD Sensitivity

If one uses the analogy of the rain buckets collecting rain drops, it becomes obvious that the size of the bucket becomes extremely important. As such, the diameter of the bucket, as well as depth of the bucket, greatly influences the volume of water that it can hold. This is exactly the same as the size of the pixel (and thus the potential well) when considering how many photons each can accept before becoming saturated, also known as the dynamic range. This is a delicate balance for a few reasons which all come back to the relationship of speed, sensitivity, and resolution introduced earlier. Consider the overall goal of being able to image two cells in a field or two vesicles in a cell for the purpose of detecting extremely small differences in intensity, possibly due to more or less total expression in the case of the cells or the abundance of a protein in the two vesicles. Forgetting resolution for the moment, let us assume that one pixel represents one vesicle's intensity and another pixel represents the second vesicle's intensity, with the two being extremely close. After exposure, the full wells under each pixel have accumulated electrons and the charge is read out by the ADC. In scenario 1, the full well is very shallow and a maximum of only 1,000 electrons are collected. In scenario 2, the wells are deeper and a maximum of 60,000 electrons are collected. As mentioned before, the bit depth of the ADC defines the ability to divide a given well into different intervals or shades of gray and thus intensity on a computer screen. For example, an 8-bit ADC could convert the potential well's charge to give 2^8 or 256 gray levels and a 16-bit ADC could convert the well's charge to give 2^{16} or 65,536 shades of gray. Since it is already known that the intensity is extremely close between the two, one might be inclined to have a higher bit depth ADC to be able to resolve this difference. However, in scenario 1, this would be ineffective as there is not enough dynamic range to be able to divide up differences so finely. In this case, it would not be possible to detect the difference in intensity. In scenario 2, given the higher well capacity, one could make use of a higher bit depth ADC in order to detect this difference. These concepts of bit depth, dynamic range, and linear full well all contribute to how sensitive a particular chip is to light and the ability to detect small changes once the light is collected.

3.2. CCD Resolution

Larger pixel dimensions and full well capacities would seem like an obvious choice to achieve adequate sensitivity, however, pixel size has an inversely proportional effect on resolution. Similar to cones on the retina of the eye, pixels are essential in being able to distinguish two objects as being separate entities and not one blurred object. The ability to distinguish these two objects is the essence of spatial resolution. The more pixels there are in between these two objects, the easier it is to resolve them. At the core of this concept is the Nyquist theorem, which proposes the amount of sampling necessary to be able to distinguish signals from one another (15, 16). Depending on the source, it is postulated that between 2 and 3 pixels (after optical projection on the chip) are needed as the minimum sampling rate between objects. Pixels that are too small for a particular application and/or magnification suffer from oversampling and unnecessary insensitivity to light. What results is empty magnification and higher exposure times. Pixels that are too large suffer from inability to resolve fine detail that the objective would otherwise permit and "grain" or "pixelation" may result. It can be safely assumed that there is no one pixel size that satisfies all scenarios, but careful consideration of what techniques and conditions warrant the most priority is recommended.

3.3. CCD Speed

The speed of a camera is most importantly linked to a given exposure time (and thus sensitivity), but also the ability of the camera to read out the data from that exposure quickly. The more pixels there are on a chip, the more data needs to be read and thus more time is required to properly digitize this information at the ADC. As the temporal resolution demands of researchers intensify, the cameras get pushed to their limit and certain sacrifices inevitably need to be made. The great majority of everyday microscopic needs in the laboratory do not require ultra-high sensitivity (large pixel) cameras. For this reason, higher resolution chips are preferred in many instances for both image quality as well as cost considerations. As a result, when dynamic situations are encountered and the sensitivity characteristics are not present, high exposure times limit the ability to capture dynamic events. Cameras built for speed typically have lower pixel counts, but more importantly are much more sensitive due to the larger pixel dimensions. Moreover, many specialty cameras also have built-in electron multiplication that amplifies the charge for each pixel in the serial register before it hits the ADC. This serves to make the camera much more sensitive to low light, and exposure times become much smaller enabling faster sampling. Now that electron multiplication has become a linear process, it has become much more valuable in the world of quantitative fluorescence microscopy. In addition, most electron multiplication-type cameras have a different manufacturing process in which the chips are back thinned, as opposed to front thinned, which has beneficial impacts on sensitivity.

Many attributes impact the imaging capabilities of a camera, including noise, quantum efficiency, cooling, readout speed, and vibration. These all can be equally important when deciding what camera is best for a particular application and readers are encouraged to research these characteristics when considering different manufacturer's products.

3.4. TIRF Applications

TIRF is being used for a wide variety of applications involving analysis of the cell membrane and extracellular interface. Vesicle fusion at the membrane, endo/exocytosis, adhesion, cellular motion/cytoskeletal dynamics, among many others represents areas which TIRF has illuminated in the field of cellular biology (17–19). Translational research and systems biology have gained insights into many cellular responses, such as contact inhibition, activation/deactivation of phosphorylation cascades, and morphological transitions, which can be governed by events at the surface (20). Infectious disease research has also gained new insight into host–pathogen interactions through TIRF (21). Not only do these contributions serve to increase our understanding of cellular function and physiology, they also identify potential drug targets. As is the case with many breakthrough technologies, TIRF has been itself a key technique to develop cutting-edge methods in microscopy. Correlational fluorescent speckle microscopy, photoactivated localization microscopy (PALM), and stochastic optical reconstruction microscopy (STORM) represent just a few of the new techniques that have been accelerated by TIRF microscopy (22–24). With the computational power of current workstations, these methods are tracking molecules past the optical limits of the microscopes themselves with resolutions as low as 18 nm being claimed by some researchers (25). Amazingly, these techniques and levels of detection are being applied to a dynamic environment at this molecular level (26). As we see the line between the computational scientist and the molecular biologist blur and collaborations grow, these capabilities surely only increase with high-end users such as these leading the way.

3.5. Experimental Considerations with TIRF

The biggest limitation that becomes immediately apparent is that for the most part TIRF is limited to the coverslip interface and is not aimed at exciting probes deeper within the sample, such as the internal cellular molecules. Thus, the degree to which this tool is useful is critically dependant on the nature of the research. Setting up a microscope to be able to accomplish TIRF imaging is much more straightforward than it has been in the past, but still is not an ubiquitously adaptable or inexpensive add-on from most manufacturers. In most (but not all) TIRF configurations, lasers are used as the excitation source providing much of the associated cost. A specialized illuminator (and compatible base stand) that is designed to bring laser (or epi-fluorescent) light into the back aperture of high NA objectives at controlled angles is also a critical component for TIRF imaging. These illuminators are either automated or manual in their ability to switch illumination sources and to alter the angle of the incoming beam to control evanescent field wave depth. While achieving and surpassing critical angle is a seemingly simple concept, there are many considerations with a complete TIRF system. One such consideration is the requirement for stringent tolerances on torque or unevenness of any reflecting or refracting surface, especially the filters and dichroic mirrors. Special metal and strain-free cubes have been developed to address this issue. In addition, if refractive indices and wavelength are not optimized for TIRF, interference patterns and inability to achieve critical angles can be problematic. This is of special note when setting up experiments using multiple-wavelength excitation beams as the critical angle of each beam almost certainly varies at the coverslip. Automated TIRF systems have addressed this problem by adding the ability to memorize beam angles that may be wavelength dependent, a key feature for live-cell, time-lapse imaging in multiple channels. While not an absolute requirement, TIRF microscopy for the most part is a live-cell technique, so it carries with it all the inherent challenges that go along with live-cell microscopy, some of which have been described above. In this regard, third-party hardware for live-cell microscopy needs to be considered. Depending on the timescale of an experiment, this hardware may include a stage-top or enclosure-type device to control environmental conditions, such as temperature, carbon dioxide, and humidity, as well as mechanisms to control focus through time which until recently has been a major limitation in long-term, time-lapse microscopy/imaging.

3.6. Conclusion

Through the phenomenon of evanescent wave field illumination, TIRF microscopy enables researchers to detect cellular entities down to the level of single-molecule fluorescence in an extremely dynamic environment. While this detection is for the most part limited to fields immediately associated with the cell surface, it provides researchers, such as those studying membrane interactions, a unique look into cellular dynamics. Fields, including systems biology, translational research, and personalized medicine, are increasingly in need of new ways to monitor dynamics of cells not only within the cytosolic boundaries, but also within the extracellular environment. TIRF-based imaging provides another avenue to shed light on this previously obscure boundary.

4. Notes

1. A complete procedure for TIRF microscopy is presented in Total Internal Reflection Fluorescence (TIRF) Microscopy by Trache et al. (7).
2. Nikon provides a Web resource for all types of microscopy applications. The MicroscopyU Web site provides tutorials, images, as well as historical information related to all types of microscopy, http://www.microscopyu.com/.

References

1. Moore, G. (1965) Cramming more components onto integrated circuits. *Electronics* **38**, 114–17.
2. Liotta, L. A., and Kohn, E. C. (2001) The microenvironment of the tumour-host interface. *Nature* **411**, 375–9.
3. Aplin, A. E., Howe, A., Alahari, S. K., and Juliano, R. L. (1998) Signal transduction and signal modulation by cell adhesion receptors: the role of integrins, cadherins, immunoglobulin-cell adhesion molecules, and selectins. *Pharmacol Rev* **50**, 197–263.
4. Ward, Y., Wang, W., Woodhouse, E., Linnoila, I., Liotta, L., *et al.* (2001) Signal pathways which promote invasion and metastasis: critical and distinct contributions of extracellular signal-regulated kinase and Ral-specific guanine exchange factor pathways. *Mol Cell Biol* **21**, 5958–69.
5. Hanahan, D., and Weinberg, R. A. (2000) The hallmarks of cancer. *Cell* **100**, 57–70.
6. DeClerck, Y. A., Mercurio, A. M., Stack, M. S., Chapman, H. A., Zutter, M. M., *et al.* (2004) Proteases, extracellular matrix, and cancer: a workshop of the path B study section. *Am J Pathol* **164**, 1131–9.
7. Trache, A., and Meininger, G. A. (2008) Total internal reflection fluorescence (TIRF) microscopy. *Curr Protoc Microbiol* **Chapter 2**, Unit 2A 2 1-2A 2 22.
8. Kahraman, A., and Inal, M. E. (2002) Protective effects of quercetin on ultraviolet A light-induced oxidative stress in the blood of rat. *J Appl Toxicol* **22**, 303–9.
9. Stryer, L. (1995) DNA Structure, Replication, and Repair, *in* Biochemistry, W.H. Freeman and Company, New York, NY, pp. 811.
10. Lodish, H., Berk, A., Kaiser, C., Kreiger, M., Scott, M., *et al.* (2007) *Molecular Cell Biology*. W.H. Freeman and Company, New York, NY.
11. Serway, R., and Faughn, J. (1999) *College Physics*. Saunders College Publishing, Philadelphia, PA.
12. Ross, S., Schwartz, S., Fellers, T., and Davison, M. (2000) *Total Internal Reflection Fluorescence (TIRF) Microscopy*. September 27, 2009, http://www.microscopyu.com/articles/fluorescence/tirf/tirfintro.html
13. Byrne, G. D., Pitter, M. C., Zhang, J., Falcone, F. H., Stolnik, S., *et al.* (2008) Total internal reflection microscopy for live imaging of cellular uptake of sub-micron non-fluorescent particles. *J Microsc* **231**, 168–79.
14. Joselevitch, C., and Zenisek, D. (2009) Imaging exocytosis in retinal bipolar cells with TIRF microscopy. *J Vis Exp* **28**, e1395, doi:10.3791/1305.
15. Young, I. T. (2001) Calibration: sampling density and spatial resolution. *Curr Protoc Cytom* **Chapter 2**, Unit 2 6.
16. Young, I. T. (1988) Sampling density and quantitative microscopy. *Anal Quant Cytol Histol* **10**, 269–75.
17. Blow, N. (2007) Cell migration: our protruding knowledge. *Nature Methods* **4**, 589–94.
18. Cohen, M., Kam, Z., Addadi, L., and Geiger, B. (2006) Dynamic study of the transition from

hyaluronan- to integrin-mediated adhesion in chondrocytes. *Embo J* **25**, 302–11.

19. Huang, S., Lifshitz, L. M., Jones, C., Bellve, K. D., Standley, C., *et al.* (2007) Insulin stimulates membrane fusion and GLUT4 accumulation in clathrin coats on adipocyte plasma membranes. *Mol Cell Biol* **27**, 3456–69.
20. Huang, C., Rajfur, Z., Borchers, C., Schaller, M. D., and Jacobson, K. (2003) JNK phosphorylates paxillin and regulates cell migration. *Nature* **424**, 219–23.
21. Ewers, H., Smith, A. E., Sbalzarini, I. F., Lilie, H., Koumoutsakos, P., *et al.* (2005) Single-particle tracking of murine polyoma virus-like particles on live cells and artificial membranes. *Proc Natl Acad Sci U S A* **102**, 15110–5.
22. Hu, K., Ji, L., Applegate, K. T., Danuser, G., and Waterman-Storer, C. M. (2007) Differential transmission of actin motion within focal adhesions. *Science* **315**, 111–5.
23. Betzig, E., Patterson, G. H., Sougrat, R., Lindwasser, O. W., Olenych, S., *et al.* (2006) Imaging intracellular fluorescent proteins at nanometer resolution. *Science* **313**, 1642–5.
24. Rust, M. J., Bates, M., and Zhuang, X. (2006) Sub-diffraction-limit imaging by stochastic optical reconstruction microscopy (STORM). *Nat Methods* **3**, 793–5.
25. Eisenstein, M. (2006) Thinking big, seeing small. *Nature* **443**, 1019–20.
26. Shroff, H., Galbraith, C. G., Galbraith, J. A., and Betzig, E. (2008) Live-cell photoactivated localization microscopy of nanoscale adhesion dynamics. *Nat Methods* **5**, 417–23.

Chapter 20

Construction and Hyperspectral Imaging of Quantum Dot Lysate Arrays

Kevin P. Rosenblatt, Michael L. Huebschman, and Harold R. Garner

Abstract

The emerging field of proteomic molecular profiling will be driven by new technologies that can measure dozens to hundreds of proteins from a small sample input from a patient's biopsy. Lysate arrays, or reverse-phase protein microarrays, provide a platform for complex mixtures of proteins extracted from cells and tissues to be directly immobilized onto a solid support (such as a biochip with protein binding capacity) in diminutive volumes (picoliter-to-nanoliter). The proteins are spotted using precision robotics and then quantitatively assayed using primary antibodies; important posttranslational modifications, such as phosphorylations that are important for protein activation, may also be assayed to provide an estimate of the regulation of cellular signaling. Until recently, chromogenic signals and fluorescence (using organic fluorophores) detection were two strategies relied upon for signal detection. Emerging regents such as quantum dots (Qdot® nanocrystals; QD) are now employed for improved performance. QD embody a more versatile detection system because the robust signals may be time averaged and the narrow spectral emissions enable many protein targets to be quantified within the same lysate spot. Previously, we found that commercially available pegylated, streptavidin-conjugated QD were effective detection agents, with low-background affinities to spurious components within heterogeneous protein mixtures. Hyperspectral imaging allows the simultaneous detection of the different colored QD reagents within a single lysate spot. Here, we described the construction and imaging of QD lysate arrays. This technology is an emerging, enabling tool within the exciting, clinically oriented field of clinical tissue proteomics.

Key words: Quantum dots, Protein microarray, Lysate arrays, Hyperspectral imaging

1. Introduction

Lysate arrays were developed for the molecular profiling of cellular signal pathways in vivo from microscopic biopsy specimens (1). The physiological state of these pathways may herald cellular disease processes, such as an indolent malignancy. The morphological and biochemical expression of malignant transformation comprises

Virginia Espina and Lance A. Liotta (eds.), *Molecular Profiling: Methods and Protocols*, Methods in Molecular Biology, vol. 823, DOI 10.1007/978-1-60327-216-2_20, © Springer Science+Business Media, LLC 2012

the invasion of malignant cells into the surrounding stroma and the complex tissue interactions cancerous cells, blood vessels, extracellular matrix, and stroma (2). An understanding of these complex interactions may be gained from the molecular profiling of activated signal transduction networks in the component cells of the lesion. While transcriptional profiling using nucleic acid microarrays may reveal activated genes in key pathways that drive disease signaling, the activation (e.g., phosphorylation, cleavage, methylation, etc.) signaling checkpoints are not discerned directly by gene expression alone. Although immunohistochemistry may elucidate activations/inactivations of regulating proteins and localize them to subcellular compartments, this technique lacks quantitative precision – the variability in tissue preprocessing and the inherent semiquantitative nature of measuring the binding of target antibody at single dilutions prevent precise determination of analyte abundance within cells and tissues. Unfortunately, a large number of the most interesting regulatory proteins will be of such low abundance in tissues that it is difficult to detect, much less quantify, these proteins. This difficulty arises in part from our inability to enzymatically amplify the proteome as is indeed possible for DNA and mRNA using PCR and RT-PCR. Lysate arrays were originally constructed out of the need for sensitive, quantitative, and precise protein measuring tools that could monitor changes in the dynamic protein circuitry regulating cell growth, survival, and microenvironment interactions present within diminutive clinical tissue samples, such as human biopsies (3, 4).

Distinct from antibody arrays, wherein the detecting antibodies are immobilized on a slide or other solid support (5), for the lysate arrays, the subject of this chapter, the entire protein content of a procured cell population (e.g., specifically via laser microdissection) from a tissue or lesion of interest is immobilized on a lysate array substrate (3, 6, 7). The analyte-loaded surfaces are then probed with antibodies specific for individual candidate proteins (or their activated counterparts) of key signaling pathways, or other targets of interest. A typical analysis using human tissue begins with precision laser capture microdissection (LCM) of the cells of interest, cell lysis in a suitable buffer for the types of proteins desired (e.g., membrane proteins vs. cytosolic, nuclear or mitochondrial), and subsequent spotting of only a few nanoliters or hundreds of picoliters of the extract, onto a substrate – often nitrocellulose-coated or chemically modified glass slides – in defined locations arrayed with a pin-, quill-, or noncontact-based microarrayer. Our group has arrayed up to 9,000 individual cellular lysates on a single slide with spot diameters as small as 50 μm. Each microarray slide may then be probed with an antibody or other affinity reagent and the binding typically detected by fluorescent or colorimetric assays. The precision, specificity, and dynamic range of lysate arrays are quantifiable and reproducible (3, 4, 8–12). Numerous serial dilutions of samples, replicates, standard curves, and positive and negative controls, all arrayed on the same slide, make this technology very useful for high-throughput yet quantitative applications.

Most methodologies widely in use today for protein detection are limited in terms of their sensitivity, dynamic range, durability, speed, safety, and their utility for multiplexing. Though many protein array detection systems previously relied upon chromogenic and organic flourophore reporter technologies, which are limited for their dynamic range, sensitivity, spectral overlap, and reagent robustness, newer reagents, such as quantum dot (Qdot® nanocrystals; QD) labeled antibodies, and other such labeled affinity reagents, have improved the performance of array detection systems (13). In order for sufficient signal to be generated for chromogenic systems, an amplification step, such as biotinyl tyramide deposition, is usually required (3, 4). Chromogenic detection methodologies only allow a single reporter output (such as the intensity of chromogen deposition) per protein spot, thus negating multiplexed measurements that allow for internal standardization and higher accuracy and precision. Fluorescent reporter molecules permit a broader dynamic range, which is more suitable for array-based assays. Unfortunately, most organic fluorophores are not robust (i.e., they exhibit significant photobleaching with prolonged excitation exposure, or quenching from the environment), and extensive time averaging for increased sensitivity and scanning optimization is not possible. The broad emission, yet narrow excitation spectra, of most fluorescent reporters necessitates the use of expensive scanners with multiple lasers and there is a severe limitation to the numbers of filters that can be used. Thus, a limit on the number of discriminating colors reduces the ability to perform multiplex analysis with different reporter molecules. Quantum dots (Qdot® nanocrystals; QD) are an attractive alternative detection system because they resist photobleaching (14–17). Their wide excitation but narrow-emission spectra enable the use of filter sets with much less spectral overlap and a single bandwidth excitation source is suitable for a low cost imager for extensive multiplexing of signal outputs (13).

A newer imaging technique referred to as hyperspectral imaging permits broad spectral measurements that span the numerous emission wavelengths of an assortment of fluorophores regardless of their spectral overlap (18–21). These imaging systems may be adapted using a spectrum generator that affords a single light source that can be dialed to the different excitation frequencies needed to excite multiple organic fluorophores. Quantum dots permit a single excitation wavelength in hyperspectral imaging measurements, but no expensive filter sets are required. Thus, a lower cost imaging system may be fashioned for multiplexed measurements. Hyperspectral imaging imbues the lysate array system described here with the proper set of features for high-throughput, quantifiable, multiplexed measurements for rapid screening of numerous clinical samples in parallel. In the following pages, we describe our methodology for the construction and hyperspectral imaging of Qdot® nanocrystal-dependent, protein lysate arrays.

2. Materials

2.1. Samples

1. Laser capture microdissected cells or cell culture pellets.
2. 2× SDS Tris–glycine sample buffer (Invitrogen, cat. no. LC2676).
3. T-PER® tissue protein extraction reagent (Pierce, cat. no. 78510).
4. β-Mercaptoethanol (Sigma cat. no. M-6250).
5. Protease inhibitors (e.g., Complete™ Protease Inhibitor Tablets, Roche), optional (see Note 1).
6. Phosphatase inhibitors (e.g., PhosSTOP Phosphatase Inhibitor, Roche), optional (see Note 2).

2.2. Reverse Phase Protein Microarray Construction

1. 2× Tris–glycine SDS sample buffer (Invitrogen, cat. no. LC2676).
2. T-PER® tissue protein extraction reagent (Pierce, cat. no. 78510).
3. Nitrocellulose-coated glass slides (Grace Bio-Labs, Schott Nexterion, or Kera-Fast).
4. SpotArray™ 24 Microarray Printing System (Perkin Elmer) or BioOdyssey™ Calligrapher™ MiniArrayer (BioRad Laboratories).

2.3. Reverse Phase Protein Microarray Immunostaining

1. Phosphate-buffered saline (PBS) without calcium or magnesium.
2. Reblot Mild 10× (Millipore). Dilute to 1× with dH_2O immediately prior to use. Add 5 mL Reblot Mild to 45 mL dH_2O in a 50-mL conical tube. Mix well by inverting the tube.
3. I-Block Solution (Applied Biosystems/Life Technologies): 1 g I-Block powder, 500 mL PBS, 500 μL Tween-20 (0.1%). In a 500 mL beaker, add 1 g I-Block powder to 500 mL PBS without calcium or magnesium. Add a magnetic stir bar. Heat at medium setting, with constant stirring, until the powder dissolves and solution becomes slightly cloudy. Particulate matter should not be visible. Store at 4°C for up to 2 weeks.
4. Primary antibodies to protein(s) of interest.
5. Biotinylated anti-IgG secondary antibody (species matched to primary antibody).

6. Catalyzed Signal Amplification kit (CSA kit, Dako cat. # K1500).
7. Biotin Blocking System (Dako cat. # X0590).
8. Tris-buffered saline with tween (TBST) (Dako cat. #S3306).
9. Antibody diluent (Dako cat. # S3022).
10. Mozaic IHC Autostainer (Zymed/Invitrogen), optional.
11. Qdot® 655-PEG-streptavidin or non-pegylated-Qdot® 655-streptavidin (Invitrogen).
12. 2% (w/v) bovine serum albumin (BSA) in PBS. Weigh 1 g of BSA. Add the BSA to 50 mL of PBS. Mix well.

2.4. Imaging

1. ScanArray (Perkin Elmer) or LS Reloaded (Tecan) with 655-nm, 605-nm, 565-nm, 525-nm, and 705-nm narrow band-width emission filters (Omega® Optical).
2. Hyperspectral imaging system (see Note 3).

3. Methods

Lysate array construction consists of four basic steps (Figs. 1 and 2): (a) Extraction of proteins from cells that have been procured either by LCM or from cell cultures, (b) application of the extracted proteins upon the solid support of the arrays, (c) immunostaining, and

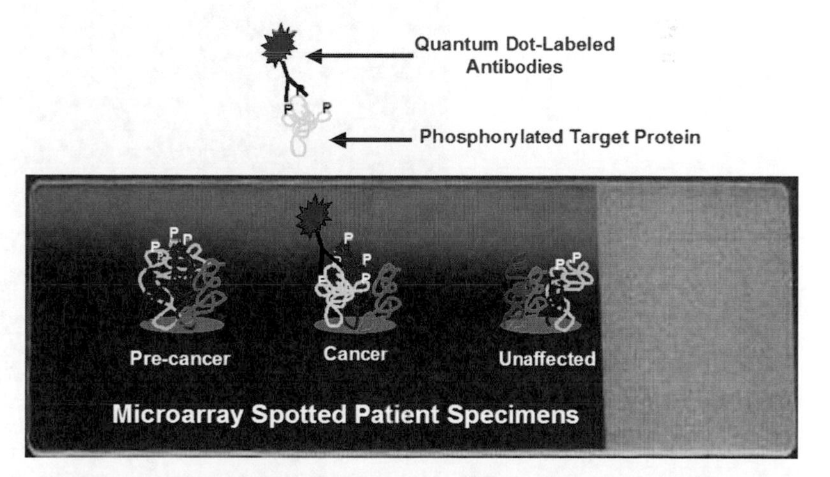

Fig. 1. Schematic outline of the steps involved in constructing a reverse phase microarray. In the lysate arrays, proteins extracted from cellular lysates are arrayed onto a nitrocellulose substrate and probed with a primary antibody. A biotinylated secondary antibody recognizes the presence of the primary antibody. The biotinyl groups are then detected by streptavidin linked to reporter molecules, such as quantum dots instead of an enzyme such as horseradish peroxidase.

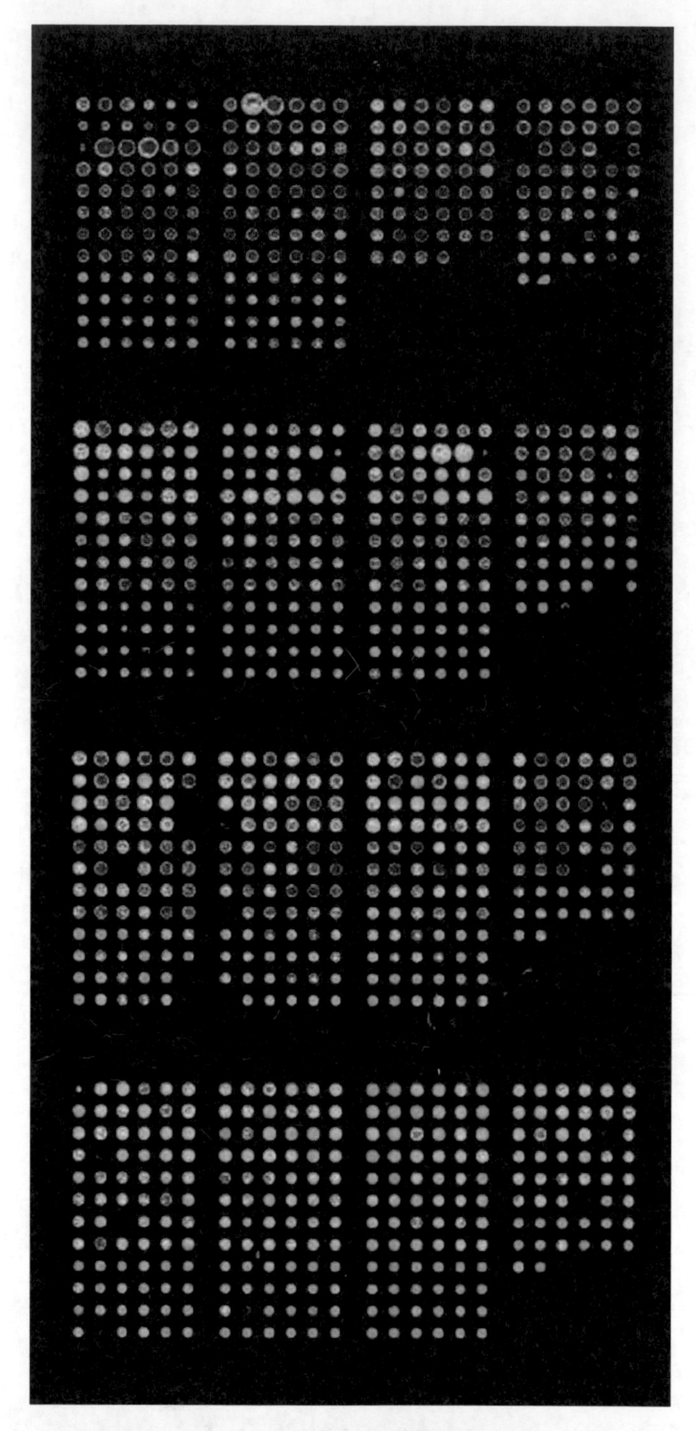

Fig. 2. Example of a lysate array labeled using quantum dot (QD) conjugates. Protein extracts from cancer cell lines cultured under hundreds of different conditions were harvested and arrayed on FAST® nitrocellulose backed glass slides. The protein extracts were arrayed upon nitrocellulose slides (in this case without serial dilutions as a quick screen for differences in Akt target protein phosphorylation). The array surfaces were then exposed to an antibody directed toward a phosphorylated (activated) isoform of the signaling protein and then subjected to the biotinyl tyramide amplification procedure. QDot® 655-streptavidin reagents (20 nM) were incubated on the arrays. For visualization, the microarrays were imaged with a ScanArray (Perkin Elmer) fitted with a 655-nm narrow-bandwidth (20 nm) emission filter. The intensity of the signals is pseudo-colored, with the lowest intensities in the blue spectrum and the highest intensities in the red – saturated signals *appear white.*

(d) imaging. The tissue source of the protein determines the proper protocol for sample lysis and preparation. Microarray spotting of proteins onto the array substrates, immunostaining, and imaging may be done in a fully automated fashion.

3.1. Protein Extraction from Microdissected Tissues

The extraction buffer for liberating proteins from cells that have been procured by LCM consists of a detergent, a denaturing agent, and a buffer (see Note 4).

1. To prepare the protein extraction buffer, prepare a 5% solution of 2-mercaptoethanol (BME) using the 2× Tris–glycine SDS sample buffer, yielding BME–SDS buffer. Example: add 50 μL of BME to 950 μL of 2× Tris–glycine SDS sample buffer.
2. Add equal amounts of BME–SDS buffer and T-PER in order to make the extraction buffer. Example: add 1 mL of BME–SDS buffer solution to 1 mL of T-PER (see Note 1).
3. If the proteins of interest are phosphorylated, phosphatase inhibitors may be added to the extraction buffer solution after the lysate is prepared (see Note 2).
4. Add the desired quantity of the final extraction buffer into the bottom of a 500-μL Eppendorf tube. The minimum volume of extraction buffer required to cover the surface of an LCM cap is 15 μL. The minimum number of cells for protein microarrays is approximately 10,000 cells/15 μL extraction buffer.
5. Thaw each LCM cap at room temperature and remove all traces of condensation from the edges and rim of the cap with a Kimwipe tissue.
6. Place the CapSure transfer film cap containing cells captured by LCM securely onto the tube. Invert and, using a slinging motion, deposit the buffer onto the surface of the cap. Check for leaks. Mix well. Do not vortex.
7. Place the inverted tube with the extraction buffer resting on the surface of the cap into a 75°C ± 2 oven for 30 min to 2 h.
8. After a 15-min incubation at 75°C, gently mix the cap so that the extraction buffer moves over the cap surface. Place the cap back in the oven.
9. At the end of the incubation, mix the inverted tube again.
10. Place the tube into a microcentrifuge and spin the sample for approximately 1 min at approximately 8,200 × *g*.
11. Remove the supernatant and place it in a clean, labeled 1.5-mL screw cap tube. Discard the LCM cap.
12. Store the lysate at −20°C until time to run the downstream assay.

3.1.1. Protein Extraction from Cell Culture Lysates

Proteins may also be extracted from cells cultured under various conditions using the denaturing extraction buffer as described for tissue collected using LCM.

1. To prepare the protein extraction buffer, prepare a 5% solution of BME using the 2× Tris–glycine SDS sample buffer, yielding a BME–SDS buffer. Example: Add 50 μL BME to 950 μL of 2× Tris–glycine SDS sample buffer.
2. Add equal amounts of BME–SDS buffer and T-PER in order to make the extraction buffer. Example: Add 1 mL of BME–SDS buffer solution to 1 mL of T-PER (see Note 1).
3. If the proteins of interest are phosphorylated, phosphatase inhibitors may be added to the extraction buffer solution (see Note 2).
4. Cells from cultures are harvested and centrifuged at 1,000 × *g* for 10 min at 4°C. The supernatant is removed and the cells are washed several times in PBS and centrifuged at 1,000 × *g* for 10 min at 4°C each time.
5. The cell pellet is then resuspended in extraction buffer. Vortex thoroughly to mix.
6. Incubate the lysate on ice for 20 min.
7. Centrifuge the lysate at 10,000 rpm for 4 min at 4°C.
8. Store the supernatant (the protein extract) at –80°C.

3.2. Spotting of Reverse Phase Protein Microarrays

1. Protein extracts can be arrayed on nitrocellulose-coated glass slides using a SpotArray 24 Microarray Printing System or BioOdyssey™ Calligrapher™ MiniArrayer.
2. The lysates are arrayed in serial dilutions in duplicate or triplicate. As a background control, protein extraction buffer alone is arrayed as well.
3. Approximately 1.5 nL per spot is arrayed.
4. Spatial densities of 6,000 spots/slide and greater can easily be accommodated on a 20–30-mm slide.
5. The slides may be stored at –20°C with a desiccant (Drierite, W.A. Hammond) or used immediately.

3.3. Immunostaining Procedure

Each reverse phase protein microarray is probed with a single primary antibody. Signal amplification occurs via horseradish peroxidase-mediated biotinyl tyramide deposition at the site of the antigen–antibody complex. The antibody complex is detected using streptavidin-conjugated quantum dots (Qdot® nanocrystals).

1. Allow the microarrays to reach room temperature in a desiccated environment.
2. Place the arrays to be stained in a shallow container.

3. Add 1× Reblot to cover the arrays. Incubate with constant rocking for 15 min.
4. Wash the arrays twice for 5–10 min with Ca^{2+}- and Mg^{2+} -free PBS.
5. Place the slides in I-Block solution for at least 2 h. I-Block powder is dissolved in calcium- and magnesium-free PBS, with 0.01% Tween 20. Do not remove the I-Block prior to immunostaining the arrays. Do not allow the array surface to dry.
6. Slides may be immunostained using an automatic slide stainer such as the Mozaic IHC Autostainer (Zymed-Invitrogen) or may be stained manually.
7. Incubate the arrays for 5 min with hydrogen peroxide solution from the CSA kit (Dako) and then rinse with high-salt Tris-buffered saline supplemented with 0.1% Tween-20 (TBST, Dako).
8. Block the slides with avidin block solution for 10 min, rinse with TBST, and then incubate with biotin block solution for 10 min.
9. Wash the arrays with TBST again and incubate for 5 min with protein block solution. Air-dry the slides.
10. The arrays are then incubated with either a specific primary antibody diluted in Dako antibody diluent or, as a control, only Dako antibody diluent (time must be optimized individually for each application) (see Note 5).
11. Wash the array with TBST.
12. Incubate the array with a secondary biotinylated antibody (time must be optimized individually for each application) (see Note 5).
13. Prepare the streptavidin–biotin complex by adding one drop of reagent A and one drop of reagent B to 1 mL of the streptavidin buffer solution. Mix well.
14. Wash the array with TBST and incubate with streptavidin–biotin complex (Dako) for 15 min. This is the first step of the signal amplification process.
15. Wash with TBST.
16. Add the amplification reagent and incubate for 15 min.
17. Wash with TBST.
18. Add streptavidin-conjugated Qdot® diluted in 2% albumin in PBS. Incubation times and Qdot® concentrations are experimentally determined (see Note 6).
19. Wash the arrays with TBST. Immediately wash the arrays with dH_2O. Do not allow the arrays to dry out after washing with TBST.

3.4. Imaging

A combination of contrast and brightness are key aspects of fluorescent image quality. Contrast is the ratio of emission intensity to background light (21). Brightness is intrinsic to each fluorophore and the excitation wavelength. Filters are required to block the excitation wavelength in order to produce fluorescent images with high contrast. Narrow band pass filters may reduce the brightness of the image by reducing a portion of the emission spectrum that reaches the detector (21).

Hyperspectral imaging allows an entire emission spectrum to be captured, for each pixel in a microarray spot (scanned image), by illuminating the microarray with a continuously variable excitation system (20, 21). A noncommercial hyperspectral imaging microscope is a superior method for detecting proteins labeled with QDs without the use of catalyzed reporter amplification (13, 18). Hyperspectral imaging with Quantum Dot probes permits multiplexing of numerous proteins in parallel (e.g., activated vs. inactivated forms) without the need for specially designed filters. If a hyperspectral image system is not available, the microarray slides probed with QDs can be imaged using a ScanArray (Perkin Elmer) or LS Reloaded (Tecan) imaging system. The emission signal is visualized using a 655-nm narrow bandwidth emission filter (Omega Optical) (see Note 7).

1. After washing the arrays with water, air dry the slides and analyzed on the hyperspectral imaging system using an imaging spectrometer coupled to a CCD camera (18, 20, 21).&&
2. The slit width of the spectrometer is 72 μm across, and a 100-W Hg lamp is used for excitation. The excitation band pass filter is 330–385 nm and the emission long pass filter is 420 nm.
3. Continuous spectra between 420 and 778 nm are collected for each array, but this range can be expanded depending on the wavelength of the QDot® used in the assays (see Note 8).
4. The hyperspectral imaging system acquires the continuous spectra for each pixel point of the array slide imaged by the spectrometer slit. In a "push broom like" motion, a motorized stage shifts the area viewed by the slit along the array slide acquiring the spectra at each stop. Computer control moves the stage one spectrometer width at a time to cover the array region without overlap; that is a 1.8-μm shift for 72 μm slit width and 40× microscope objective.
5. The data are stored as two spatial dimensions, one wavelength dimension, and the emission count is called an image cube. The image cube is analyzed by applying a least squares fitting algorithm that compares the data spectra to linear combinations of standard spectra and a measured background spectrum. Figure 3 is an example of the output of the software

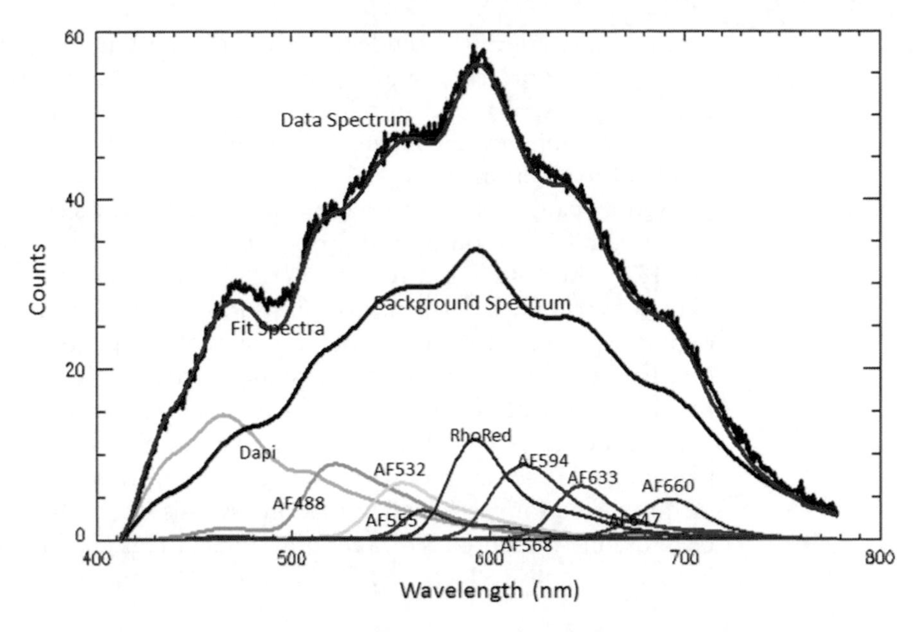

Fig. 3. Example of the analysis software display during a linear fit for ten standard spectra markers for one pixel. The labels indicate the data, the background, standards, and fit spectra. The standards and background are displayed relative to their fit-coefficient. The fit spectrum is the sum of all the standards and the background multiplied by their fit-coefficients. Ten different quantum dot markers in an array spot could be quickly de-convolved similarly since their individual spectra have less overlap with each other than do fluorescent dyes.

program when analyzing a cell stained with ten fluorophores rather than quantum dots, an illustration that more than one marker could also be used in a spot on the array.

6. The analysis software stores the coefficients of each marker as image layers in a TIFF-format file along with an image of all the intensity values of the image cube summed for each pixel, i.e., grayscale image. Depending on the objective power employed in the scan, more than one scan down the array will be needed to scan the entire array. Mosaics of the grayscale images and/or the marker coefficients can be assembled for visual interpretation.
7. In addition to calculating relative amount of the marker in each spot, false color images of the markers and heat map images of the concentrations can be displayed.

3.5. Conclusion

We described herein protocols for constructing lysate protein arrays using streptavidin-conjugated Qdot® nanocrystals and imaging them using both filter-dependent and filter-independent imaging systems. Qdot® nanocrystals overcome many performance limitations of traditional fluorescent tags. Their unique physicochemical properties address a number of issues with standard fluorescent tags, including (a) lack of quenching in signal output during repeated excitation, (b) tendency of tissues and liquids to quench non-Qdot® fluorescent signals, (c) tunable wavelength that can bypass regions of tissue autofluorescence, and (d) limited multiplexing capabilities secondary to broadly overlapping emission bandwidths of different tags.

Qdot® nanocrystals withstand the physical stresses (heat, light, X-rays, and irradiation) that destroy organic fluorophores (13, 14, 16, 17). Their stability presents an advantage over thermo- and light-sensitive tags. Their narrow emission bandwidths and broad excitation bandwidths make the very suitable reagents for multiplexing, because a single light source may excite a collection of QD tags while their narrow emissions permit their separation in charge-coupled device (CCD)-based systems using well-designed narrow pass filters. However, the hyperspectral imaging system itself obviated the need for specially designed filters; this system has recently been used to measure 13 different fluorescent tags (12 plus DAPI) in tissue sections to localize different molecular markers (19).

4. Notes

1. If the proteins of interest are particularly subject to degradation, protease inhibitors can be added to the T-PER® prior to adding the 2× SDS Tris–glycine sample buffer. Dissolve one Complete™ mini protease inhibitor cocktail tablet (Roche, cat. no. 1836153) in 10 mL of T-PER. Alternatively, add 1 mL deionized water to one vial of Calbiochem protease inhibitor cocktail set I. Dilute this 100× solution to 1× (10 μL protease inhibitor cocktail for each 1,000 μL of extraction buffer) for use in the extraction buffer. Then proceed with the preparation of the extraction buffer.
2. Phosphatase inhibitors may be added to the sample lysate after the addition of the denaturing extraction buffer. If kinases are active, addition of phosphatase inhibitors may cause false elevation of phosphorylated proteins (22) after the sample lysate has been denatured by heating at 100°C for 5–8 min. Dissolve one PhosSTOP tablet in 1 mL water or buffer (e.g., 100 mM phosphate buffer, pH 7.0) via swirling. The 10× stock solution is stable for more than 1 month if stored at +2°C to +8°C, or for at least 6 months if stored at −20°C.
3. A hyperspectral microscope is available at the BioCenter at University of Texas Southwestern Medical Center.
4. It is not necessary to extract proteins immediately after microdissection. The caps containing microdissected cells can be stored in a clean microcentrifuge tube at −80°C until all samples have been collected.

5. Typically, RPMA are incubated with primary antibodies for 30 min. Secondary antibody incubation is 15 min. The antibody concentration must be determined empirically for each primary antibody and secondary antibody. Typical primary antibody dilutions are 1:100 to 1:1,000 with a secondary antibody dilution of 1:10,000.
6. To decrease nonspecific binding, a modified form of streptavidin-conjugated QDs, QDot 655-PEG-Sav, may be used as the detection molecule in the RPMA format. This modification involves the introduction of polyethylene glycol groups onto the streptavidin-QD conjugates. In contrast to the nonpegylated form of the streptavidin-QD conjugate, QDot 655-PEG-Sav markedly improves the assay by decreasing the intrinsic, nonspecific binding characteristics of the streptavidin QDs to the protein spots.
7. In this example, the Qdot-probed array was scanned using a ScanArray (Perkin Elmer) imaging system and emission measured at 655 nm. Among the several scanning modalities employed, the most effective signal-to-noise ratio using CCD cameras and filters was achieved using an Omega Optical filter made for assessment of 655 nm fluorescent emissions (20 nm bandwidth centered at 655 nm with added heavy blocking of wavelengths in the infrared and ultraviolet ranges). Hyperspectral systems preclude issues associated with filters and provide the greatest sensitivity.
8. The stability of Qdots®, in addition to their resistance to photobleaching, allows them to be imaged multiple times, for extended periods of time, in order to capture the linear portions of markedly different dilution curves located on the same array surface. Qdots® are available as streptavidin conjugates with emission wavelengths from 525 to 800 nm (Invitrogen).

References

1. Espina, V., Geho, D., Mehta, A. I., Petricoin, E. F., 3rd, Liotta, L. A., Rosenblatt, K. P. (2005) Pathology of the future: molecular profiling for targeted therapy. *Cancer Invest* 23, 36–46.
2. Liotta, L. A., Kohn, E. C. (2001) The microenvironment of the tumour-host interface. *Nature* 411, 375–9.
3. Liotta, L. A., Espina, V., Mehta, A. I., Calvert, V., Rosenblatt, K., Geho, D. et al. (2003) Protein microarrays: meeting analytical challenges for clinical applications. *Cancer Cell* 3, 317–25.
4. Paweletz, C. P., Charboneau, L., Bichsel, V. E., Simone, N. L., Chen, T., Gillespie, J. W. et al. (2001) Reverse phase protein microarrays which capture disease progression show activation of pro-survival pathways at the cancer invasion front. *Oncogene* 20, 1981–9.
5. Haab, B. B. (2001) Advances in protein microarray technology for protein expression and interaction profiling. *Curr Opin Drug Discov Devel* 4, 116–23.
6. Bobrow, M. N., Harris, T. D., Shaughnessy, K. J., Litt, G. J. (1989) Catalyzed reporter deposition, a novel method of signal amplification. Application to immunoassays. *J Immunol Methods* 125, 279–85.
7. Bobrow, M. N., Litt, G. J., Shaughnessy, K. J., Mayer, P. C., Conlon, J. (1992) The use of

catalyzed reporter deposition as a means of signal amplification in a variety of formats. *J Immunol Methods* 150, 145–9.

8. Sheehan, K. M., Calvert, V. S., Kay, E. W., Lu, Y., Fishman, D., Espina, V. et al. (2005) Use of reverse phase protein microarrays and reference standard development for molecular network analysis of metastatic ovarian carcinoma. *Mol Cell Proteomics* 4, 346–55.
9. Silvestri, A., Colombatti, A., Calvert, V. S., Deng, J., Mammano, E., Belluco, C. et al. (2010) Protein pathway biomarker analysis of human cancer reveals requirement for upfront cellular-enrichment processing. *Lab Invest* 90, 787–96.
10. Petricoin, E. F., 3rd, Espina, V., Araujo, R. P., Midura, B., Yeung, C., Wan, X. et al. (2007) Phosphoprotein pathway mapping: Akt/mammalian target of rapamycin activation is negatively associated with childhood rhabdomyosarcoma survival. *Cancer Res* 67, 3431–40.
11. Kornblau, S. M., Tibes, R., Qiu, Y., Chen, W., Kantarjian, H. M., Andreeff, M. et al. (2009) Functional proteomic profiling of AML predicts response and survival. *Blood* 113 (1);154–64.
12. Korf, U., Derdak, S., Tresch, A., Henjes, F., Schumacher, S., Schmidt, C. et al. (2008) Quantitative protein microarrays for time-resolved measurements of protein phosphorylation. *Proteomics* 8, 4603–12.
13. Geho, D., Lahar, N., Gurnani, P., Huebschman, M., Herrmann, P., Espina, V. et al. (2005) Pegylated, steptavidin-conjugated quantum dots are effective detection elements for reverse-phase protein microarrays. *Bioconjug Chem* 16, 559–66.
14. Michalet, X., Pinaud, F. F., Bentolila, L. A., Tsay, J. M., Doose, S., Li, J. J. et al. (2005) Quantum dots for live cells, in vivo imaging, and diagnostics. *Science* 307, 538–44.
15. Pinaud, F., Michalet, X., Bentolila, L. A., Tsay, J. M., Doose, S., Li, J. J. et al. (2006) Advances in fluorescence imaging with quantum dot bio-probes. *Biomaterials* 27, 1679–87.
16. Bruchez, M., Jr., Moronne, M., Gin, P., Weiss, S., Alivisatos, A. P. (1998) Semiconductor nanocrystals as fluorescent biological labels. *Science* 281, 2013–6.
17. Chan, W. C., Nie, S. (1998) Quantum dot bioconjugates for ultrasensitive nonisotopic detection. *Science* 281, 2016–8.
18. Huebschman, M. L., Schultz, R. A., Garner, H. R. (2002) Characteristics and capabilities of the hyperspectral imaging microscope. *IEEE Eng Med Biol Mag* 21, 104–17.
19. Huebschman, M. L., Rosenblatt, K. P., Garner, H. R. (2009) Hyperspectral microscopy imaging to analyze pathology samples with multicolors reduces time and cost. *Proc. SPIE 7182* 7182F doi:10.1117/12.809277.
20. Katari, S., Wallack, M., Huebschman, M., Pantano, P., Garner, H. (2009) Fabrication and evaluation of a near-infrared hyperspectral imaging system. *J Microsc* 236, 11–7.
21. Schultz, R. A., Nielsen, T., Zavaleta, J. R., Ruch, R., Wyatt, R., Garner, H. R. (2001) Hyperspectral imaging: a novel approach for microscopic analysis. *Cytometry* 43, 239–47.
22. VanMeter, A. J., Rodriguez, A. S., Bowman, E. D., Jen, J., Harris, C. C., Deng, J. et al. (2008) Laser capture microdissection and protein microarray analysis of human non-small cell lung cancer: differential epidermal growth factor receptor (EGPR) phosphorylation events associated with mutated EGFR compared with wild type. *Mol Cell Proteomics* 7, 1902–24.

Chapter 21

Microarray Data Analysis: Comparing Two Population Means

JiangHong Deng, Valerie Calvert, and Mariaelena Pierobon

Abstract

Scientists employing microarray profiling technology to compare sample sets generate data for a large number of endpoints. Assuming the experimental design minimized sources of bias, and the analytical technology was reliable, precise, and accurate, how does the experimentalist determine which endpoints are meaningfully different between the groups? Comparison of two population means for individual analysis measurements is the most common statistical problem associated with microarray data analysis. This chapter focuses on the hands-on procedures using SAS software to describe how to choose statistical methods to find the statistically significantly different endpoints between two groups of data generated from reverse phase protein microarrays. The four methods outlined are. (a) two-sample *t*-test, (b) Wilcoxon rank sum test, (c) one-sample *t*-test, and (d) Wilcoxon signed rank test. Two sample *t*-test is used for two independently normally distributed groups. One-sample *t*-test is used for a normally distributed difference of paired observations. Wilcoxon rank sum test is considered a nonparametric version of the two-sample *t*-test, and Wilcoxon signed rank test is considered a nonparametric version of the one-sample *t*-test.

Key words: Two-sample *t*-test, Wilcoxon rank sum test, One-sample *t* test, Wilcoxon signed rank test, Parametric, Nonparametric, Normal distribution

1. Introduction

Although protein arrays are often portrayed as an extension of transcript profiling arrays, in reality the two technologies provide information that is distinct and complementary (1, 2). The relative level of a transcript often bears no relationship to the subsequent encoded protein level, and of course, cannot measure posttranslational events. Since posttranslational events (e.g., phosphorylation) drive cellular signaling, proteomic analysis using protein microarrays provides critical information not attainable by gene arrays (3).

The state of protein networks is embodied in posttranslational modifications and protein–protein interactions (4). Following a stimulus, cellular proteins coalesce into clusters and networks.

Virginia Espina and Lance A. Liotta (eds.), *Molecular Profiling: Methods and Protocols*, Methods in Molecular Biology, vol. 823, DOI 10.1007/978-1-60327-216-2_21, © Springer Science+Business Media, LLC 2012

The network may effectively dissolve after the stimulus is removed or following feedback down-regulation. Information flow through the networks takes place through reversible posttranslational modifications, such as protein phosphorylation, and irreversible modifications, such as enzymatic cleavage of target proteins (e.g., caspase cleavage in apoptosis), and specific protein–protein or protein–DNA binding events. Defective, hyperactive, or dominating signal pathways may drive cancer progression, growth, survival, invasion, and metastasis. Pathogenic alternations in protein networks extend outside the cancer cell to the tumor–extracellular matrix and tumor–host interface.

Reverse phase protein microarrays represent a unique format in which an individual test sample is immobilized in each array spot, such that an array comprises hundreds of different patient samples or cellular lysates (3, 5–7). Each array is incubated with one detection antibody and a single endpoint is measured and directly compared across multiple samples.

Reverse phase protein microarrays are currently being applied to cancer research for: (a) discovery of novel ligands that bind to specific bait molecules on the array and (b) profiling the state of specific members of known signal pathways and protein networks. This chapter describes common statistical methods for comparing groups of microarray data, using reverse phase protein microarray data as an example. These statistical comparisons can be applied to gene microarray data as well. Statistical comparison examples will be used to infer which protein endpoints or pathways are meaningful and/or different between groups of patients/samples (8, 9).

2. Materials

1. Software: Statistical analysis software is SAS 9.1.3 from SAS Institute Inc. (see Note 1).
2. Microarray data: Data generated from any experiment in which the experimental design minimized sources of bias, and the analytical technology was reliable, precise, and accurate can be used (10, 11). Data must be annotated with clinical features or other biologically relevant information in order to discriminate between groups (see Note 2).

3. Statistical Methods

Microarray data analysis includes comparing two groups, comparing multiple groups, survival analysis, classification, and clustering. Comparing the means of two groups is the most common statistical function (Fig. 1). Comparing group means allows the researcher to

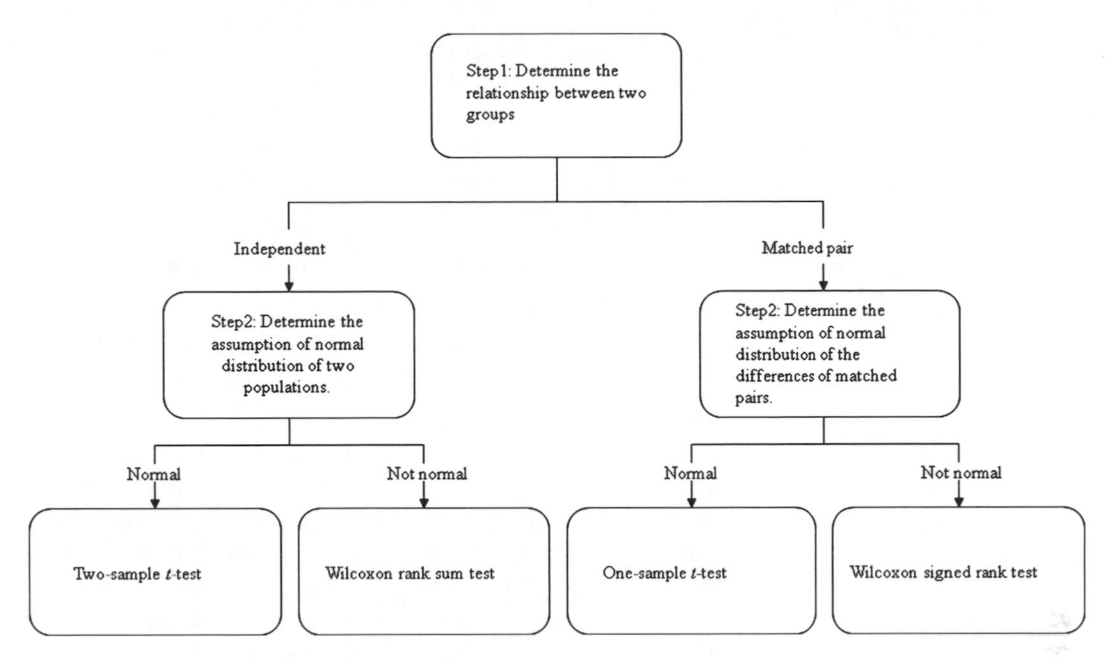

Fig. 1. Flowchart for selecting the appropriate statistical method.

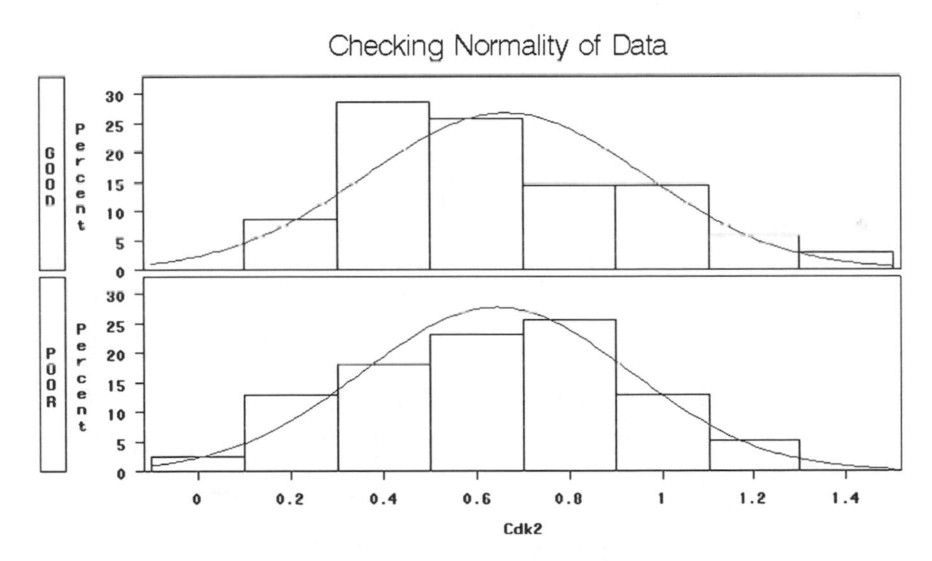

Fig. 2. Histograms of example 1 normal distribution test for two groups, good or poor responders.

answer questions, such as are the microarray data intensities of endpoints from patients who respond to a drug statistically different from the endpoint intensities from patients who do not respond to the drug?

3.1. Data Independence and Normal Distribution Tests

The normal distribution is the basis for many statistical inferences. Normal distribution assumes that the data is centered around the data mean (Gaussian distribution) (Fig. 2) (12). If the means of two groups are compared the distributions of two groups of data should be checked. The two-sample *t*-test is based on the assumption that each group of data follows normal distribution. If the data is not normal, the method cannot be employed. If distributions of both data sets are normal, then a two-sample *t*-test can be used. Otherwise, Wilcoxon rank sum test should be used instead of a two-sample *t*-test. An example test of normal data distribution is shown below in Example 1.

1. The samples of each group of data should be independent.
2. If the data of two groups are independent, the next step is to check if the data of two groups are normally distributed.

   Example 1: Normal Distribution Test: Prednisone therapy is used to treat children with B-cell derived acute lymphoblastic leukemia (B-ALL) and patient response can predict positive or negative treatment outcomes. Patients were enrolled from two groups: 39 prednisone resistant (POOR responders) and 35 prednisone sensitive (GOOD responders). Differences in protein signaling between these two groups were measured by reverse phase protein microarray analysis. One of the endpoints studied was Cdk2, with signal intensities returned for each sample.

   Question: Do the signal intensities of the two groups follow normal distribution?

SAS Code for Example 1

```
/*********************************************
EXAMPLE 1 CHECKING NORMALITIY OF DATA SAS CODE
*********************************************/

DATA ENDPOINT1;
      INPUT RESPONSE $ Cdk2 @@;
      DATALINES;
      POOR  0.307278739  POOR  0.215671255  POOR  0.261845669  POOR
0.251830257
      POOR  0.091446608  POOR  0.414782912  POOR  0.295230167  POOR
0.501576069
      POOR  0.726149037  POOR  0.199488238  POOR  0.511708578  POOR
0.778800783
      POOR  0.532591801  POOR  0.414782912  POOR  0.379083038  POOR
0.771051586
      POOR  1.150273799  POOR  0.618783392  POOR  0.458406011  POOR
0.600495579
      POOR  0.878095431  POOR  0.427414932  POOR  0.748263568  POOR
0.618783392
      POOR  1.02020134   POOR  0.618783392  POOR  0.980198673  POOR
0.778800783
      POOR  0.697676326  POOR  1.02020134   POOR  0.860707976  POOR
0.625002268
      POOR  0.472366553  POOR  0.860707976  POOR  0.878095431  POOR
1.25860001
      POOR  0.826959134  POOR  1.061836547  POOR  0.923116346  GOOD
0.402524224
      GOOD  0.238830783  GOOD  0.313486181  GOOD  0.48190899   GOOD
0.440431655
```

```
        GOOD  0.436049286  GOOD  0.444858066  GOOD  0.339595526  GOOD
0.39455371
        GOOD  0.755783741  GOOD  0.48190899   GOOD  0.511708578  GOOD
0.267135302
        GOOD  1.25860001   GOOD  0.600495579  GOOD  0.778800783  GOOD
0.904837418
        GOOD  0.690734331  GOOD  1.150273799  GOOD  0.207214664  GOOD
1.447734615
        GOOD  0.631283646  GOOD  0.554327285  GOOD  0.980198673  GOOD
0.65704682
        GOOD  0.886920437  GOOD  0.960789439  GOOD  0.496585304  GOOD
0.548811636
        GOOD  0.625002268  GOOD  0.625002268  GOOD  0.990049834  GOOD
0.818730753
        GOOD 1.040810774 GOOD 0.835270211
        ;

TITLE1 "Checking Normality of Data";
PROC UNIVARIATE NORMAL PLOTS DATA = ENDPOINT1;
        CLASS RESPONSE;
        VAR Cdk2;
        HISTOGRAM Cdk2/NORMAL;
        PROBPLOT Cdk2;
RUN;
```

SAS Result for Example 1

```
                         Checking Normality of Data
                       The UNIVARIATE Procedure
                           Variable:  Cdk2
                           RESPONSE = GOOD

                                 Moments

N                          35    Sum Weights                 35
Mean               0.66280845    Sum Observations    23.1982956
Std Deviation       0.2990591    Variance            0.08943634
Skewness           0.69167706    Kurtosis            0.07763775
Uncorrected SS     18.4168619    Corrected SS        3.04083569
Coeff Variation     45.119989    Std Error Mean      0.05055021


                       Basic Statistical Measures

          Location                     Variability

     Mean     0.662808     Std Deviation            0.29906
     Median   0.625002     Variance                 0.08944
     Mode     0.481909     Range                    1.24052
                           Interquartile Range      0.44649

NOTE: The mode displayed is the smallest of 2 modes with a count of 2.


                       Tests for Location: Mu0=0

          Test           -Statistic-     -----p Value------

          Student's t    t  13.11188     Pr > |t|     <.0001
          Sign           M      17.5     Pr >= |M|    <.0001
          Signed Rank    S       315     Pr >= |S|    <.0001
```

```
                         Tests for Normality

Test                  --Statistic---     -----p Value------

Shapiro-Wilk          W     0.958502     Pr < W      0.2061     1
Kolmogorov-Smirnov    D     0.113405     Pr > D     >0.1500
Cramer-von Mises      W-Sq   0.07474     Pr > W-Sq   0.2394
Anderson-Darling      A-Sq  0.443237     Pr > A-Sq  >0.2500

                     Checking Normality of Data
                      The UNIVARIATE Procedure
                          Variable:  Cdk2
                          RESPONSE = POOR

                               Moments

N                              39    Sum Weights                 39
Mean                   0.64197661    Sum Observations    25.0370878
Std Deviation          0.28810243    Variance            0.08300301
Skewness               0.09084353    Kurtosis            -0.7243486
Uncorrected SS         19.2273393    Corrected SS         3.1541145
Coeff Variation        44.8774036    Std Error Mean      0.04613331


                     Basic Statistical Measures

          Location                    Variability

     Mean     0.641977     Std Deviation            0.28810
     Median   0.618783     Variance                 0.08300
     Mode     0.618783     Range                    1.16715
                           Interquartile Range      0.44593


                     Tests for Location: Mu0=0

          Test           -Statistic-    -----p Value------

          Student's t    t  13.91568    Pr > |t|    <.0001
          Sign           M      19.5    Pr >= |M|   <.0001
          Signed Rank    S       390    Pr >= |S|   <.0001


                        Tests for Normality

Test                    --Statistic---    -----p Value------

Shapiro-Wilk            W     0.983504    Pr < W      0.8263          2
Kolmogorov-Smirnov      D     0.061953    Pr > D     >0.1500
Cramer-von Mises        W-Sq  0.032711    Pr > W-Sq  >0.2500
Anderson-Darling        A-Sq  0.206375    Pr > A-Sq  >0.2500
```

3. Interpreting the SAS Result of Example 1:
   SAS UNIVARIATE procedure provides a variety of methods to describe data. Before data is analyzed, data exploration is necessary. HISTOGRAM, PROBPLOT options of UNIVARIATE procedure in SAS create histogram and probability-normal plot, which are very good tools to visualize the data and determine if data is normally distributed or not.

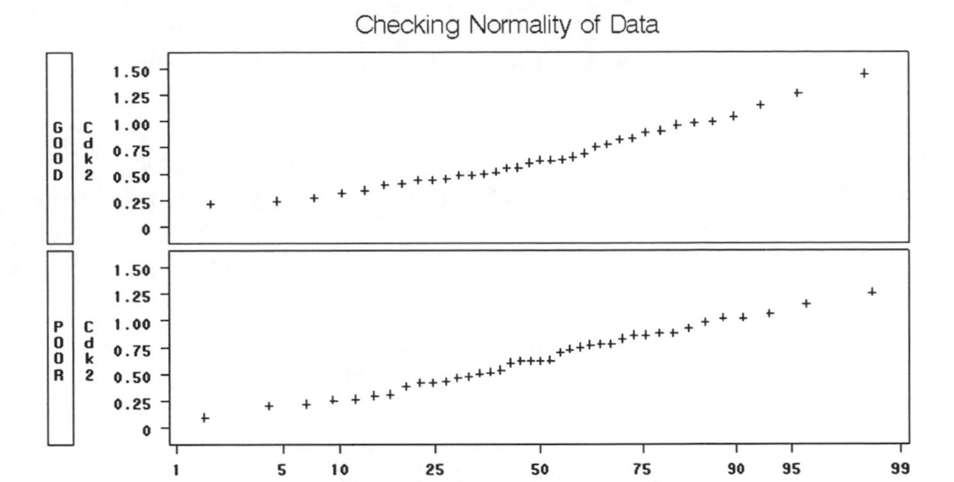

Fig. 3. Quantile-normal plots of example 1 normal distribution test.

The PROBPLOT procedure compares ordered variable values with the percentiles of a specified theoretical distribution. If the data distribution matches the theoretical distribution, the points on the plot form a linear pattern (13).
The characteristics of normal distribution are:

1. The histogram of normally distributed data looks like a "bell-shaped" curve (Fig. 2).
2. The points of quantile-normal plot of normally distributed data would like a diagonal line (Fig. 3).
3. *P*-value of *Shapiro–Wilk* test or *Kolmogorov–Smirnov* test should be *greater* than the statistically significant level. In general, the statistically significant level is 0.05 (see Note 3).

   For example 1, both histograms are "bell-shaped" and symmetric (Fig. 2). Both shapes of quantile-normal plots are diagonal lines (Fig. 3). The normality test is determined by the *Shapiro–Wilk* test or *Kolmogorov–Snirnov* test. *P*-value of *Shapiro–Wilk* test for GOOD response group is 0.2061 1 (see SAS Result for Example 1) which is greater than 0.05. *P*-value of *Shapiro–Wilk* test for POOR response group is 0.8263 2 (see SAS Result for Example 1). So both data of GOOD and POOR response groups follow the normal distribution. From all these characteristics, both GOOD and POOR response group microarray spot intensity data follows a normal distribution. If researchers want to compare the means of these two groups, the two-sample *t*-test should be chosen, since the assumptions of normal distribution of two groups are satisfied (see Note 4).

3.2. Comparison of Data Groups Means

1. If two groups of data are independent and normal, two-sample *t*-test should be chosen. Details are given at Subheading 3.2.1.
2. If two groups of data are independent and not normal, Wilcoxon rank sum test should be used. Details are presented at Subheading 3.2.2.
3. If the data are matched pairs, the difference of the matched pair should be analyzed. The null hypothesis of the mean difference of the matched pair is zero and is rejected only when there is convincing evidence that it is false. When the null hypothesis is true, there is no difference between two groups. If there is not enough evidence to conclude the null hypothesis is correct, that indicates there is difference between two groups. If the difference is a normal distribution, one-sample *t*-test will be chosen. Details are presented at Subheading 3.2.3. Otherwise, Wilcoxon signed rank test will be used (Fig. 1) (see Note 5). Details are presented at Subheading 3.2.4.

3.2.1. Independent Two-Sample t-Test

The two-sample *t*-test is used to compare the means of two populations. The prerequisite for using the two-sample *t*-test is that the data of both groups are normally distributed. If any group of data is not normal, the two-sample *t*-test cannot be used. In this case, Wilcoxon rank sum test should be chosen (see Subheading 3.2.2).

Basic Procedure

1. The assumption of normal distribution of two groups of data should be checked (see Subheading 3.1).
2. If both data are normal, the assumption of equal variances of two groups should be verified. The assumption of equal variances is checked by *F*-test (see Note 6) (16).
3. Find the proper *P*-value from SAS output based on Step 2 result.

   Example 2: Two-Sample t-Test: Prednisone therapy is used to treat children with B-ALL and patient response can predict positive or negative treatment outcomes. Patients were enrolled from two groups: 39 prednisone resistant (POOR responders) and 35 prednisone sensitive (GOOD responders). Differences in protein signaling between these two groups were measured by reverse phase protein microarray analysis. One of the endpoints studied was Cleaved Caspase 3 Asp315, with signal intensities returned for each sample.

   Question: Are the means of endpoint intensities of the two groups (POOR and GOOD) statistically different?

   SAS Code for Example 2

```
/***********************************
EXAMPLE 2 TWO-SAMPLE T-TEST SAS CODE
***********************************/

DATA ENDPOINT2;
      INPUT RESPONSE $ ClCasp9D315 @@;
      DATALINES;
      POOR   0.723973854   POOR   0.531527682   POOR   0.490661892   POOR
0.876340995
      POOR   0.955997482   POOR   1.010050167   POOR   0.734915318   POOR
1.344470157
      POOR   0.771823023   POOR   0.322065609   POOR   1.31127507    POOR
1.20321844
      POOR   0.815462371   POOR   0.706805328   POOR   1.862649618   POOR
1.166490887
      POOR   1.091988122   POOR   0.672334026   POOR   1.05865581    POOR
0.930530896
      POOR   1.076806805   POOR   1.015113065   POOR   1.016128685   POOR
0.804125442
      POOR   1.0010005     POOR   1.337764584   POOR   1.117394907   POOR
1.11071061
      POOR   1.403543345   POOR   1.436198942   POOR   1.306040446   POOR
1.320486197
      POOR   1.328432931   POOR   1.223848008   POOR   0.930530896   POOR
1.168826203
      POOR   1.321807344   POOR   1.212882794   POOR   1.515885869   GOOD
0.875465092
      GOOD   0.488703164   GOOD   0.70468809    GOOD   0.622507253   GOOD
0.661000951
      GOOD   0.913017711   GOOD   1.268709193   GOOD   1.421908524   GOOD
0.879853379
      GOOD   0.901225297   GOOD   1.061836547   GOOD   0.852143789   GOOD
1.029424594
      GOOD   1.302128196   GOOD   0.684545613   GOOD   0.879853379   GOOD
0.921271959
      GOOD   0.836942423   GOOD   1.157196188   GOOD   0.953133787   GOOD
1.074655344
      GOOD   0.620022197   GOOD   0.737123374   GOOD   1.044982355   GOOD
0.97628571
      GOOD   0.922193691   GOOD   1.094174284   GOOD   0.813833076   GOOD
1.168826203
      GOOD   0.866754069   GOOD   0.578104865   GOOD   0.898525673   GOOD
1.099658855
      GOOD 0.815462371 GOOD 0.954087398
      ;

TITLE1 "Checking Normality of Data";
PROC UNIVARIATE NORMAL PLOTS DATA = ENDPOINT2;
      CLASS RESPONSE;
      VAR ClCasp9D315;
      HISTOGRAM ClCasp9D315/NORMAL;
      PROBPLOT ClCasp9D315;
RUN;

TITLE1 "The Two-Sample T-Test";
PROC TTEST DATA = ENDPOINT2;
      CLASS RESPONSE;
      VAR ClCasp9D315;
RUN;
```

SAS Result for Example 2

```
                              The Two-Sample T-Test

                              The TTEST Procedure

                                   Statistics

                                   Lower CL            Upper CL  Lower CL            Upper CL
Variable     RESPONSE        N       Mean      Mean      Mean    Std Dev   Std Dev   Std Dev   Std Err

ClCasp9D315  GOOD           35     0.8443    0.9166    0.9888    0.1701    0.2103    0.2755    0.0355
ClCasp9D315  POOR           39     0.9568    1.0571    1.1575     0.253    0.3095    0.3989    0.0496
ClCasp9D315  Diff (1-2)            -0.265    -0.141    -0.017    0.2299    0.2673    0.3194    0.0622


                                    T-Tests

             Variable      Method           Variances      DF    t Value    Pr > |t|

             ClCasp9D315    Pooled            Equal          72      -2.26      0.0269     1
             ClCasp9D315   Satterthwaite   Unequal      67.3     -2.30     0.0243          2


                              Equality of Variances

               Variable      Method       Num DF    Den DF    F Value    Pr > F

               ClCasp9D315   Folded F        38        34       2.17    0.0244        3
```

Interpreting the SAS Result for Example 2

1. The normality of two groups data are checked like example 1. The conclusion of example 2 is that two data are normal.
2. After checking normality, two-sample *t*-test can be implemented. If you check the *P*-value of two-sample *t*-test, there are two *P*-values in the SAS output 1 2 (*see* SAS Result of Example 2).
3. Before determining which *P*-value should be chosen, *F*-test of equality of variances needs to be conducted. *P*-value of *F*-test is 0.0244 3 (*see* SAS Result of Example 2), which indicates the null hypothesis test of equal variances is rejected. So the two variances are not equal.
4. Now, going back to the SAS output of two-sample *t*-test, *P*-value of two-sample *t*-test with equal variances is chosen. It is 0.0243 2 (*see* SAS Result of Example 2).

3.2.2. Wilcoxon Rank Sum Test

The Wilcoxon rank sum test is the nonparametric version of two-sample *t*-test. It is used to compare the means of two populations when the distributions of data are not normal. The calculation of *P*-value is based on the sum of the ranks within the combined two groups of the observations. If data is highly skewed, the Wilcoxon rank sum test is more powerful than two sample *t*-test.

Basic Procedure

1. Check the assumption of normal distribution of two data (see Subheading 3.1).

2. If any one of two data sets is not normal, Wilcoxon rank sum test is applied.

   Example 3: Wilcoxon Rank Sum Test: Prednisone therapy is used to treat children with B-ALL and patient response can predict positive or negative treatment outcomes. Patients were enrolled from two groups: 39 prednisone resistant (POOR responders) and 35 prednisone sensitive (GOOD responders). Differences in protein signaling between these two groups were measured by reverse phase protein microarray analysis. One of the endpoints studied was CREB Ser133, with signal intensities returned for each sample.

   Question: Are the means of endpoint intensities of the two groups (POOR and GOOD) statistically different?

   SAS Code for Example 3

```
/*************************************
EXAMPLE 3 WILCOXON RANK SUM TEST SAS CODE
*************************************/

DATA ENDPOINT;
      INPUT RESPONSE $ pCREBS133 @@;
      DATALINES;
      POOR  0.037217  POOR  0.022038  POOR  0.085435  POOR  0.241714  POOR
0.180866
      POOR  0.205975  POOR  0.606531  POOR  0.548812  POOR  0.055911  POOR
0.012613
      POOR  0.559898  POOR  0.310367  POOR  0.343009  POOR  0.251579  POOR
0.066073
      POOR  0.343009  POOR  0.364219  POOR  0.107959  POOR  0.346456  POOR
0.382893
      POOR  0.496585  POOR  0.077227  POOR  0.205975  POOR  0.046888  POOR
0.180866
      POOR  0.472367  POOR  0.477114  POOR  0.329559  POOR  0.414783  POOR
0.382893
      POOR  0.246597  POOR  0.22313   POOR  0.755784  POOR  0.398519  POOR
0.307279
      POOR  0.410656  POOR  0.436049  POOR  0.771052  POOR  1.185305  GOOD
0.143704
      GOOD  0.108284  GOOD  0.170333  GOOD  0.115325  GOOD  0.594521  GOOD
0.423162
      GOOD  0.427415  GOOD  0.516851  GOOD  0.182684  GOOD  0.463013  GOOD
0.71177
      GOOD  0.18452   GOOD  1.030455  GOOD  0.548812  GOOD  0.11475   GOOD
0.133989
      GOOD  0.543351  GOOD  0.491644  GOOD  0.516851  GOOD  0.100761  GOOD
0.22091
      GOOD  0.0845    GOOD  0.249075  GOOD  0.559898  GOOD  0.436049  GOOD
0.210136
      GOOD  0.571209  GOOD  0.289384  GOOD  0.353455  GOOD  0.131336  GOOD
0.131336
      GOOD 0.218712 GOOD 0.075698 GOOD 0.216536 GOOD 0.313486
      ;
```

```
TITLE1 "Checking Normality of Data";
PROC UNIVARIATE NORMAL PLOTS DATA = ENDPOINT;
      CLASS RESPONSE;
      VAR pCREBS133;
      HISTOGRAM pCREBS133/NORMAL;
      PROBPLOT pCREBS133;
RUN;

TITLE1 "The Wilcoxon Rank Sum Test";
PROC NPAR1WAY WILCOXON DATA = ENDPOINT;
      CLASS RESPONSE;
      VAR pCREBS133;
RUN;
```

SAS Result for Example 3

```
                         The Wilcoxon Rank Sum Test

                         The NPAR1WAY Procedure

         Wilcoxon Scores (Rank Sums) for Variable pCREBS133
                    Classified by Variable RESPONSE

                          Sum of      Expected       Std Dev          Mean
RESPONSE        N         Scores      Under H0      Under H0         Score
ƒƒƒƒƒƒƒƒƒƒƒƒƒƒƒƒƒƒƒƒƒƒƒƒƒƒƒƒƒƒƒƒƒƒƒƒƒƒƒƒƒƒƒƒƒƒƒƒƒƒƒƒƒƒƒƒƒƒƒƒƒƒƒƒƒƒƒƒƒƒƒ
POOR           39        1442.50       1462.50     92.358610     36.987179
GOOD           35        1332.50       1312.50     92.358610     38.071429

                  Average scores were used for ties.


                      Wilcoxon Two-Sample Test

                  Statistic                 1332.5000

                  Normal Approximation
                  Z                            0.2111
                  One-Sided Pr >  Z            0.4164
                     Two-Sided Pr > |Z|           0.8328                 1

                  t Approximation
                  One-Sided Pr >  Z            0.4167
                  Two-Sided Pr > |Z|           0.8334

             Z includes a continuity correction of 0.5.


                        Kruskal-Wallis Test

                  Chi-Square                   0.0469
                  DF                                1
                  Pr > Chi-Square              0.8286
```

SAS Result for Example 3 of Checking Normality

```
                         Checking Normality of Data

                          The UNIVARIATE Procedure
                           Variable:  pCREBS133
                             RESPONSE = GOOD

                                  Moments

N                                 35    Sum Weights                 35
Mean                        0.330969    Sum Observations     11.583915
Std Deviation             0.22055113    Variance             0.0486428
Skewness                  1.05086216    Kurtosis            1.24050701
Uncorrected SS            5.48777199    Corrected SS        1.65385523
Coeff Variation           66.6380023    Std Error Mean      0.03727995


                        Basic Statistical Measures

              Location                    Variability

          Mean     0.330969     Std Deviation            0.22055
          Median   0.249075     Variance                 0.04864
          Mode     0.131336     Range                    0.95476
                                Interquartile Range      0.38286

NOTE: The mode displayed is the smallest of 2 modes with a count of 2.


                        Tests for Location: Mu0=0

             Test           -Statistic-    -----p Value------

             Student's t    t  8.877937    Pr > |t|    <.0001
             Sign           M      17.5    Pr >= |M|   <.0001
             Signed Rank    S       315    Pr >= |S|   <.0001


                           Tests for Normality

        Test                  --Statistic---    -----p Value------

          Shapiro-Wilk          W     0.889482    Pr < W       0.0021

        Kolmogorov-Smirnov    D      0.176831    Pr > D      <0.0100
        Cramer-von Mises      W-Sq   0.183546    Pr > W-Sq    0.0082
        Anderson-Darling      A-Sq   1.147569    Pr > A-Sq   <0.0050
```

```
                         Checking Normality of Data

                          The UNIVARIATE Procedure
                            Variable:  pCREBS133
                             RESPONSE = POOR

                                 Moments

N                               39    Sum Weights                 39
Mean                    0.33054364    Sum Observations     12.891202
Std Deviation           0.23945864    Variance            0.05734044
Skewness                1.29185487    Kurtosis             3.0521545
Uncorrected SS           6.4400416    Corrected SS        2.17893675
Coeff Variation         72.4438808    Std Error Mean      0.03834407


                         Basic Statistical Measures

            Location                        Variability

        Mean     0.330544     Std Deviation            0.23946
        Median   0.329559     Variance                 0.05734
        Mode     0.180866     Range                    1.17269
                              Interquartile Range      0.25518

NOTE: The mode displayed is the smallest of 4 modes with a count of 2.


                         Tests for Location: Mu0=0

             Test           -Statistic-    -----p Value------

             Student's t    t  8.620463    Pr > |t|    <.0001
             Sign           M      19.5    Pr >= |M|   <.0001
             Signed Rank    S       390    Pr >= |S|   <.0001


                           Tests for Normality

        Test                  --Statistic---      -----p Value------

          Shapiro-Wilk            W      0.908624     Pr < W       0.0039   3

        Kolmogorov-Smirnov    D       0.106087    Pr > D      >0.1500
        Cramer-von Mises      W-Sq    0.084852    Pr > W-Sq    0.1798
        Anderson-Darling      A-Sq    0.680293    Pr > A-Sq    0.0738
```

Interpreting the SAS Result for Example 3

1. The normalities of two groups are checked (*see* SAS Result for Example 3 of Checking Normality and Figs. 4 and 5).
2. *P*-value of GOOD group of normality is 0.0021 2 and *P*-value of POOR group of normality is 0.0039 3. The histograms of two data are not symmetric and the shapes of quantile-normal plots look like two curves. There is insufficient evidence to conclude that both data are normal. Two-sample *t*-test cannot be used now.

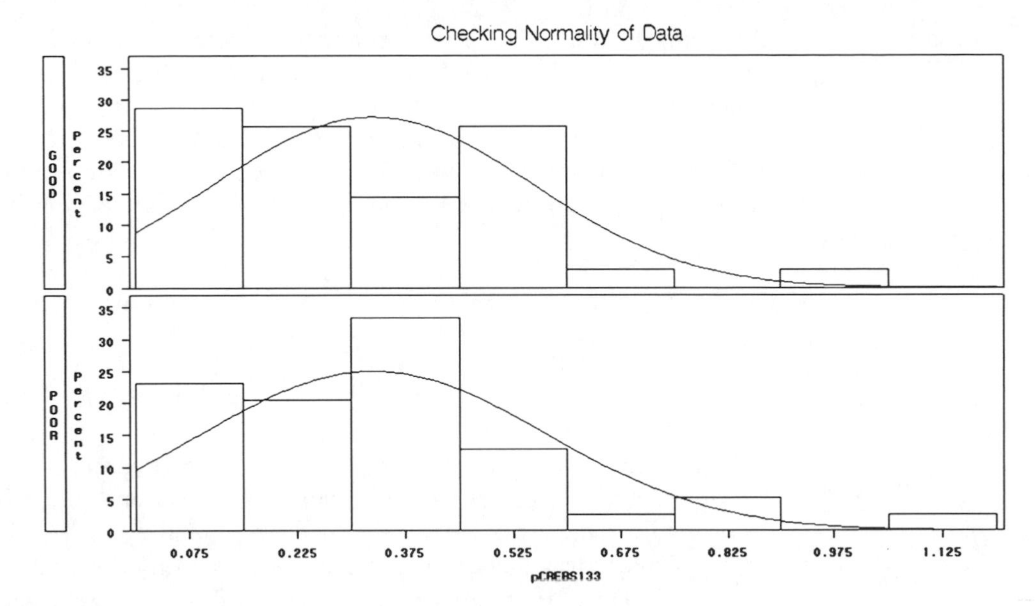

Fig. 4 Histograms of example 3 Wilcoxon rank-sum test of non-normally distributed data.

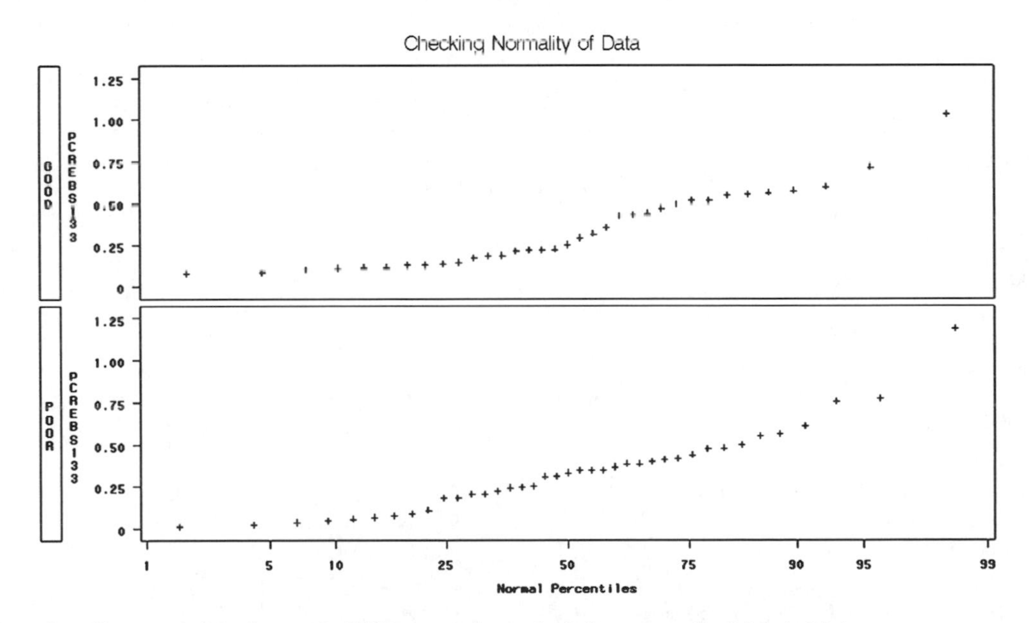

Fig. 5 Quantile-normal plots of example 3 Wilcoxon rank-sum test of non-normally distributed data.

3. In this case, Wilcoxon rank sum test is a more powerful test to compare the means of these two groups. *P*-value of Wilcoxon rank sum test is 0.8328 1, which points out that the intensity mean of GOOD response group is the same as that of POOR response group.

3.2.3. One Sample t-Test

One-sample t-test is used to determine if the mean of population is equal to a given value (17). Mostly, one-sample t-test is used to find if there is any difference between paired observations. In this case, the given value is zero. The assumption of normal distribution of the differences of paired data should be satisfied. If the assumption of normality is not satisfied, Wilcoxon signed rank test will be applied.

Basic Procedure

1. Calculate the difference of each paired observations.
2. Check the normality assumption of the differences.
3. Perform one-sample t-test with SAS when the distribution of the differences is normal.

   Example 4: One Sample t-Test: The phosphoproteomic profile of patients affected by primary colorectal cancer and synchronous liver metastases has been analyzed in order to identify molecules that characterize the metastases from the primary lesions. Thirty-four patients affected by primary colorectal cancer and liver metastases have been enrolled in this study. For each case, the primary colon cancer and its synchronous liver metastasis were collected from the same patient at the same time of surgery. Multiple endpoints between these two groups were measured by reverse phase microarray analysis.

   Question: Is there any difference of endpoint CREB Ser133 intensities between colorectal cancer and liver cancer?

   SAS Code for Example 4

```
/**********************************
EXAMPLE 4 One Sample T Test SAS CODE
**********************************/

DATA ENDPOINT4;
      INPUT ID $ pCREBS133_COLON pCREBS133_LIVER @@;
      pCREBS133_Diff = pCREBS133_COLON - pCREBS133_LIVER;        1
      DATALINES;
1492 3.320116923 2.459603111 2099 0.367879441 0.496585304
2760 0.40656966  1           2764 0.496585304 0.904837418
3046 0.090717953 0.009657698 3386 0.332871084 0.367879441
3539 0.301194212 0.449328964 3703 0.40656966  0.740818221
3904 0.670320046 0.165298888 3952 0.182683524 1.491824698
3997 0.060810063 0.740818221 4096 0.182683524 1
4215 0.818730753 0.60653066  4216 2.718281828 0.904837418
4230 0.40656966  0.367879441 4254 0.904837418 0.740818221
4319 0.548811636 0.818730753 4337 1.221402758 1.105170918
4424 0.670320046 0.60653066  443 0.818730753  1.105170918
4444 1.349858808 0.818730753 5007 0.449328964 0.60653066
5029 0.904837418 1           5092 0.367879441 0.044600955
```

```
5210 0.740818221 0.367879441 5225 0.301194212 0.904837418
5359 0.367879441 0.22313016  5514 0.332871084 1
5527 1.491824698 0.100258844 5647 0.201896518 0.039163895
5700 0.496585304 1.349858808 6661 1           0.22313016
6896 0.182683524 1.221402758 810  1.349858808 1.024660894
;
RUN;

TITLE1 "Checking Normality of Data and One-Sample T-Test";
PROC UNIVARIATE NORMAL PLOTS DATA = ENDPOINT4;
      VAR pCREBS133_Diff;
      HISTOGRAM pCREBS133_Diff/NORMAL;
      PROBPLOT pCREBS133_Diff;
RUN;
```

SAS Result for Example 4

```
        Checking Normality of Data and One Sample T-Test

                    The UNIVARIATE Procedure

                  Variable:  pCREBS133_Diff

                            Moments

N                          34    Sum Weights                 34
Mean               -0.0159501    Sum Observations     -0.542303
Std Deviation      0.64927896    Variance            0.42156317
Skewness           0.60250655    Kurtosis             1.1603074
Uncorrected SS     13.9202345    Corrected SS        13.9115847
Coeff Variation    -4070.6919    Std Error Mean      0.11135042


                  Basic Statistical Measures

        Location                    Variability

    Mean     -0.01595     Std Deviation            0.64928
    Median    0.00184     Variance                 0.42156
    Mode       .          Range                    3.12259
                          Interquartile Range      0.73153


                  Tests for Location: Mu0=0
```

```
Test             -Statistic-      -----p Value------

 Student's     t  -0.14324     Pr > |t|     0.8870      2
Sign           M         0    Pr >= |M|   1.0000
Signed Rank    S     -19.5    Pr >= |S|   0.7442


                     Tests for Normality

Test                    --Statistic---      -----p Value------

Shapiro-Wilk            W     0.968287     Pr < W       0.4159      3
Kolmogorov-Smirnov      D     0.098131     Pr > D      >0.1500
Cramer-von Mises        W-Sq  0.053071     Pr > W-Sq   >0.2500
Anderson-Darling        A-Sq  0.358829     Pr > A-Sq   >0.2500
```

Interpreting the SAS Result for Example 4

The difference of paired observations is calculated by SAS 1. The *P*-value of normality of the difference is 0.4159 (>0.05) with *Shapiro–Wilk test* 3. That indicates the data of difference is normal. It is reasonable to use one sample *t*-test, since the assumption of normality is followed. The *P*-value of the hypothesis test of difference equal to zero is 0.887 2. There is not strong evidence to conclude that two means are different.

3.2.4. Wilcoxon Signed Rank Test

Wilcoxon signed rank test is the nonparametric version of one-sample *t*-test (15). It is for testing that the mean of a population is equal to a hypothesized value, based on the sum of rank of each observation. If data is highly skewed and not symmetric, that shows the data is not normal. In this case, the Wilcoxon signed rank test is more powerful than the one-sample *t*-test.

Basic Procedure

1. Calculate the difference of each pair of observations.
2. Check the normal assumption of the differences.
3. Perform Wilcoxon signed rank test with SAS when the distribution of the differences is not normal.

   Example 5: Wilcoxon Signed Rank Test: The phosphoproteomic profile of patients affected by primary colorectal cancer and synchronous liver metastases has been analyzed in order to identify molecules that characterize the metastases from the primary lesions. Thirty-four patients affected by primary colorectal cancer and liver metastases have been enrolled in this study. For each case, the primary colon cancer and its synchronous liver metastasis were collected from the same patient at the same time of surgery. Multiple endpoints between these two groups were measured by reverse phase protein microarray analysis.

   Question: Is there any difference of endpoint BAD Ser112 intensities between colorectal cancer and liver cancer?

SAS Code for Example 5

```
/****************************************
EXAMPLE 5 Wilcoxon Signed Rank Test SAS CODE
****************************************/

DATA ENDPOINT5;
     INPUT ID $ pBADS112_COLON pBADS112_LIVER @@;
     pBADS112_Diff = pBADS112_COLON - pBADS112_LIVER;
     DATALINES;
1492 1.648721271 1.491824698 2099 1           1.221402758
2760 0.496585304 1           2764 0.904837418 0.818730753
3046 0.246596964 0.044600955 3386 0.740818221  0.548811636
3539 0.496585304 0.670320046 3703 0.496585304 0.740818221
3904 0.740818221 0.301194212 3952 0.40656966  1.8221188
3997 0.165298888 0.548811636 4096 0.548811636 1.349858808
4215 1.105170918 0.670320046 4216 2.225540928 0.904837418
4230 1.105170918 0.60653066  4254 1.491824698 1
4319 1.105170918 1           4337 2.013752707 1
4424 1.105170918 0.740818221 443  0.740818221 0.740818221
4444 1.105170918 0.548811636 5007 1           1.221402758
5029 1.491824698 1           5092 1.221402758 0.131335521
5210 0.740818221 0.449328964 5225 1           0.548811636
5359 0.60653066  0.301194212 5514 0.40656966  0.818730753
5527 2.459603111 0.135335283 5647 0.496585304 0.126185782
5700 0.904837418 1           6661 0.904837418 0.548811636
6896 0.22313016  0.740818221 810  0.904837418 0.995773438
;
RUN;

TITLE1 "Checking Normality of Data and Wilcoxon Signed Rank Test";
PROC UNIVARIATE NORMAL PLOTS DATA = ENDPOINT5;
     VAR pBADS112_Diff;
     HISTOGRAM pBADS112_Diff/NORMAL;
     PROBPLOT pBADS112_Diff;
RUN;
```

SAS Result for Example 5

```
Checking Normality of Data and Wilcoxon Signed Rank Test

              The UNIVARIATE Procedure
              Variable:  pBADS112_Diff

                      Moments

N                         34    Sum Weights                34
Mean              0.19007762    Sum Observations   6.46263923
Std Deviation     0.65474058    Variance           0.42868522
Skewness          0.70226045    Kurtosis            3.0391196
Uncorrected SS    15.3750154    Corrected SS       14.1466123
Coeff Variation   344.459574    Std Error Mean     0.11228708
```

```
                      Basic Statistical Measures

           Location                       Variability

        Mean      0.19008     Std Deviation            0.65474
        Median    0.19700     Variance                 0.42869
        Mode     -0.22140     Range                    3.73982
                              Interquartile Range      0.67259

NOTE: The mode displayed is the smallest of 2 modes with a count of 2.


                      Tests for Location: Mu0=0

           Test           -Statistic-     -----p Value------

           Student's t    t  1.692783     Pr > |t|     0.0999
           Sign           M       4.5     Pr >= |M|    0.1628
           Signed Rank    S      93.5     Pr >= |S|    0.0952          1


                         Tests for Normality

      Test                  --Statistic---     -----p Value------

      Shapiro-Wilk          W     0.935165     Pr < W       0.0443      2
      Kolmogorov-Smirnov    D     0.171663     Pr > D       0.0120
      Cramer-von Mises      W-Sq   0.11905     Pr > W-Sq    0.0618
      Anderson-Darling      A-Sq  0.736957     Pr > A-Sq    0.0495
```

Interpreting the SAS Result for Example 5

P-value of normality test is 0.0443 (<0.05) 2, which indicates data is not normal. Wilcoxon signed rank test is then chosen. *P*-value of Wilcoxon signed rank test is 0.0952 1. All these tests can be performed by SAS UNIVARIATE procedure (13).

4. Notes

1. If one wants to learn basic SAS procedures, a book titled "The Little SAS Book a primer", by Lora D. Delwiche and Susan J. Slaughter (ISBN-10: 1599947250), is very helpful.
2. Data examples in this chapter were culled from actual clinical research studies. The most important aspect of a translational research project is the availability of robust, in-depth clinical/biological parameters associated with the samples.
3. Significant level and *P*-value. Significant level α is also called type I error. Type I error α is the probability of incorrectly rejecting the null hypothesis. In general, significant level is set

as 0.05. Sometimes, it is set as 0.01 or 0.001. *P*-value is the probability of obtaining extreme part of rejection area. The smaller the *P* value is, the more likely the null hypothesis H_0 is rejected. If *P*-value is less than significant level α (0.05), there is insufficient evidence to accept the hypothesis test.

4. After these four methods have been presented, some researchers would ask, "why not use Wilcoxon rank sum test instead of two-sample *t*-test and Wilcoxon signed rank test instead of one-sample *t*-test without checking the normality of data?" In general, parametric test, such as two-sample *t*-test and one-sample *t*-test, is more powerful than the nonparametric test when the sampled population is approximately normally distributed. For instance, when sample size is very small, parametric methods allow us to reject hypothesis test when the statistical level is low. However, in such cases, nonparametric methods will not allow us to do that. When the data is far from normal distribution and highly skewed, nonparametric test is a better choice.

5. Hypothesis test of two-sample *t*-test and Wilcoxon rank sum test is:

   The null hypothesis
   $$H_0 : \mu_1 = \mu_2.$$
   The alternative hypothesis
   $$H_a : \mu_1 \neq \mu_2,$$
   where μ_1, μ_2 are means of two sampled populations.

   Hypothesis test of one-sample *t*-test and Wilcoxon signed rank test is:

   The null hypothesis
   $$H_0 : \mu = \mu_0.$$
   The alternative hypothesis
   $$H_a : \mu \neq \mu_0,$$
   where μ is the mean of the data, μ_0 is the particular value you want to test.

6. Hypothesis test of *F*-test is:

   The null hypothesis
   $$H_0 : \sigma_1^2 = \sigma_2^2.$$
   The alternative hypothesis
   $$H_a : \sigma_1^2 \neq \sigma_2^2,$$
   where σ_1^2 and σ_2^2 are the variances of the populations.

References

1. Celis, J. E., Gromov, P. (2003) Proteomics in translational cancer research: toward an integrated approach. *Cancer Cell* 3, 9–15.
2. Celis, J. E., Gromov, P., Gromova, I., Moreira, J. M., Cabezon, T., Ambartsumian, N. et al. (2003) Integrating proteomic and functional genomic technologies in discovery-driven translational breast cancer research. *Mol Cell Proteomics* 2, 369–77.
3. Liotta, L. A., Espina, V., Mehta, A. I., Calvert, V., Rosenblatt, K., Geho, D. et al. (2003) Protein microarrays: meeting analytical challenges for clinical applications. *Cancer Cell* 3, 317–25.
4. Hanahan, D., Weinberg, R. A. (2000) The hallmarks of cancer. *Cell* 100, 57–70.
5. Paweletz, C. P., Charboneau, L., Bichsel, V. E., Simone, N. L., Chen, T., Gillespie, J. W. et al. (2001) Reverse phase protein microarrays which capture disease progression show activation of pro-survival pathways at the cancer invasion front. *Oncogene* 20, 1981–9.
6. Petricoin, E. F., 3rd, Espina, V., Araujo, R. P., Midura, B., Yeung, C., Wan, X. et al. (2007) Phosphoprotein pathway mapping: Akt/mammalian target of rapamycin activation is negatively associated with childhood rhabdomyosarcoma survival. *Cancer Res* 67, 3431–40.
7. Gulmann, C., Sheehan, K. M., Conroy, R. M., Wulfkuhle, J. D., Espina, V., Mullarkey, M. J. et al. (2009) Quantitative cell signalling analysis reveals down-regulation of MAPK pathway activation in colorectal cancer. *J Pathol* 218, 514–9.
8. Liu, W., Bretz, F., Hayter, A. J., Wynn, H. P. (2009) Assessing Nonsuperiority, Noninferiority, or Equivalence When Comparing Two Regression Models Over a Restricted Covariate Region. *Biometrics* 65 (4); 1279–1287. ISSN 1541–0420.
9. Orina, J. N., Calcagno, A. M., Wu, C. P., Varma, S., Shih, J., Lin, M. et al. (2009) Evaluation of current methods used to analyze the expression profiles of ATP-binding cassette transporters yields an improved drug-discovery database. *Mol Cancer Ther* 8, 2057–66.
10. Ball, C. A., Sherlock, G., Parkinson, H., Rocca-Sera, P., Brooksbank, C., Causton, H. C. et al. (2002) Standards for microarray data. *Science* 298, 539.
11. Brazma, A., Hingamp, P., Quackenbush, J., Sherlock, G., Spellman, P., Stoeckert, C. et al. (2001) Minimum information about a microarray experiment (MIAME)-toward standards for microarray data. *Nat Genet* 29, 365–71.
12. Hamilton, C. Regression with Graphics: A Second course in Applied Statistics, Duxbury Press, Belmont, CA, 1992.
13. SAS (2006) Base SAS 9.1.3 Procedures Guide, in *SAS Procedures Guide*ed.), SAS Institute, Cary, NC, pp.
14. Shapiro, S., Wilk, M. (1965) An Analysis of Variance Test for Normality (Complete Samples). *Biometrika* 52, 591–611.
15. Hayter, A. J. Probability and Statistics for Engineers and Scientists, 2nd. Duxbury Press, Pacific Grove, CA, 2002.
16. Lomax, R. (2007) One-Factor Analysis of Variance-Fixed-Effects Model, in *Statistical Concepts: A Second Course* 3rd ed., Lawrence Erlbaum Associates, Inc., Mahwah, NJ, pp. 10–12.
17. Fisher Box, J. (1987) Guinness, Gosset, Fisher and Small Samples. *Statistical Science* 2, 45–52.

Chapter 22

Bioinformatics/Biostatistics: Microarray Analysis

Gabriel S. Eichler

Abstract

The quantity and complexity of the molecular-level data generated in both research and clinical settings require the use of sophisticated, powerful computational interpretation techniques. It is for this reason that bioinformatic analysis of complex molecular profiling data has become a fundamental technology in the development of personalized medicine. This chapter provides a high-level overview of the field of bioinformatics and outlines several, classic bioinformatic approaches. The highlighted approaches can be aptly applied to nearly any sort of high-dimensional genomic, proteomic, or metabolomic experiments. Reviewed technologies in this chapter include traditional clustering analysis, the Gene Expression Dynamics Inspector (GEDI), GoMiner (GoMiner), Gene Set Enrichment Analysis (GSEA), and the Learner of Functional Enrichment (LeFE).

Key words: Bioinformatics, Biostatistics, Clustering, Genomics, Microarray

1. Introduction

Drawing meaningful conclusions from the analysis of microarray data remains a challenge for many researchers. The difficulty frequently lies in determining which tools and methods are most appropriate for answering the biological question at hand. Answering that question is not simple since microarray technology can be used for a wide range of analyses and the development of new computational techniques remains an active area of bioinformatics research that will continue to expand over the coming years. Nevertheless, this chapter focuses on providing an overview of the fundamental computational approaches for performing the most commonly encountered analyses of microarray data.

Before delving into a description of microarray analysis, it is necessary to provide a more formal description of the results of a microarray experiment. The dataset that emerges from such an

Virginia Espina and Lance A. Liotta (eds.), *Molecular Profiling: Methods and Protocols*, Methods in Molecular Biology, vol. 823, DOI 10.1007/978-1-60327-216-2_22,

experiment is in the form of a matrix quantifying the presence of M molecules (proteins or mRNAs) measured in N samples. Often, a microarray data matrix is also associated with a list of annotations for each sample. The sample's annotations often describe a high-level state, class, or category of the samples and can be categorical, ordinal, or continuous. Sample annotations can be useful for some analyses discussed in this chapter.

It is important to draw attention to the fact that the term *microarray* can have various meanings depending on its context. For example, microarrays can refer to gene expression microarrays, which detect tens of thousands of mRNA transcripts, or protein microarray technologies used to measure tens or hundreds of polypeptides. Single nucleotide polymorphism (SNP) (1) arrays and comparative genomic hybridization arrays (aCGH) (2) are also sometimes referred to as microarray technologies. However, SNP and aCGH technologies are used to study the structure and sequence of genomic DNA, respectively, and therefore they pose unique analytical challenges not addressed by the methods discussed here. Instead, this chapter focuses on the analysis of data originating from gene expression and protein microarray systems.

It is useful to categorize clinical microarray experiments into two broad nonexclusive classes: (1) sample classification and (2) molecular analysis. The objective of the first category, sample classification, is to identify or characterize molecularly distinctive sample subclasses (e.g., therapeutic responders versus nonresponders). If the sample classes are not known a priori then the primary objective is to identify the unknown sample subclasses. Alternatively, if the sample classes are known a priori, the goal of the analysis is to understand the biological difference between the subclasses or perhaps create a predictive *molecular classifier* that can classify unannotated samples encountered in the future. After a molecular classifier is *trained*, it can be used to predict the class of unseen samples and to aid clinicians in making treatment decisions. As such, the popularity of creating molecular classifiers as a diagnostic tool has grown dramatically in recent years and the first generation of multiplexed molecular classifiers have been approved by the FDA for clinical use. Molecular classifiers will likely become increasingly widespread in clinical practice.

The second category of microarray experimentation, molecular analysis, attempts to identify the molecular subsystems (e.g., molecular pathways or functional gene categories) that systematically differ according to the sample's annotations (e.g., healthy versus diseased). Frequently, molecular analyses are carried out using gene set-based approaches that interpret the experimental data within the context of *biological knowledge* contained in independent databases. Gene set-based approaches to molecular analysis are discussed in the second part of this chapter.

2. Sample Classification from Gene Microarray Data

2.1. Microarray Study Design

The first step in identifying sample classes starts with the design of the study itself. It is imperative that all samples be handled in a uniform manner and that the process of sample handling must not be concordant with the already known or predicted sample classes. For example, if two different disease subtypes of samples are being collected, it would be improper for the samples to be collected and processed in two different laboratories, each handling a single sample class. That flaw in sample handling can lead to serious issues in the downstream analysis since there is often a detectable bias originating from minute differences in the sample processing procedures. Generally speaking, all samples should be handled in a manner that is as consistent as possible. Furthermore, if batch processing is necessary, the procedure should be designed to process an equal number of samples from each class in each batch. Likewise spurious molecular differences can be introduced if the microarrays are processed in a manner that temporally or spatially distinguishes the sample classes (3).

2.2. Hierarchical Clustering

Often the first technique applied to a new set of microarray data is a clustering analysis which groups the samples based on the similarity of their molecular expression profiles. Clustering analysis can also be used to identify whether any of the samples may be outliers (i.e., technical or biological anomalies) and should be removed from the analysis. Since clustering is among the class of unsupervised algorithms it operates without using the samples' annotations. Therefore, it can be used to discover new classes of samples or to validate the existence of previously identified classes.

One of the most popular clustering methods, *hierarchical clustering*, creates a tree-like *dendrogram* visualization of the relationships between the samples. Similar to a family tree, dendrograms are drawn so that the height of the branch points is in proportion to the similarities of the connected subgroups. Frequently, dendrograms are visualized in conjunction with a two-dimensional heatmap that uses small colored squares to represent the abundance of each gene or protein in each sample.

Clustering analysis is both an art and a science since there are many parameter choices associated with a clustering analysis and choices about parameter configurations are not straightforward (4). One of the most fundamental clustering parameters is the *distance metric* that quantifies the distance between a pair of samples in M-dimensional space.

Below are some frequently used distance metrics:

Euclidean Distance: The square root of the sum of the squares of each of the N expression values.

Manhattan Distance: The sum of linear distances (similar to counting blocks while walking in Manhattan).

1-Correlation Coefficient: Though technically not a distance metric, this can be useful for comparing samples in which differences in the scales are irrelevant. It can be useful to consider the absolute value of the correlation coefficient in situations when a negative correlation is as informative as a positive correlation.

In addition to computing the distance between samples with a distance metric, clustering also requires a method of computing the distance between groups of samples, known as the *linkage*. Typically, one of the following three linkage methods is used, although there is no strong rationale for using one linkage type over another.

Complete Linkage: The maximal distance between the two groups of samples.

Average Linkage: The average pairwise distance between the two groups of samples.

Single Linkage: The smallest pairwise distance between the two groups of samples.

The actual implementations of hierarchical clustering algorithms can differ, although most software packages usually perform the following steps:

1. Determine the distance between all pairs of samples using the chosen distance metric.
2. Merge the closest pair of samples into a single group.
3. Update the computed distances to include the distance between all samples and the newly formed group using the distance metric and linkage type.
4. Merge the closes pair of samples, groups and samples.
5. Repeat steps 3 and 4 until all of the samples are unified into one single large group.

The choices of linkage and distance metric can produce very different clustering results. Figure 1 shows a comparison of the exact same example dataset in which three a priori known sample classes, labeled as ClassA, ClassB, and ClassC, are clustered using a distance metric of either Euclidean or Manhattan Distance and a linkage type of Average Linkage, Complete Linkage, or Single Linkage. Each of the resultant dendrograms is equally valid yet their results would imply markedly different relationships between the samples. For example, all three of the dendrograms using Euclidean distance were able to clearly separate the three groups. But, the dendrogram that was created using single linkage and Manhattan distance failed to clearly distinguish the three groups. Those differences in results due to choices of input parameters underscore the difficulties with applying hierarchical clustering

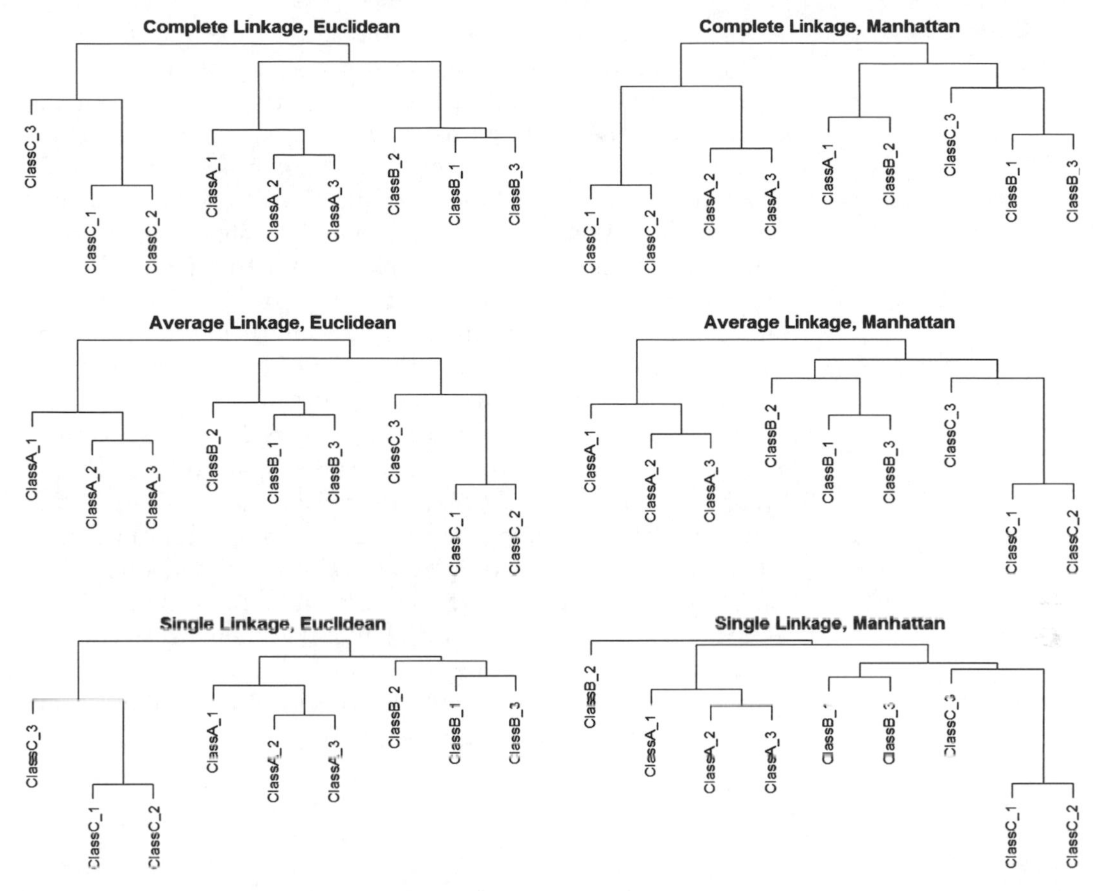

Fig. 1. Results of clustering a dataset using different distance metrics. The same example data set comprised of samples, labeled as ClassA, ClassB, and ClassC, were clustered using a distance metric of either Euclidean or Manhattan Distance and a linkage type of Complete Linkage, Average Linkage, or Single Linkage. Each of the resultant dendrograms is equally valid yet their results imply markedly different relationships between the samples. All three Euclidean distance dendrograms were able to clearly separate the three groups. The single linkage, Manhattan distance dendrogram, failed to clearly distinguish the three groups. These differences in results due to choices of input parameters underscore the difficulties with applying hierarchical clustering techniques to microarray data sets.

techniques. However, sometimes that variability between the dendrograms is used for qualitative assessment of the sample associations' robustness by noting which sample classes frequently co-cluster using several clustering parameter configurations.

2.3. Alternative Clustering Approaches

There are many alternative clustering methods that may be used instead of hierarchical clustering although most methods rely on a distance metric. One useful alternative clustering technique is *k*-means, which separates the samples into *k* distinct groups where *k* is chosen by the operator. The concreteness by which samples are assigned to clusters can improve the interpretability of the analysis since some human subjectivity is required to separate sample groups in a hierarchical clustering analysis. Many other clustering methods are available in a variety of freely available open source software platforms (see Note 1).

2.4. Multidimensional Scaling

Multidimensional Scaling (MDS) (5) is another useful technique for visualizing the relationships between samples. Similar in result to *Principle Component Analysis* (PCA) (6), MDS organizes samples on a two-dimensional plane based on a distance metric computed between all pairs of samples. The placement of the samples on the 2D plane is done to preserve the relative pairwise distances between each sample (point). A useful way to think of MDS analysis is as if there are springs connecting each pair of samples and the length of each spring is in proportion to the distance between the pair of samples it connects. The MDS plot organizes the samples in such a way that it minimizes the strain on all of the springs. MDS plots are useful if it is possible to place the samples with less than 15% *strain*, otherwise, the visualization may be misleading. Figure 2 shows two example MDS plots of the same three-class dataset as above. The distinct clustering of sample classes on the 2D plane visualizes the distinctiveness of the three sample classes when using a Euclidean distance metric. However, similar to the results of hierarchical clustering, the Manhattan distance does not clearly distinguish the sample classes.

Even though the above description of clustering, MDS, and PCA has focused on applications to the analysis of microarray samples, the three techniques can also be used to elucidate the relationships and subclasses found within the gene or protein dimensions of the data.

Manhattan Distance

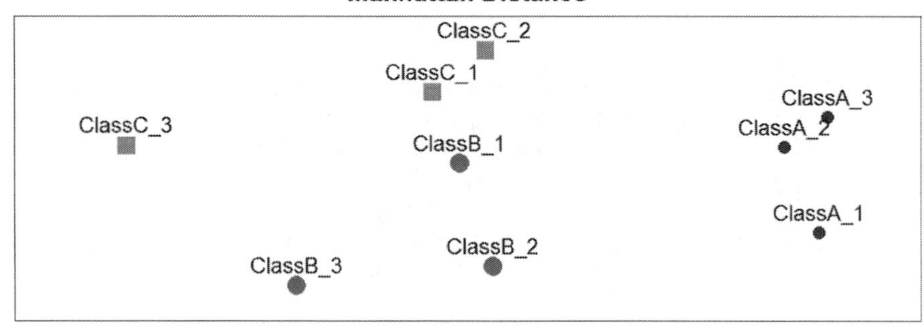

Euclidean Distance

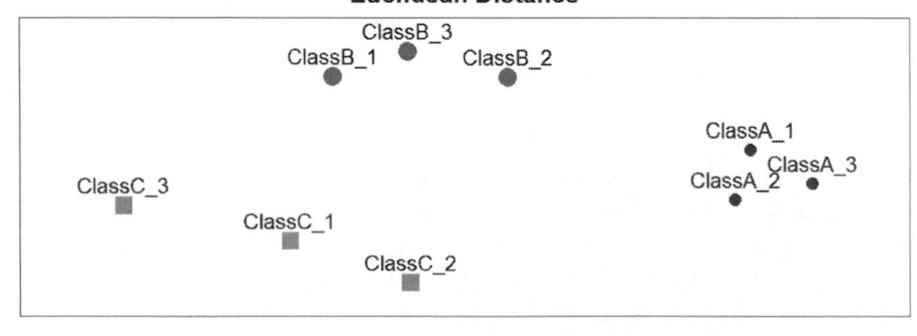

Fig. 2. Multidimensional scaling organizes samples on a two-dimensional plane based on a distance metric computed between all pairs of samples. Two example MDS plots of a three-class dataset are shown using either a Euclidean distance metric or a Manhattan distance metric. Similar to the results of hierarchical clustering, the Manhattan distance does not clearly distinguish the sample classes.

However, since the number of genes is so vast in that dimension, the dendrograms can become excessively complex and difficult to interpret. Therefore, Subheading 3 describes alternative methods for finding relationships within the gene dimension of a microarray dataset.

2.5. Gene Expression Dynamics Inspector and other tools

Another data visualization technique that can be used to elucidate the relationships between samples is the Gene Expression Dynamics Inspector (GEDI) (7). The GEDI method is based on an algorithm called a Self-organizing Map (SOM), which is similar in spirit to MDS, assigning the genes to a two-dimensional map (array) of tiles such that genes with similar expression patterns across the samples are placed on the same or nearby tiles. Once the SOM has been trained, a heatmap-like *mosaic* is created for each sample, by coloring the SOM's tiles according to a color spectrum representing the expression of each of the genes assigned to each tile. That process creates a set of *N* colored mosaics representing the gene expression of each of the samples. Users can use GEDI mosaics to understand the relationships between the different samples. Furthermore, users can click on the mosaic's tiles to reveal the genes underlying the graphical patterns. GEDI provides an intuitive representation of the gene expression data and allows the powerful visual perception of the human mind to identify the relationships between samples.

Additional information on sample analysis resources:

BRB Array Tools: A freely available software system that integrates with Microsoft Excel. The software provides many resources for assessing the relationships among biological samples including many types of clustering, MDS and PCA. It can be found at http://linus.nci.nih.gov/BRB-ArrayTools.html.

The Gene Expression Dynamics Inspector: Learn more about GEDI and its Java-based software package at http://www.childrenshospital.org/research/ingber/GEDI/gedihome.htm.

R and Bioconductor: Both the open-source R programming language (http://www.r-project.org) and its free add-on package called Bioconductor (http://www.bioconductor.org) can be extremely useful for advanced users to perform both supervised and unsupervised sample classification. Some programming knowledge is required.

3. Molecular Analysis of Microarray Data

Molecular analysis of microarray data can provide insights into the samples' underlying biology and is one of the best means of generating novel hypotheses. In order to apply most molecular analysis approaches, each sample must be associated with an annotation that describes high-level information of the sample. The annotation may be a sample classification (e.g., disease versus normal) or a continuous value (e.g., disease-free survival time, response to therapy). The goal of a molecular analysis is to identify the biological subsystems that may be responsible for the differences in annotations of the sample.

Molecular analysis of microarray data has been an evolving research domain in bioinformatics. The first applications of molecular analysis to microarray data usually focused on the identification of individual genes that were most strongly associated with the sample annotations. Unfortunately, that type of analysis was usually dominated by unreliable or outlying probes and frequently highlighted un-interpretable results. In addition, transcript variation among a population, such as splice variants, may confound the results (8) (see Note 2). Given that shortcoming, more recent and successful molecular analysis methods have used an integrative algorithm that analyzes the microarray experiment's data in the context of *knowledge* contained within an independent database of pathways or a set of functionally related genes often called *gene sets*. The objective of that integrative approach is to identify plausible hypotheses that are congruent with previously established findings. Typically, the gene set-based approaches quantify the association between the microarray experiment's annotation data and each gene set and then highlight the most significantly associated gene sets.

Gene set databases have unique ways of organizing their genes. For example, the Kyoto Encyclopedia of Genes and Genomes (KEGG) (9) organizes genes into functional pathways. In contrast to that database, the Gene Ontology (GO) (10) database categorizes several thousand genomes into three distinct but overlapping hierarchical organizations: Biological Process, Molecular Function, and Cellular Component. The precise experimental objectives of the study may help determine which database of gene sets is best to use for a molecular analysis.

There are many software implementations that support molecular analysis of gene microarray data. Three of the major software systems and how they perform their molecular analyses are discussed below.

3.1. GoMiner

As one of the first gene set-based methods for performing molecular analysis, GoMiner (11), as well as several subsequently published approaches, performs a test to detect what statisticians refer to as *enrichment*. Prior to inputting data into GoMiner it must first be preprocessed by computing a univariate statistic that quantifies the strength of association between each gene with the experiment's sample annotations. The choice of statistic is usually chosen by the researcher, however, if the sample annotations are categorical, users typically perform a conventional t-statistic, and if the annotations are a continuous value, a correlation is usually employed. Once the univariate statistics are computed between each gene and the annotation, the user identifies a reasonable threshold value in the distribution of the univariate statistics, above which there is a meaningful association between each gene and the annotation data. The gene symbols of all of the genes that surpass that threshold are put in a *changed file*. Similarly, a *total file* is created using the gene symbols of all of the genes for which a univariate statistic has been computed.

Once the two files have been created, GoMiner uses a Fisher's exact test (12) to identify the GO categories that are overrepresented in the changed file. In other words, GoMiner determines which GO categories contain more genes in the changed list than would otherwise be expected by chance. The most statistically enriched GO categories are presumed to have the strongest association with the experiment's annotations.

One advantage of GoMiner's use of the changed and total files is that they provide a very flexible interface for GoMiner. Should the user have a different means of identifying genes that are strongly associated with the sample annotations, GoMiner can still be used to determine enrichment. However, one drawback of GoMiner is the requirement to make the somewhat arbitrary choice of statistical threshold for determining changed genes. A solution to that shortcoming was addressed by the following method.

3.2. Gene Set Enrichment Analysis

This method was designed to use a special, continuous statistical method. Like GoMiner, Gene Set Enrichment Analysis (GSEA) (13) first computes univariate associations between each gene and the sample's associations. However, the two methods then begin to differ. Rather than relying on a binary classification of each gene as changed or unchanged, GSEA uses all of the genes in the experiment by considering their ranked order of the strength of their univariate associations. The algorithm uses a weighted Kolmogorov–Smirnov statistic (K–S test) to identify the gene sets that are overrepresented at the top or bottom (i.e., a strong positive or negative association, respectively) of the rank order gene list. Since all of the genes are used in the analysis, it is unnecessary to set a threshold differentiating changed from unchanged genes, as is required in GoMiner. One of the drawbacks of the GSEA method is that gene sets with less than 20 genes cause unreliable statistics and therefore must be ignored. GSEA is widely used and remains one of the most popular gene set analysis approaches.

An important issue with most gene-set-based enrichment methods is that they use univariate statistics computed between each gene and the sample's annotations. Though those statistics provide a convenient means to apply classical statistical methods such as the Fisher's exact test, the simplicity of the one-to-one univariate associations may fail to capture more complex many-to-one relationships that exist between multiple genes and a higher-level biological process described by the sample's annotations. That shortcoming was addressed by the LeFE method that is described next.

3.3. Learner of Functional Enrichment

Learner of Functional Enrichment (LeFE) (14) differs from most other methods in that it relies on multivariate, nonlinear machine learning models and does not use univariate statistics. Briefly, for each gene set, LeFE builds a machine learning model of the sample's annotations based on both the genes in a gene set and a comparably sized set of background genes not in the gene set. The machine learning model uses both the genes in the gene set and the background genes to *learn* the sample annotations. LeFE then determines whether the machine learner relied more on the genes in the gene set than the genes in the background set. The most significantly biologically associated gene sets are those in which LeFE's machine learning model relied primarily on the genes in the gene set instead of the background set of genes.

LeFE has several advantages over the previously published methods. One primary advantage is that LeFE uses complex machine learning models that are capable of capturing complex biological phenomena that may be overlooked by simple univariate statistics. Comparisons of LeFE and GSEA have found that former identifies biologically meaningful results that are overlooked by the latter method (14). Furthermore, LeFE's analysis is valid on gene sets with as few as two genes, making it useful for assessing the statistical significance of practically all meaningful types of gene sets. Both of those advantages set LeFE apart from GoMiner and GSEA. One drawback, however, is that since LeFE operates a complex machine learning model, it work best on datasets with at least seven to ten samples. LeFE is ideally suited for experiments with more than 20 microarrays.

Additional information on molecular analysis resources:

The Gene Ontology Consortium (*GO*): Provides information and access to the GO Hierarchies. http://www.geneontology.org/.

Kyoto Encyclopedia of Genes and Genomes (*KEGG*): The pathways database contains annotations for several thousand genes. http://www.genome.ad.jp/kegg/pathway.html.

GoMiner: The GoMiner Application http://discover.nci.nih.gov/gominer/index.jsp as well as a web-based online version called GoMiner which contains additional features for processing multiple different experiments simultaneously. http://discover.nci.nih.gov/gominer/htgm.jsp.

Gene Set Enrichment Analysis (*GSEA*): Contains several versions of GSEA that are all freely accessibly to registered users. http://www.broad.mit.edu/gsea/.

The Learner of Functional Enrichment (*LeFE*): Provides web-based access to the LeFE algorithm. http://discover.nci.nih.gov/lefe/.

3.4. Conclusion

This chapter has reviewed the fundamental approaches to analyzing microarrays. Nevertheless, the analytical techniques discussed are a mere snapshot of a rapidly advancing bioinformatic frontier. As the quality of the experimental platforms improves, so will the computational techniques for interpreting the complex datasets. All signs suggest that microarrays will have a bright future as they become increasingly used in the clinical decision-making process. The new wave of individually tailored molecular medicine will require the integration of high-dimensional molecular profiling information into the clinical decision making process. Microarray technologies can provide those data and thereby empower practical approaches for successful individualized medicine.

4. Notes

1. Example open source clustering software is available at the following websites:
   (a) *k*-means clustering, hierarchical clustering, and self-organizing maps in a single multipurpose open-source library (15) http://bonsai.ims.u-tokyo.ac.jp/mdehoon/software/cluster.
   (b) ScanAlyze for Microarray Image Analysis (16) http://rana.lbl.gov/EisenSoftware.htm.
   (c) Cluster (17) and TreeView for visualization and analysis of complex microarray data. http://rana.lbl.gov/EisenSoftware.htm.
   (d) TimeClust (18) for analyzing data from DNA array time-course experiments http://aimed11.unipv.it/TimeClust.
2. SpliceCenter, http://discover.nci.nih.gov/splicecenter, provides a means of assessing the impact of transcript variation on the design and interpretation of genomic and proteomic data generated from RT-PCR, gene expression microarrays, antibody based arrays, and mass spectrometry proteomics.

References

1. Fan, J. B., Chee, M. S., and Gunderson, K. L. (2006) Highly parallel genomic assays. *Nat Rev Genet* 7, 632–44.
2. Cho, E. K., Tchinda, J., Freeman, J. L., Chung, Y. J., Cai, W. W., *et al.* (2006) Array-based comparative genomic hybridization and copy number variation in cancer research. *Cytogenet Genome Res* **115**, 262–72.
3. Yuan, D. S., and Irizarry, R. A. (2006) High-resolution spatial normalization for microarrays containing embedded technical replicates. *Bioinformatics* **22**, 3054–60.
4. D'Haeseleer, P. (2005) How does gene expression clustering work? *Nat Biotechnol* **23**, 1499–501.
5. Gower, J. C. (1966) Some distance properties of latent root and vector methods used in multivariate analysis. *Biometrika* **53**, 325–28.
6. Pearson, K. (1901) On lines and planes of closest fit to systems of points in space. *Philosophical Magazine* **2**, 559–72.
7. Eichler, G. S., Huang, S., and Ingber, D. E. (2003) Gene Expression Dynamics Inspector (GEDI): for integrative analysis of expression profiles. *Bioinformatics* **19**, 2321–2.

8. Ryan, M. C., Zeeberg, B. R., Caplen, N. J., Cleland, J. A., Kahn, A. B., *et al.* (2008) SpliceCenter: a suite of web-based bioinformatic applications for evaluating the impact of alternative splicing on RT-PCR, RNAi, microarray, and peptide-based studies. *BMC Bioinformatics* **9**, 313.
9. Kanehisa, M. (1997) A database for post-genome analysis. *Trends Genet* **13**, 375–6.
10. Ashburner, M., Ball, C. A., Blake, J. A., Botstein, D., Butler, H., *et al.* (2000) Gene ontology: tool for the unification of biology. The Gene Ontology Consortium. *Nat Genet* **25**, 25–9.
11. Zeeberg, B. R., Feng, W., Wang, G., Wang, M. D., Fojo, A. T., *et al.* (2003) GoMiner: a resource for biological interpretation of genomic and proteomic data. *Genome Biol* **4**, R28.
12. Fisher, R. (1922) On the interpretation of X^2 from contingency tables, and the calculation of P. *Journal of the Royal Statistical Society* **85**, 87–94.
13. Subramanian, A., Tamayo, P., Mootha, V. K., Mukherjee, S., Ebert, B. L., *et al.* (2005) Gene set enrichment analysis: a knowledge-based approach for interpreting genome-wide expression profiles. *Proc Natl Acad Sci U S A* **102**, 15545–50.
14. Eichler, G. S., Reimers, M., Kane, D., and Weinstein, J. N. (2007) The LeFE algorithm: embracing the complexity of gene expression in the interpretation of microarray data. *Genome Biol* **8**, R187.
15. de Hoon, M. J., Imoto, S., Nolan, J., and Miyano, S. (2004) Open source clustering software. *Bioinformatics* **20**, 1453–4.
16. Ball, C. A., Awad, I. A., Demeter, J., Gollub, J., Hebert, J. M., *et al.* (2005) The Stanford Microarray Database accommodates additional microarray platforms and data formats. *Nucleic Acids Res* **33**, D580-2.
17. Eisen, M. B., Spellman, P. T., Brown, P. O., and Botstein, D. (1998) Cluster analysis and display of genome-wide expression patterns. *Proc Natl Acad Sci U S A* **95**, 14863–8.
18. Magni, P., Ferrazzi, F., Sacchi, L., and Bellazzi, R. (2008) TimeClust: a clustering tool for gene expression time series. *Bioinformatics* **24**, 430–2.

Chapter 23

Structure-Based Functional Design of Drugs: From Target to Lead Compound

Amy C. Anderson

Abstract

Proteomic and genomic discoveries have identified vast numbers of new drug targets for investigation. In the quest to discover drugs that modulate the function of these targets, identification of small-molecule drug leads is one of the earliest steps. Structure-based drug design has emerged as a valuable, inexpensive, and rapid computational resource that identifies lead compounds that are complementary to the structure of the target. Leads identified through this process are biologically evaluated and "hit compounds" with affinity and activity are further optimized. This chapter introduces the process of structure-based drug design, including preparation of the ligand database, preparation of the target structure, docking and scoring, and evaluation.

Key words: Docking, Virtual screening, Structure-based drug design, *In silico* library, Flexibility

1. Introduction

Recently, there has been an explosion of genomic, proteomic, and structural information for new drug targets and consequently, hundreds of opportunities for new drug lead discovery. Structure-based drug design (SBDD) continues to play a critical role in this process. The advantages of SBDD are many: hundreds of thousands of ligands can be described and virtually screened as potential drug leads without the need for initial purchase or synthesis, the speed of the SBDD process is rapid relative to in vitro screening, and the cost of the process is relatively low.

The process of SBDD is iterative and fits nicely within the context of a larger drug discovery program (1–3). In this process, software is used to identify optimal binding modes of small-molecule ligands in the structure of a target (docking); these binding modes are then scored for their noncovalent interactions (see Note 1).

Virginia Espina and Lance A. Liotta (eds.), *Molecular Profiling: Methods and Protocols*, Methods in Molecular Biology, vol. 823, DOI 10.1007/978-1-60327-216-2_23, © Springer Science+Business Media, LLC 2012

The ranked list of ligands is then visually evaluated in the complex with the target and top-scoring molecules are often purchased or synthesized. Of these, some are considered "hits" and exhibit affinity for the target. Often, 50% inhibition constants (IC_{50}) for hits are in the range of 10–100 μM. These hits can be optimized toward a higher affinity interaction (IC_{50} = 10–100 nM) by undergoing additional cycles of SBDD using focused databases comprising analogs of the hit scaffold. The optimized hit is a lead that must then be developed for drug-like properties, such as bioavailability, stability, and efficacy.

There are several docking and scoring programs available. Each program has its own strengths and weaknesses (1–6) as well as its own procedures and nuances. The choice of program depends on priorities placed on requirements for flexibility of the target and ligand, virtual screening of whole molecules or de novo construction of a molecule from docked functional groups, and, lastly, purchase price. To describe the exact procedures for even a few of the available programs would be prohibitively lengthy in this chapter; therefore, a more general approach with overall considerations is presented.

2. Materials

2.1. In Silico Ligands

1. A ligand database can consist of compounds that are commercially available, a private collection of compounds, or a collection of functional groups (see Note 2). One example of a collection of commercially available compounds is the ZINC database ((7) and http://zinc.docking.org/).
2. Software for converting the two-dimensional representation of ligands in the database to three-dimensional representations. The programs, CONCORD (Tripos, Inc.) and CORINA, are common examples of software to perform this function.

2.2. Target Structure

The macromolecular structure can often be obtained from the protein database ((8) and http://www.rcsb.org/). Structures determined with X-ray diffraction data are most commonly used for drug design, although solution structures determined with NMR methods and homology models can also be effective.

2.3. Docking Software

Several programs are available and each has key features (Table 1). A few programs are available free of charge to academic users (DOCK and Autodock are two examples), and others are associated with a fee.

Table 1
Ten common docking programs and their characteristics

Program	Description	References
DOCK 6	Docks small molecules, includes solvent effects, uses incremental construction for ligands. Available to academic users without charge	(24)
Autodock	Uses an interaction grid to account for receptor conformations and simulated annealing to account for ligand conformations. Available free of charge	(25)
Glide	Performs complete conformational, orientational, and positional search for ligand	(26)
GOLD	Uses genetic algorithms. Allows partial protein flexibility	(22)
FlexX	Uses incremental construction for ligands	(27)
Flo	Allows protein flexibility and ligand flexibility	(20)
Surflex	Uses incremental construction with fragment assembly	(28)
SLIDE	Allows protein side chain and ligand flexibility	(21)
LUDI	Calculates interaction energy for small-molecule fragments	(29)
GRID	Calculates interaction energy of functional groups	(30)

2.4. Software for Scoring Corrections (Optional)

Most docking software packages have associated scoring functions; however, additional scoring functions to assess the contribution of solvent and the free energy of the target:molecule interaction may increase the predictive accuracy of the docking process.

3. Methods

The process of SBDD begins with preparation of the *in silico* library of ligands and structure of the target. Using docking software, the ligands are then positioned in the target and scored and ranked for noncovalent interactions with the target.

3.1. Ligand Preparation

1. Ligands in the database are usually represented as "strings" that describe the two-dimensional connectivity of atoms. These strings are automatically converted to three-dimensional, minimized representations for docking with software available within the docking package or as a stand-alone utility.
2. The ligand library can be initially filtered to select compounds that are more likely to be bioavailable in later stages. Several criteria may apply, including molecular weight, number of

rotatable bonds, and number of hydrogen bond donor and acceptor groups (9, 10).

3. Ligands are checked for proper geometry, including reasonable bond distances and angles. The conformation of the ligands can be minimized if necessary to achieve proper geometry.
4. Ligands with stereocenters are examined as independent enantiomers.
5. Ligands are appropriately protonated for the pH of the target solution.

3.2. Target Preparation

1. Hydrogens, usually absent from crystal structures determined with data at resolutions lower than 1 Å, are added to macromolecular structures.
2. Charges are calculated and assigned for individual residues.
3. The general docking site is defined. Ideally, the docking site is a pocket in which small-molecule ligands can interact. The ligand docking site can be the active site of an enzyme or an assembly site with another macromolecule. RNA secondary structural elements can also form good ligand docking sites since they have available functional groups arranged in a specific fashion (see Note 3).
4. The specific docking site is defined within the software using the procedures specific to the program (see Note 4). One approach is to define individual residues within the general docking site, and another approach uses a 3.5–6 Å radius around a preexisting ligand.
5. A decision to keep metals and cofactors that may be bound in the docking site is made at this point. Metals and cofactors may form an integral part of the binding interaction with the ligand and, if so, are considered part of the docking site. However, if displacement of the metal or cofactor is desired, they should be removed in order to make the functional groups to which they bind available to the ligand.
6. A decision to keep or remove ordered water molecules that exist in the docking site is made at this point. The water molecules can remain in the docking site if they are critical to ligand binding; they may be removed if the ligand is designed to replace them.
7. If the docking program allows target flexibility in order to accommodate conformational changes induced by the ligand, the number and identity of flexible residues and the degree of flexibility are defined.

3.3. Docking

1. There are individual procedures associated with different docking software packages. For instance, the format of the ligands in

the ligand database may change, the ligand may be docked in entirety or in fragments, or the algorithms for ligand placement may differ. However, the general principles for docking software packages are similar: compounds are first positioned in the target site using an algorithm specific to the software and then evaluated for noncovalent interactions using a scoring algorithm available in the package. Most packages use an automatic format to place and score individual ligands from the database.

2. If the orientation of the ligands in the site is known a priori, this information can be used during and after the docking process. In some programs that allow an anchor fragment, ligand placement can be guided during the docking process. After the docking process, knowledge of binding orientation can be an important filter for evaluating results. As an example, once the orientation of a hit compound has been established by a crystal structure with the target, analogs of that lead can be assumed to dock in a similar fashion.

3.4. Optional Scoring Corrections

Docking scores are inherently approximations to the true binding constant, based primarily on the noncovalent interactions between the ligand and target (11). Postdocking software that attempts to narrow the difference between the docking score and true binding constant can be useful. Three postdocking corrections are presented here: estimating the contribution of solvent, consensus scoring, and calculating a free energy of binding.

1. Solvent plays an important role in ligand binding both in forming specific interactions with ligands and in affecting the dielectric constant and therefore the strength of electrostatic interactions. Some docking algorithms account for solvent effects, others do not and in this case, increased accuracy can be achieved by including a solvation correction to the score (12, 13). In order of increasing accuracy, (a) docking and scoring operations can be performed without modeling the effect of solvent, (b) simple approaches estimate a dielectric constant (either fixed or distance dependent), and (c) explicit solvation models are employed.
2. Consensus scoring, in which the top hits from the docking exercise are rescored with a variety of different scoring algorithms, has proven valuable in achieving greater accuracy (14). Hits appearing at the top of multiple lists are then selected for further investigation.
3. Free energy perturbation (FEP) calculations provide a rigorous measure of the changes in free energy between the unbound and bound complexes in solvent (15).

3.5. Interpretation of Results

1. The docking and scoring software provides a ranked list of ligands (see Note 4). Using computer graphics, these are visually

evaluated in complex with the target to assess goodness of fit, formation of key hydrogen bonds or hydrophobic interactions, surface complementarity, and stability of the bound conformation of the ligand compared to the free conformation. Several top-scoring ligands are then purchased or synthesized and evaluated in the laboratory using in vitro assays with the target.

2. Compounds that succeed as hits in the laboratory assays are then subjected to further rounds of *in silico* screening using more focused libraries of potential analogs. The structure of the complex comprising the target and hit molecules is often experimentally determined in order to validate the binding mode.

4. Notes

1. The choice of target structure is often critical to achieving accurate results in docking. Crystal structures are preferred over solution structures or homology models. Specifically, the best targets for docking are crystal structures determined with data at a resolution better than 2.5 Å, R-factors below 30%, low coordinate error (the Luzzati coordinate error should be in the range 0.2–0.3 Å), and good stereochemistry. The structure should also place at least 90% of the backbone φ and ψ angles in the most favored regions of the Ramachandran plot. Analyses of structure evaluations are available from the PDB. Solution structures determined with NMR data can also be used as docking targets, but care must be exercised to ensure accuracy of the docking results (16). Solution structures are presented as ensembles of structures, all of which satisfy the experimental data; individual ensemble members vary widely and may not be accurate receptors for docking. This problem may be alleviated by docking against each individual member of the ensemble and combining the scores (see below). Homology models generally represent poor targets for docking since they lack the accuracy necessary to predict the proper noncovalent interactions with ligands.
2. Ligands may adopt different conformations with different proteins, and therefore ligand flexibility has been incorporated in many docking algorithms (all programs listed in Table 1 allow ligand flexibility). Ligand conformations can either be precalculated and docked individually or may be "incrementally constructed" during the docking process; the choice between these options is dependent on the docking program. Incremental construction docks an anchor fragment first and then builds up the remainder of the ligand by exploring the potential interactions for various placements of the next fragment across a

rotatable bond. If ligand flexibility is available as an option, it should be utilized.

3. Ligands can induce targets, especially proteins, to undergo significant conformational changes. Incorporating protein flexibility can be critical to achieving accurate docking results (16–19). Some docking programs, such as Flo (20), SLIDE (21), and GOLD (22), directly incorporate protein flexibility and permit the user to identify key residues that are allowed to move during the docking process. Another approach to simulate protein flexibility is to create an ensemble of target structures that represent potential conformations (16, 18, 19, 23). The ensemble can be generated using molecular dynamics simulations, amino acid rotamer libraries, or multiple experimental structures. Ligands can either be docked to individual members of the ensemble and the scores combined or ligands can be docked to a single unified version that represents the individual ensemble members.

4. Evaluation of docking programs is often based on "enrichment" or the identification of known hits from a large database containing decoys. High enrichment values are valuable in selecting initial hits from a diverse database. Nevertheless, accurately ranking a set of subtly different analogs with a con served scaffold is important during lead optimization in order to properly guide new synthetic chemistry efforts. Advances in docking software progress toward the goal of accurate ranking, but the goal remains elusive. In order to judge whether novel compounds are properly ranked, the docking scores of several compounds with known affinity should be compared.

Acknowledgments

The author thanks Dr. Erin Bolstad for her critical review and comments. This work was supported by NIGMS (GM 067542).

References

1. Anderson, A. C. (2003) The process of structure-based drug design. *Chem Biol* 10, 787–97.
2. McInnes, C. (2007) Virtual screening strategies in drug discovery. *Curr Opin Chem Biol* 11, 494–502.
3. Moitessier, N., Englebienne, P., Lee, D., Lawandi, J., Corbeil, C. R. (2008) Towards the development of universal, fast and highly accurate docking/scoring methods: a long way to go. *Br J Pharmacol* 153 Suppl 1, S7-26.
4. Cummings, M. D., DesJarlais, R. L., Gibbs, A. C., Mohan, V., Jaeger, E. P. (2005) Comparison of automated docking programs as virtual screening tools. *J Med Chem* 48, 962–76.
5. Warren, G. L., Andrews, C. W., Capelli, A. M., Clarke, B., LaLonde, J., Lambert, M. H. et al. (2006) A critical assessment of docking programs and scoring functions. *J Med Chem* 49, 5912–31.

6. Wang, R., Lu, Y., Wang, S. (2003) Comparative evaluation of 11 scoring functions for molecular docking. *J Med Chem* 46, 2287–303.
7. Irwin, J. J., Shoichet, B. K. (2005) ZINC--a free database of commercially available compounds for virtual screening. *J Chem Inf Model* 45, 177–82.
8. Berman, H. M., Westbrook, J., Feng, Z., Gilliland, G., Bhat, T. N., Weissig, H. et al. (2000) The Protein Data Bank. *Nucleic Acids Res* 28, 235–42.
9. Lipinski, C. A., Lombardo, F., Dominy, B. W., Feeney, P. J. (2001) Experimental and computational approaches to estimate solubility and permeability in drug discovery and development settings. *Adv Drug Deliv Rev* 46, 3–26.
10. Veber, D. F., Johnson, S. R., Cheng, H. Y., Smith, B. R., Ward, K. W., Kopple, K. D. (2002) Molecular properties that influence the oral bioavailability of drug candidates. *J Med Chem* 45, 2615–23.
11. Huang, S. Y., Grinter, S. Z., Zou, X. (2010) Scoring functions and their evaluation methods for protein-ligand docking: recent advances and future directions. *Phys Chem Chem Phys* 12, 12899–908.
12. Shoichet, B. K., Leach, A. R., Kuntz, I. D. (1999) Ligand solvation in molecular docking. *Proteins* 34, 4–16.
13. Liu, H.-Y., Kuntz, I. D., Zou, X. (2004) Pairwise GB/SA scoring function for structure-based drug design. *J Phys Chem B* 108, 5453–5462.
14. Hattotuwagama, C. K., Davies, M. N., Flower, D. R. (2006) Receptor-ligand binding sites and virtual screening. *Curr Med Chem* 13, 1283–304.
15. Plount Price, M. L., Jorgensen, W. L. (2000) Analysis of Binding Affinities for Celecoxib Analogues with COX-1 and COX-2 from Combined Docking and Monte Carlo Simulations and Insight into the COX-2/COX-1 Selectivity. *J Am Chem Soc* 122, 9455–9466.
16. Bolstad, E. S., Anderson, A. C. (2008) In pursuit of virtual lead optimization: the role of the receptor structure and ensembles in accurate docking. *Proteins* 73, 566–80.
17. B-Rao, C., Subramanian, J., Sharma, S. D. (2009) Managing protein flexibility in docking and its applications. *Drug Discov Today* 14, 394–400.
18. Paulsen, J. L., Anderson, A. C. (2009) Scoring ensembles of docked protein:ligand interactions for virtual lead optimization. *J Chem Inf Model* 49, 2813–9.
19. Totrov, M., Abagyan, R. (2008) Flexible ligand docking to multiple receptor conformations: a practical alternative. *Curr Opin Struct Biol* 18, 178–84.
20. McMartin, C., Bohacek, R. S. (1997) QXP: powerful, rapid computer algorithms for structure-based drug design. *J Comput Aided Mol Des* 11, 333–44.
21. Schnecke, V., Swanson, C. A., Getzoff, E. D., Tainer, J. A., Kuhn, L. A. (1998) Screening a peptidyl database for potential ligands to proteins with side-chain flexibility. *Proteins* 33, 74–87.
22. Jones, G., Willett, P., Glen, R. C., Leach, A. R., Taylor, R. (1997) Development and validation of a genetic algorithm for flexible docking. *J Mol Biol* 267, 727–48.
23. Knegtel, R. M., Kuntz, I. D., Oshiro, C. M. (1997) Molecular docking to ensembles of protein structures. *J Mol Biol* 266, 424–40.
24. Moustakas, D. T., Lang, P. T., Pegg, S., Pettersen, E., Kuntz, I. D., Brooijmans, N. et al. (2006) Development and validation of a modular, extensible docking program: DOCK 5. *J Comput Aided Mol Des* 20, 601–19.
25. Morris, G. M., Huey, R., Lindstrom, W., Sanner, M. F., Belew, R. K., Goodsell, D. S. et al. (2009) AutoDock4 and AutoDockTools4: Automated docking with selective receptor flexibility. *J Comput Chem* 30, 2785–91.
26. Friesner, R. A., Banks, J. L., Murphy, R. B., Halgren, T. A., Klicic, J. J., Mainz, D. T. et al. (2004) Glide: a new approach for rapid, accurate docking and scoring. 1. Method and assessment of docking accuracy. *J Med Chem* 47, 1739–49.
27. Rarey, M., Kramer, B., Lengauer, T., Klebe, G. (1996) A fast flexible docking method using an incremental construction algorithm. *J Mol Biol* 261, 470–89.
28. Jain, A. N. (2003) Surflex: fully automatic flexible molecular docking using a molecular similarity-based search engine. *J Med Chem* 46, 499–511.
29. Boehm, H.-J. (1992) The computer program LUDI: A new method for the de novo design of enzyme inhibitors. 6, in *Journal of Computer-Aided Molecular Design*. Springer Netherlands, 61–78.
30. Goodford, P. J. (1985) A computational procedure for determining energetically favorable binding sites on biologically important macromolecules. *J Med Chem* 28, 849–57.

Chapter 24

Personalized Medicine: Changing the Paradigm of Drug Development

Robin D. Couch and Bryan T. Mott

Abstract

Despite an increased investment in research and development, there has been a steady decline in the number of drugs brought to market over the past 40 years. The tools of personalized medicine are refining diseases into molecular categories, and future therapeutics may be dictated by a patient's molecular profile relative to these categories. The adoption of a personalized medicine approach to drug development may improve the success rate by minimizing variability during each phase of the drug development process. This chapter describes the current paradigm of drug development and then discusses how molecular profiling/personalized medicine might be used to improve upon this paradigm.

Key words: Clinical trials, Drug development, High-throughput screening, Personalized medicine, Theranostic

1. Introduction

Personalized medicine, the use of biomarker-based diagnostics and tailored therapies derived from a patient's molecular profile, does not only revolutionize the approach to patient care, but also profoundly impacts the process of drug development. This chapter provides an overview of the process of drug development and then addresses the influence that molecular profiling and personalized medicine may have on this process.

Virginia Espina and Lance A. Liotta (eds.), *Molecular Profiling: Methods and Protocols*, Methods in Molecular Biology, vol. 823, DOI 10.1007/978-1-60327-216-2_24, © Springer Science+Business Media, LLC 2012

2. The Process of Drug Development

Serendipity largely contributed toward the drug development process well into the 1960s, as pharmaceutical companies randomly tested molecules for desirable activity, in large part, using in vivo model systems. While drugs and drug leads were successfully identified, this approach generally suffered from molecular libraries of insufficient size and limited chemical diversity. Furthermore, the mechanism of action of lead molecules was often unknown, thereby hindering the development of alternatives if the lead molecule failed due to toxicity. To address these shortcomings, a more rational approach to drug discovery was adopted, utilizing structural information (e.g., of a receptor bound to its ligand) and knowledge acquired by studying biological systems and processes. In vitro assays generated valuable information on structure–activity relationships and were used to guide the optimization of lead molecules. Thus, if a lead molecule failed, there was sufficient data to pinpoint and remove the cause from the molecule. The dawn of molecular biology, in conjunction with developments in synthetic chemistry and screening technologies, now permits a combination of both rational and random approaches for drug discovery. As illustrated in Fig. 1, the current process of drug discovery can be divided into five phases: target identification, target validation, lead identification, lead optimization, and preclinical and clinical trials (1).

2.1. Target Identification

The identification of new molecular targets is of paramount importance to the process of drug discovery and development. Modern techniques such as genomics, metabolomics, and proteomics have become essential tools in the search for druggable targets. Genomics involves the study of an organism's entire genome, and includes efforts to determine the complete nucleotide sequence. *Haemophilus influenzae* was the first free-living organism to be sequenced (2), and through advances in automation and informatics the complete genomic sequence of numerous other organisms is being determined at a rapid pace. As of October 2011, the complete genomes of approximately 1,691 bacteria, 144 eukaryotes, and 2,592 viruses have been sequenced (http://www.ebi.ac.uk/genomes/index.html). A bioinformatic comparison of whole genomes enables the

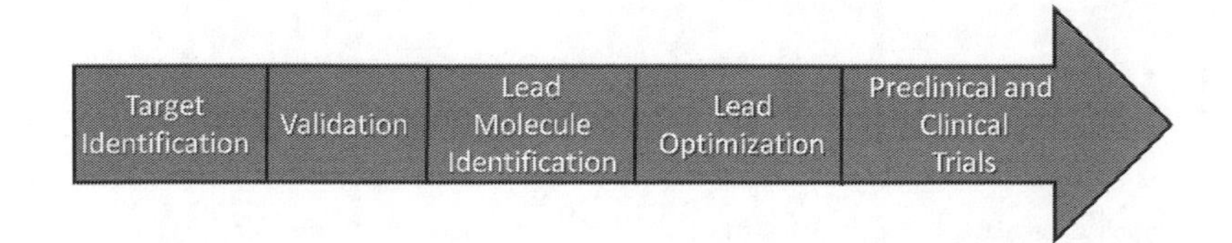

Fig. 1. The current, linear process of drug development.

identification of novel drug targets. For example, a comparison of the human genome with that of the pathogenic bacterium *Francisella tularensis* reveals that each utilizes a different biochemical pathway for the biosynthesis of isoprenes. Isoprenoids are a family of molecules fundamentally involved in a variety of crucial biological functions, including electron transport (quinones), cell wall biosynthesis (dolichols), signal transduction (prenylated proteins), and the regulation of membrane fluidity (hopanoids and sterols). Despite their structural and functional diversity, all isoprenoids are derived from two building blocks, isopentenyl diphosphate and dimethylallyl diphosphate, which originate from either the mevalonic acid (MVA) or methylerythritol phosphate (MEP) pathway (3). Genomic examination reveals that humans exclusively utilize the MVA pathway and are devoid of MEP pathway genes, whereas *F. tularensis* uses the MEP pathway and is devoid of MVA pathway genes (3). Thus, genomic comparison suggests that the MEP pathway enzymes would make attractive targets for the development of novel antimicrobials.

Knowledge of genomic sequences has also permitted functional genomic studies concerned with gene expression patterns of various conditions. For example, using gene chip technology with high-density arrays of oligonucleotide probes, thousands of human genes can be analyzed simultaneously analyzed, permitting a comparison of gene expression between healthy and diseased. Differences in gene expression may implicate pathways or proteins that are involved in the pathophysiology of disease, thereby identifying potential targets for drug development.

The metabolome, the collection of small-molecule metabolites in a biological sample, contains indicators of underlying cellular processes. Metabolomics is a technology concerned with the collection and analysis of multiple metabolites within a sample. The global metabolomic profiling of small molecule metabolites can lead to enhanced understanding of disease mechanisms, the discovery of new diagnostic biomarkers, and the identification of targets for drug discovery. Changes in small molecule abundance have been demonstrated in biological samples such as urine and blood. Global metabolomic profiles can become altered by a variety of physiological and pathological processes, and therefore changes in global profiles may also be indicative of a particular disease state. Hence, differential metabolomic comparison of healthy and diseased can provide insight into gene function and implicate processes, pathways, and proteins for drug development.

But genomics and metabolomics alone do not permit the complete understanding of biological complexity. Proteins are the workhorses of the cell, and many diseases reflect a dysfunction at the protein level. Accordingly, proteins are the primary targets of most drugs (4). Proteomics involves the large-scale characterization of all proteins within a biological system (in contrast to traditional biochemistry which investigated a small number of isolated proteins) and is often used for target identification by comparing protein expression levels in normal and diseased tissues. Complete proteomes can be separated by column chromatography or two-dimensional polyacrylamide gel electrophoresis (2D-PAGE) and individual proteins identified using liquid chromatography tandem mass spectrometry (LC-MS/MS). Differences in global proteome composition or protein abundance may be used to identify proteins involved in the pathophysiology of a disease, which may serve as potential drug targets.

2.2. Target Validation

Target validation involves demonstrating the role of the target protein in disease and serves to illustrate its therapeutic potential. Target validation is often accomplished with genetic knockout and knock-in animal models of disease (5–7) or by using neutralizing antibodies (8), existing small molecules (such as known agonists or antagonists of the target protein) (9), RNAi (10), or ribozymes (11). Furthermore, knowledge of a protein's three-dimensional structure may also assist in validating a target for drug development. For example, protein structure information is central to ascertaining the feasibility of rationally designing drug-like molecules to bind and modulate a protein.

2.3. Lead Identification

In this phase of drug development, rational (structure-based design) or random (screening) approaches are used to identify compounds that interact with the target protein and influence its activity. Structure-based drug design aims to produce rational compounds using the resolved three-dimensional structure of the target protein. Through an iterative process of protein structure determination, *in silico* docking, scoring of ligand molecules from a molecular database, biochemical evaluation of the best compounds, and structure determination of the protein–ligand complex, further *in silico* optimization of the ligand structure can be determined. Structure-based drug design seeks to generate optimized lead compounds with marked improvement in binding and specificity for the target (see Note 1) (12).

The random approach to lead identification relies upon efficient screening of a large number of compounds with the goal of identifying candidate molecules that are amenable to further optimization. High-throughput screening (HTS) is used to evaluate large numbers of compounds, 10,000–100,000 per day, for their ability to modulate the target protein (see Note 2) (13). The molecular libraries that are screened often consist of natural products or their derivatives (14), or are fully synthetic in origin, generated via combinatorial chemistry and parallel synthesis approaches (15). A critical concern for a reliable HTS is the quality and robustness of the assay. HTS assays often consist of biochemical assays with purified, isolated components or cell-based assays. Biochemical assays may be

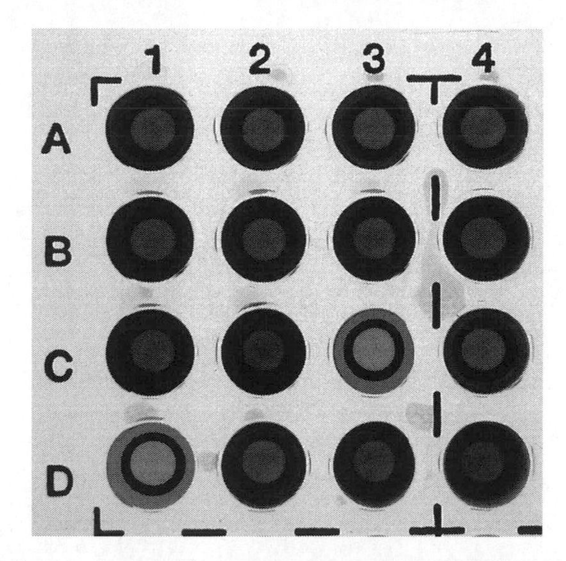

Fig. 2. A colorimetric assay used to screen a molecular library for inhibitors of an enzyme. The enzyme is inhibited in the wells that appear lighter in color (C3 and D1).

colorimetric enzyme assays (Fig. 2), scintillation proximity assays (16), or fluorescence-based assays (17) and are designed to evaluate the interaction of the molecular library with the protein target. Cell-based HTS assays are often reporter gene assays evaluating cellular response, signal transduction assays, or cell proliferation assays evaluating stimulation or inhibition of cell growth.

2.3.1. High-Throughput Screening: Potential Pitfalls

HTS is not immune to analytical interferences from nonspecific chemical reactivity, autofluorescence, or colloidal aggregation (31, 32). Autofluorescence is an inherent property of many small-molecule compounds that affects fluorescent-based assays (33). Autofluorescence can be detected by noting the initial fluorescence intensity compared to signal intensity after 1 min (32). Colloidal aggregation can act on enzyme targets causing reproducible, yet meaningless, inhibition (false positives) in the assay. Aggregation can be abrogated by adding a minimal amount of detergent (0.01% Triton-X) to assay buffers (32). Although a detergent cannot reduce all aggregation, analysis of the Hill coefficients in the dose-dependent response curves assists in identifying false-positive screening hits (32).

2.4. Lead Optimization

The lead optimization phase of drug development involves the simultaneous optimization of pharmacodynamic (efficacy, potency, and selectivity), pharmacokinetic (absorption, distribution, metabolism, elimination), and physicochemical properties to enable the small-molecule to reach and bind its protein target and exhibit its effects over a desirable duration. During lead optimization, structure–activity relationships are explored by chemically modifying the lead molecule and subsequently characterizing the derivatives using biochemical or cell-based assays. This typically involves multiple rounds of synthesis, where hundreds of compounds are prepared using combinatorial or parallel chemistry, colloquially termed "library synthesis" (18). Computational approaches that dock small molecules into the structures of macromolecular targets and score their potential complementarity to binding sites are also widely used in hit identification and lead optimization. The physicochemical properties of a molecule relate to dissolution, oral absorption, distribution, plasma protein binding, ability to cross the blood–brain barrier, and metabolism. Thus, methods of physicochemical profiling and absorption estimation, such as turbidimetric or kinetic solubility approaches (19), determination of lipophilicity (20), and drug absorption models (21–24), are used during the lead optimization phase to guide a project toward a small set of desirable drug candidates. These candidates are then evaluated in preclinical (in vitro disease-model assays and animal studies) and clinical trials (trials with humans). The lead optimization process is exemplified in studies of compounds that inhibit cysteine proteases in *Trypanosoma cruzi*, the causative agent of Chagas' disease (25, 26). *T. cruzi* relies on cruzain, a cysteine protease, for survival. A quantitative high-throughput screen of cruzain identified triazine nitriles, which are known inhibitors of other cysteine proteases, as reversible inhibitors of the enzyme (25, 26). Lead optimization as a tool for rational drug design is also being used to identify modulators for a wide array of proteins, including NF-κB signaling (27), cdc-2 like kinases (Clk) (28), human immunodeficiency virus (HIV) integrase (29), and thyroid-stimulating hormone (TSH) (30).

2.5. Preclinical and Clinical Trials

In preclinical studies, compounds are introduced to a cell line or animal to obtain preliminary efficacy and pharmacokinetic information, which assists in decisions regarding further development of the compound. Clinical trials are commonly classified into five different phases. Phase 0 clinical trials are first-in-human trials designed to evaluate if a compound behaves in human subjects as was anticipated from preclinical studies. In phase 1 clinical trials, a small group of volunteers (20–80) is used to evaluate the tolerability, pharmacokinetics, and pharmacodynamics of a therapy and to determine dosage for clinical use. Phase 2 clinical trials continue phase 1 safety assessments, but with a larger group size (20–300 individuals). Phase 3 clinical trials are randomized control trials on large patient groups (hundreds to thousands) to assess the efficacy of the new therapy, in comparison with standard therapy. Once a compound is proven acceptable and is approved by the FDA, phase 4 clinical trials evaluate postlaunch drug safety and detect rare or long-term adverse effects over a large patient population and timescale.

3. Personalized Medicine and Drug Development

Historically, most pharmaceutical companies have sought to create "blockbuster" treatments exhibiting a high therapeutic index for the majority of patients with a particular disease. However, since 1980, the US pharmaceutical companies have nearly doubled their spending on research and development every 5 years without a corresponding increase in new chemical entities brought to the market (see Note 3) (34). A new drug today requires, on average, a $750 million to $1.1 billion investment for research and development (35). About 75% of this cost is due to failure (36), with most (~80%) drug candidates failing phase 1 clinical trials (37). Clinical trials are designed to evaluate differences between treatments and controls, not differences in individual responses within the treatment group. Hence, agents that are effective in a subgroup of the test population, due to differences in individual molecular profiles, often produce nonstatistically significant results. Thus, drug candidates that may have been of value to a smaller percentage of the population are deemed a failure in clinical trials. Furthermore, agents that cause a specific serious side effect in a subgroup of the test population may lead to the conclusion that the compound is too toxic for overall use. Personalized medicine promises to refine the drug development process by incorporating new diagnostics to identify individual predictive characteristics/molecular profiles to better control or even exploit this variability.

The goal of personalized medicine is to define disease at the molecular level and then utilize molecular profiling for diagnostic and therapeutic decisions. Examples of the use of personalized medicine for disease treatment are illustrated with the anticancer agents, Herceptin® and Iressa, and with the use of Ziagen for the treatment of HIV/AIDS.

3.1. Herceptin®

Herceptin® (developed clinically as trastuzumab by Genentech) is a monoclonal antibody used to treat breast cancer. Herceptin® acts upon the human epidermal growth factor receptor (HER-2) in metastatic breast cancer, but is only effective when this receptor is overexpressed (38). Overexpression occurs in 25–30% of early-stage breast cancer patients (38, 39). While Herceptin could simply be administered to all patients diagnosed with breast cancer, there are several side effects associated with its use, most notably cardiac dysfunction (40), and the treatment is cost prohibitive, as a full course can cost upward of $70,000. Hence, accurate determination of HER-2 levels is desirable to identify patients that respond to Herceptin® before treatment is initiated. Several FDA-approved tests have been developed, based upon immunohistochemistry and in situ hybridization methods of determining the levels of HER-2 (38), which predict a patient's response to Herceptin®. Thus, through a personalized medicine approach, only those patients that are likely to benefit from Herceptin® will undergo treatment.

3.2. Iressa

Iressa, an epidermal growth factor receptor (EGFR) tyrosine kinase inhibitor (developed clinically as gefitinib by AstraZeneca), is used for the treatment of non-small-cell lung carcinoma, which accounts for approximately 85% of all lung cancer cases (41–44). Mutations in the EGFR gene, which occur at a frequency as low as 2% of patients in the USA to 26% of patients in Japan, are associated with enhanced tumor response to Iressa (42, 45). In a manner similar to that described for Herceptin®, a biomarker assay (EGFR mutation assay and fluorescent in situ hybridization (FISH)) is used to justify and predict the outcome of Iressa treatment.

3.3. Ziagen

Ziagen (clinically developed as abacavir by GlaxoSmithKline) is a nucleoside analog reverse transcriptase inhibitor used for the treatment of HIV-1 infections and AIDS (46). It is usually not administered as a monotherapy, but rather as a cocktail with one or more other anti-HIV therapeutic agents (e.g., AZT, 3TC). Hypersensitivity to Ziagen has been observed in up to 10% of patients, with symptoms that include fever, rash, fatigue, nausea, and vomiting, along with mild to severe respiratory distress and even death. The symptoms usually appear within 6 weeks of first dose, disappear after discontinuation of use, and return within hours of resuming Ziagen treatment (46–48). The human leukocyte antigen (HLA) allele HLA-B*5701 was found to be the most dominant risk factor linked with hypersensitivity toward Ziagen (47). Personalized medicine, through genotyping the HLA-B allele, has nearly eliminated the occurrence of hypersensitivity by identifying those patients at risk before initiating treatment (47–49).

3.4. The New Paradigm of Drug Development

In addition to having an impact on cancer and AIDS therapies, the tools of personalized medicine (genomics, proteomics, imaging technologies, bioinformatics, etc.) are further refining other diseases into molecular categories, and future treatments for these diseases will be dictated by a patient's molecular profile relative to these categories. Thus, the personalized medicine approach to patient care necessitates an evolution in the process of drug development. The linear paradigm must transform into a more cyclic and integrated process (Fig. 3). The revised drug development process revolves around a molecular profile of a disease. In essence, the target identification phase uses the tools of personalized medicine to define a molecular "fingerprint" of a disease. The protein chosen for drug development is viewed in relation to the presence (or absence) of other molecular markers. As a consequence, to minimize variability, the subsequent in vivo assays used in the lead

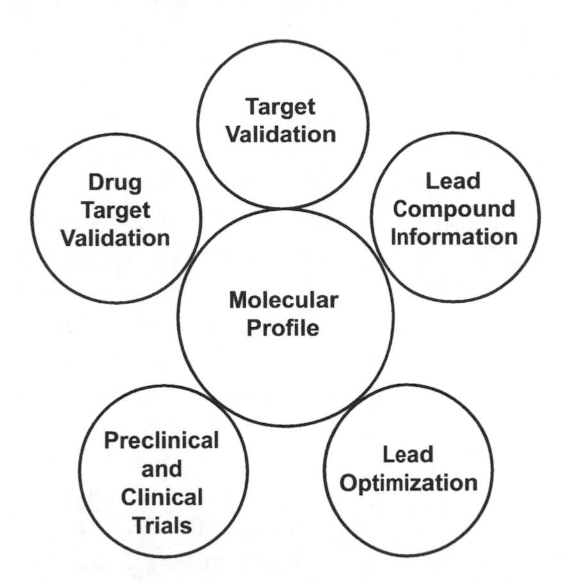

Fig. 3. The new paradigm of drug development. Due in part to the personalized medicine approach to patient care, the linear process of drug development must evolve into a cyclic and integrated process that revolves around a molecular profile of a disease. Each phase in the new paradigm generates data that may impact all other phases of development.

identification and optimization phases should be selectively performed with specific cell lines that also exhibit the molecular fingerprint, as should the test subjects used in clinical trials, to strengthen the predictive value of the earlier phases for treatment response and substantially improve the risk/benefit analysis during clinical trials.

Clearly, it is economically desirable for the molecular profile to be reflective of that found in the majority of patients. Thus, a statistically acceptable number of samples need to be profiled and analyzed before drug development can proceed. While one or two key proteins identified in the molecular profile become the targets for drug development, the target validation phase expands beyond exclusively validating the role of the prospective drug target in the disease, and includes validating other molecular markers for use as theranostics (diagnostic tests that identify patients most likely to be helped or harmed by a new medication).

In essence, the tools of personalized medicine are used to evolve the drug development process from one that seeks a "one-size-fits-all" therapy to a one that produces a more individual treatment for a molecular subset of a disease. While this new paradigm of drug development may demand an even greater investment of time and resources and the resulting product may serve a smaller target market, a reduction in the overall failure rate of drug development should economically justify this approach and consequently get more drugs to the market.

4. Notes

1. *In silico* screening can save thousands of dollars in the drug discovery process and is gaining utility and popularity as an initial screening phase of potential compounds.
2. HTS can be very costly, yet it provides information related to on-target as well as off-target drug effects. Adverse drug reactions and off-target effects must be taken into consideration as part of the rational drug design process (50–53). The National Institutes of Health Molecular Libraries Probe Production Centers are a network of the US national laboratories that perform HTS, secondary screening, and medicinal chemistry service for the scientific community (http://mli.nih.gov/mli/). The NIH Molecular Libraries provides publicly available compound structure and screening results.
3. The white paper published by the US Food and Drug Administration discussing issues affecting the drug pipeline can be found at http://www.fda.gov/oc/initiatives/criticalpath/whitepaper.html. Reports on the status of drug development, including funding, resources, and new compounds, are available from the Pharmaceutical Research and Manufacturers of America (http://www.phrma.org/profiles_%26_reports/).

Acknowledgments

Special thanks to Arthur Tsang for supplying the colorimetric plate assay results.

References

1. Hillisch, A., and Hilgenfeld, R. (2003) *Modern methods of drug discovery.* Birkhauser Verlag, Basel, Switzerland.
2. Fleischmann, R. D., Adams, M. D., White, O., Clayton, R. A., Kirkness, E. F., *et al.* (1995) Whole-genome random sequencing and assembly of Haemophilus influenzae Rd. *Science* **269**, 496–512.
3. Jawaid, S., Seidle, H., Zhou, W., Abdirahman, H., Abadeer, M., *et al.* (2009) Kinetic characterization and phosphoregulation of the Francisella tularensis 1-deoxy-D-xylulose 5-phosphate reductoisomerase (MEP synthase). *PLoS One* **4**, e8288.
4. Drews, J. (2000) Drug discovery: a historical perspective. *Science* **287**, 1960–4.
5. Abuin, A., Holt, K. H., Platt, K. A., Sands, A. T., and Zambrowicz, B. P. (2002) Full-speed mammalian genetics: in vivo target validation in the drug discovery process. *Trends Biotechnol* **20**, 36–42.
6. Tornell, J., and Snaith, M. (2002) Transgenic systems in drug discovery: from target identification to humanized mice. *Drug Discov Today* **7**, 461–70.
7. Sanseau, P. (2001) Transgenic gene knockouts: a functional platform for the industry. *Drug Discov Today* **6**, 770–71.
8. Tse, E., Lobato, M. N., Forster, A., Tanaka, T., Chung, G. T., *et al.* (2002) Intracellular antibody capture technology: application to selection of intracellular antibodies recognising

the BCR-ABL oncogenic protein. *J Mol Biol* **317**, 85–94.

9. Ivy, S. P., Wick, J. Y., and Kaufman, B. M. (2009) An overview of small-molecule inhibitors of VEGFR signaling. *Nat Rev Clin Oncol.*
10. Bartz, S., and Jackson, A. L. (2005) How will RNAi facilitate drug development? *Sci STKE* **2005**, pe39.
11. Zaffaroni, N., Pennati, M., and Folini, M. (2007) Validation of telomerase and survivin as anticancer therapeutic targets using ribozymes and small-interfering RNAs. *Methods Mol Biol* **361**, 239–63.
12. Davies, J. W., Glick, M., and Jenkins, J. L. (2006) Streamlining lead discovery by aligning in silico and high-throughput screening. *Curr Opin Chem Biol* **10**, 343–51.
13. Bender, A., Bojanic, D., Davies, J. W., Crisman, T. J., Mikhailov, D., *et al.* (2008) Which aspects of HTS are empirically correlated with downstream success? *Curr Opin Drug Discov Devel* **11**, 327–37.
14. Lam, K. S. (2007) New aspects of natural products in drug discovery. *Trends Microbiol* **15**, 279–89.
15. Musonda, C. C., and Chibale, K. (2004) Application of combinatorial and parallel synthesis chemistry methodologies to antiparasitic drug discovery. *Curr Med Chem* **11**, 2519–33.
16. Glickman, J. F., Schmid, A., and Ferrand, S. (2008) Scintillation proximity assays in high-throughput screening. *Assay Drug Dev Technol* **6**, 433–55.
17. Liu, B., Li, S., and Hu, J. (2004) Technological advances in high-throughput screening. *Am J Pharmacogenomics* **4**, 263–76.
18. Thomas, G. (2007) *Medicinal chemistry: an introduction.* John Wiley and Sons, Hoboken, NJ.
19. Lipinski, C. A., Lombardo, F., Dominy, B. W., and Feeney, P. J. (2001) Experimental and computational approaches to estimate solubility and permeability in drug discovery and development settings. *Adv Drug Deliv Rev* **46**, 3–26.
20. Barton, P., Davis, A. M., McCarthy, D. J., and Webborn, P. J. (1997) Drug-phospholipid interactions. 2. Predicting the sites of drug distribution using n-octanol/water and membrane/water distribution coefficients. *J Pharm Sci* **86**, 1034–9.
21. Artursson, P., Palm, K., and Luthman, K. (2001) Caco-2 monolayers in experimental and theoretical predictions of drug transport. *Adv Drug Deliv Rev* **46**, 27–43.
22. Bachmann, K. A., and Ghosh, R. (2001) The use of in vitro methods to predict in vivo pharmacokinetics and drug interactions. *Curr Drug Metab* **2**, 299–314.
23. Gres, M. C., Julian, B., Bourrie, M., Meunier, V., Roques, C., *et al.* (1998) Correlation between oral drug absorption in humans, and apparent drug permeability in TC-7 cells, a human epithelial intestinal cell line: comparison with the parental Caco-2 cell line. *Pharm Res* **15**, 726–33.
24. Irvine, J. D., Takahashi, L., Lockhart, K., Cheong, J., Tolan, J. W., *et al.* (1999) MDCK (Madin-Darby canine kidney) cells: A tool for membrane permeability screening. *J Pharm Sci* **88**, 28–33.
25. Mott, B. T., Ferreira, R. S., Simeonov, A., Jadhav, A., Ang, K. K., *et al.* (2010) Identification and optimization of inhibitors of Trypanosomal cysteine proteases: cruzain, rhodesain, and TbCatB. *J Med Chem* **53**, 52–60.
26. Ferreira, R. S., Simeonov, A., Jadhav, A., Eidam, O., Mott, B. T., *et al.* (2010) Complementarity between a docking and a high-throughput screen in discovering new cruzain inhibitors. *J Med Chem* **53**, 4891–905.
27. Miller, S. C., Huang, R., Sakamuru, S., Shukla, S. J., Attene-Ramos, M. S., *et al.* (2010) Identification of known drugs that act as inhibitors of NF-kappaB signaling and their mechanism of action. *Biochem Pharmacol* **79**, 1272–80.
28. Mott, B. T., Tanega, C., Shen, M., Maloney, D. J., Shinn, P., *et al.* (2009) Evaluation of substituted 6-arylquinazolin-4-amines as potent and selective inhibitors of cdc2-like kinases (Clk). *Bioorg Med Chem Lett* **19**, 6700–5.
29. Marinello, J., Marchand, C., Mott, B. T., Bain, A., Thomas, C. J., *et al.* (2008) Comparison of raltegravir and elvitegravir on HIV-1 integrase catalytic reactions and on a series of drug-resistant integrase mutants. *Biochemistry* **47**, 9345–54.
30. Neumann, S., Huang, W., Eliseeva, E., Titus, S., Thomas, C. J., *et al.* (2010) A small molecule inverse agonist for the human thyroid-stimulating hormone receptor. *Endocrinology* **151**, 3454–9.
31. Thorne, N., Auld, D. S., and Inglese, J. (2010) Apparent activity in high-throughput screening: origins of compound-dependent assay interference. *Curr Opin Chem Biol* **14**, 315–24.
32. Jadhav, A., Ferreira, R. S., Klumpp, C., Mott, B. T., Austin, C. P., *et al.* (2010) Quantitative analyses of aggregation, autofluorescence, and reactivity artifacts in a screen for inhibitors of a thiol protease. *J Med Chem* **53**, 37–51.
33. Simeonov, A., Jadhav, A., Thomas, C. J., Wang, Y., Huang, R., *et al.* (2008) Fluorescence spectroscopic profiling of compound libraries. *J Med Chem* **51**, 2363–71.
34. Frantz, S. (2004) FDA publishes analysis of the pipeline problem. **3**, 379.

35. Gilbert, J., Henske, P., and Singh, A. (2003) Rebuilding big pharma's business model. *In Vivo The Business & Medicine Report* **21**, 1–4.
36. Dimasi, J. A. (2001) Risks in new drug development: approval success rates for investigational drugs. *Clin Pharmacol Ther* **69**, 297–307.
37. Bouchie, A. (2006) Clinical trial data: to disclose or not to disclose? *Nat Biotechnol* **24**, 1058–60.
38. Bange, J., Zwick, E., and Ullrich, A. (2001) Molecular targets for breast cancer therapy and prevention. *Nat Med* **7**, 548–52.
39. Nahta, R., and Esteva, F. J. (2003) HER-2-targeted therapy: lessons learned and future directions. *Clin Cancer Res* **9**, 5078–84.
40. Seidman, A., Hudis, C., Pierri, M. K., Shak, S., Paton, V., *et al.* (2002) Cardiac dysfunction in the trastuzumab clinical trials experience. *J Clin Oncol* **20**, 1215–21.
41. Marko-Varga, G., Ogiwara, A., Nishimura, T., Kawamura, T., Fujii, K., *et al.* (2007) Personalized medicine and proteomics: lessons from non-small cell lung cancer. *J Proteome Res* **6**, 2925–35.
42. Paez, J. G., Janne, P. A., Lee, J. C., Tracy, S., Greulich, H., *et al.* (2004) EGFR mutations in lung cancer: correlation with clinical response to gefitinib therapy. *Science* **304**, 1497–500.
43. Pao, W., Miller, V., Zakowski, M., Doherty, J., Politi, K., *et al.* (2004) EGF receptor gene mutations are common in lung cancers from "never smokers" and are associated with sensitivity of tumors to gefitinib and erlotinib. *Proc Natl Acad Sci U S A* **101**, 13306–11.
44. Kim, K. S., Jeong, J. Y., Kim, Y. C., Na, K. J., Kim, Y. H., *et al.* (2005) Predictors of the response to gefitinib in refractory non-small cell lung cancer. *Clin Cancer Res* **11**, 2244–51.
45. Lynch, T. J., Bell, D. W., Sordella, R., Gurubhagavatula, S., Okimoto, R. A., *et al.* (2004) Activating mutations in the epidermal growth factor receptor underlying responsiveness of non-small-cell lung cancer to gefitinib. *N Engl J Med* **350**, 2129–39.
46. Hetherington, S., McGuirk, S., Powell, G., Cutrell, A., Naderer, O., *et al.* (2001) Hypersensitivity reactions during therapy with the nucleoside reverse transcriptase inhibitor abacavir. *Clin Ther* **23**, 1603–14.
47. Mallal, S., Nolan, D., Witt, C., Masel, G., Martin, A. M., *et al.* (2002) Association between presence of HLA-B*5701, HLA-DR7, and HLA-DQ3 and hypersensitivity to HIV-1 reverse-transcriptase inhibitor abacavir. *Lancet* **359**, 727–32.
48. Zucman, D., Truchis, P., Majerholc, C., Stegman, S., and Caillat-Zucman, S. (2007) Prospective screening for human leukocyte antigen-B*5701 avoids abacavir hypersensitivity reaction in the ethnically mixed French HIV population. *J Acquir Immune Defic Syndr* **45**, 1–3.
49. Rauch, A., Nolan, D., Martin, A., McKinnon, E., Almeida, C., *et al.* (2006) Prospective genetic screening decreases the incidence of abacavir hypersensitivity reactions in the Western Australian HIV cohort study. *Clin Infect Dis* **43**, 99–102.
50. Bender, A., Scheiber, J., Glick, M., Davies, J. W., Azzaoui, K., *et al.* (2007) Analysis of pharmacology data and the prediction of adverse drug reactions and off-target effects from chemical structure. *ChemMedChem* **2**, 861–73.
51. Crisman, T. J., Parker, C. N., Jenkins, J. L., Scheiber, J., Thoma, M., *et al.* (2007) Understanding false positives in reporter gene assays: in silico chemogenomics approaches to prioritize cell-based HTS data. *J Chem Inf Model* **47**, 1319–27.
52. Scheiber, J., Jenkins, J. L., Sukuru, S. C., Bender, A., Mikhailov, D., *et al.* (2009) Mapping adverse drug reactions in chemical space. *J Med Chem* **52**, 3103–7.
53. Scheiber, J., Chen, B., Milik, M., Sukuru, S. C., Bender, A., *et al.* (2009) Gaining Insight into Off-Target Mediated Effects of Drug Candidates with a Comprehensive Systems Chemical Biology Analysis. *J Chem Inf Model* **49**, 308–17.

Chapter 25

Grant Writing Tips for Translational Research

Lindsay Wescott and Michael Laskofski

Abstract

All investigators face the same challenge – the highly competitive nature of the grant review process. Innovation alone is not enough to ensure grant supported funding. Applied clinical research requires a diverse collaborative team of investigators with specialized skills, a supportive clinical research environment, and access to clinical material. In addition, ethical limitations, and lack of animal models for many diseases, prevent direct mechanistic experiments that are possible using in vitro systems or animal models. Therefore, specific granting mechanisms and program initiatives target translational research studies. This chapter provides grant writing tips and lists resources that may prove helpful for new investigators seeking research funding in support of translational research, biobanking, and research utilizing molecular biomarkers.

Key words: Budget, Clinical research, Clinical trial protocol, Funding, Grant, Human subjects' protection, Institutional review board, Translational research, Writing

1. Introduction

Young investigators who conduct translational research face two challenges as they compete for grant funding. The first challenge is the same that all investigators face: the highly competitive nature of the grant review process. Innovation alone is not enough. Investigators setting out to discover novel genomic or proteomic clinical biomarkers or proposing to develop new technology must present the work as hypotheses driven in order to be appreciated by study sections experienced in clinical research. Thus, applicants planning to conduct proteomic or genomic surveys, exploratory studies, or open-ended biomarker discovery studies may be interpreted by the reviewers as unfocused or as a "fishing expedition." You must frame the biomarker discovery project around a testable hypothesis. If you propose a new technology, you must show its

Virginia Espina and Lance A. Liotta (eds.), *Molecular Profiling: Methods and Protocols*, Methods in Molecular Biology, vol. 823, DOI 10.1007/978-1-60327-216-2_25, © Springer Science+Business Media, LLC 2012

advantages compared to prior work in the field and you must convince the reviewers that the technology will lead to the discovery of mechanistic or therapeutic insights that would not otherwise be possible.

The second challenge relates to the nature of translational research. Applied clinical research requires a diverse collaborative team of investigators with specialized skills, a supportive clinical research environment in a medical school or community hospital, and access to clinical material (tissue, blood, body fluids, and clinical data) (1). Translational research often involves time delays due to institutional review board (IRB) assessment, patient recruitment and accrual, time span between treatment and clinical end points, and specimen collection. Moreover, ethical limitations, and lack of animal models for many diseases, prevent the type of direct mechanistic experiments that are possible using in vitro systems or animal models.

The National Institutes of Health (NIH) extramural program has recognized the critical need for translational research, and appreciates the challenges imposed on young clinical investigators. For this reason, specific granting mechanisms and program initiatives are made available by the NIH to assist investigators proposing to conduct clinical research. The reader is encouraged to access the NIH Guide LISTSERV describing the current funding opportunities, NIH-Wide Initiatives, Requests for Applications (RFA), Requests for Proposals (RFP), and Program Announcements (PA) notices at http://grants.nih.gov/grants/guide/listserv.htm.

2. Grant Writing Tips

This section provides grant writing tips and lists helpful resources for investigators who are seeking research funding to support translational research and research on molecular biomarkers (2).

2.1. Get Advice Before and During the Writing Process

Study successful proposals by colleagues in your field and request help from more experienced colleagues to read your grant and offer suggestions (3). Ask two or three senior colleagues to act as your friendly mini "grant committee." Discuss your ideas for the application with them before starting the writing process. Write 3–5 specific aims and discuss these with your colleagues before beginning to write the body of the application. Review your experimental design with colleagues who are familiar with the methods you propose. By the time you tackle the bulk of the writing, the logical development and content of your proposal should have undergone detailed scrutiny and critical consideration. For translational research, it is vitally important to consult with a clinician and a diagnostic pathologist who can help you frame your clinical question and design your specimen collection protocols. Make sure that you consult with a biostatistician and an epidemiologist if you are planning to use clinical samples. Be familiar with the requirements for research involving human subjects. Assume that the members of the review committee assigned to your application are intimately familiar with the disease topic that you are addressing. As such, they will be looking carefully at the control populations you choose. Therefore, it is wise to speak to clinicians who work in the disease topic you have chosen.

2.2. Contact the Granting Agency

1. Contact the NIH Program Directors or Scientific Review Administrator (SRA) in your research field to discuss the proposed research. Ask the SRA if he/she will accept the application for review. The SRA has several roles in the granting process.
   (a) He/she reviews the application and assigns it to a study section.
   (b) He/she selects members of the community to participate in the study section panel.
   (c) He/she interacts with the section chair.
   (d) He/she assigns applications to specific reviewers.
   (e) He/she assists in coordination of logistical issues related to study section reviews.
2. Carefully study the specialized topics covered by the different Integrated Review Groups (IRG). The IRG is a cluster of study sections responsible for the review of grant applications in scientifically related areas, http://cms.csr.nih.gov/peerreview meetings/csrirgdescriptionnew/.
3. Contact the NIH by Web and phone to reach people who want to help you:
   (a) NIH – http://www.nih.gov/
   (b) Center for Scientific Review – http://www.csr.nih.gov/
   (c) National Institute of General Medical Sciences – http://www.nigms.nih.gov/
   (d) Computer Retrieval of Information on Scientific Projects (CRISP, a searchable database of federally funded biomedical research projects) – http://report.nih.gov/crisp.aspx
4. Hard and fast submission deadlines: Applications submitted in response to an RFA or RFP must be received by NIH on or before the deadline or you have to wait until the RFA or RFP is advertised again. Guidelines for submission deadlines can be accessed at http://grants.nih.gov/grants/guide/listserv.htm.
   (a) Read the entire RFA or RFP carefully. Follow the instructions exactly (see Note 1).

(b) Do not exceed page limits. NIH grant applications require each section of the application to be uploaded electronically via Grants.gov. Strict adherence to page limits, page formatting, and fonts is required. Failure to follow the instructions results in truncated or illegible text resulting in immediate rejection of the application.

(c) Find out your institutional deadlines for proposal submissions. Most organizations require at least 2 days for review and processing of proposal submissions through Grants.gov.

(d) Meet all deadlines.

2.3. Key Criteria

Essential attributes that generate excitement in the eyes of a study section reviewer are (a) innovation and novelty within a significant and timely field and (b) a realistic experimental plan that is not overambitious (see Note 2).

1. Establish a reasonable and testable hypothesis and a strong research plan. The elements used by reviewers in scoring grant applications are significance, approach, innovation, investigator, and environment.
2. Make sure that you anticipate pitfalls and propose a realistic timeline for the work.

2.4. Preliminary Data Is Very important

The grant instructions from the funding agency may state that preliminary data is not required. Nevertheless, in this competitive funding climate, preliminary data is often crucial to convince the reviewers that your hypothesis has merit or your experimental model is feasible.

2.5. Imagine Yourself in the Reviewer's Shoes

While the reviewer is a recognized expert, he/she may not be familiar with the details of your specific field or your specialized experimental models and technology. Consequently, it is imperative that all sections of your grant application are written clearly with subheadings and introductory sections that respectfully educate the reviewer. On the other hand, you do not want to use elementary terms and lay explanations that could irritate a reviewer who is a direct expert in your field.

2.6. Keep the Plan Focused

It is essential to focus the application. Be enthusiastic. Do not include lengthy appendices or use a series of obscure abbreviations. Spell everything out (within reason) and avoid all but the most common acronyms. Keep the reviewer interested and involved, and make every effort to keep the format consistent and readable. Carefully review the application for figure legend, spelling, or format errors. An annoyed or bored reviewer is not likely to give a favorable review.

2.7. Prove That You Can Do It

Your publication track record, the budget you propose, your collaborators, and the facilities for your work are all critical to prove that you can actually accomplish the proposed studies.

2.8. Do Not Give Up

If it is not funded the first time, your grant can be revised and resubmitted under NIH guidelines. Many grants are funded on the second round. There was an NIH policy change effective from 1/25/09 that limits the number of allowable submissions to one original submission and one revision.

2.9. Do Not Put All Your Eggs in One Basket

Submit many grant applications to a variety of agencies, funding institutions, RFA, and PA. Some Federal agencies may ask you to disclose if you are submitting the same application simultaneously to other agencies. You may have to submit ten or more high-quality applications before one is funded.

2.10. IRB Approval for Human Subjects' Research

NIH policy was recently changed to allow applications that use human subjects to be submitted without prior IRB approval, http://grants.nih.gov/grants/guide/notice-files/NOT-OD-00-031.html.

If your grant receives a good score upon review, you should begin the process to obtain approval of the IRB. No application can be funded without IRB approval. The principal investigator (PI) and all key personnel must submit a description of education they have completed in the protection of human subjects, http://grants.nih.gov/grants/guide/notice-files/NOT-OD-00-039.html.

For human subjects, your grant application must include documentation of six items described in Section C of the instructions for the application:

1. Characteristics of the subjects
2. The sources of research materials
3. Recruitment plans and consent procedures
4. Potential risks
5. Procedures for protecting against or minimizing potential risks
6. Potential benefits to the subjects and mankind

NIH requires that clinical research subjects include minorities and women, and that wherever possible children are also to be included.

http://grants.nih.gov/grants/guide/notice-files/NOT-OD-02-001.html.

http://grants2.nih.gov/grants/funding/SBIRConf2000/Scharke/.

http://grants.nih.gov/grants/guide/notice-files/not98-024.html.

IRB processes and requirements may vary by institution; it is important to become familiar with the specifics of your IRB.

2.11. IACUC Approval for Animal Subjects' Research

Applications with animal research subjects must include documentation of Institutional Animal Care and Use Committee (IACUC) approval. For animal subjects, you must provide a detailed description of the proposed use of the animals, a justification for the choice of species and number of animals in the experiment, information on the veterinary care of the animals, an explanation of procedures to ensure that the animals will not experience unnecessary discomfort, distress, pain, or injury, and a justification for any euthanasia method to be used.

IACUC processes and requirements may vary by institution; it is important to become familiar with the specifics of your IACUC.

2.12. Collaborators and Coinvestigators

Translational research requires collaboration between clinicians and basic scientists. You should obtain a signed letter from each collaborator listing the contribution he or she intends to make and his or her enthusiasm for the proposed study. These letters should be included in the grant application and are often the primary assurance to the reviewers that this work can and will be done. It is essential that the support letters document the availability of clinic or hospital facilities, describe the clinical laboratory environment, and refer to the roles of the clinical collaborators.

2.13. Developing the Budget

Start with personnel. You need to fully explain the role of each person on the grant. Review the modular grant writing format at http://grants.nih.gov/grants/funding/modular/modular.htm.

NIH modular grant rules specify that you must request funds in $25,000 modules and do not permit increments for inflation in the "out-years." Out-years are the yearly periods after the initial year of the grant. Although you may not have to detail budgetary needs, keep in mind that the reviewers judge your competence in part by how well your funding request matches the scope of the project. Be careful not to underestimate the cost of performing the work; underestimation can be as negative as overestimation.

1. Personnel
   - (a) Describe all personnel, whether professional or nonprofessional, by name, position, and proposed time and effort even if no salary is involved.
   - (b) Justify job descriptions.
   - (c) List dollar amounts separately for each individual.
   - (d) Request only consultants who have agreed to participate.
2. Equipment and supplies
   - (a) Request and justify all equipment necessary for the completion and performance of the proposed research.
   - (b) Request and itemize supplies needed for completion of the proposed research.

   (c) Justify species, number, and cost of all animals.
   (d) Request costs for patient care where appropriate.
   (e) Add subtotals and check calculations.
3. Travel
   (a) Consider requesting travel money to encourage dissemination of knowledge gained through your research.
   (b) Request costs for flights and conference registration.
   (c) Describe the purpose of the travel.
4. Other
   (a) Include accurate rates, such as fringe benefits and facilities and administrative costs (indirect) (see Note 3).

2.14. Budget Reminders

Do not forget to check your budget for mathematical errors.

1. Personnel
   (a) Do not exceed 100% for the collective sum of percentages in time and effort proposed for each individual.
   (b) Do not request consultants that cannot be justified due to lack of expertise or level of effort for the proposed research.
2. Equipment and supplies
   (a) Do not request equipment purchases that appear to be duplicative (for example, investigator in collaborative institution requests equipment available in PI's institution).
   (b) Do not add supplies indiscriminately.
   (c) Do not propose to use animal species that cannot be correlated to human data or that are not appropriate for the proposed area of research.
   (d) Do not request funds for coverage of laboratory tests that are routinely provided as part of a patient's basic tests or standard of care.

3. Special Grant Application Features of Clinical Studies

Clinical studies have several features that are not typically found in basic research studies. The grant application should clearly state the significance of the biological problem and likelihood of altering current clinical practice. In addition, the following features must be fully justified and addressed in the application:

1. Demonstrated critical assessment of preliminary clinical data and literature

2. Well-defined, justified use of human subjects compared with the value of similar studies in animals
3. Full consideration of human subject protection issues
4. Consideration of any potential medical problems or side effects arising from the new procedures
5. Appropriate and accurate methods for clinical and biological patient assessment; if methods utilized are somewhat simplistic and possibly unsophisticated, explain how the technology to be utilized is powerful enough to answer the questions posed

3.1. Data Quality

Data must meet clinical data standards that include:

1. Criteria for collecting, analyzing, and assessing clinical and laboratory data; define the accuracy of all data collected, including intra- and interoperator errors
2. Compliance monitoring for adverse events
3. Monitoring procedural rigor for issues, such as randomization and dosage modification
4. Monitoring by statistical unit (and PI where appropriate) for treatment benefits and side or toxic effects.
5. Collaborative, multi-institutional studies require special organizational structures to ensure the uniformity of testing at each clinical site

3.2. Principal Investigator Qualifications and Experience

1. Strong qualifications as a biological, basic science, or clinical investigator.
2. PIs of clinical research studies are often required to be licensed medical doctors.
3. Prior experience in clinical studies.
4. Basic science skills may not translate to clinical studies, so choose collaborators that complement your expertise.

3.3. Statistical Design

1. Statistical methods should be appropriate for data to be collected.
2. If possible and appropriate, consider randomized, double-blind prospective study designs.
3. Carefully estimate the incidence of the observed effect on normal or control populations.
4. Show proper selection of control populations, including matching for age, sex, race, color, clinical conditions, etc.
5. Provide the statistical rationale and methodology to determine the number of subjects required for each cohort; dropout and withdrawal must be considered; and statistical power assumptions should be well-defined.
6. Typically, statistical consultants are required in the preparation of applications and the performance of studies; often, independent

data units or centers are established to implement the study and allow the PI to remain unaware of patient assignment to treatment or control groups.

3.4. Patient Population

1. Well-defined, homogeneous subject population and control.
2. Availability of a suitable patient population, as well as any prior experience with that group of individuals.
3. Clearly defined trial entry/inclusion criteria.
4. Clearly defined exclusion criteria based on current or past diseases, utilization of specific medications, or the results of laboratory and clinical tests.
5. Clearly defined recruitment procedures to be followed.
6. Prior experience in recruiting and enrolling patients in a clinical study.

3.5. Summary

You are not funded unless you apply! Almost no one is funded with the first application; it is the repeated effort that succeeds. Persistence pays off; continue to apply for grants and reapply for those which you received good scores. Specifically address the criticisms of your proposal and resubmit in time for the next grant submission cycle.

NIH funding, while difficult to obtain, is the "gold standard" of peer review. It is less encumbered than funding from industry, much more prestigious, and permits you to develop your research concepts in an atmosphere of scientific freedom.

4. Notes

1. Most grant submission forms are entered electronically. Failure to follow formatting guidelines may result in the application being unreadable. Grant reviewers appreciate quality figures, formatting, and absence of typographical and grammatical errors. Attention to the details of your application creates a sense of quality and commitment to the project.
2. There are numerous publications and electronic resources available for every step of the grant writing and application process (4). Particularly helpful resources are listed below.
   (a) NIH Office of Extramural Research, Electronic Submission of Grant Applications: Training library, including videos and pdf resources, http://era.nih.gov/ElectronicReceipt/training.htm.
   (b) NIH Application Transition Process and A Walk through the SF424 (R&R): Grants.gov: How to Complete an

Application Package Demo, http://www.grants.gov/applicants/apply_for_grants.jsp#demo.

(c) eRA Commons Registration: Training demo, http://era.nih.gov/ElectronicReceipt/training.htm.

(d) Check out the "Tips and Tools" resources to avoid common application submission errors, http://era.nih.gov/ElectronicReceipt/avoiding_errors.htm.

(e) Finding help/get support for NIH grants: NIH has several support teams ready to answer the questions and concerns not covered on the Web site, http://era.nih.gov/ElectronicReceipt/support.htm.

(f) Monitor announcements. Monitor the NIH Guide for Grants and Contracts, http://grants2.nih.gov/grants/grant_tips.htm.

(g) Social Science Research Council (SSRC): The Art of Writing Proposals, http://www.ssrc.org/publications/view/7A9CB4F4-815F-DE11-BD80-001CC477EC70/.

(h) National Institute of Allergy and Infectious Diseases (NIAID): Tutorials all about grants, http://www.niaid.nih.gov/ncn/grants/default.htm.

(i) National Science Foundation (NSF): A Guide for Proposal Writing, http://www.nsf.gov/pubs/1998/nsf9891/nsf9891.htm.

(j) GrantsNet: Resource for funding and training in the sciences, http://sciencecareers.sciencemag.org/funding.

(k) American Association For the Advancement of Science (AAAS), http://www.aaas.org/.

(l) Society of Research Administrators (SRA) International Resource Guide: Grant Seeking Publications and Funding Information/Opportunities, http://www.srainternational.org/sra03/resourceguide/.

(m) How to Prepare a Winning Proposal: National Sciences and Engineering Research Council of Canada (NSERC), http://www.nserc-crsng.gc.ca/Professors-Professeurs/PoliciesProf-PolitiqueCorpProf_eng.asp.

(n) Canadian Research Funding Opportunities (Queen's University, Canada), http://queensu.ca/ors/funding-sources.html.

(o) NSERC Hints for Discovery Grant Applicants (University of Manitoba, Canada), http://www.umanitoba.ca/research/funding/tips/nserc_grant_tips.pdf.

(p) Canadian Institutes of Health Research (CIHR) grant writing tips from McGill University.

(q) Proposal Writing Short Course (The Foundation Center), http://foundationcenter.org/getstarted/tutorials/shortcourse/index.html.

3. Fringe rates and facility/administrative costs typically help universities defray the cost of administering the grants while supporting university infrastructure. Each institution develops its own rate structure. Fringe rates may also vary based on personnel classification, such as staff versus faculty.

References

1. Lowe, B., Hartmann, M., Wild, B., Nikendei, C., Kroenke, K. et al. (2008) Effectiveness of a 1-year resident training program in clinical research: a controlled before-and-after study. *J Gen Intern Med* **23**, 122–8.
2. Agarwal, R., Chertow, G. M., Mehta, R. L. (2006) Strategies for successful patient oriented research: why did I (not) get funded? *Clin J Am Soc Nephrol* **1**, 340–3.
3. Koren, G. (2005) How to increase your funding chances: common pitfalls in medical grant applications. *Can J Clin Pharmacol* **12**, e182–5.
4. Reif-Lehrer, L. (2000) Applying for grant funds: there's help around the corner. *Trends Cell Biol* **10**, 500–4.

Chapter 26

Inventions and Patents: A Practical Tutorial

J. Lille Tidwell and Lance A. Liotta

Abstract

Patents are designed to protect and encourage creativity and innovation. Patenting a biomedical discovery can be a requirement before a pharmaceutical company or biotech entity will invest in the lengthy and costly clinical testing necessary to achieve patient benefit. Although scientists and clinicians are well versed in research publication requirements, patent descriptions and claims are formatted in a manner quite different from a research paper. Patents require (a) a series of logical statements clearly delineating the boundaries of the novel aspects of the invention and (b) sufficient disclosure of the invention so that it can be reproduced by others. Patents are granted only for inventions that meet three conditions: novelty, nonobviousness, and usefulness. This chapter provides basic guidelines and definitions of technology transfer: inventions, inventorship, and patent filing, which are summarized using a question and answer format.

Key words: Disclosure, Discovery, Invention, Nondisclosure agreement, Patent, Technology transfer

1. Introduction

Patenting biomedical discoveries is an important requirement for commercialization. A pharmaceutical company, a diagnostic company, or a biotechnology firm will not spend the considerable resources to develop a new product that can benefit patients suffering from disease, if the product is not covered by an issued or pending patent (1). Even experienced academic scientists may not understand that inventorship, an invention, and a patent are quite different from a scientific publication. In many ways, obtaining a granted patent can be much harder and longer than publishing a research study in a peer reviewed journal. This is because of the requirement for absolute novelty and utility, as judged by the lengthy examining process by the US Patent and Trademark Office (USPTO, http://www.uspto.gov) (1, 2). Patent claims must be written in a form of an

Virginia Espina and Lance A. Liotta (eds.), *Molecular Profiling: Methods and Protocols*, Methods in Molecular Biology, vol. 823, DOI 10.1007/978-1-60327-216-2_26, © Springer Science+Business Media, LLC 2012

independent clause or multiple-dependent clauses that logically delineate the novel features claimed under the invention. Licensing and assignment of intellectual property rights (inventions) provide employers/inventors with a means to effectively sell or "rent" the invention under specific conditions (3).

Molecular profiling and individualized therapy are providing new insights into disease treatment while at the same time providing new technologies and therapies. Although the concept of genomic and proteomic analysis is not new, the wealth of patentable information gleaned from these molecular insights is constantly challenging current health and patent laws (1, 4, 5). It is our intent in this chapter to explain basic guidelines applicable to technology transfer for biomedical inventions and patents, including a tutorial for writing a patent application (based on the Patent Cooperative Treaty and United States Patent and Trademark Office requirements).

2. Types of Patents

Question: What constitutes an invention that can be patented?

Answer: Inventions can be patented if they are novel, nonobvious, and useful (1, 5).

Meaning of novel, nonobvious, and useful in patent terms:

New and Novel: For a US patent the invention must never have been disclosed in public in any way, anywhere in the world, more than 1 year before the date on which the patent application is filed. In other countries, the inventor does not have a 1-year grace period.

Original and nonobvious: An invention involves a creative, inventive step. When compared with what is already known, it would not be obvious to someone experienced in the subject matter, or would be unexpected or contrary to established theories or findings.

Useful: This means that the invention must take the practical form of a machine, apparatus, device, diagnostic kit, pharmaceutical compound; it has to accomplish something of practical value to society.

A utility invention can fulfill any of the following definitions: a Kit for accomplishing a useful goal, a Method or Process of synthesis or production, a Machine, an Article of Manufacture, a Composition of Matter (such as a chemical compound), or an improvement of any of the above categories. Design patents are for the new ornamental design of an article of manufacture. Plant patents provide patent protection for any asexually reproduced distinct and new variety of plant.

2.1. Issued Patents

Question: What is an issued patent and what protection does it afford to the inventor?

Answer: The term "patent" is derived from "letters of patents patent"; an open letter by which a sovereign entity conferred a special privilege or right on subject. The first recorded patent was granted to Filippo Brunelleschi in 1421 in Florence, Italy, for an industrial invention (6). Since then countries have set their own rules to grant patents, including the duration, types of patents, and filing rules.

An invention is a property right (an owned article of property that comes into existence the instant it is invented) for an invention granted by a government to the inventor. A US patent gives inventors the right "to exclude others from making, using, offering for sale, or selling their invention throughout the United States or importing their invention into the United States" for a limited time.

Utility and plant patents are granted for a term that begins with the date of the grant and usually ends 20 years from the date the applications were filed. You must make the timely payment of the appropriate maintenance fees.

Design patents last 14 years from the date you are granted the patent. No maintenance fees are required for design patents.

The patent is a personal property: so it can be sold, assigned, or transferred as determined by the owner. As such there can be disputes, in which case the authority or jurisdiction concerned has to mediate and investigate infringement. If infringement is found then a determination must be made to grant penalties to the violator and award damages to the rightful owner.

In the 1990s the establishment of World Trade Organization set forth a common minimum set of rights that should be granted to all patent owners by governments, as well as a period of 20 years (from the date the application filed) as the term of the patent.

Table 1 lists links to patent offices/organizations/procedural guideline sources for selected countries, world bodies, and interest groups.

2.1.1. Nonpatentable Articles

Question: What cannot be patented?

Answer: Unmodified pre-existing articles of nature cannot be patented. You cannot patent an unmodified natural chemical, gene, protein, or animal, or plant species (1, 5). However, you can patent a modified form of an article of nature if the modification serves a useful purpose. You can patent the use of existing articles of nature in devices, compounds, or diagnostic kits that are useful.

A 2010 court case ruling highlights the controversy concerning the patenting of genes and proteins (7, 8). A US district court ruled in March 2010 that the claims were not valid in seven patents covering genetic testing using breast cancer susceptibility genes. The ruling followed a lawsuit against the company Myriad Genetics

Table 1
Resources for patent offices/organizations/procedural guidelines for selected countries, world bodies, and interest groups

Country/entity	Web site resources
USA	How to get a Patent in USA http://www.tutorial-reports.com/innovation/patent/howtogetus.php
India	How to get a Patent in India http://www.tutorial-reports.com/innovation/patent/howtogetindia.php
Europe	european-patent-office.org http://www.european-patent-office.org/wbt/pi-tour/tour.php
Japan	http://www.jpo.go.jp/
Korea	http://www.kipo.go.kr
Italy	http://www.info-brevetti.org/
Canada	http://strategis.ic.gc.ca/sc_mrksv/cipo/
Australia	http://www.ipaustralia.gov.au/
African Region	http://www.aripo.wipo.net/
New Zealand	http://www.iponz.govt.nz/
Singapore	http://www.ipos.gov.sg/
UK	http://www.ukpats.org.uk
WIPO (World Intellectual Property Organization)	http://www.wipo.int/patentscope/en/
Patent Cooperation Treaty (PCT) World Trade Organization (WTO)	http://www.wto.org/english/tratop_e/trips_e/trips_e.htm http://www.pctlearningcenter.org/
The Public Patent Foundation (PUBPAT)	http://www.pubpat.org/index.html
Intellectual Property Owners Association	http://www.ipo.org/

and the University of Utah Research Foundation, which hold the patents on the *BRCA1* and *BRCA2* breast cancer susceptibility genes (7). A woman who tests positive has on average an 82% risk of developing breast cancer in her lifetime and a 44% risk of developing ovarian cancer, according to Myriad. The plaintiffs who brought the lawsuit were the Association for Molecular Pathology and the American College of Medical Genetics and they were represented by the American Civil Liberties Union (ACLU) and the New York-based Public Patent Foundation. Judge Robert Sweet of the US District Court for the Southern District of New York ruled that both Myriads' composition and method claims are invalid under the law, because a product of nature, in this case a gene, even if isolated, cannot be patented as an invention (7). While this case is being appealed, many experts worry that the ruling will have a chilling impact on the biotechnology industry. Nevertheless, if the ruling is upheld it does not prevent the patenting of diagnostic tests using genes or proteins to predict disease or guide therapy. The take home message is to craft patent claims around the nonobvious practical use of a gene or protein, or its modified form, instead of trying to patent the gene or protein itself as a composition of matter.

You cannot patent: laws of nature, physical phenomena, abstract ideas, literary, dramatic, musical, and artistic works. These can be copy write protected. You cannot patent inventions that are considered not useful or physically impossible by the USPTO (for example perpetual motion machines) or considered offensive to public morality.

2.2. Persons Qualifying as Inventors or Co-inventors

Question: If two people or groups make the same invention around the same time, who gets the patent?

Answer: *First to invent rule*: The USA grants a patent to the first inventor who conceives and reduces the invention to practice, e.g., a working prototype or well-written description. Other countries use the *first to file rule* granting a patent and all rights to the first person who files a patent application for an invention (9, 10) (see Note 1). As of September 2011, the USA changed its patent regulations so as to switch from a first to invent to a first to file rule under the Americans Invent Act of 2011 inacted by President Obama. "Patent Reform Act of 2011," Pub.L. 112–29, H.R. 1249.

Clause 101 of US Code 35 states:

"Whoever invents or discovers any new and useful process, machine, manufacture, or composition of matter, or any new and useful improvement thereof, may obtain a patent therefore, ..."(11). Effective January 1, 1996, clause 104 of US Code 35 was changed with the effect that residents of World Trade Organization member countries can now rely on the law of "first to invent" in establishing invention priority in the USA (11) (see Note 2).

2.2.1. Co-inventors

Question: Who qualifies to be a co-inventor?

Answer: Each co-inventor must have made an independent creative contribution to at least one of the claims specified in the patent. This definition is quite different from co-authorship on a scientific publication. Thus, those who creatively and directly generated the invention, qualifying to be co-inventors, may be only a small subset of co-authors listed on a research publication relevant to the subject invention (12).

2.3. Provisional Patent Application

Question: When should I file a Provisional application?

Answer: You may be in a hurry to get a patent because you see immediate commercial application or you want to beat a competitor (see Note 3). In the USA, you can take the option of filing a Provisional

Application for Patent. The Provisional application discloses detailed information about the invention, but not to the depth required in the regular application. You must file a regular patent application on the invention of your Provisional application within 1 year.

The Provisional application establishes a registration date for your invention that is much earlier in time than the ultimate date of patent issue after filing the regular application. The Provisional in no way resembles a regular utility patent, as it expires in a year's time, cannot be searched, and "it does not start a 20-year patent term running."

Provisional applications are usually filed for reasons of urgency to establish priority. However, sometimes there are many good reasons not to file a Provisional application, such as higher overall costs and extra time delay before the patent is granted (prosecution will begin only on the utility application). US patent laws have been amended to allow upgrading of the Provisional application into a Utility patent application.

An alternative means of proving that you were the first to think of the idea is a Disclosure Document. A Disclosure Document is an "evidence of conception" of an idea or an invention. In no way does it substitute for a Provisional application or a regular utility patent application. For a fee of $10, it enables the applicant to have a recorded proof of date of conception provided it is followed up by the regular patent application within 2 years of receipt of the disclosure document at the USPTO (see Note 4). Unlike a Provisional application, the date of the Disclosure Document cannot be considered an effective filing date. Because the Provisional application serves the purpose of providing an earlier filing date, most intellectual property offices recommend filing the Provisional application and not bothering with the Disclosure Document.

2.4. Nondisclosure Agreements

Question: What are the proper components of a Nondisclosure Agreement (see Note 5)?

Answer: Nondisclosure Agreement.

A Nondisclosure Agreement is an agreement under which a party (the "Recipient") agrees not to disclose proprietary and confidential information ("Confidential Information") that it receives from another party (the "Owner").

(a) *No warranty.* There is a possibility that the Confidential Information could contain mistakes or errors, or be based on assumptions that later prove to be incorrect. Therefore, it is common for Owners to include a "no warranty" provision that specifies that the Owner will not be responsible for any damages that the Recipient might incur from using the Confidential Information.

(b) *Risk of accepting disclosure.* The Owner may also want to provide that any disclosure made by the Recipient of any information is at the Recipient's risk. Because the Owner has already stated that it will not warrant the accuracy of the information, the Owner can further provide that the Recipient will bear the risk of using the information in violation of the agreement. For example, if the Recipient acts on some of the information and the information was inaccurate, the Recipient cannot hold the Owner responsible for the harm caused by the inaccurate information.

(c) *Limited license.* Generally, the Owner and the Recipient intend that the Confidential Information will only be used by the Recipient for the limited purpose of reviewing the information to determine whether the parties might have interest in future transactions. A "limited license" provision specifies that the Recipient is not acquiring the right to use the Confidential Information for any other than the specified purpose.

(d) *General provisions.* A Nondisclosure Agreement (*see* **Note 5**) should include provisions that (1) provide a detailed description of the Confidential Information and the purpose of disclosing, (2) require all Confidential Information be identified as such and all oral disclosures reduced to writing within a specific time frame, (3) include a termination date (i.e., 2–5 years) and return policy, (4) acknowledge an obligation under Federal and State Freedom of Information Act (FOIA), (5) require amendments (changes) to the agreement to be in writing and signed by both parties, (6) specify the state whose laws will govern and interpret disputes between the parties regarding the matters covered by the agreement, and (7) prohibit the parties from assigning their obligations under the agreement to third parties. Generally, the state whose laws should govern the agreement should be the state of the Owner.

3. Preparing a Patent Application

3.1. Records of the Invention

Question: What is the proper format for an inventor's notebook record?

Answer: The inventor's logbook ideally should be a separate book or witnessed highlighted pages or entries in an ongoing laboratory notebook (see Note 6). Detailed records of the concepts, test results, and other information related to making an invention should be kept in a logbook (3, 9). Look for sequentially preprinted numbered pages, fade-away backgrounds, spaces for you and a witness to sign and date. Never use a loose leaf notebook or a 3-ring binders as a log book. Never use a legal pad or any glued together notebook. Use a notebook with bound or sewn pages. The pages must be bound so that you can prove in a legal patent dispute that you did not simply add the notebook record later and back date it (9, 10).

3.2. Patent Specifications

Question: What are the essential features of the patent specification and the sections of the application?

Answer: The specification is a written detailed description of the invention and how to make and use the invention. The specification must be written such that a person that is skilled in the technology could make and use your invention. The components of the application are listed below (11, 13, 14):

- Title of invention.
- Cross-reference to related applications.
- Statement regarding federally sponsored research or development.
- Reference to a sequence listing, a table, or a computer program, listing compact disc appendix.
- Background of the invention.
- Brief summary of the invention.
- Brief description of the several views of the drawing.
- Detailed description of the invention.
- Claim or claims.
- Abstract of the disclosure.
- Drawings (when necessary).
- Oath or declaration.
- Sequence listing (when necessary).

3.2.1. Title of Invention

The title of the invention may have up to 500 characters, and should be as short and specific as possible.

3.2.2. Cross-Reference to Related Applications

Any nonprovisional utility patent application claiming the benefit of one or more prior filed co-pending nonprovisional applications (or international applications) under laws 120, 121, or 365 (c) must contain in the first sentence of the specification following the title, a reference to each prior application, identifying it by the application number or international application number and international filing date, and indicating the relationship of the applications, or include the reference to the earlier application (11). Cross-references to other related patent applications may be made when appropriate.

3.2.3. Statement Regarding Federally Sponsored Research or Development

The application should contain a statement as to rights to inventions made under federally sponsored research grants or intramural programs if applicable.

3.2.4. Background of the Invention

This section should include a statement of the field of endeavor to which the invention pertains. This section may also include a

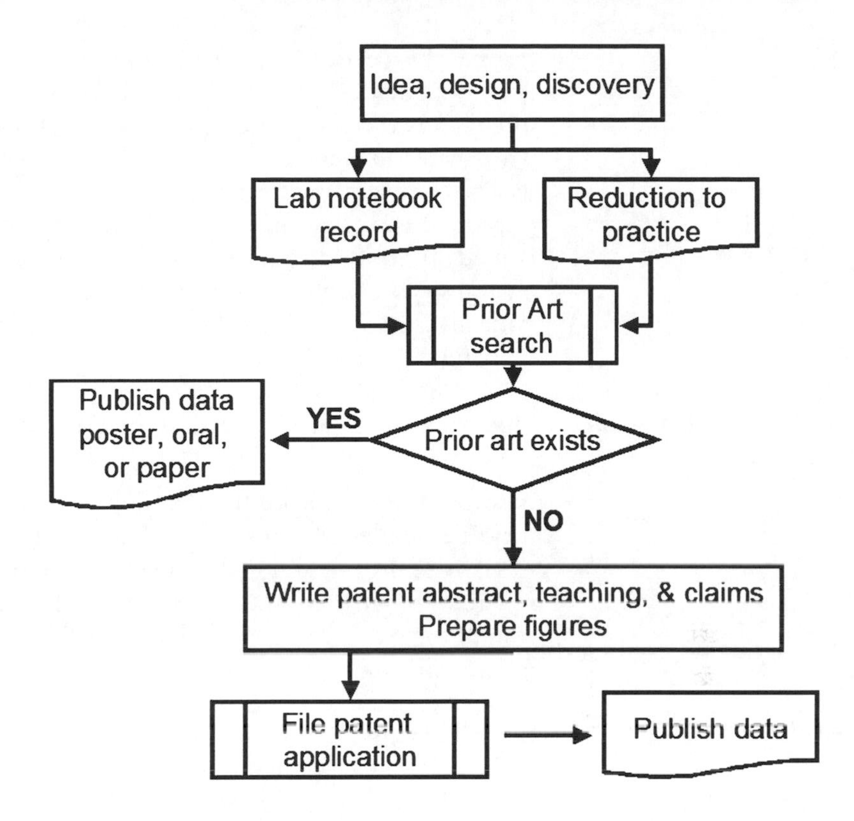

paraphrasing of the applicable US patent Classification Definitions or the subject matter of the claimed invention. In the past, this part of this section may have been titled "Field of Invention" or "Technical Field." This section should also contain a description of information known to you, including references to specific documents, which are related to your invention. It should contain, if applicable, references to specific problems involved in prior art or missing gaps or needs in existing technology.

3.2.5. Brief Summary of the Invention (Different from the Abstract)

This section should present the substance, objective, or general idea of the claimed invention in summarized form. The summary may point out the advantages of the invention and how it solves previously existing problems identified in the Background of the Invention.

3.2.6. Brief Description of the Several Views of the Drawing

Where there are drawings, you must include a listing of all figures by number (e.g., Fig. 1a) and with corresponding statements explaining what each figure depicts.

3.2.7. Detailed Description of the Invention

The specification is the description along with the claims. In this section, the invention must be explained along with the process of making and using the invention in full, clear, concise, and exact terms. This section should distinguish the invention from other inventions and prior art. The description for biomedical patents often takes the form of experimental examples that are presented with materials and methods, results, and conclusions. You must write a complete and thorough, yet broad, description because you cannot add any new information to your patent application once it is filed. If you are required by the patent examiner to make any changes, you can only make changes to the subject matter of your invention that could be reasonably inferred from the original drawings and description.

3.3. Writing the Patent Claims

Question: What are the claims of the patent and why are they important?

Answer: The claims are the defining features of the invention and form the legal basis for protection. The claim or claims must particularly point out and distinctly claim the subject matter that you regard as the invention. The claims define the scope of the protection of the patent. Whether a patent will be granted is determined, in large measure, by the choice of wording of the claims.

The most important part of the patent is the claims. The claims must be crafted to encompass the invention as broadly as possible while not overstepping the requirements of novelty and utility. The material covered in the patent application is often referred to as "teaching" because an important requirement of the patent is that it "teaches" others how to carry out the invention. The terms in the claims must refer back to words used in the teaching (body) of the patent. Importantly, a claim in the patent cannot extend very far beyond what is reduced to practice in the data and examples given in the patent. The breadth of a claim becomes a balance between what can be argued is novel beyond the prior art, and what is disclosed or logically inferred from the teaching of the patent. This is best explained as an example of an invention and a discovery in the field of molecular diagnostics.

Imagine that you have discovered that the phosphorylation of a specific site on the molecule Beclin-1 shows a strong correlation with disease-free survival in patients with multiple myeloma who have been treated with dexamethasone. You could file a patent with the following claim:

"We claim a method of selecting therapy for a patient with multiple myeloma comprising the steps (a) measuring the phosphorylation level of Beclin-1 in a cell sample derived from the patient, and (b) determining if the level of phosphorylation is above a defined threshold, and (c) administering dexamethasone to the patient." This claim would not cover the use of phosphorylated Beclin-1 to choose the therapy for any other type of disease, or any other type of therapy besides dexamethasone. A broader claim therefore, might be as follows, if there is data or rationale provided in the patent teaching:

"We claim a method of selecting therapy for a patient with a hematologic malignancy, comprising the steps of (a) measuring the phosphorylation level of Beclin-1 in a cell sample derived from the patient, (b) determining if the level of phosphorylation is above a defined threshold, and (c) administering a steroid alone or in combination with another treatment modality." This broader claim can be argued because (a) dexamethasone is classified as a steroid, and (b) multiple myeloma is one of many types of hematologic malignancies including myelodysplasias and leukemias.

3.3.1. Dependent and Independent Claims

The claims section must begin with the statement, "What I claim as my invention is..." or "I (We) claim..." followed by the statement of what you regard as your invention. One or more claims may be presented in dependent form, referring back to and further limiting another claim or claims in the same application. All dependent claims should be grouped together with the claim or claims to which they refer to the extent practicable. Any dependent claim that refers to more than one other claim shall refer to such other claims in the alternative only.

3.3.2. In "Claims" Every Word Is Important

Claims are the parts of a patent that define the boundaries of patent protection. Patent claims are the legal basis for your patent protection. They form a protective boundary line around your patent that lets others know when they are infringing on your rights. The limits of this line are defined by the words and phrasing of your claims (see Note 7).

(a) *Scope.* Each claim should have only one meaning which can be either broad or narrow, but not both at the same time. In general, a narrow claim specifies more details than a broader claim. Having many claims, where each one is a different scope, allows you to have legal title to several aspects of your invention.

(b) *Important characteristics.* Three criteria to take note of when drafting your claims are that they should clear, complete, and supported in the application. Any terms you use in the claims must be either found in the description or clearly inferred from the description.

(c) *Structure of a claim.* A claim is a single sentence composed of three parts: the introductory phrase, the body of the claim, and the link that joins the two. The introductory phrase identifies the category of the invention and sometimes the purpose for example, "a diagnostic test kit," or "a composition for treating cancer." The body of the claim is the specific legal description of the exact invention that is being protected. The linking consists of words and phrases such as "which comprises," "which consists of," "including," "consisting of," or "consisting essentially of." The linking word or phrase describes how the body of the claim relates to the introductory phrase. The linking words are also important in assessing the scope of the claim, as they can be restrictive or permissive in nature.

(d) *Merit of the claim.* Each claim will be evaluated by the Patent Examiner on its own merit. It is important to make claims on all aspects of your invention to ensure that you receive the most protection possible. One way of ensuring that specific inventive features are included in several or all claims is to write an initial claim and refer to it in claims of narrower scope. Thus, all the elements in the first claim are also included in the subsequent claims. As more features are added the claims become narrower in scope.

3.4. Patent Abstract

Question: What are the essential features of the Patent Abstract?

Answer: Abstracts are limited to 150 words and are used primarily for searching patents. They should be written in a way to make the invention easily understood by those with a background in the field. The abstract should summarize your invention and how it is useful, but does not discuss the scope of your claims.

3.5. Patent Drawings and Sequence Tables

Question: What are the essential features of the Patent Drawings and Sequence tables?

Answer: A patent application is required to contain drawings if drawings are necessary for the understanding of the subject matter to be patented (see Note 8). The drawings must show every feature of the invention as specified in the claims. Omission of drawings may cause an application to be considered incomplete. All patent drawings must show every feature of the invention specified in the claims, and is required to be in a particular form. The reason for specifying the standards in detail is that the patent drawings are printed and published in a uniform style when the patent issues, and patents must provide enough information so that someone with proper expertise can reproduce the invention (14).

3.6. Amino Acid and Nucleotide Sequence Listing

If they apply to your invention, amino acid and nucleotide sequences must be included as they are considered part of the description. You must prepare this section, for the disclosure of a nucleotide and/or amino acid sequence, with a listing of the sequence that complies with the following patent rules: 1.821, 1.822, 1.823, 1.824, and 1.825 [37 CFR 1.821 Nucleotide and/or amino acid sequence disclosures in patent applications and WIPO Standard ST.25 (1998)) (13, 14).

3.7. International Agreement for Patent Application Filing

Question: How do I obtain patent protection in a foreign country? What is a PCT Application?

Answer: To be protected in any country you must attain an issued patent in that country. For this reason, most new patents are filed under an international agreement for sharing patent application among countries (http://www.pctlearningcenter.org/). The Patent Cooperation Treaty or PCT is an international agreement for filing patent applications having effect in up to 117 countries. Although the PCT system does not provide for the grant of an international patent, the system is designed to (a) simplify the process of filing patent applications across multiple countries, (b) delay the expenses associated with applying for patent protection in other countries, and (d) allow the inventor more time to evaluate the commercial viability of his/her invention. Under the PCT, an inventor can file one international patent application in one language with one patent office in order to simultaneously seek protection for an invention in up to 117 countries.

Filing a PCT application does not mean that a separate application is automatically filed in all the countries covered in the agreement. The invention must still be individually filed in each country and the inventor must follow the rules specific to that country. The specifications of the PCT patent application are similar to the US patent application, but each country has its unique requirements. Filing fees for the PCT application are considerable and the inventor still has to pay additional fees for each foreign country filing. In the end the invention may be granted in some countries but not others and the patent claims allowed for the same invention may be different in each country.

3.8. Publication of the Patent Application

Question: When is the application published?

Answer: Publication (making copies available to the public) of patent applications is required for most plant and utility patent applications. Publication of patent applications is one of the functions of the WIPO (World Intellectual Property Organization) and the USPTO. On filing of a US plant or utility application, an applicant may request that the application not be published, but only if the invention has not been and will not be the subject of an application filed in a foreign country that requires publication 18 months after filing (or earlier claimed priority date) or under the Patent Cooperation Treaty (PCT).

Publication occurs after expiration of an 18-month period, both by the USPTO and WIPO/PCT patent applications, following the earliest effective filing date or priority date claimed by an application. Following publication, the application for patent is no longer held in confidence by the US PTO or WIPO Office and any member of the public may request access to the entire file history of the application.

3.9. Patent Examination of the Patent Office

Question: What happens after the patent application is filed?

Answer: Once the patent is filed your attorney will receive a notice of the filing and the identifying registration number. If the resolution of the figures, the size of the tables, or the format of the application has any errors, the patent office may request corrections to be made (see Note 8). They will provide a deadline for filing the complete patent, and they will charge a fee. When you make the corrections you cannot change any of the text or the data. Once the application is in order it will then be reviewed by an Examiner who has expertise in the field of the invention. The Examiner will evaluate every claim for novelty and utility. Your attorney will eventually receive a report from the Examiner rejecting or accepting one or more Claims. For any Claim that is rejected, the Examiner will cite the reason for rejection and will provide a citation of literature that the Examiner believes anticipate the specific Claim or Claims. Your attorney can appeal the decision in a written rebuttal to each of the Examiner's rejections. In response to the rebuttal, the Examiner may, or may not, agree with your arguments. Your attorney may request a meeting with the Examiner to discuss your rebuttal in person. At some point in the exchange the Examiner will declare their decision as "final." Any Claims that are allowed at this stage will become the Claims of the published patent.

3.10. Benefits of Publication for the Inventor

As a result of publication, an applicant may assert provisional rights. These rights provide a patentee with the opportunity to obtain a reasonable royalty from a third party that infringes a published application claim. Thus, damages for prepatent grant infringement by someone violating one or more of the claims of the invention are now available.

3.10.1 Patent Assignment

Question: What is a patent assignment?

Answer: Assigning your patent is like selling your house. You do not own it any longer. Licensing your patent is like renting your house, you can evict the renter's if they violate the terms of the lease.

Patent law provides for the transfer or sale of a patent by a written agreement called an "assignment" that can transfer the entire interest in the patent. The assignee, when the patent is assigned to him or her, becomes the owner of the patent and has the same rights that the original patentee had. Patent law also provides for the assignment of a part interest, that is, a half interest, a fourth interest, etc., in a patent. An assignment can be granted for a particularly specified part of the invention or a specific field of use.

The US Patent Office records assignments, grants, and similar instruments sent to it for recording, and the recording serves as notice. If an assignment, grant, or conveyance of a patent or an interest in a patent (or an application for patent) is not recorded in the US Patent Office within 3 months from its date, there can be no subsequent purchaser(s).

3.10.2. Patent Licensing and Joint Ownership

Patents may be owned jointly by two or more persons as in the case of a patent granted to joint inventors, or in the case of the assignment of a part interest in a patent. Any joint owner of a patent, no matter how small the part interest, may make, use, offer for sale and sell and import the invention for his or her own profit provided they do not infringe another's patent rights, without regard to the other owners, and may sell the interest or any part of it, or grant patent licensing to others, without regard to the other joint owner.

A patent licensing agreement is in a promise by the licensor not to sue the licensee for patent infringement. No particular form of license is required; a license is a written contract and may include whatever provisions the parties agree upon, including the payment of royalties, etc.

3.11. Patent Infringement

Question: What do I do if someone is infringing my patent?

Answer: Patent infringement consists of the "unauthorized making, using, offering for sale or selling any patented invention within the United States or United States Territories, or importing into the United States of any patented invention during the term of the patent" (35 United States Code 271; http://www.uspto.gov/web/offices/pac/mpep/consolidated_laws.pdf) (11).

When patent infringement happens, the patentee may sue for relief in the appropriate Federal court. The patentee may ask the court for an injunction to prevent the continuation of the patent infringement and may also ask the court for an award of damages because of the patent infringement (15).

The defendant usually challenges the validity of the patent, which is then decided by the court. An invalid patent, for example, could be wrongly granted because the Examiner did not know about a prior art reference, produced by the defendant that anticipated the invention. The defendant may also try to say that what is being done does not constitute infringement. Infringement is determined primarily by the specific language of the claims. The defendant will argue that the specific terms of the claims are not violated.

Suits for infringement of patents follow the rules of procedure of the Federal courts. From the decision of the district court, there is an appeal to the Court of Appeals for the Federal Circuit. The Supreme Court may thereafter take a case by *writ of certiorari.* If the US Government infringes a patent, the patentee has a remedy for damages in the US Court of Federal Claims. The Government may use any patented invention without permission of the patentee, but the patentee is entitled to obtain compensation for the use by or for the Government.

3.12. Product Patent Documentation

Question: If I buy a product or kit and there is no patent number listed on the product, can it still be covered by a patent?

Answer: No. Anyone who sells patented articles is required to mark the articles with the word "Patent" and the number of the patent. The penalty for failure to mark is that the patentee may not recover damages from an infringer unless the infringer was duly notified of the infringement and continued to infringe after the notice. The marking of an article as patented when it is not in fact patented is against the law and subjects the offender to a penalty.

3.12.1. Patent Pending

Articles can be sold with the terms "Patent Applied For" or "Patent Pending." These phrases have no legal effect, but only give information that an application for patent has been filed in the US Patent and Trademark Office or WIPO. The protection afforded by a patent does not start until the actual grant of the patent. False use of these phrases or their equivalent is prohibited.

4. Notes

1. A common pitfall encountered by inventors is failure to actually make your invention. You can't sell ideas – you can only sell inventions. You must create a working prototype, provide example data demonstrating utility, or at least describe you invention in enough detail so that it would be expected to work by someone knowledgeable in the field (2).
2. Pursuing an invention that has no commercial market can be costly and time-consuming. Before using your children's college funds to heavily invest in your invention, assess the true value in the market compared to the existing state of the art. Does your invention offer true advantages in performance or economy? Know when it's time to move on to your next great idea (4).
3. Revealing your invention prematurely is another common pitfall encountered by inventors. In the USA, a 1-year countdown begins the instant you reveal your invention to the public or anybody that has not signed a confidentiality agreement with you (see definition of novelty above). You only have 1 year to patent your invention in the USA. For other countries, if you reveal the invention, this is no 1-year grace period and you lose the rights. At any point in time, scientists all over the world are exposed to similar information and perceive similar needs for new products or treatments. There is a good chance that someone is thinking of, or has already patented, your new idea. Conduct a search for prior art to see if anybody else has already patented an invention similar or identical to yours. This can be done in the scientific literature as well as patent databases. Under the "Patent Reform Act of 2011," Pub.L. 112–29, H.R. 1249, USA inventors now obtain priority based on the "first to file" rule. Inventors are urged to study the new provisions of this law that strongly influence the timing of a patent application.
4. Filing a patent is very expensive, and this cost must be weighed in the decision to file a patent and where to file a patent. If the inventor files a patent only in her/his home country the costs

incurred will include the (a) filing fee, (b) additional fees depending on the number of claims, (c) maintenance fees, and (d) additional fees required during the prosecution of the case. In addition to the cost of a PCT application, if the invention is filed in a foreign country, each country will have its own set of fees. The inventor will have to pay fees to translate the patent into the appropriate language for each country. On top of all the filing fees for the application, the inventor or the sponsor must factor in the legal fees paid to the patent attorney. Filing a patent therefore can range from $2,000 to $10,000, and the total cost for an issued patent may be much greater than $20,000.

5. A typical confidential disclosure agreement, addresses proprietary information, and how long must the information be kept secret. To be considered "Proprietary Information" all such information must clearly and conspicuously be identified as "confidential" or "proprietary" to the Disclosing Party, and all oral proprietary information must be reduced to writing or other tangible form and delivered identified as confidential or proprietary within 20 days of disclosure. A university may have an obligation under Federal and state or province statutes to disclose certain information in the possession of University to the public. Unless terminated by either party, the obligations set forth herein shall remain in full force and effect for a period of 2 years from the date hereof. *Termination of agreement and return of proprietary information:* Either party may terminate this agreement without cause by giving written notice of such termination by certified mail, express or overnight mail, or by telephone facsimile. Such termination shall be effective immediately upon receipt of the notification by the other party. Upon termination the parties shall immediately (1) return to the Disclosing Party all items of Proprietary Information (including all copies thereof) of the Disclosing Party, upon written request or (2), at the option of the disclosing party, destroy any notes or personal memoranda which include or make reference to such Proprietary Information.

6. You should keep detailed records of the concepts, data, and other information related to making an invention in a logbook. Start a logbook entry series from the very first moment you think of an idea. Proper record keeping can be used as proof of the conception date of an invention. The best way to prove that an idea is yours is by maintaining an inventor's journal or logbook, recording the experiments and discussions including with whom you discussed the invention, and having a witness who can testify that you made the invention (3, 9).

7. Each individual claim is evaluated by the patent examiner. Each claim is either allowed or not allowed based on its individual merit. For this reason, the language of the claim is often redundant in stating the novelty of the invention.

8. Patent figures or tables must be drawn following specific guidelines. The figure must be on A4 paper, with one inch margins, 32 mm text (14-point font) and must be in black and white. The resolution must be adequate for reproduction (300 dpi). A listing of a nucleic acid sequence must comply with the following patent rules: 1.821, 1.822, 1.823, 1.824, and 1.825, and may be in paper or electronic form (see: 35 United States Code 271; http://www.uspto.gov/web/offices/pac/mpep/consolidated_laws.pdf.) (11)

Acknowledgment

The authors thank Richard Peet for invaluable discussion regarding patent law and intellectual property.

References

1. Jones, B. W. (2006) Broadening the Scope of Inherent Anticipation and Its Impact on the Patentability of Chemical Structures, Comment, SmithKline v. Apotex. *5J MARSHALL REV INTELL PROP L* **456**, 455–476.
2. Gholz, C. L. (2007) A Critique of Recent Opinions in Patent Interferences. *J Pat & Trademark Office Society* **89**, 1–43.
3. Merges, R. P. (1999) The Law and Economics of Employee Inventions. *Harvard Journal of Law and Technology* **13**, 1–53.
4. Kelton, T. (2007) Pharmacogenomics:The rediscovery of the concept of tailored drug therapy and personalized medicine. *The Health Lawyer* **19**, 1–10.
5. Shi, Q. (2005) Patent System Meets New Sciences: Is the Law Responsive to Changing Technologies and Industries? *NYU Annual Survey of American Law* **61**, 317–346.
6. (2011) *General information about patents* in *Patent Information Tour*. Accessed March 8, 2011, http://www.european-patent-office.org/wbt/pi-tour/tour.php.
7. Ledford, H. (2010) *US government wants limits on gene patents* in *Nature News*. Accessed March 6, 2011, http://www.nature.com/news/2010/101102/full/news.2010.576.html.
8. Wadman, M. (2010) *Breast cancer gene patents judged invalid* in *Nature News*. Accessed March 6, 2011, http://www.nature.com/news/2010/100330/full/news.2010.160.html.
9. Dolak, L. A. (1999) Patents Without Paper: Proving a Date of Invention With Electronic Evidence. *Houston Law Review* **36**, 472–530.
10. Gholz, C. L. (2000) First-to-File or First-to-Invent? *J Pat & Trademark Office Society* **82**, 891–895.
11. (2007) *Title 35 Patent Laws* in *United States Code Title 35-Patents*.
12. Matt, J. (2002) Searching for an Efficacious Joint Inventorship Standard. *Boston College Law Review* **44**, 245–287.
13. (2011) *Title 37 Patents, Trademarks, and Copyrights* in *United States Code Title 37 Parts 1–199*.
14. (2010) *Manual of Patent Examining Procedure (MPEP)*March 6, 2011, http://www.uspto.gov/web/offices/pac/mpep/index.htm.
15. Adamo, K. R., McCrum, R. B., Gerber, S. M. (2007) The Curse of "Copying". *7J MARSHALL REV INTELL PROP L* **296**, 296–324.

Chapter 27

Regulatory Approval Pathways for Molecular Diagnostic Technology

Lance A. Liotta and Emanuel F. Petricoin III

Abstract

This chapter describes the basic categories for regulatory approval to sell/market a molecular profiling technology. The US Food and Drug Administration regulates and provides guidance, for marketing in vitro diagnostic devices (IVD). Three different paths currently exist for obtaining Food and Drug Administration (FDA) approval of an IVD: (a) If the new test can be shown to be substantially equivalent to an existing predicate test on the market, then the 510(k) is the regulatory path for new device approval. (b) If your new diagnostic technology cannot be considered substantially equivalent to an existing technology, and will be used to make a critical medical decision concerning the diagnosis, treatment, or medical management, then the premarket approval (PMA) is the regulatory path of choice. (c) If no predicate device exists and the test is of low or moderate risk, it may be eligible for a de novo reclassification. If the test is done "in house," in the designated laboratory only, for a patient sample that is sent to the laboratory from an outside physician's office or medical facility, then the test can be potentially marketed under "home brew" guidelines (also known as laboratory developed tests) regulated under the Clinical Laboratory Improvement Amendments (CLIA). The Centers for Medicare and Medicaid Services (CMS) assumes primary responsibility for financial management operations of the CLIA program, but the categorization of commercially marketed in vitro diagnostic tests under CLIA is the responsibility of the FDA. Definitions, guidelines, information sources, and instructions for data requirements are outlined for each regulatory pathway.

Key words: 510(k), Accuracy, Approval, Biomarker, Code of Federal Regulations, Devices, Diagnostic, Molecular profiling, Precision, Premarket approval application

1. Introduction

While the urgent goal of translational research is public health benefit, regulatory guidelines exist to insure that molecular profiling technology, before it is marketed, is accurate, safe, sensitive, and specific for the intended use. A patent application, or a publication describing a

Virginia Espina and Lance A. Liotta (eds.), *Molecular Profiling: Methods and Protocols*, Methods in Molecular Biology, vol. 823, DOI 10.1007/978-1-60327-216-2_27, © Springer Science+Business Media, LLC 2012

candidate biomarker or diagnostic technology, is just the first step in a long process to ultimately market a product to physicians and their patients. A set of guidelines, good manufacturing processes, and regulatory approval steps, must all be followed to insure the safety and effectiveness of the new product. Finally, you must establish that the product can be manufactured reliably and reproducibly with an accurate and designated shelf life.

An in vitro diagnostic device (IVD) is any system used to diagnose a disease or other condition. The US Food and Drug Administration (FDA) regulates IVDs mainly through the Office of In Vitro Diagnostic Device Evaluation and Safety (OIVD); Center for Devices and Radiological Health (CDRH); with consultations or joint review from the Center for Biologics Evaluation and Research (CBER), and Center for Drug Evaluation and Research (CDER), depending on the assay. The intended use of the device determines how much data, type of data, and approval route to use. Four main pathways for regulatory approval of new diagnostic devices are (1) a 510(k) pre-market submission, (2) PMA submission, (3) de novo reclassification, or (4) marketing under "Home Brew," or in-house developed assay.

The OIVD encourages pre-investigational devise exemption (pre-IDE) meetings with in vitro diagnostic manufacturers prior to submission of regulatory approval documents. These meetings have been termed "pre-IDE" for pre-investigational device exemption. 510(k) Submissions generally have abundant documentation available to the public on the FDA website, therefore pre-IDE meetings are generally not cost-effective. PMA applications are better suited to pre-IDE meetings compared to 510(k) submissions. Pre-IDE meetings should only be arranged if the devise manufacturer has a study plan, population/study set, statistical plan, clear definition of intended use, and preclinical testing plan (1).

This chapter summarizes the definitions and qualifications to support the marketing of a candidate molecular profiling test under each of these regulatory categories. The following summary material provides general guidelines. The reader is encouraged to explore the cited documents and web sites for detailed information and guidance.

2. Regulatory Approvals for New Diagnostic Devices

Three routes of regulatory approval currently exist for IVDs. Each route requires general requirements to be met such as (a) availability of adequate specimens with associated clinical data, (b) samples relevant to the intended market/question, (c) compliant with informed consent and/or applicable regulatory issues, and (d) lack of legal or ethical issues (2, 3).

1. 510(k) is a rapid regulatory path for the marketing of a diagnostic test or device that qualifies for approval if the new test can be shown to be substantially equivalent to an existing test already on the market. Generally, the 510(k) process is used for tests that monitor disease progression.
2. PMA is required if your new diagnostic technology cannot be considered substantially equivalent to an existing technology, and will be used to make a critical medical decision concerning the diagnosis, treatment, or medical management of the patient. PMA approval is required for screening tests.
3. De novo reclassification is appropriate for a low to moderate risk device or test that does not have a predicate device for comparison. Usually, a 510(k) is submitted for the devise and the FDA rules that it is not substantially equivalent. At this point, a petition for de novo reclassification can be instituted (2).
4. "Home brew," in-house developed, tests are diagnostic laboratory tests brought to market by a specific medical center or reference laboratory. The test is done in-house, in the designated laboratory only, for a patient sample that is sent to the laboratory from an outside physician's office or medical facility. Marketing a diagnostic test under the home brew category does not apply to a test kit that is sold to multiple laboratories. While the FDA does not specifically regulate these types of assays, it is in charge of the categorization of any assay brought to commercialization in the USA.

2.1. Sample Procurement to Support Clinical Trials

2.1.1. Prospective Sample Collection

In vitro diagnostics intended for marketing require clinical trial data that support the claims of the device. Prospective studies are often designed for development of new biomarkers or new intended uses (3). Clinical trial data generated from samples collected prospectively allows the researcher greater control over the collection process and study design. The disadvantage of prospective sample collection for in vitro diagnostic marketing is the extended time required to correlate the diagnostic test measurement with a subsequent disease recurrence, stabilization, or remission. Clinical performance of the device or biomarker may not be immediately evident due to diseases that progress or recur (3).

2.1.2. Retrospective Sample Collection

Retrospective studies, which use archived or banked samples, may also be used to support 510(k) or PMA requirements. Retrospective samples present a quandary because the patients may not be available to provide consent. The FDA has issued a guidance document allowing the use of anonymized, leftover tissue without obtaining informed consent "Guidance on Informed Consent for In Vitro Diagnostic Device Studies Using Leftover Human Specimens that Are Not Individually Identifiable" (3). Clinical data associated with leftover tissue, not required for diagnosis, must also be de-identified to qualify under this exemption. The first FDA approved test based on retrospective tissue collection with prospective analysis was Agendia's MammaPrint 70-gene signature assay for prognosis of breast cancer (4–6). MammaPrint data was derived from fresh frozen tumor tissue without prior knowledge of the genes involved in breast cancer progression/metastasis (4–6).

2.2. US Federal Documents/Websites for Medical Devices

Government regulatory agencies provide a variety of electronic resources to assist with the approval process. Records are updated periodically and should be consulted prior to submitting approval documents for your device/test.

1. The Code of Federal Regulations (CFR) Title 21: CFR Title 21 consists of parts 1 – 1499. Each part details specific regulations governing a particular topic related to foods and drugs. CFR is updated annually and issued quarterly. CFR citations are identified by title, part, and section, e.g., 21CFR11.1 (see Note 1).
2. Product Classification Database: the database search provides the name of the device, classification, and a link to the CFR, if any. The CFR provides the device type name, identification of the device, and classification information (see Note 2).
3. Clinical Laboratory Improvement Amendments (CLIA) are federal regulations to ensure quality laboratory testing of human samples, except for research only activities (see Note 3).

   Clinical trials conducted under CLIA compliance meet the requirements for submission to the FDA.

3. Regulatory Processes for New Device Approval

Companies should contact the FDA for feedback/advice when they are almost finished developing their clinical plan. The clinical plan must have an intended use statement, statistical plan, and clinical statement. The intended use statements determine how much data and data type are required as well as determining whether a 510(k) or PMA will be required. It is important to include sales and marketing input when you are developing the intended use statement (7). Seemingly inconsequential wording such as "rule in" or "rule out" can change the intended use of a device. Examples of subtle changes in wording are (a) tests to measure prostate-specific antigen (PSA) to *screen* for prostate cancer have required a PMA while the same test needs a 510(k) if intended to *monitor* PSA levels in patients diagnosed with prostate cancer (7) and (b) an ovarian cancer screening test requires a PMA. An IVD intended to help stratify the risk of ovarian cancer in women who have pelvic masses and are scheduled to have surgery may be reviewed via the 510(k) or de novo process (7).

3.1. 510(k) Submission

510(k) is a rapid regulatory path for the marketing of a diagnostic test or device that qualifies for approval if the new test can be shown to be substantially equivalent to an existing test already on the market. For example, if you develop a new diagnostic assay technology that could be used to measure PSA, you could seek approval under 510(k) to establish that your new test was substantially equivalent to existing immunoassay kits (the "predicate"), which measure PSA for the intended purpose of monitoring prostate cancer. The 510(k) application will consist of a presentation of data verifying the precision, accuracy, specificity, and stability of your new test system compared to the predicate test for the same set of unknown serum samples, and the same intended use.

The first step is for the applicant to determine whether or not the device falls under an exemption category. The FDA lists a series of 510(k) exemptions and Good Manufacturing Practice (GMP)/ Quality System exemptions listed by device class (see Note 4). If your devise is nonexempt and is to be used in humans, a 510(k) submission must be made to the FDA for any device that falls under the following description. A 510(k) is a "pre-marketing submission made to FDA to demonstrate that the device to be marketed is as safe and effective, that is, substantially equivalent (SE), to a legally marketed device that is not subject to PMA (see Note 5). Applicants must compare their 510(k) device to one or more similar devices currently on the US market and make and support their substantial equivalency claims. A legally marketed device is a device that was legally marketed prior to May 28, 1976 (pre-amendments device), or a device which has been reclassified from Class III to Class II or I, a device which has been found to be substantially equivalent to such a device through the 510(k) process, or one established through Evaluation of Automatic Class III Definition. The legally marketed device(s) to which equivalence is drawn is known as the 'predicate' device(s)." (8–11).

Applicants must submit descriptive laboratory data demonstrating performance for the intended use to establish that their device is substantially equivalent to a predicate device. The data in a 510(k) is to show comparability, that is, substantial equivalency (SE) of a new device to a predicate device. Data types vary based on the assay and/or device.

The applicant may not proceed to market the device, unless they receive an order declaring a device SE. Once the device is determined to be SE, it can then be marketed in the USA. The SE determination is usually made within 90 days and is made based on submitted information. The manufacturer should be prepared for an FDA quality system (21CFR820) inspection at any time after 510(k) clearance.

3.1.1. What is Substantial Equivalence?

According to the FDA "a device is substantially equivalent if, in comparison to a predicate it: has the same intended use as the predicate; and has the same technological characteristics as the predicate; or has the same intended use as the predicate; and has different technological characteristics and the information submitted to FDA; does not raise new questions of safety and effectiveness; and demonstrates that the device is at least as safe and effective as the legally marketed device." (2, 8).

A claim of substantial equivalence does not mean the new and predicate devices must be identical. Substantial equivalence is established with respect to intended use, design, energy used or delivered, materials, chemical composition, manufacturing process, performance, safety, effectiveness, labeling, biocompatibility, standards, and other characteristics, as applicable.

A device may not be marketed in the USA until the submitter receives a letter declaring the device substantially equivalent. If FDA determines that a device is not substantially equivalent, the applicant may (a) resubmit another 510(k) with new data, (b) a Class I or II designation through the de novo process (2), (c) file a reclassification petition, or (d) submit a PMA application (12).

3.2. Premarket Approval

PMA is the FDA process of scientific and regulatory review to evaluate the safety and effectiveness of Class III medical devices. Class III devices are those that support or sustain human life, are of substantial importance in preventing impairment of human health, or which present a potential, unreasonable risk of illness or injury (2).

PMA is the most stringent type of device marketing application required by FDA. The applicant must receive FDA approval of its PMA application prior to marketing the device. PMA approval is based on a determination by the FDA that the PMA contains sufficient valid scientific evidence to assure that the device is safe and effective for its intended use(s). An approved PMA is, in effect, a private license granting the applicant (or owner) permission to market the device.

FDA regulations provide 180 days to review the PMA and make a determination. In reality, the review time is normally longer. Before approving or denying a PMA, the appropriate FDA advisory committee may review the PMA at a public meeting and provide FDA with the committee's recommendation on whether FDA should approve the submission. After FDA notifies the applicant that the PMA has been approved or denied, a notice is published on the Internet (1) announcing the data on which the decision is based, and (2) providing interested persons an opportunity to petition FDA within 30 days for reconsideration of the decision.

The regulation governing PMA is located in 21CFR814, Premarket Approval. A class III device that fails to meet PMA requirements is considered to be adulterated under section 501(f) of the FD&C Act and cannot be marketed. Device product classifications can be found by searching the FDA Product Classification Database (see Note 6). The database search provides the name of the device, classification, and a link to the CFR, if any. The CFR provides the device type name, identification of the device, and classification information. If it is unclear whether the unclassified device requires a PMA, use the three letter product code to search the PMA database and the Premarket Notification 510(k) database. If there are 510(k) documents cleared by FDA, and the new device is substantially equivalent to any of these cleared devices, then the applicant should submit a 510(k).

3.2.1. PMA Data Requirements

In addition to the administrative data, good science and lucid scientific writing are the keys to the approval of PMA application. If a PMA application lacks elements listed in the administrative checklist provided, the FDA will refuse to file a PMA application and will not proceed with the in-depth review of scientific and clinical data. If a PMA application lacks valid clinical information and scientific analysis on sound scientific reasoning, it will delay FDA's review and approval (1, 2).

Nonclinical Laboratory Studies Section: nonclinical laboratory studies section includes information on microbiology, toxicology, immunology, biocompatibility, stress, wear, shelf life, and other laboratory or animal tests. Nonclinical studies for safety evaluation must be conducted in compliance with 21CFR Part58 (Good Laboratory Practice for Nonclinical Laboratory Studies).

Clinical Investigations Section: clinical investigations section includes study protocols, safety and effectiveness data, adverse reactions and complications, device failures and replacements, patient information, patient complaints, tabulations of data from all individual subjects, results of statistical analyses, and any other information from the clinical investigations. Any investigation conducted under an Investigational Device Exemption (IDE) must be identified as such (1, 2).

A variety of device-specific FDA guidance documents that describe data requirements are available (see Note 7) (13).

3.2.2. Instructions for Reviewing PMA and 510(k) Approved Applications

Once a PMA or a 510(k) is approved, the associated application documents and review are made publicly available by the FDA. Review of the regulatory documents associated with previously approved diagnostic devices provides examples for new applicants (see Note 8) (14–16).

By example, the steps to access the records for the PMA approval documents describing the Dako HercepTest™ (17–21) are as follows:

1. Access the FDA data base using the URL http://www.accessdata.fda.gov/scrIpts/cdrh/cfdocs/cfPMA/pmasimplesearch.cfm.
2. The screen will show a search field box "Search Pre-market Approval (PMA) Database." Search the company name "Dako."

3. A list of records by company and Approval Date are displayed in a table. Scroll through the Approval Date for the record from Dako on September 25, 1998. The record provides the PMA number P980018. Be sure to note the date of approval because PMA records are stored by approval year.
4. Go to the FDA home page http://www.fda.gov/cdrh/ and enter the PMA number into the search field.
5. Go to the FDA Medical Devices page to access the PMA record. Click on "Premarket Approvals" http://www.fda.gov/MedicalDevices/ProductsandMedicalProcedures/DeviceApprovalsandClearances/PMAApprovals/default.htm.
6. Select the "Devices approved in…" for the year that the device was approved.
7. Scroll through the list to find the approval document of interest.
8. The same process can be used to search 510(k) approved applications, starting with the CDRH home page listed above.

3.3. In-house Developed Assays or Home Brew Tests

In-House Developed Assays, or "Home brew" tests, are diagnostic laboratory tests brought to market by a specific medical center or reference laboratory. The test is done in-house, in the designated laboratory only, for a patient sample that is sent to the laboratory from an outside physician's office or medical facility. Marketing a diagnostic test under the home brew category does not apply to a test kit that is sold to multiple laboratories.

Home brew tests are configured using analyte-specific reagents (ASRs) (22). An ASR are "antibodies, both polyclonal and monoclonal, specific receptor proteins, ligands, nucleic acid sequences, and similar reagents which, through specific binding or chemical reactions with substances in a specimen, are intended for use in a diagnostic application for identification and quantification of an individual chemical substance or ligand in biological specimens." (21 CFR 864.4020(a)). ASRs are subject to GMPs (21 CFR Part 820) (see Note 9).

While the components that comprise home brew tests are regulated by the FDA, especially regarding features such as quality control and labeling of reagents, the agency does not regulate how the tests are put together. Instead, home brew tests are regulated under CLIA. To qualify for "in-house developed assay" status, a diagnostic test must be conducted in a clinical testing laboratory that is compliant with CLIA laboratory guidelines, and is subject to a CLIA site inspection and approval. The CLIA regulatory framework insures that the test follows stringent standards for quality assurance, reliability and reproducibility, precision and accuracy for the intended use (23).

An example of a test marketed under the home brew category is the Breast Cancer Gene Expression Ratio offered by Quest

Diagnostics Nichols Institute (Breast Cancer Gene Expression Ratio; real-time PCR of HOXB13 and IL-17BR gene expression ratio) (24, 25) (see Note 10).

This is a gene expression test applied to tissue samples. The originators of the test examined the association between the ratio of the homeobox 13 (HOXB13) to interleukin-17B receptor (IL-17BR) expression and the clinical outcomes of relapse and survival in women with ER-positive breast cancer enrolled in a North Central Cancer Treatment Group adjuvant tamoxifen trial (NCCTG 89-30-52) (24–26). Tumor blocks were obtained from 211 of 256 eligible patients, and quantitative reverse transcription-PCR profiles for HOXB13 (27) and IL-17BR were obtained from 206 patients. The cut point for the two-gene log 2(expression ratio) that best discriminated clinical outcome (recurrence and survival) was selected and identified women with significantly worse relapse-free survival (RFS), disease-free survival (DFS), and overall survival (OS), independent of standard prognostic markers. This cut point was used as the basis for marketing a genomic test for therapeutic sensitivity to adjuvant Tamoxifen for ER-positive breast cancer patients (24–26).

3.3.1. Disadvantages of the "Home Brew" Regulatory Pathway

The first disadvantage of the home brew regulatory route is that the specific test will not automatically qualify for medical payment reimbursement. It is worth noting that FDA approval also does not automatically qualify tests for reimbursement. The second disadvantage is that the test cannot be sold as a product/kit to multiple laboratories. Thus, the entity marketing the in-house developed assay test must establish a fully operation CLIA inspected clinical laboratory to conduct the test in its entirety. This would include the laboratory facility, the technical staff, the laboratory director, and all the required equipment. The testing facility would be responsible for receiving, tracking, and accurately reporting the sample test results to the physician sending in the patient sample. Only tests that are certified under CLIA can be offered on the laboratory test menu. To date, the FDA has not regulated "home brew" clinical and genetic tests offered by laboratories as clinical services, but regulation remains an option. The FDA does provide a standard of measurement for regulating certain testing reagents as ASRs.

3.3.2. Quality Assurance for Genetic Tests Marketed Under Home Brew

Many state agencies regulate laboratories performing genetic testing through licensure of facilities and personnel. New York requires laboratories to submit clinical validation data for approval prior to offering patient testing. Many states specifically require that genetic tests be performed in a laboratory that is accredited by a program approved by the Department of Health and Human Services and enrolls in a proficiency-testing program such as those offered through the College of American Pathologists.

4. Conclusions

The information provided herein is provided as an educational resource for a scientist or entrepreneur who is contemplating the development and commercialization of a diagnostic test in the field of molecular profiling. The reader is urged to work with a qualified professional for definitive information concerning a specific test platform or analyte, and to consult with the FDA early and often during the development process. No matter what regulatory path is chosen, the requirements are designed to ensure the safety and effectiveness of the project when marketed for a clear and specific intended use.

5. Notes

1. Code of Federal Regulations website is http://www.gpoaccess.gov/cfr/index.html.
2. Device product classifications can be found by searching http://www.accessdata.fda.gov/scripts/cdrh/cfdocs/cfPCD/classification.cfm.
3. Clinical Laboratory Improvement Amendment website is http://www.cms.hhs.gov/clia/.
4. Device classification and 510(k) submission information is available at the following websites:

   New Section 513(f)(2) – Evaluation of Automatic Class III Designation

   http://www.fda.gov/MedicalDevices/DeviceRegulationandGuidance/GuidanceDocuments/ucm080195.htm

   510(k) Substantial Equivalence Decision-Making Process Flowchart

   http://www.fda.gov/MedicalDevices/DeviceRegulationandGuidance/HowtoMarketYourDevice/PremarketSubmissions/PremarketNotification510k/ucm134783.htm

   Premarket Notification Review Program 6/30/86 (K86-3) (blue book memorandum) http://www.fda.gov/MedicalDevices/DeviceRegulationandGuidance/GuidanceDocuments/ucm081383.htm

   Deciding When to Submit a 510(k) for a Change to an Existing Device Deciding (K97-1) http://www.fda.gov/MedicalDevices/DeviceRegulationandGuidance/GuidanceDocuments/ucm080235.htm

Pre-amendment Status http://www.fda.gov/cdrh/comp/preamend.html

5. Substantial equivalence (SE) means that the new device is at least as safe and effective as the predicate device.
6. FDA product classifications can be viewed at: http://www.accessdata.fda.gov/scripts/cdrh/cfdocs/cfPCD/classification.cfm.
7. FDA guidance documents for medical devices can be found at: http://www.accessdata.fda.gov/scripts/cdrh/cfdocs/cfGGP/Search.cfm.
8. 510(k) and PMA applications may be heavily redacted prior to publication under the US Freedom of Information act (7).
9. ARSs regulations can be found in the Code of Federal Regulations, 21 CFR 809.10(e), 809.30 and 864.4020.
10. The HOXB13 gene information can be found at: http://www.gene-profiles.org/gene/hoxb13-homo-sapiens-10481.
11. The following private sector organizations have established guidelines for the safety and effectiveness of genetic testing relevant to molecular profiling:

    American College of Medical Genetics (ACMG).

    College of American Pathologists (CAP).

    American Academy of Pediatrics.

    American College of Obstetricians and Gynecologists (ACOG).

    American Society of Gene Therapy (ASGT).

    National Society of Genetic Counselors (NSGC).

    Association of Molecular Pathologists (AMP).

    Clinical Laboratory Standards Institute (CLSI).

References

1. Gibbs, J. N. (2009) Making the most of pre-IDE meetings. *Genetic Engineering & Biotechnology News* **29**, 5.
2. Gibbs, J. N. (2008) Regulatory pathways for molecular Dx. *Genetic Engineering & Biotechnology News* **28**, 14.
3. Gibbs, J. N. (2009) Banked-Specimen retrospective studies: A clinical validation shortcut or a lengthy detour? *Genetic Engineering & Biotechnology News* **29**, 12.
4. van't Veer, L. J., Dai, H., van de Vijver, M. J., He, Y. D., Hart, A. A., Mao, M. et al. (2002) Gene expression profiling predicts clinical outcome of breast cancer. *Nature* **415**, 530–6.
5. van de Vijver, M. J., He, Y. D., van't Veer, L. J., Dai, H., Hart, A. A., Voskuil, D. W. et al. (2002) A gene-expression signature as a predictor of survival in breast cancer. *N Engl J Med* **347**, 1999–2009.
6. Knauer, M., Mook, S., Rutgers, E. J., Bender, R. A., Hauptmann, M., van de Vijver, M. J. et al. (2010) The predictive value of the 70-gene signature for adjuvant chemotherapy in early breast cancer. *Breast Cancer Res Treat* **120**, 655–61.
7. Gibbs, J. N. (2009) Adroit Crafting of "Intended Use" Critical
8. Regulatory Fate of a New IVD May Well Depend on a Few Dozen Words. *Genetic Engineering & Biotechnology News* **28**, 20.
9. Food and Drug Administration, U. S. (2009) Premarket notification (510k).

10. Food and Drug Administration, U. S. (2009) New Section 513(f)(2) - Evaluation of Automatic Class III Designation, Guidance for Industry and CDRH Staff.
11. Food and Drug Administration, U. S. (2009) Deciding When to Submit a 510(k) for a Change to an Existing Device (K97-1).
12. Food and Drug Administration, U. S. (2009) Guidance on the CDRH Premarket Notification Review Program 6/30/86 (K86-3).
13. Food and Drug Administration, U. S. (2009) 510(k) "Substantial Equivalence" Decision Making Process.
14. Food and Drug Administration, U. S. (2009) Guidance documents (Medical Devices).
15. Food and Drug Administration, U. S. (2002) The Least Burdensome Provisions of the FDA Modernization Act of 1997: Concept and Principles; Final Guidance for FDA and Industry.
16. Food and Drug Administration, U. S. (2009) Code of Federal Regulations Title 21 21CFR814.
17. Food and Drug Administration, U. S. (2009) Manual of Compliance Policy Guides.
18. Wolff, A. C., Hammond, M. E., Schwartz, J. N., Hagerty, K. L., Allred, D. C., Cote, R. J. et al. (2007) American Society of Clinical Oncology/College of American Pathologists guideline recommendations for human epidermal growth factor receptor 2 testing in breast cancer. *Arch Pathol Lab Med* **131**, 18–43.
19. Wolff, A. C., Hammond, M. E., Schwartz, J. N., Hagerty, K. L., Allred, D. C., Cote, R. J. et al. (2007) American Society of Clinical Oncology/College of American Pathologists guideline recommendations for human epidermal growth factor receptor 2 testing in breast cancer. *J Clin Oncol* **25**, 118–45.
20. Slamon, D. J., Clark, G. M., Wong, S. G., Levin, W. J., Ullrich, A., McGuire, W. L. (1987) Human breast cancer: correlation of relapse and survival with amplification of the HER-2/neu oncogene. *Science* **235**, 177–82.
21. Birner, P., Oberhuber, G., Stani, J., Reithofer, C., Samonigg, H., Hausmaninger, H. et al. (2001) Evaluation of the United States Food and Drug Administration-approved scoring and test system of HER-2 protein expression in breast cancer. *Clin Cancer Res* **7**, 1669–75.
22. Bartlett, J. M., Going, J. J., Mallon, E. A., Watters, A. D., Reeves, J. R., Stanton, P. et al. (2001) Evaluating HER2 amplification and overexpression in breast cancer. *J Pathol* **195**, 422–8.
23. Food and Drug Administration, U. S. (2007) Guidance for Industry and FDA Staff commercially distributed analyte specific reagents (ASRs):Frequently Asked Questions.
24. Department of Health & Human Services, U. S. (2009) Clinical Laboratory Improvement Amendments (CLIA).
25. Goetz, M. P., Suman, V. J., Ingle, J. N., Nibbe, A. M., Visscher, D. W., Reynolds, C. A. et al. (2006) A two-gene expression ratio of homeobox 13 and interleukin-17B receptor for prediction of recurrence and survival in women receiving adjuvant tamoxifen. *Clin Cancer Res* **12**, 2080–7.
26. Ma, X. J., Hilsenbeck, S. G., Wang, W., Ding, L., Sgroi, D. C., Bender, R. A. et al. (2006) The HOXB13:IL17BR expression index is a prognostic factor in early-stage breast cancer. *J Clin Oncol* **24**, 4611–9.
27. Ma, X. J., Salunga, R., Dahiya, S., Wang, W., Carney, E., Durbecq, V. et al. (2008) A five-gene molecular grade index and HOXB13:IL17BR are complementary prognostic factors in early stage breast cancer. *Clin Cancer Res* **14**, 2601–8.

Chapter 28

Small Business Development for Molecular Diagnostics

Anthanasia Anagostou and Lance A. Liotta

Abstract

Molecular profiling, which is the application of molecular diagnostics technology to tissue and blood specimens, is an integral element in the new era of molecular medicine and individualized therapy. Molecular diagnostics is a fertile ground for small business development because it can generate products that meet immediate demands in the health-care sector: (a) Detection of disease risk, or early-stage disease, with a higher specificity and sensitivity compared to previous testing methods, and (b) "Companion diagnostics" for stratifying patients to receive a treatment choice optimized to their individual disease. This chapter reviews the promise and challenges of business development in this field. Guidelines are provided for the creation of a business model and the generation of a marketing plan around a candidate molecular diagnostic product. Steps to commercialization are outlined using existing molecular diagnostics companies as learning examples.

Key words: Business development, Industry, Market, Molecular diagnostic, Personalized medicine

1. Introduction

1.1. Molecular Diagnostics Industry and Market Overview

Diagnostic methods are vital to the entire field of medicine and are a commonplace part of routine practice. The term, "diagnostic," is widely encompassing and can be as simple as a patient interview, or the measurement of a pulse. Conversely, it can be as complex as the determination of cytochrome P450 gene expression prior to warfarin treatment or the process of predicting recurrence of breast cancer through a 21 gene signature (1–3). Today, within the over 5,500 hospitals in the USA (4), the outcome of a diagnostic test influences 60–70% of the clinical decisions made by treating physicians (5).

Since the mapping of the human genome was completed more than 10 years ago, innovative technologies and tests have emerged

Virginia Espina and Lance A. Liotta (eds.), *Molecular Profiling: Methods and Protocols*, Methods in Molecular Biology, vol. 823, DOI 10.1007/978-1-60327-216-2_28, © Springer Science+Business Media, LLC 2012

based on an improved understanding of the role that genes, and gene products, play in disease and influence therapeutic outcome. As the molecular understanding of disease continues to advance, molecular diagnostics will become increasingly important (6–9).

1.2. Applications of Molecular Diagnostics

Molecular diagnostics (MDx) is the testing or analysis of an organism to obtain molecular information, (i.e., at genetic, biochemical, proteomic, or biosystems level) (10). The practical application of MDx already spans a wide spectrum of diseases, including cardiology, oncology, obesity, diabetes, aging, and sepsis. Indeed, MDx plays an important role at many points along the entire disease management cycle. This includes risk assessment, diagnostic staging, presymptomatic disease detection and identification, drug response prediction, therapeutic monitoring, and relapse prediction & management (11).

Molecular diagnostics has the potential to go beyond immediate disease diagnosis to become a roadmap of disease prevention and overall lifestyle management (8). Futurists have pointed to a day when personal genetic testing and biochemical metabolic analysis will impact on everyday dietary and lifestyle habits (6, 8). Unfortunately, personal genetic information can be abused and there are many concerns about protecting the privacy of molecular test data: "If genomic innovations have great power to help, they also have power to harm, and patients and the public deserve a rigorous evaluation of what scientists bring to the table" cautions Khoury et al. (12). In May of 2008, President Bush signed into law the Genetic Information Nondiscrimination Act (GINA), which will protect Americans against discrimination based on their genetic information in regard to health insurance and employment. In 2010 Francis Collins, former head of the Human Genome Project, and NIH Director, declared that ethical protection of genetic information is of the highest priorities to prevent abuse of personal genetic information (8).

1.3. Industry Characteristics and Market Size

From a financial perspective, diagnostics have historically been the poor cousin of the biotech industry. Many of the current clinical diagnostic tests analyzed through reference labs are essentially commodities. It is no surprise that low profit margins, reimbursement issues, and the difficulty of market penetration have made conventional diagnostics companies unattractive to institutional investors. The global diagnostic market, valued at $36.5 billion in 2005 (prior to the downturn in the economy), was much smaller than the $602 billion projected for the pharmaceutical industry. Drug companies may seem to have the inside track in the industry, but there's usually a crack in every drug pipeline. After 10 years and over $1 billion dollars for each new product's development, failure rates for new drugs are climbing to as high as 7 out of 8 (5). Further, drugs are subject to a stringent, burdensome regulatory process. The drug industry's forecast is worsening as drug attrition and FDA approval rates have steadily declined over the past few years while intellectual property protections for the so-called "blockbuster" drugs are imminently running out. Many of these obstacles are nonexistent or largely diminished within the diagnostics segment that has lower costs of product development, has lower failure rates, and takes less time to commercialize.

Based on economic arguments, molecular diagnostics is actively infusing optimism into the biotechnology industry. The molecular diagnostics market segment, with an Annual Growth Rate of 10.4%, was valued at $13.8 billion in 2005 and is expected to nearly double, reaching $22.7 billion in 2010 (13). Within the USA alone, molecular diagnostics was valued at approximately $2.7 billion in 2006 and is expected to reach $5 billion by 2010 (AGR 15%) (14). Cancer molecular diagnostics, in particular, is the fastest growing sector and is predicted to increase by 30% each year tripling from its 2005 level of $315 million to more than $1.35 billion by 2010 (14). Another extension of molecular diagnostics is the mechanical hardware that performs the tests: the clinical laboratory equipment sector, which is also experiencing a high level of growth at 9.3%. The USA is the world's largest single market for diagnostic equipment, with a value of $13.5 billion in 2002. This segment is specifically comprised of laboratory instruments, point-of-care diagnostics and kits. Laboratory instrumentation & equipment alone was valued at $7.76 billion in 2007, most of which is sold to clinical reference laboratories in hospitals and represents a distinctively different business than service-based molecular diagnostics.

The MDx market is diversified with some participants selling equipment, reagents, and kits, others point-of-care diagnostics, and still others providing in-house, or "home-brew" services. The path to commercialization and the regulatory environment differ depending on the type of molecular diagnostic. If performed as a laboratory test, diagnostics must be conducted in a CLIA (Clinical Laboratory Improvement Amendments) approved lab but are rendered exempt from the regulations required for in vitro diagnostic kits. Application Specific Reagents that are sold to "high complexity" labs must be regulated under good manufacturing processes (GMP), and cannot market claims of clinical utility. "Kits" must necessarily prove clinical utility and be approved by the Food and Drug Adminsitration (FDA) as a medical device through a PMA or 510 k before being marketed. The FDA is actively analyzing and debating exactly how to regulate the space; a variety of draft guidances and papers on the topics of molecular diagnostics have been issued, including a draft guidance on In Vitro Diagnostic Multivariate Index Assays issued on July 26, 2007, an Analyte Specific Reagent draft guidance on September 14, 2007, and a draft Drug/Diagnostic Codevelopment paper issued in April of 2005. Meanwhile, the regulatory path is usually decided on a case-by-case basis by each company and largely determined by the technology and product delivery type.

1.4. The Evolution of Molecular Profiling

Over the last 15 years molecular diagnostics has rapidly advanced from microscopy, chemistry and immununoassays to measure whole new classes of molecules. The current molecular analytes now encompass DNA, RNA, and mitochondrial DNA sequence, DNA methylation patterns, gene expression profiles, proteins and protein expression, and combination biomarkers (13).

In 1989 molecular diagnostic assays were principally Southern Blots (15). Polymerase chain reaction (PCR) technologies had not yet made their way into clinical laboratories. Molecular diagnostics was esoteric, centralized in specialized laboratories, and all tests that were brought online were that way. In the early 1990s, an inflection point in molecular diagnostics was reached when Roche bought rights to PCR from Cetus and came out with a *Chlamydia trachomatis* (16)/*Neisseria gonorrhoeae* (17).

The first-generation molecular diagnostics were single analyte tests, detecting a single viral or bacterial gene or protein. Indeed the blockbuster diagnostic, Dako's widely used and publicized HercepTest™ (18–20), measures the presence of only one protein (Her2/neu) in tissue samples. The emerging second generation of products utilizes multiplex platforms such as DNA and of protein microarrays, performing multiple biomarker analyses in parallel. Innovation over time has rendered this multiplexed analysis possible by increased automation, quantitation, and sensitivity of these techniques, along with reduced costs (13). The frontrunners of the new generation of molecular diagnostics such as Genomic Health's onco*type* DX® and Agendia's Mammaprint tests (21, 22), evaluate 21 and 70 genes, respectively. These indexes, used to predict disease recurrence, are widely applied in cancer management. Third-generation molecular diagnostics of the future will integrate many classes of analytes (e.g., proteins, DNA, RNA, metabolites) into a "systems biology portrait" and will directly apply this information to optimize therapy for an individual patient

What should be called third-generation molecular diagnostics will not only integrate the newest technological developments to form a clearer portrait of disease through the systems biology perspective but also directly apply this information toward treatment.

The number of companies involved in the MDx field has increased remarkably during the past few years. Today, more than 500 companies are developing molecular diagnostics (23, 24). This trend has been primarily due to the realization of the value of integrating drug development with complementary diagnostics or "companion diagnostics." Companion diagnostics have gained more acceptance as the field of medicine has shifted toward targeted or personalized treatment methods. It is then no coincidence that the biopharmaceutical industry itself is witnessing two trends in innovations that are in line with the trend in personalized medicine and theranostics. Targeted therapies are being developed to act on very specific disease mechanisms that are present in subpopulations of patients. Predictive tests are being codeveloped with these therapies to determine in advance how likely a patient is to respond to a certain drug. The next step will be combination therapies of these targeted drugs, based on third-generation molecular diagnostics that can quantitatively analyze multiple analytes at once.

2. Business Considerations in Molecular Diagnostics

2.1. Overview of Business Models

The business considerations for any particular molecular diagnostic product are largely dependent on the characteristics of the technology itself, its product delivery method and its intended use (25). There are two key MDx customers to consider: pharmaceutical companies and physicians. Pharmaceutical companies use a variety of molecular diagnostic methods within their preclinical drug discovery programs to, for example, characterize the mechanism of action of their drug candidates and optimize their leads. These techniques are also applied in animal models in the translational medicine space and are increasingly being implemented within clinical trial development. In the case where an MDx implemented in drug development goes on to be approved as a companion diagnostic test for a new drug, the ultimate customer will be the physician. The ideal companion diagnostic codevelopment model has yet to have been clearly articulated, from both a business development and regulatory standpoint, and the best that can be done is to analyze case studies as examples. However, the schema below demonstrates the breadth of molecular diagnostic applications within the drug development pipeline (Fig. 1).

Molecular diagnostic techniques can be performed in-house, where drug development companies build their own capabilities for testing and analysis, or be run through pharmaceutical service providers at centralized laboratories. After its clinical launch, a diagnostic test will consequently either be analyzed through the same centralized service provider, or the capability will be established at a hospital reference laboratory. Meanwhile, independent from any drug, clinical molecular diagnostic products continue to be developed to diagnose disease, predict recurrence, and even provide treatment information relating to existing drugs. The customers for these clinical molecular diagnostics are primarily physicians within hospitals and private practices.

2.2. Business Considerations

The expected path from biomarker discovery to molecular diagnostic commercialization for a molecular diagnostics company is somewhere between 2 and 8 years. This represents a much shorter

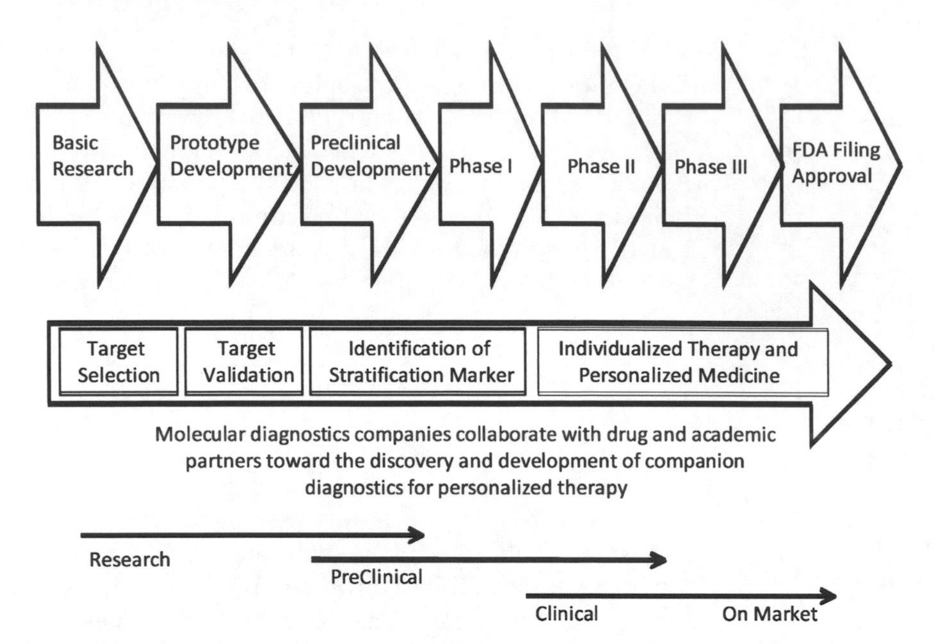

Fig. 1. Molecular diagnostic development stages. All phases of the product/drug development pipeline provide opportunities to develop a commercial molecular diagnostic product.

timeline than that for a pharmaceutical product, but nevertheless presents its own set of obstacles and complexities. In molecular diagnostics development, as opposed to other sectors of the biopharmaceutical industry, the questions that beg to be answered for early-stage technologies have not changed much in the last decade. There are fundamental questions about the technology, the intellectual property it is based on, the competition, and the scale of the commercial impact, that must be answered for prospective investors:

2.2.1. The Technology

1. Is it more than a concept?
2. What is the evidence that it can work?
3. What is its stage of development?
4. Do data indicate that it is much simpler, faster, cheaper, more accurate, or better in some way than existing technology? Is the technology qualitatively or quantitatively different from the prior art?
5. Are the data reproducible?
6. Are there – or could there be – alternative ways to achieve similar results?
7. What commercial products, or fields would the technology create or impact?

2.2.2. Patentability

1. Is the technology new, useful, and not obvious?
2. Was it publicly divulged? If so, when and how?

3. Would patent infringement be detectable?
4. Could a patent be enforced?
5. Would the available patent claims protect a developer's investment in marketing a product?
6. Who holds the rights to the technology?
7. Are the large investments in international patents necessary to derive value from the technology?

2.2.3. Economic Value

1. What commercial needs are addressed by this technology?
2. Does it have multiple applications?
3. Will it create a new market or is it an improvement on an existing product?
4. What is the size of the current and future markets (23, 24)?
5. How many products that incorporate the technology will be needed to satisfy market demand, both now and in the future?
6. What competes with the proposed product?
7. What are the competitive advantages of the proposed product?
8. What is the likely difference between the manufacturing cost and the price for which the product can be sold?
9. Who are the potential suitors, licensors, and strategic relationships?

2.2.4. Risk Factors

1. Will a licensee or other investor find this technology more attractive than other investments?
2. What are the legal, regulatory, and physical barriers to market entry? (The reader is directed to the chapter in this volume describing FDA review pathways for molecular diagnostics).
3. Can the technology be scaled up or proven safe and effective in the anticipated end use?
4. How much financial commitment is necessary to bring the technology to market?
5. Are there sources of capital to develop technologies in this scientific area?
6. What are the company milestones for commercialization? How much time exists before the next milestone?
7. How much time will be needed to bring a product to market?

With the challenges of ever-growing competition, successful molecular diagnostic companies are working to find unexploited niches and also address these considerations.

2.3. The Commercialization Process

In the case where an early-stage technology is patentable, has reached proof of concept and will overcome the abovementioned risk factors while providing economic value, there are additional

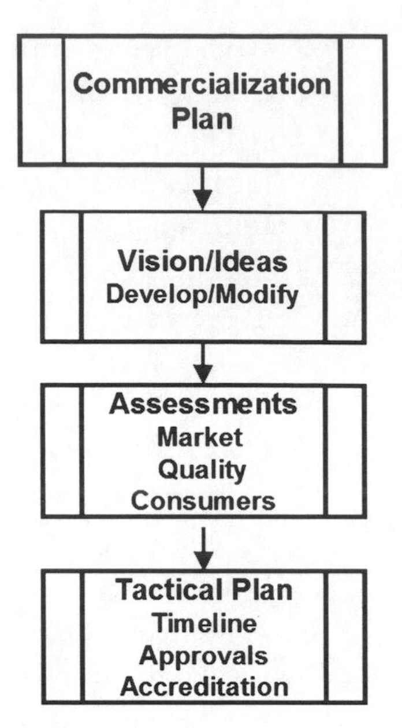

Fig. 2. Pathway to commercialization.

preliminary strategic and tactical considerations that must be addressed, as outlined below (Fig. 2).

System: Understand and create the intended use and the delivery system for your product – what are analytes how will you create a fully functional product from sample acquisition to customer report? What is the clinical context in which the product is applied how will it be used in a clinical decision, how will it be delivered, and how it will it be performed in a particular environment; For example, a MDx could be performed in a hospital, primary-care provider's or general physician's office.

Mission: Define your target and how you will reach your target. Decide what is the mission of that particular diagnostic application. Who are you looking to target? For example, what disease condition in what population? Do you have a well-defined test group with a sufficient disease incidence?

Structure: What structure is necessary to implement? Do you need outsourced resources, a full marketing sales team, or a contract manufacturer? Decide if you need outsourced relationships with marketing and sales. Is there additional infrastructure required for this technology or is a relationship required with a contract manufacturer? Do you need to educate your customers? Decide on your product delivery model: will it conducted in centralized lab, or decentralized lab for example. From this, create a tactical plan with timelines.

Standards: Quality control and quality assurance are essential components of molecular diagnostic assays (see Note 1). Aim to tackle Quality Assurance (QA) and reimbursement issues up front to become a best-in-class product.

2.4. Identifying Market Opportunity and Gaining Intelligence about a Tactical Plan

2.4.1. Identifying Market Opportunity

1. Assess your target market: Who will be your customers? In the case of molecular diagnostics, they are likely to be drug development companies, academic researchers and/or physicians. There may also be a necessity to assess patients as a potential target market, depending on the type of diagnostic.
2. Define the market segment that you will work within, for example (see Note 2):
   (a) What size drug developer: large pharmaceutical or mid-size biotech
   (b) What are the disease categories or therapeutic modalities? Cancer, neurological disorders, cardiovascular disease, etc.
   (c) What type of products would you help drug companies to develop? For example, small-molecule drugs, antibody therapeutics, vaccines, or antibiotics may be several examples.
   (d) If your customers will be physicians, determine what type you will work with: both by specialty and business, i.e., private practice vs. hospital network
3. Assess the size of your market space: This is in terms of economic size and number of customers. For example, an entrepreneur considering marketing a diagnostic for breast cancer would discover that:
   (a) Oncology molecular diagnostics was worth $300 million in 2011.
   (b) It is estimated that there are 12,000 physicians within the USA that treat breast cancer.
   (c) There are 220,000 new breast cancer patients a year (26).
4. Feedback: Once you have identified your target customer base, start to communicate with its constituents. There are several methods to gain initial feedback into product planning; two specific suggestions include the following:
   (a) Surveys: these can be conducted in-house or outsourced to a specialized company.
   (b) Focus Groups: a focus group of up to ten individuals, gathered within the context of an interactive setting, may provide valuable feedback from your potential customers about your product concept and will inform product planning and development.

2.4.2. Market Intelligence

1. Market intelligence involves gathering and synthesizing detailed information about potential customers, competing companies and competing products, as well as market risks and opportunities. This information is used to make informed decisions about business strategy and planning. The result of useful market intelligence may be concrete actionable items for your business.

2. Methods: There are a variety of methods for gathering market intelligence, such as simply visiting company Web sites, reading product brochures, news, journal articles, and press releases, attending events and listening to lectures and actually speaking with people from other companies. What is most significant is the synthesis of this information and ability to link it to possible tangible effects on your product.
3. Ethical Considerations: The gathering of "competitive intelligence" about competing companies and products must be conducted in a completely ethical way that does not misrepresent you, your company, or your intentions.
4. Protecting your products and company: Be aware that other companies will also be seeking competitive intelligence about you!

2.4.3. Think About Formulating Your Brand and Message

1. Alongside the process of clearly defining your product, customers, and market segment, the language and image that you will create and eventually disseminate to articulate and define your product to customers and to the market at large will necessarily have to be carefully crafted.
2. Branding Overview: Typically you start with a logo design. In that you strive to incorporate not just the character and ideals of the product and company (i.e., two things coming together) but also shapes, colors, and textures that then will be used to build the style guidelines around all other subsequent graphic design pieces, be they for print, Web, onscreen presenting, video, or other type.
3. Logo: is vital. The designs elements contained therein determine a great deal and should be reflected in every other branded piece. If successful, the brand will identify your company in an instant and anywhere it is used it will speak directly and solely to your identity. Nike for example is the ultimate brand. That logo stands alone. In this case, it is the shape of the logo that really carries over, but in other cases it can be color schemes, typography, and shape combinations.
4. Develop "branding" for the product: with a graphic logo and design system for all materials that incorporates this logo and overall corresponding color palette and design concept. Moving forward, all materials, including print and Web, should incorporate this brand identity and possibly be governed by a style guide that you develop.
5. Messaging: Set out to build a strong corporate message platform and continuous dialog with your customers. Define the technology/product messages, with specifics for each product and customer category. This will include both defining and

articulating the product features, the benefits and advantages that it will convey to customers, and the defining characteristics that make your product different than other MDx products: for example, technical advantages: sample type, assay sensitivity or precision (10), product delivery type or ease of use, price, value of information derived from the test for both drug developers and medical practitioners, or potential economic outcomes and health-care savings.

6. Identify key customers and overall audiences: For example, the audiences for a molecular diagnostic may not just be pharmaceutical or physician customers, but patients, patient interest groups, payors, or legislators (27, 28). Create messages that will differentiate and define you as a leader in the field and plan message dissemination tactics to align with these audiences and their interests.

7. Develop a communications plan and program with both your key audiences in mind and events and means to target your key audiences (see Note 3). Develop a calendar of upcoming milestones and events: opportunities that will build your visibility and credibility.

3. Case Studies in MDx Business Development

3.1. The Economic Value of Strategic Alliances for Companion Diagnostic Development

A standard method of business development in companion diagnostics is a strategic alliance between diagnostic company and pharmaceutical drug developer. Drug and Diagnostic companies have increasingly forged alliances for companion drug and diagnostic development (14). Nevertheless, complications in their relationships have often clouded the process of working together. Diagnostics often demand up-front payments from Pharmaceutical companies for development of tests for therapeutic products that could take up to 10 years to reach the market. Pharmaceutical companies may not be willing to pay, being accustomed to this long development cycle, whereas diagnostic companies look to bring products to market within a much shorter period of 2–5 years (29).

Despite these obstacles, there have been several hallmark examples of strategic treatment/diagnostic alliance. The prime example of companion diagnostics is Dako's HercepTest™, which is used to select patients for treatment with Genentech's Herceptin® (trastuzumab) (18–20). This has grown into a $1.2 billion a year product and has paved the way for companion diagnostics. Over 300,000 patients have been treated with Herceptin®. The HercepTest™ has created over $35 million in savings in clinical trials (30, 31).

Another case study is the Pfizer/Monogram Biosciences alliance. The alliance between these companies was forged in May of 2006 through a $25 million investment on the part of Pfizer, and a global cooperation agreement for comarketing outside of the USA. The success of clinical trials for Pfizer's novel antiretroviral HIV drug, Selzentry®, was largely dependent on the Monogram Trofile assay™ being able to qualify patients for treatment (29). Trofile™, which has been determined to be 90% accurate, determines which of the 2 coreceptors HIV utilizes to enter CD4 cells: CXCR4 or CCR5 (32, 33). Most antiretrovirals block entry into the CXCR4 receptor, while Selzentry® (maraviroc) is the first in class to block entry into CCR5 (34). In September of 2007, Pfizer gained approval for Selzentry®. Today, patients are only qualified for Selzentry treatment through Trofile™ assays.

Large pharmaceutical companies are increasingly integrating business practices based on molecular diagnostics coupled to their therapeutics (23, 24). Companion diagnostics can accelerate and improve the drug development process by applying biomarkers to weed out unsuccessful drug candidates and refine and restratify the clinical trial population. This is a method to both rescue drugs from failure and reduce the costs of clinical trials (13, 35).

Commercial Need: Diagnostic companies can create economic value for pharmaceutical companies and the health-care system as a whole by actively addressing the commercial need for increased efficiency in drug development, and personalizing therapy.

Applications: Diagnostic companies are applying their technologies at all stages of the drug development pipeline, from preclinical drug discovery through clinical development to product launch and beyond (see Note 4).

Market Creation: Not only can companion diagnostics improve the chances of success for new drugs but they can also create new markets for existing drugs.

Market Size, Number of Products: Once marketed, the most exciting prospect for a companion diagnostic is that more diagnostic tests will have to be given than actual drugs prescribed. This means that diagnostics as an entire market should grow in proportion to the therapeutics market.

Pricing: The traditional issues of commodity pricing of diagnostics will likely be eclipsed by a new pricing model based on the actual economic value conveyed by MDx: this is the new era of value-based pricing, where tests will be sold at a level comparable to the prices of the drugs they select for.

The competition within technologies in this space is fierce, but the key for a successful molecular diagnostic assay will be the choice of an optimal companion for commercialization.

3.2. How to Market: Genomic Health as an Example of a Successful Marketing Campaign

Genomic Health, Inc., was founded in August of 2000 and is located in Redwood City, California (http://www.genomichealth.com). The company's primary product is the onco*type* DX® breast cancer assay, which has been marketed since 2004 for diagnosis of early-stage breast cancer patients (1–3). Onco*type* DX® paved the way for other molecular diagnostic companies in various ways:

Novel technology applied to a specific disease and clinical decision: A multiple analyte test of 21 genes that combine into a score that predicts recurrence of disease and response to chemotherapy for breast cancer (1–3).

Insights from customers guided product development: Genomic Health consulted with physicians and payors throughout the entire process of product planning, conception, and development. This information was used to inform extensive clinical studies that would establish the clinical validity necessary for physician adoption of the test (29).

Aggressive marketing: A key element of the Genomic Health strategy was identifying and targeting customers: oncology physicians. Once Genomic Health determined who its key customers were, it mustered the resources to reach and influence them: investing heavily in marketing: In 2006, the company spent over $24.6 million in sales and marketing (29). This not only went toward programs for advertising, promotion, and public relations but also was put toward hiring an experienced oncology sales team of over 60 people who began by targeting key opinion leaders to drive adoption. In addition, customer service and a full billing and reimbursement infrastructure was put in place (36). *Value-Based Pricing*: The onco*type* DX® test costs about $3,500 and its price tag was determined based on the value that it provides: reducing the administration of unnecessary chemotherapy regimens which can cost up to $40,000 per patient per year, thereby saving the health-care system significant amounts of money by reducing systemic inefficiencies (37).

3.3. Building Molecular Diagnostic Capacity and Addressing Risk Factors Through a Novel Business Model

The Sequenom® Center for Molecular Medicine started in 2007 as an advanced molecular pathology laboratory located in Grand Rapids, Michigan (http://www.scmmlab.com). The Center was set within the backdrop of its two parent organizations: Spectrum Health Systems, an integrated health-care system that serves over 650,000 patients annually, and the Van Andel Institute, a well-established biomedical research institute. After having established three other hospital-based DNA diagnostic labs, Daniel H. Farkas, Executive Director of the Center, opted to go out on a limb and establish the center as a unique, experimental molecular diagnostics business model (38).

The general concept of the Center was to run a molecular diagnostics laboratory with one or two successful diagnostic assays that would "underwrite other more esoteric tests that are being offered" (Daniel Farkas, personal communication). The Center brought online the most cutting edge tests that by definition were ahead of the curve: physicians did not yet understand their clinical utility, they had a low test ordering volume, and companies weren't yet reimbursing them well (see Note 5) (39). The Center speculated in the value of molecular diagnostics by offering cutting edge tests, so that when the test becomes standard of care (in several years), the Center would have already been one of the first to offer the test.

There are nonetheless, financial ramifications to this strategy that are addressed by the other half of the business model. The Center combines a wide array of genomic and proteomic technologies, including DNA and RNA extraction, DNA microarrays, multiplex protein detection, gene expression profiling, and more. Technology access, combined with bioinformatics expertise and excellent medical technologist staffing within a laboratory that has access to clinical specimens, provides all the necessary components for a one-stop-shop for clinical trials in a CLIA-certified, College of American Pathologists (CAP) accredited environment. The Center offered its services and facilities to help customers generate data demonstrating analytical and clinical validity that will help them submit new device/product applications to the FDA and ultimately get approved and reimbursed. The original business model included a straight fee for service offering, where for example an oncology group could send in several thousand samples for genotyping (personal communication (38)).

Initially the Center had forged alliances with some of the current leaders in molecular diagnostics. Currently, the Center's partnership with Sequenom® provides direct patient testing as well as on-going clinical trial studies. In 2008, Daniel Farkas, Executive Director, used an interesting baseball analogy to describe future success in molecular diagnostics (personal communication (38)): "It's hard to know what's going to be the next gram slam within companion diagnostics but if you have a baseball lineup, not everyone has to hit a grand slam home run. With nine batters in your lineup who all hit, that's a winning team."

3.4. The Future of Molecular Diagnostics

The future of the molecular diagnostics space is promising. The industry is coming of age; companion diagnostics are becoming more common and are increasingly being integrated into the corporate strategies of pharmaceutical companies and the practices of physicians. Although adoption by payers continues to be slow, the reimbursement environment is improving. Meanwhile, patients are increasingly taking a personal interest and are expected to begin participating as individual consumers. MDx stands to help physicians make accurate treatment decisions, enable pharmaceutical companies to develop more effective therapies, and give health-care providers and payors the ability to cost-effectively manage patient care. The future of molecular diagnostics aims to lead a transformation in the disease and health-care management system.

4. Notes

1. Preanalytical variability can be a source of bias during clinical testing or a source of false negatives and false positive in your commercial product. Make sure that you understand the stability of your analyte(s) in the biologic sample and know how to educate the customer with regard to sample collection, handling, and shipment.
2. For companion diagnostics a great deal of emphasis is placed on the future value of the product. It is essential to consider the patent life (15–20 years), and potential for side effects, of the companion drug or treatment. Your companion diagnostic is worthless if the drug has no patent protection, or is withdrawn due to toxicity.
3. Develop strong relationships with pathologists, medical practitioners, and reimbursement agencies (third-party payors), thought to be leaders early in product development, and keep them involved throughout the process.
4. Meet with the FDA or equivalent regulatory agencies as soon as you have your first set of data demonstrating sensitivity, specificity, and precision of your test. Describe your intended use and present your data..
5. This strategy allows companies to bet on the future value of a particular molecular diagnostic.

References

1. Mamounas, E. P., Tang, G., Fisher, B., Paik, S., Shak, S., Costantino, J. P. et al. (2010) Association between the 21-gene recurrence score assay and risk of locoregional recurrence in node-negative, estrogen receptor-positive breast cancer: results from NSABP B-14 and NSABP B-20. *J Clin Oncol* **28**, 1677–83.
2. Paik, S., Tang, G., Shak, S., Kim, C., Baker, J., Kim, W. et al. (2006) Gene expression and benefit of chemotherapy in women with node-negative, estrogen receptor-positive breast cancer. *J Clin Oncol* **24**, 3726–34.
3. Paik, S., Shak, S., Tang, G., Kim, C., Baker, J., Cronin, M. et al. (2004) A multigene assay to predict recurrence of tamoxifen-treated, node-negative breast cancer. *N Engl J Med* **351**, 2817–26.
4. American Hospital Association (2010) Fast Facts on US Hospitals. American Hospital Association: Chicago, IL, http://www.aha.org/aha/content/2010/pdf/101207fastfacts.pdf Accessed 16 February 2011
5. Batchelder, K., Miller, P. (2006) A change in the market--investing in diagnostics. *Nat Biotechnol* **24**, 922–6.
6. Goldstein, D. B. (2010) Personalized medicine. *Nature* **463**, 10.
7. Callaway, E. (2011) Cancer-gene testing ramps up. *Nature* **467**, 766–7.
8. Collins, F. (2010) Has the revolution arrived? *Nature* **464**, 674–5.
9. Hamburg, M. A., Collins, F. S. (2010) The path to personalized medicine. *N Engl J Med* **363**, 301–4.
10. Taube, S. E., Jacobson, J. W., Lively, T. G. (2005) Cancer diagnostics: decision criteria for marker utilization in the clinic. *Am J Pharmacogenomics* **5**, 357–64.
11. Schafer, J. C. (2007) Molecular Diagnostic Assays: Leading a High Value Evolutions. 2011, in *Biotechnology Focus*,
12. Khoury, M. J., Evans, J. P., Burke, W. (2010) A reality check for personalized medicine. *Nature* **464**, 680.

13. Doig, A. (2007) Molecular Diagnostics Market Assessment. *Genetic Engineering & Biotechnology News* 27, http://www.genengnews.com/gen-articles/molecular-diagnostics-market-assessment/2006/. Accessed 16 February 2011.
14. Sannes, L. (2007) Molecular Diagnostics: A Rapidly Shifting Commercial and Technology Landscape. in *Insight Pharma Reports.* Cambridge Healthtech Institute, 232.
15. Southern, E. M. (1974) An improved method for transferring nucleotides from electrophoresis strips to thin layers of ion-exchange cellulose. *Anal Biochem* **62**, 317–8.
16. Leslie, D. E., Azzato, F., Ryan, N., Fyfe, J. (2003) An assessment of the Roche Amplicor Chlamydia trachomatis/Neisseria gonorrhoeae multiplex PCR assay in routine diagnostic use on a variety of specimen types. *Commun Dis Intell* **27**, 373–9.
17. Whiley, D. M., Tapsall, J. W., Sloots, T. P. (2006) Nucleic acid amplification testing for Neisseria gonorrhoeae: an ongoing challenge. *J Mol Diagn* **8**, 3–15.
18. Shah, S. S., Ketterling, R. P., Goetz, M. P., Ingle, J. N., Reynolds, C. A., Perez, E. A. et al. (2010) Impact of American Society of Clinical Oncology/College of American Pathologists guideline recommendations on HER2 interpretation in breast cancer. *Hum Pathol* **41**, 103–6.
19. Jacobs, T. W., Gown, A. M., Yaziji, H., Barnes, M. J., Schnitt, S. J. (1999) Specificity of HercepTest in determining HER-2/neu status of breast cancers using the United States Food and Drug Administration-approved scoring system. *J Clin Oncol* **17**, 1983–7.
20. Hammond, M. E., Hayes, D. F., Dowsett, M., Allred, D. C., Hagerty, K. L., Badve, S. et al. (2010) American Society of Clinical Oncology/College of American Pathologists guideline recommendations for immunohistochemical testing of estrogen and progesterone receptors in breast cancer (unabridged version). *Arch Pathol Lab Med* **134**, e48-72.
21. Veer, L. J., Dai, H., van de Vijver, M. J., He, Y. D., Hart, A. A., Mao, M. et al. (2002) Gene expression profiling predicts clinical outcome of breast cancer. *Nature* **415**, 530–6.
22. van de Vijver, M. J., He, Y. D., Veer, L. J., Dai, H., Hart, A. A., Voskuil, D. W. et al. (2002) A gene-expression signature as a predictor of survival in breast cancer. *N Engl J Med* **347**, 1999–2009.
23. Winter, P. (2009) Let's make a deal. *The Journal of Life Sciences* 14–15.
24. Levine, D. S. (2009) A diagnosis of deals. *The Journal of Life Sciences* 12–13.
25. Meckley, L. M., Neumann, P. J. (2010) Personalized medicine: factors influencing reimbursement. *Health Policy* **94**, 91–100.
26. Allegra, C. J., Aberle, D. R., Ganschow, P., Hahn, S. M., Lee, C. N., Millon-Underwood, S. et al. (2010) National Institutes of Health State-of-the-Science Conference statement: Diagnosis and Management of Ductal Carcinoma In Situ September 22–24, 2009. *J Natl Cancer Inst* **102**, 161–9.
27. Kontos, E. Z., Viswanath, K. (2011) Cancer-related direct-to-consumer advertising: a critical review. *Nat Rev Cancer* **11**, 142–50.
28. Farkas, D. H., Holland, C. A. (2009) Direct-to-consumer genetic testing: two sides of the coin. *J Mol Diagn* **11**, 263–5.
29. Levine, D. S. (2007) Getting Personal. *The Journal of Life Sciences* 42–28.
30. Elkin, E. B., Weinstein, M. C., Winer, E. P., Kuntz, K. M., Schnitt, S. J., Weeks, J. C. (2004) HER-2 testing and trastuzumab therapy for metastatic breast cancer: a cost-effectiveness analysis. *J Clin Oncol* **22**, 854–63.
31. Ferrusi, I. L., Marshall, D. A., Kulin, N. A., Leighl, N. B., Phillips, K. A. (2009) Looking back at 10 years of trastuzumab therapy: what is the role of HER2 testing? A systematic review of health economic analyses. *Per Med* **6**, 193–215.
32. Hunt, P. W., Harrigan, P. R., Huang, W., Bates, M., Williamson, D. W., McCune, J. M. et al. (2006) Prevalence of CXCR4 tropism among antiretroviral-treated HIV-1-infected patients with detectable viremia. *J Infect Dis* **194**, 926–30.
33. Daar, E. S., Lynn, H. S., Donfield, S. M., Lail, A., O'Brien, S. J., Huang, W. et al. (2005) Stromal cell-derived factor-1 genotype, coreceptor tropism, and HIV type 1 disease progression. *J Infect Dis* **192**, 1597–605.
34. Vandekerckhove, L., Verhofstede, C., Demecheleer, E., De Wit, S., Florence, E., Fransen, K. et al. (2011) Comparison of phenotypic and genotypic tropism determination in triple-class-experienced HIV patients eligible for maraviroc treatment. *J Antimicrob Chemother* **66**, 265–272.
35. Rader, R. A. (2010) FDA Biopharmaceutical Product Approvals and Trends: Significantly More Approvals Were Granted in 2009. Biotechnology Information Institute. http://www.biopharma.com/approvals_2009.html. Accessed 16 Feb 2011

36. Zenios, S., Chess, R. B., Denend, L. (2006) Genomic Health: Launching a Paradigm Shift... and an Innovative New Test (Case No: OIT49). Stanford Graduate School of Business: Palo Alto, CA, 1–36.
37. Ross, J. S., Hatzis, C., Symmans, W. F., Pusztai, L., Hortobagyi, G. N. (2008) Commercialized multigene predictors of clinical outcome for breast cancer. *Oncologist* **13**, 477–93.
38. Farkas, D. H. (2008) Interview with Daniel H. Farkas, PhD, Executive Director of the Center for Molecular Medicine. Anagnostou, A., Rockville, telephone interview.
39. Farkas, D. H. (2008) Diagnostic Molecular Pathology in an Era of Genomics and Translational Bioinformatics. *Diagnostic Molecular Pathology* **17**, 1–2 10.1097/PDM.0b013e31815dd481.

INDEX

Virginia Espina and Lance A. Liotta (eds.), *Molecular Profiling: Methods and Protocols*, Methods in Molecular Biology, vol. 823, DOI 10.1007/978-1-60327-216-2,

C

D

E

F

G

K

L

M

N

O

P

Q

R

S

U

V

W

X

Z